AF777909

A PHOTOGRAPHIC GUIDE TO THE

BIRDS OF JAMAICA

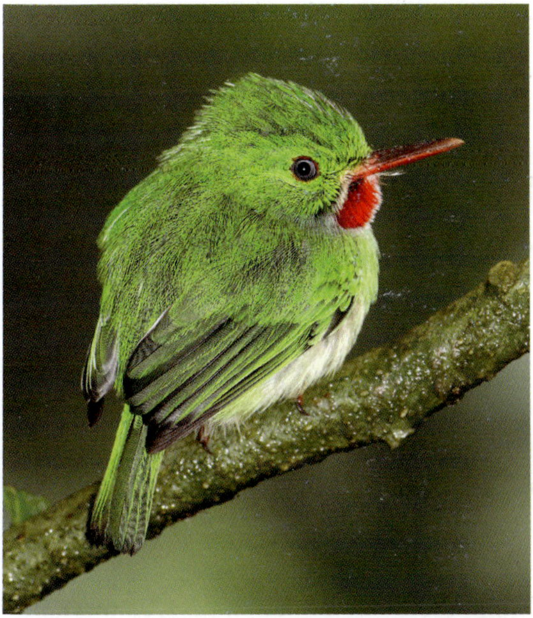

For my mother Marta and my mentor Patrick W. Fairbairn, who fostered my interest in birds; and my sister Carol, who encouraged me to complete this book. A. H-S.

For my children Richard and Jean, and my grandchildren Stephen, Natalie and Edward. A. D.

To my late wife, Anne-Marie. Y.-J. R-M.

A PHOTOGRAPHIC GUIDE TO THE

BIRDS OF JAMAICA

ANN HAYNES-SUTTON, AUDREY DOWNER & ROBERT SUTTON

PHOTOGRAPHY BY YVES-JACQUES REY-MILLET

PRINCETON UNIVERSITY PRESS
PRINCETON AND OXFORD

Published in the United States, Canada, and the Philippine Islands in 2009 by
Princeton University Press, 41 William Street, Princeton, New Jersey 08540

nathist.press.princeton.edu

Published in the United Kingdom and European Union in 2009 by
Christopher Helm, an imprint of A&C Black Publishers Ltd., 36 Soho Square, London, W1D 3QY

www.acblack.com

Library of Congress Control Number 2009925827
ISBN 978-0-691-14391-0

Commissioning Editor: Nigel Redman
Project Editor: Jim Martin

Design by Julie Dando, Fluke Art, Cornwall

Printed in China by C&C Offset Printing Co. Ltd

10 9 8 7 6 5 4 3 2 1

CONTENTS

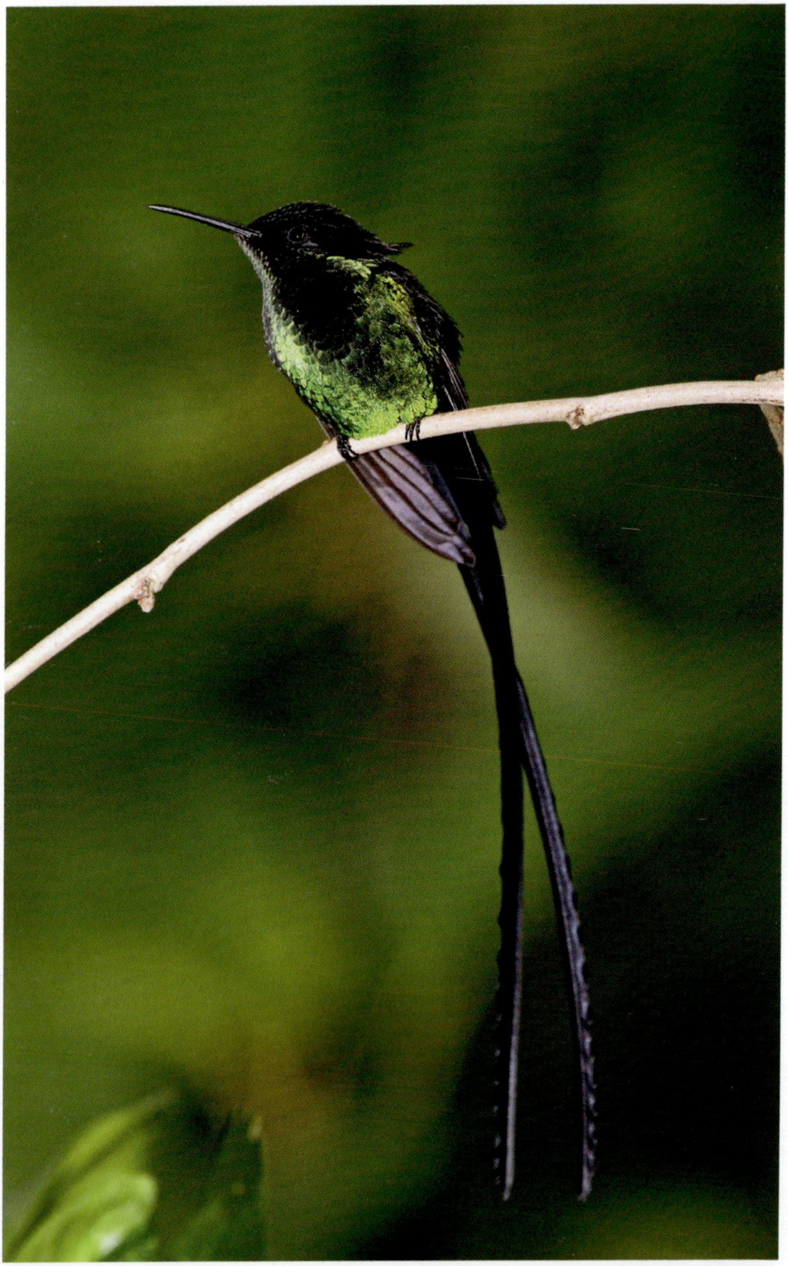

Adult Black-billed Streamertail. Port Antonio, Jamaica, November.

PREFACE

Work on this book began in 2000, when Audrey Downer, Robert Sutton, Yves-Jacques Rey-Millet and Ann Haynes-Sutton agreed on the need for a replacement for *Birds of Jamaica – a photographic field guide* (Downer and Sutton 1990). From the outset we identified the need for a comprehensive field guide, with photographs of all the species likely to be seen in Jamaica. Before it could be finished, Robert Sutton died in 2002, followed by Audrey Downer in 2006, and sadly we have had to complete this book without them. We cannot do better than to reiterate their words in the preface to the 1990 book:

> "This book is our answer to all the local and foreign ornithologists, scientists and people with a general interest in natural history, who, over the years, have had difficulty identifying some of Jamaica's unique, beautiful and fascinating birds and have requested our assistance. It is a compilation of knowledge gleaned during more than fifty years of birding, banding and observation."

It is our hope that through this book people will become more aware of and appreciative of Jamaica's birds, that they will be inspired to carry out the research that is needed, to take action to protect species that are being affected by invasive species, habitat destruction and hunting, to help others to know and understand them, and to simply enjoy a hobby that can last a lifetime, and add a new dimension to life.

Ann Haynes-Sutton and Yves-Jacques Rey-Millet
March 2009

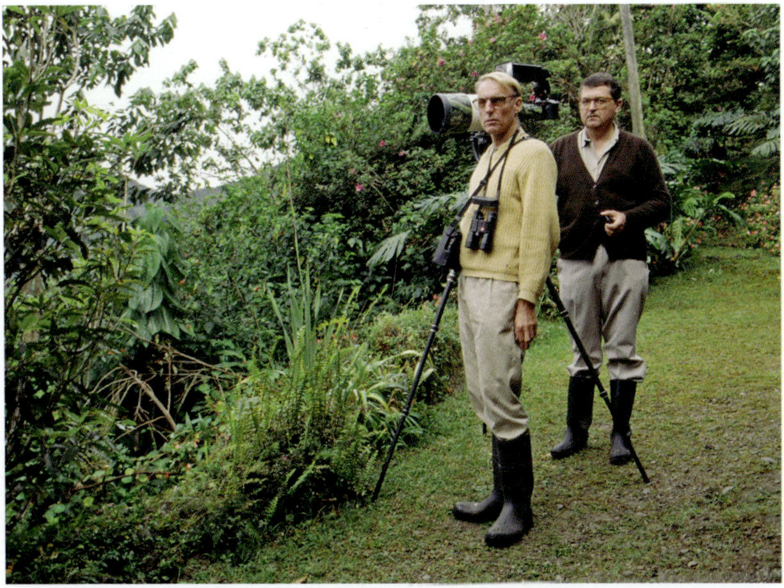

Robert Sutton and Yves-Jacques Rey-Millet.

ACKNOWLEDGEMENTS

We gratefully acknowledge the assistance of the many people without whom this book could never have been written. In particular we thank Josh Larsen whose indefatigable assistance in the field and the office was essential to the quality of the photography and the survival of the photographer.

We thank D. Brandon Hay for his assistance with finding birds and reviewing the text. We are especially grateful to Steve Schill of The Nature Conservancy for his help with the maps for the introduction, and to John Fletcher, Herlitz Davis, Leo Douglas, Rich Hoyer, Peter P. Marra and Vaughan Turland, and many other local and international scientists, for assistance with information on status, distribution and identification of various species. We also thank the following for important advice and assistance at various points in the gestation of this book: Neville Brown, Junior Campbell, Sergio Corbet, Paul Dean, Eladio Fernandez, Claude Fletcher, Mike Foley, Stuart Lacey, Mike and Alli Lubbock, Orestes 'El Chino' and Angel Martinez, Ricardo Miller, Ron Pagliaro, Hopeton Parnell, Hanna and Mark R. Roberts, Fritz and Oliver (of Rocklands Bird Sanctuary), Wayne Sutherland, and Borris Sproul.

We thank Nigel Redman and Jim Martin at Christopher Helm/A&C Black Publishers for their hard work in bringing the book's editorial and production processes to a successful conclusion, and Julie Dando for her outstanding design.

We stayed at many places around the island during the making of this book. We especially appreciate the support of Mr and Mrs Nairne of Starlight Chalets, as well as the management and staff of Strawberry Hill, The Tryall Club and Half Moon.

Ann Haynes-Sutton and Yves-Jacques Rey-Millet

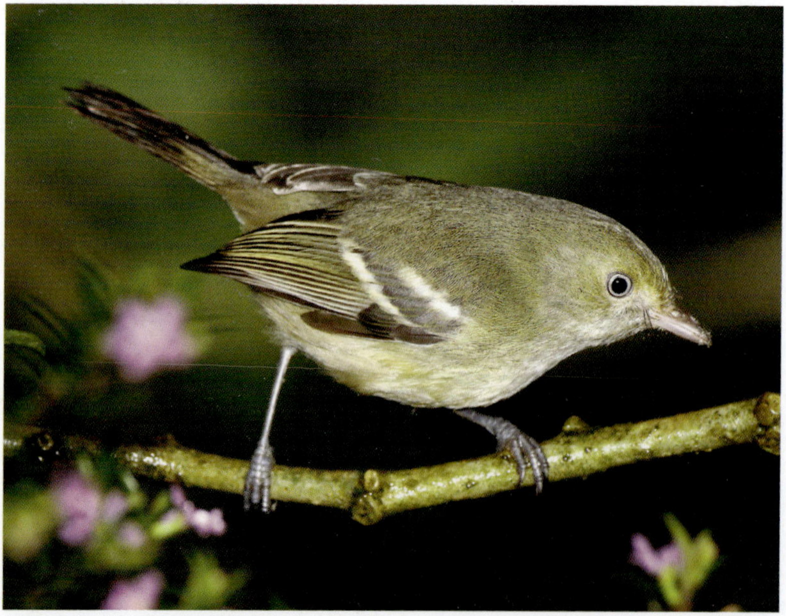

Adult Jamaican Vireo. Port Royal Mountains, Jamaica, February.

INTRODUCTION

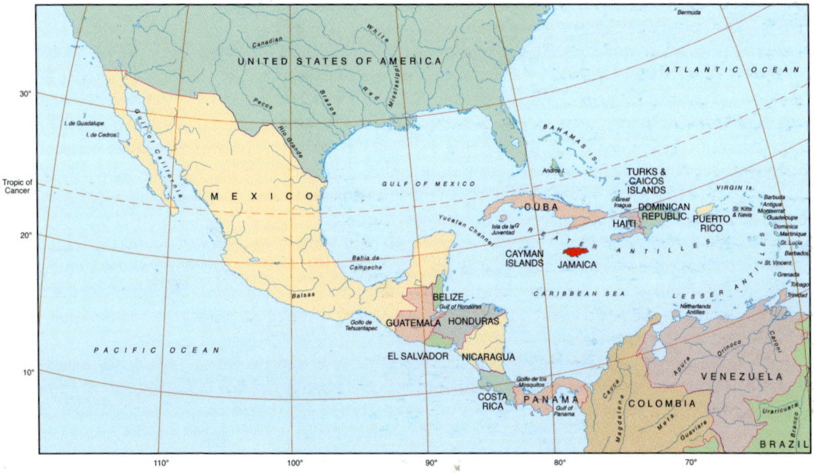

Figure 1. Jamaica and the wider Caribbean.

INTRODUCING JAMAICA

Location	18°N 77°W; 150km (95 miles) south of Cuba and 180km (110 miles) west of Hispaniola; 650km (400 miles) from Honduras – the nearest mainland (Figure 1).
Size	Jamaica is the third largest island in the Caribbean (after Cuba and Hispaniola). Area: 10,982km^2 (4,244 sq. miles). Length: 235km (146 miles). Width: 35km (22 miles) to 82km (51 miles). Highest point is Blue Mountain Peak at 2,290m (7,402ft).
Population	Approximately 2,800,000 (2008).
Government	British Commonwealth nation, independent since 1962.
Capital	Kingston.
Language	English. Most people also speak an English-based patois.

CLIMATE

The climate of Jamaica is tropical maritime. It is modified by the north or northeast trade winds and the daily pattern of land/sea breezes. The average daily temperature ranges between 27°C (80°F) in the coastal lowlands to 13°C (56°F) in the Blue Mountains. Generally there is a drop in temperature of 1°C for every 150m increase in altitude. The coolest and driest months are December, January and February. Average annual rainfall varies from less than 750mm (30in) in the south-central coastal lowlands to more than 7,500mm (300in) in the John Crow Mountains. The wettest months are usually May and October, but tropical downpours can be expected any afternoon, especially during the summer. The hurricane season is July to November. 'Northers' sometimes bring wind and rain in winter. Long term trends indicate increasing temperatures across the island.

GEOLOGY AND GEOMORPHOLOGY

Jamaica's topography and geology are complicated and diverse, the result of the alternation of periods of volcanic activity and marine submersion that are still not well understood. During the early or mid-Cretaceous (about 130 million years ago) a 'proto-Antillean arc' of volcanic islands formed on the edge of the Caribbean plate to the west, possibly in the eastern Pacific. It moved gradually westward between North and South America, 'docking' briefly between the two continents before moving on to its current position. The volcanoes that were to become Jamaica were subject to a complex sequence of lifting and inundation. About fifty million years ago the whole island was submerged and limestones were laid down over the volcanic bedrocks. The Blue Mountains of Jamaica probably emerged around five to ten million years ago, but some parts of the island were still submerged until about one or two million years ago. Once Jamaica re-emerged, erosion began and in some parts of the centre of the island the limestones were completely eroded, revealing the ancient Cretaceous rocks of the central inlier. The eroded material formed the inland valleys and coastal plains.

Today, limestone formations still cover more than two-thirds of the island. They include the John Crow Mountains in the east (running approximately northwest/southeast), the east/west ridge of central and western mountains (including the Dry Harbour Mountains, Cockpit Country and Dolphin Head), the southern coastal hills and two small, north/south ridges in the south-centre (the Don Figueroa and the Santa Cruz Mountains). In the east, between the John Crows and central Jamaica, lies the highest range, the Blue Mountains, composed of igneous shales.

Although the details remain controversial, it seems that since its re-emergence Jamaica has not been connected to any other land mass by a land bridge. However during the last Ice Age (which ended about 18,000 years ago) sea levels were up to 120m (360ft) below present levels. This exposed a series of volcanic islands between Jamaica and Nicaragua along which birds could 'island hop' to colonise the new habitats of the West Indies.

This complex geological history, as well as a long history of climate change, has created a rich diversity of habitats and thus provided the basis for the evolution of high levels of species diversity. This is reflected in many terrestrial groups of fauna and flora as well as birds. Jamaica has particularly high levels of endemism in many groups including plants, land snails, frogs and lizards. Scientific names of plant species mentioned in this book are given in Appendix 6 (page 294).

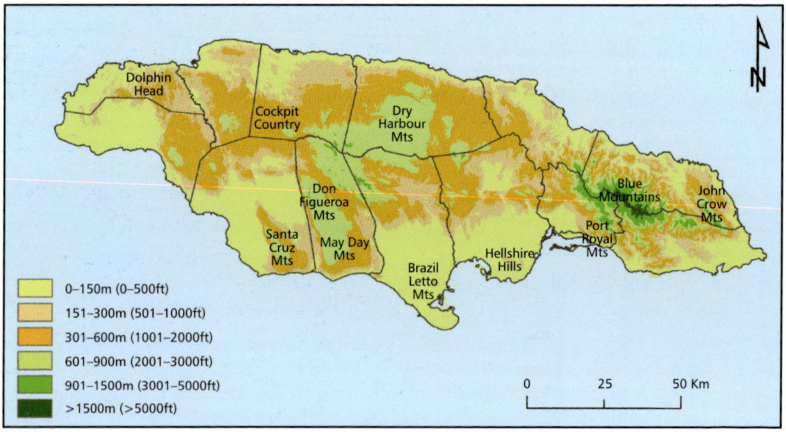

Figure 2. Jamaican topography – elevations and main mountain ranges.

HABITATS

From the spectacular coastal wetlands with their slow, meandering rivers, to the dense wet forests of the mountains with rapid streams and waterfalls via dry savannahs and rolling pastures, Jamaica boasts an astonishing diversity of climate, landscape and vegetation.

Habitat destruction began with the arrival of the first people – the Tainos – in about 800 AD, followed by Columbus and the Spanish in 1494, whose descendants were mostly ousted by the British in 1655. The lowland and riverine forests of the coastal plains of Jamaica have been entirely cleared and have vanished almost without trace, replaced by canefields and settlements. The forests of much of the coastal hills and uplands have been disturbed and less than 670km² (260 sq. miles) remains

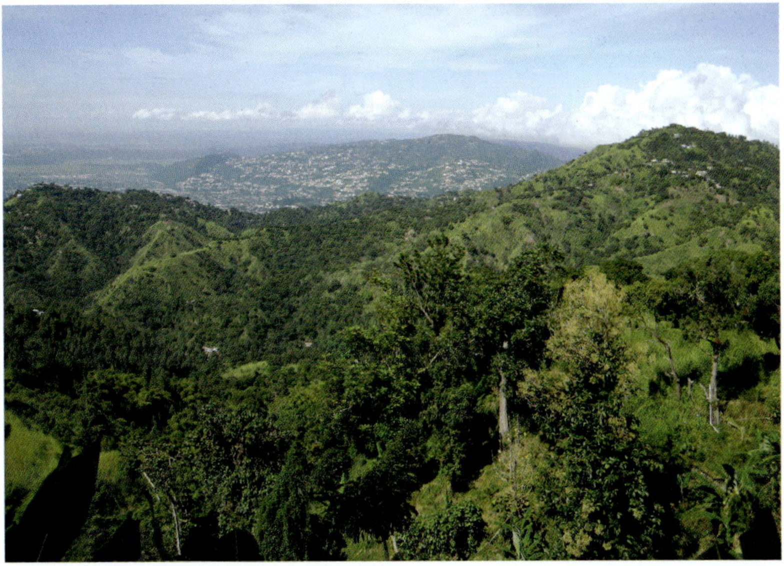

View from Strawberry Hill to the coast – a rich variety of landscapes.

as undisturbed broadleaf forests. Remaining forests can be classified into three main categories: dry limestone forest, found on the northern and southern lowlands and coastal hills including Long Mountain, Hellshire, Portland Ridge, and Canoe Valley; the forests of the central spine of Jamaica including mesic, wet and very wet limestone forests from Dolphin Head in the west, Cockpit Country, Dry Harbour and Don Figueroa Mountains of central Jamaica and the John Crow Mountains in the east; and the wet montane forests on the shale of the Blue Mountains.

Jamaica is affected by the northeast trade winds and the northeastern end of the island, with its high mountain ranges, is the wettest. The central ranges that run east/west along the axis of the island

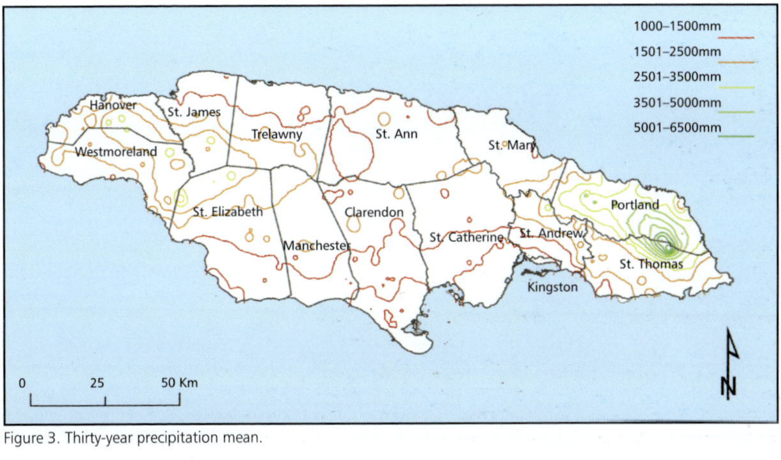

Figure 3. Thirty-year precipitation mean.

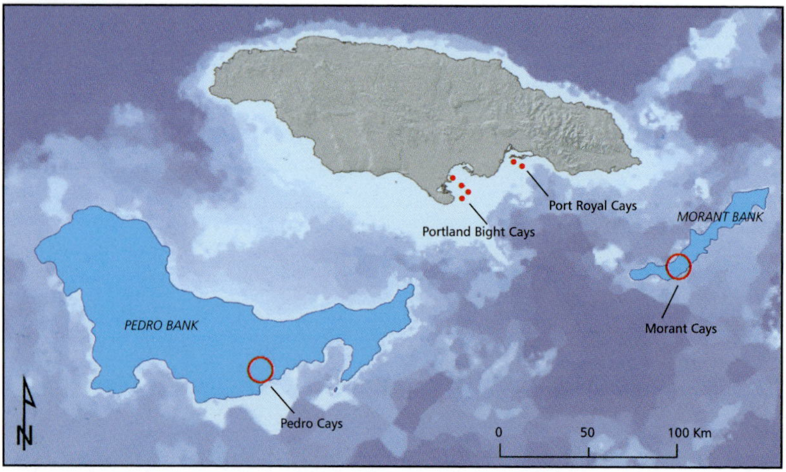

Figure 4. Locations of main groups of Jamaican Cays. These are very important for seabirds.

mean that the north side is wetter than the south. This has a distinct effect on bird life. Birds such as the Jamaican Spindalis, Orangequit and White-chinned Thrush are found at sea level in the northeast, but only in the mountains and foothills in the south except in the wettest areas (Figure 3).

Coasts and cays

Jamaica is famous for its coastline. Sandy white- and black-sand beaches, cliffs, rocky pavements, mudflats and harbours all support bird populations. But the richest areas for nesting seabirds are the small coralline islands around Jamaica known as cays. These are mostly low-lying and sandy with

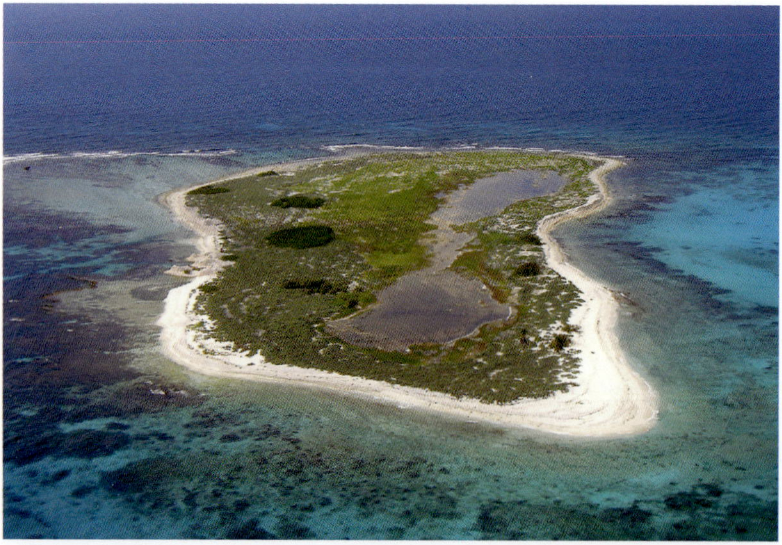

A typical seabird nesting cay, surrounded by coral reefs. This is an aerial view of Southwest Cay, Pedro.

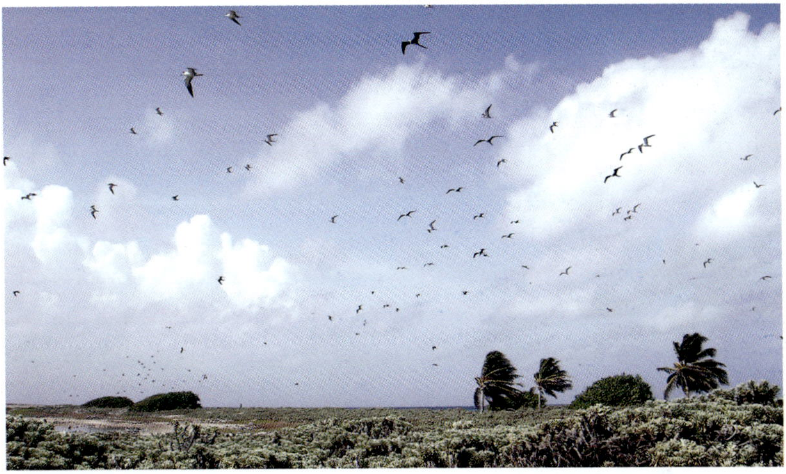

Seabirds soar above Southwest Cay, Pedro.

rocky outcrops. They support typical beach vegetation, mangroves or patches of coastal woodland and colonies of nesting seabirds. On the north coast of Jamaica the coastal shelf is narrow (about 1.6km, 1 mile) and there is a steep drop-off close to shore with spectacular and diverse reefs. On the south coast the shelf is wide – up to 24km (15 miles) across. The Port Royal Cays and Portland Bight Cays are close to the mainland. The Morant and Pedro Cays are 65km (40 miles) and 95km (60 miles) offshore respectively (Figure 4). The Morant Cays provide the breeding site for up to 100,000 pairs of Sooty Terns – one of the largest colonies in the West Indies. The Pedro Cays are the only breeding location in Jamaica for boobies, including one of the two largest nesting colonies of Masked Boobies in the region.

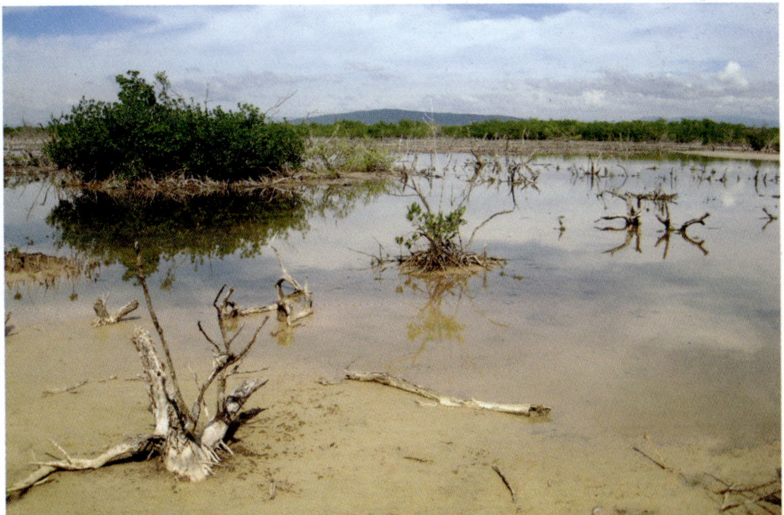

The mangroves of Portland Bigh Shallow mangrove lagoons In Portland Bight are good for shorebirds and Yellow Warbler.

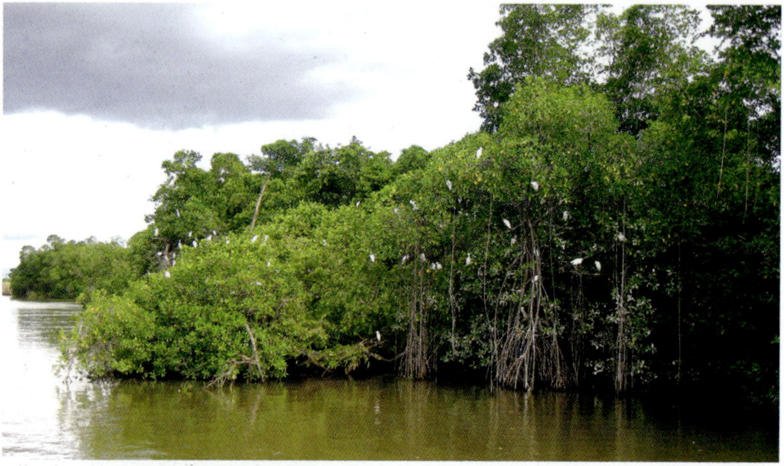

Egrets abound on the mangroves at Black River.

Cays are threatened by development of fishing camps, with associated loss of habitat, pollution and the introduction of invasive species such as cats, rats and dogs that can devastate seabird populations. They are small and low-lying, and are particularly vulnerable to hurricane damage and sea-level rise.

Wetlands, rivers and ponds

Common wetland birds on Jamaica include many species of herons and egrets. Rails, bitterns and whistling ducks are present but they are secretive and hard to see. Many species of migratory ducks and shorebirds can be seen in coastal ponds and mudflats in winter (e.g. on saline or brackish

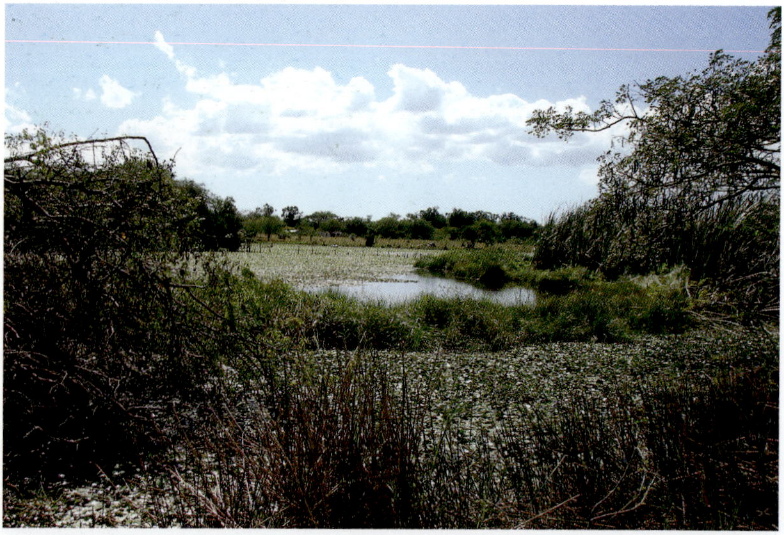

Freshwater ponds, such as this one at Black River, provide good habitat for Masked Duck and Yellow-breasted Crake.

ponds such as Great Pedro Pond and Parottee Ponds, St. Elizabeth, and Yallah's Pond, St Thomas). Jamaica's largest wetlands are the Black River Lower Morass, Negril Morass, Portland Bight and St. Thomas Great Morass. They include mangroves, shallow estuaries, lagoons, salt-pans, marshes and swamp forests. The most extensive stretches of mangrove coastline are found in the Portland Bight Protected Area.

Several of Jamaica's wetlands have been unsuccessfully drained for agriculture. Most others are threatened by conflicting land-use, including unsustainable tourism, hotel construction and industrial development. In the 1980s the Black River Lower Morass (which is very beautiful as well as ecologically important) and Negril Morass were threatened by plans to mine their peat reserves as an energy source, but these initiatives were abandoned when oil prices dropped.

Dry limestone forests

Pigeons (including the White-crowned Pigeon and Caribbean Dove), parakeets, hummingbirds, Jamaican Woodpeckers, orioles, vireos (especially the Jamaican Vireo), and Yellow Warblers are common year-round in the scrubby, spiny habitat. During the winter, migrant warblers can often be abundant.

Along the coasts, this is one of the most widely distributed habitats. The canopy is usually low (8–2m high; 20–35ft), trees are spindly and shrubs spiny and many are drought-deciduous (dropping their leaves in drought conditions). In cleared, unsheltered areas, tall upright cacti and thatch palms grow in patches. In some places there are many thorny species such as logwood and acacias. Thick growths of climbers, vines and strangler figs sometimes make the forest impenetrable, while the loose and bare honeycomb rock makes walking precarious. Distinctive trees include Red Birch, Wild Bauhinia (Bull Hoof) and agaves (Maypole; whose flowers are attractive to orioles and hummingbirds). *Broughtonia* and *Oncidium* orchids and bromeliads also occur locally. Dry limestone forest can be seen on the coastal plains and lower hills (such as the Dry Harbour Mountains, St Ann). In very dry areas the dry limestone forest is replaced by cactus thorn scrub (for example in coastal Hellshire, south Clarendon and Manchester) with endemic columnar cacti including *Cephalocereus swartzii*, *Harrisia gracilis* and *Opuntia spinossisima*. Along the coasts there may be a line of coastal woodland, which is typified by Buttonwood, Sea Grape and other salt-tolerant species such as *Capparis*.

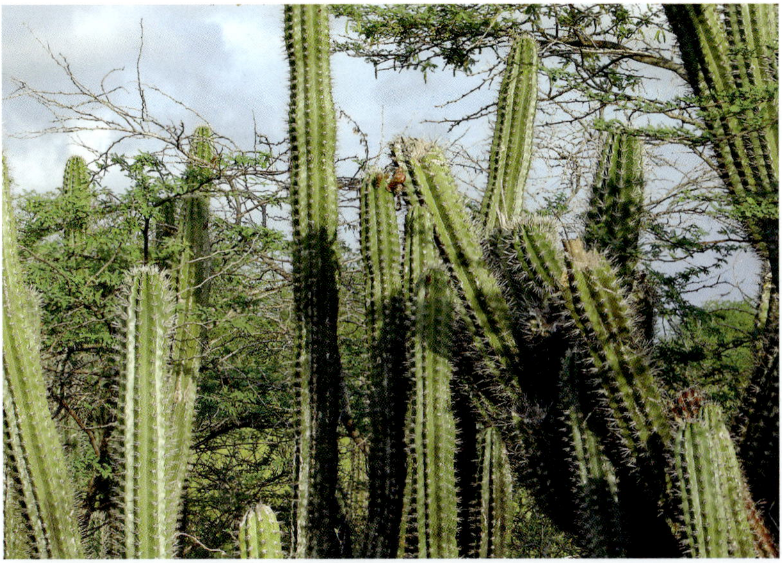

Cactus thorn scrub at Treasure Beach is good for warblers and Jamaican Mango.

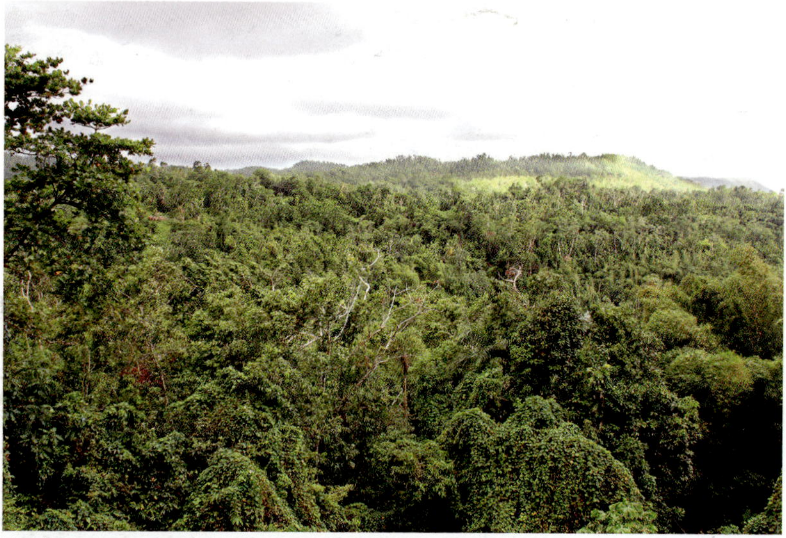

The John Crow Mountains – good habitat for parrots, and for Ring-tailed Pigeon.

Wet limestone forests

The jabbering calls of Jamaican Crows and the squawking of flocks of endemic parrots overhead are among the unique experiences of birding in this habitat. Yellow-billed and Black-billed Parrots, Jamaican Woodpeckers, Jamaican Todies and Arrowhead Warblers are common. Rarer birds such as Ring-tailed Pigeons and the Jamaican Blackbird can also be seen.

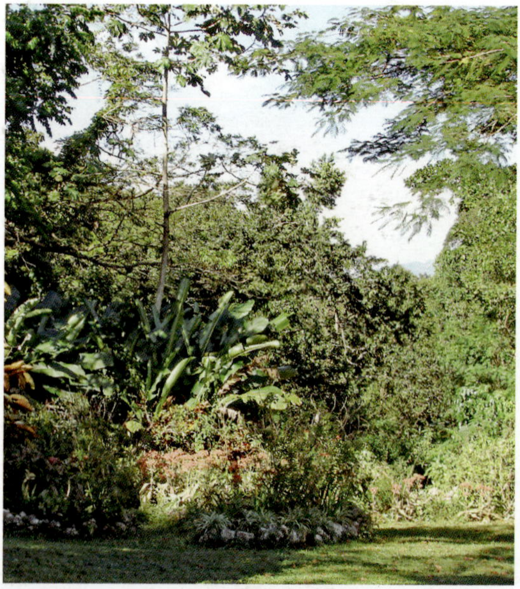

Marshall's Pen garden. A mesic limestone forest.

Wet limestone forests occur in the rugged John Crow Mountains, Dolphin Head, the Cockpit Country (for example in Trelawny at Windsor and Good Hope; or on the Burnt Hill road, from Clark's Town to Albert Town) and Mount Diablo area. The forest is layered and may be as tall as 15–20m (50–65ft) with occasional examples of West Indian Cedar, Silk Cotton and Broadleaf reaching 25–35m (80–115ft). There is a layer of shrubs and herbs that includes sweetwoods, thatch palms and many vines and bromeliads. The forest floor is dark, with many species of fern growing between outcrops of broken rock. Trees like Mountain Pride and Spanish Elm, shrubs like *Euphorbia punicea* and many orchids and bromeliads have colourful flowers that attract birds and insects. Mid-level or mesic limestone forests, such as those at Marshall's Pen in Mandeville, share many of these species but are drier.

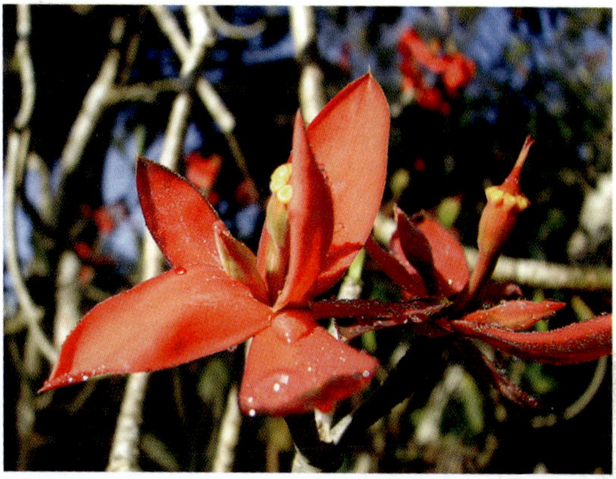

Euphorbia punicea is one of the more than 900 plants endemic to Jamaica.

Montane forests

The haunting, flute-like calls of the Rufous-throated Solitaire heard echoing across the valleys are characteristic of the Blue Mountains. Other characteristic birds include the Ring-tailed Pigeon, Crested Quail Dove, White-eyed Thrush, Greater Antillean Elaenia and Blue Mountain Vireo. The rare Jamaican Blackbird may be observed foraging for insects in the lush growth of bromeliads and tree ferns.

Hardwar Gap in the Port Royal Mountains above Kingston is the most accessible place to see this type of forest, which is restricted to the Blue Mountains and upper Port Royal Mountains. The forests are many-layered, lush, dark, dense and cool. There are many large trees, which are often heavily overgrown with bromeliads, orchids, creepers and fungi. Long strands of grey lichen trail from the branches above. Where the forest is undisturbed, the main canopy is composed of large timber trees like Santa Maria, Blue Mahoe and a huge variety of other species including Yacca. About 60% of the plant species of the

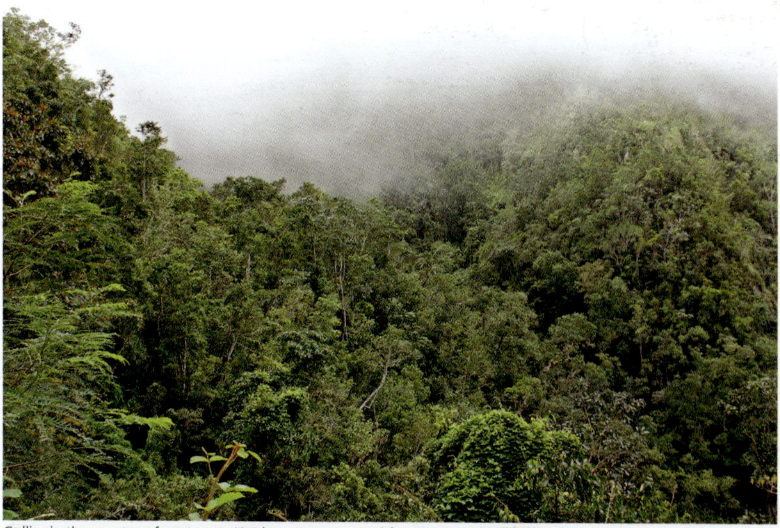

Gullies in the montane forests near Hardwar Gap are good for Jamaican Blackbird.

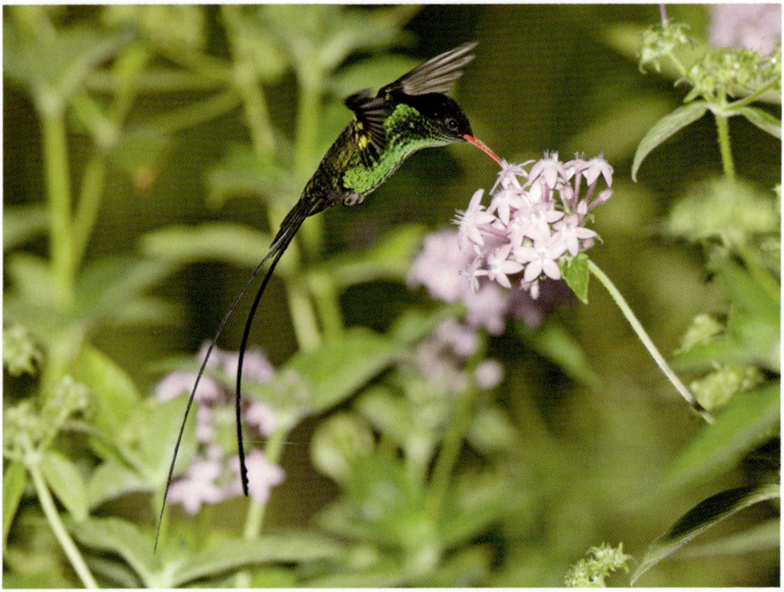

Red-billed Streamertail feeding on Pentas flowers. Marshall's Pen, Jamaica, November.

montane forest are endemic. In the sub-canopy there are several species of Sweetwoods, Waxwood, Winterberry and many other smaller tree species. There is a dense and varied shrub layer, with many species which have large, beautiful leaves and flowers. Many plants are evergreen and have dark green, waxy leaves with drip-tips which encourage water to run off to the ground. The ground may be covered with mosses and ferns. Where a fallen forest tree creates a break in the canopy, elegant groups of tree ferns may be found. Roadsides are thickly covered with exotic invasive Ginger Lilies which are nonetheless attractive to hummingbirds and Blue Mountain Vireos. Above 1,500m (5,000ft) the forests are often shrouded in mist for most of the day and the resulting elfin or cloud forest is stunted, gnarled and twisted.

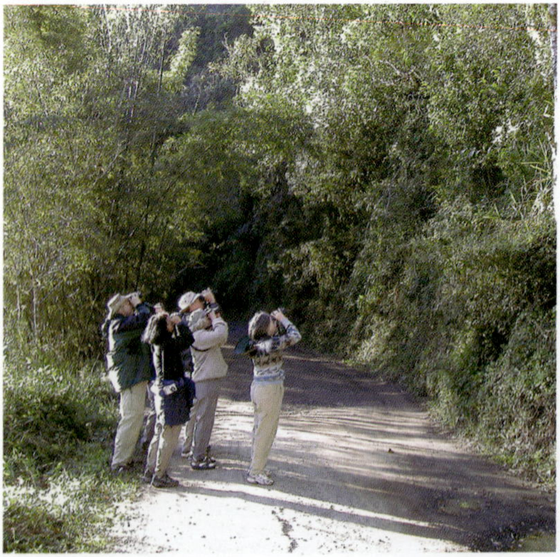

Hardwar Gap – a spot of roadside birding.

Sylviculture, coffee planting, illegal cultivation and charcoal burning are some causes of forest loss. Invasive species such as *Pittosporum undulatum* are also a serious threat even in undisturbed areas, because the seeds are spread by birds.

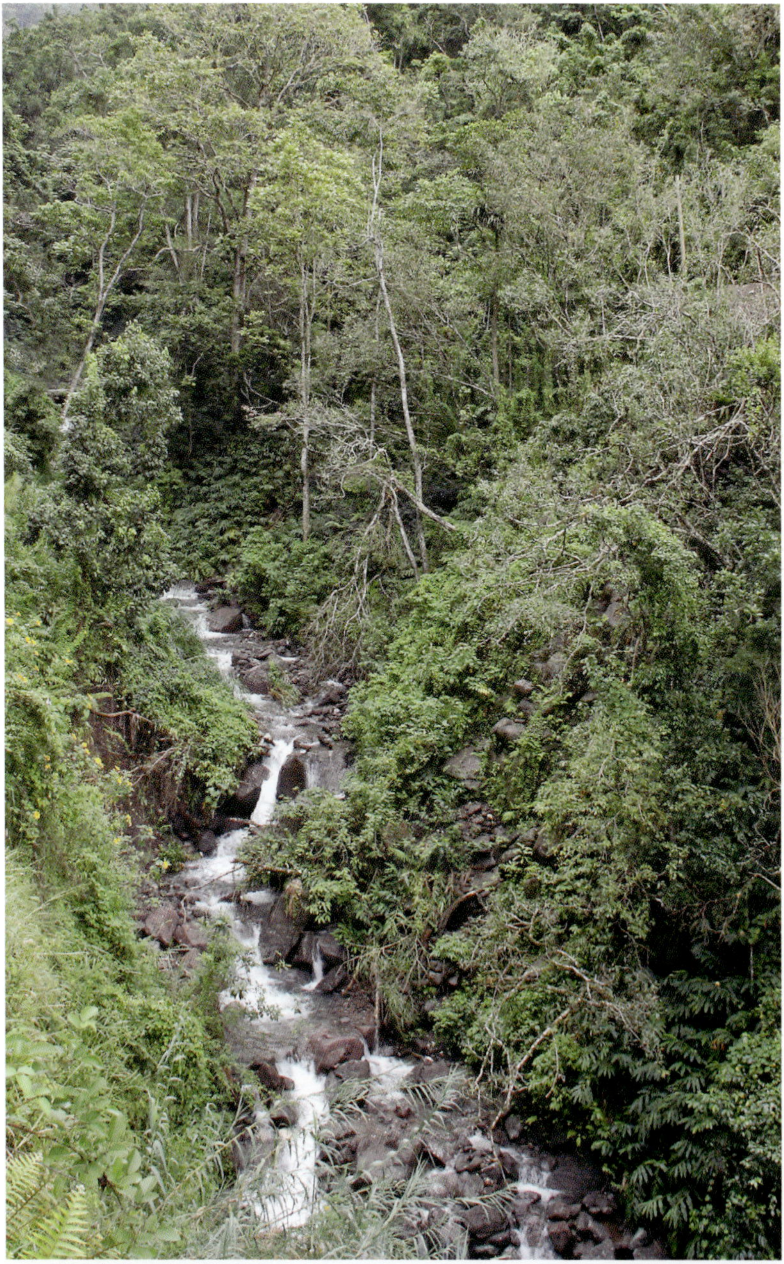

A waterfall in the heart of the Blue Mountains.

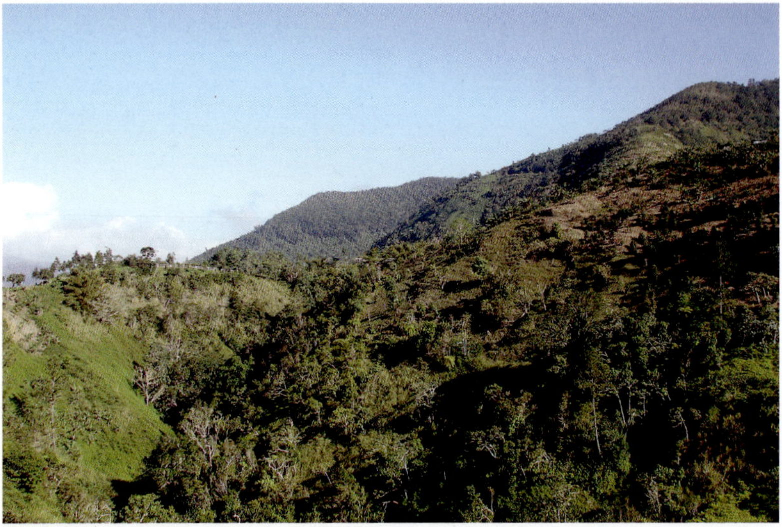

This forest at Buff Bay Valley was cleared for coffee planting and then abandoned.

Agriculture, ruinate and urban areas

More than 75% of Jamaica's land surface is settled, cultivated, mined-out or otherwise disturbed. Some cultivated areas and tracts of ruinate woodland (abandoned cultivated land which has reverted to bush) are good for birdwatching. The edges of pastures and cattle ponds, gardens which include mature trees and mixed cultivation (such as Hope Gardens and Marshall's Pen) can be pleasant and productive places to watch birds. Common birds include cuckoos, todies, grassquits, kingbirds, doves, warblers, flycatchers and vireos. There are trees in even the most urbanised areas and several species of birds are common there, including Northern Mockingbirds, White-crowned Pigeons, Common Ground Doves, Vervain Hummingbirds, grassquits, Bananaquits, grackles and Barn Owls.

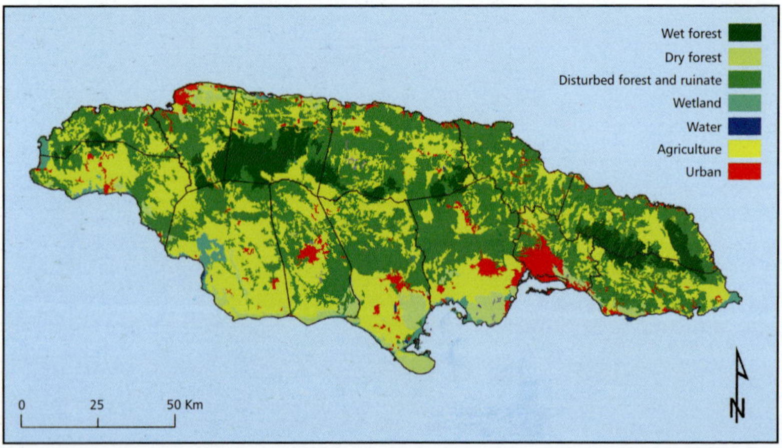

Figure 5. Landuse and habitat distribution in Jamaica.

COMPOSITION OF THE AVIFAUNA

Jamaica's avifauna is one of the most distinctive in the world. It is composed of about 307 species, including 127 breeding species and 180 migrants. Thirty species are endemic (see Appendix 3, page 289) – more than on any other West Indian island and one of the highest totals in the world for oceanic islands of comparable size. There are four endemic genera: *Trochilus* (Streamertails), *Loxipasser* (Yellow-shouldered Grassquit), *Euneornis* (Orangequit) and *Nesopsar* (Jamaican Blackbird). There are also nineteen Jamaican endemic subspecies (Appendix 4, page 290), three Caribbean endemic species and fifteen Caribbean endemic races (Appendix 5, page 293).

ORIGINS OF THE AVIFAUNA

Jamaica was never connected to the mainland, therefore the ancestors of all the native species must have arrived over water. This explains the relatively low total numbers of bird species in Jamaica (compared to a similar area in mainland Central America) as well as the high level of endemism. Possible sources of ancestral species include island-hopping from the north (from Florida via Cuba) or the south (from the Lesser Antilles via Hispaniola) or the west (from Nicaragua along a chain of islands that were exposed at times of low sea level).

Two endemic species have become extremely rare or extinct since the mid 19th century. These are the Jamaican Petrel and Jamaican Pauraque, which have not been collected since around 1880 and 1861 respectively. Searches for these species continue, fuelled by optimism because of their low detectability and the inaccessibility of their habitats. Historical accounts suggest that several other species and subspecies have become extinct in the 400 years since Columbus' second voyage brought the Spanish to Jamaica. These include two macaws – the Jamaican Red Macaw *Ara gossei* and Jamaican Green and Yellow Macaw *Ara erythrocephala*, and the nearly flightless Jamaica Wood Rail or Uniform Crake *Amaurolimnas concolor concolor*, which was a forest- and wetland-dwelling insectivore. Other species known only from the fossil record include a flightless ibis *Xenicibis xympithecus*, an unnamed large hawk and the Burrowing Owl *Athene cunicularia*.

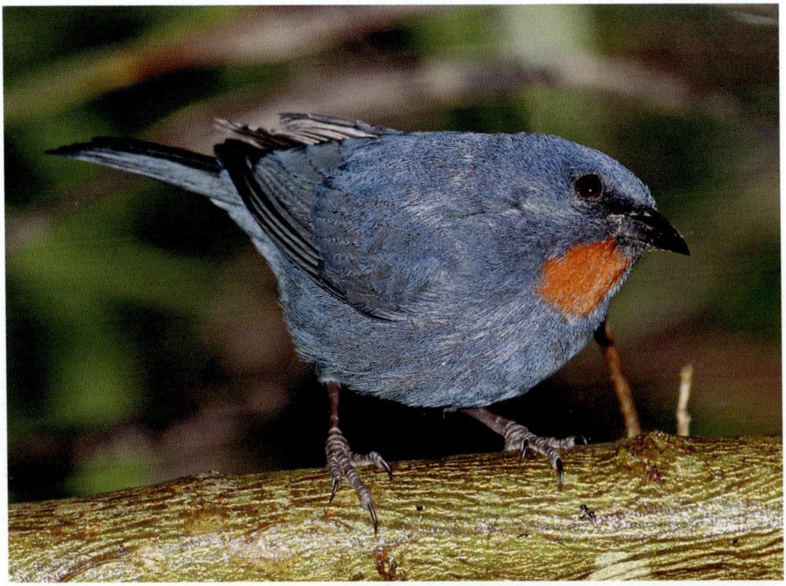

An adult male Orangequit. This species is the sole representative of an endemic genus. Rocklands, Jamaica, November.

MIGRATION

Winter migration

Winter migrants do not breed in Jamaica but pass through on their way to and from their wintering grounds, sometimes pausing only briefly at stop-over sites to refuel. Some overwinter in Jamaica. The winter migrants generally breed in North America in summer, taking advantage of the long daylight hours to feast on seasonally abundant foods such as insects and fruit. As the days get shorter and food starts to become scarce, the birds accumulate fat in preparation for the long flight south. The main flyways from eastern North America run along the Gulf Coast and from Florida through the West Indies via Cuba, Hispaniola and the Lesser Antilles. Jamaica is not on the main route of either, and thus does not receive the same number of migrants as the other Antillean islands. Many migrants, including the warblers, have high site-fidelity and individuals return to the same winter territory annually. Others, like ducks, are opportunistic and will stop for the winter at the first place they come to – if plenty of suitable habitat and food are available. Most migrants arrive in October–November and leave in March–April, but a few return as early as July and leave as late as May. The migratory period composes half or more of their life-cycles.

Migration starts with hormonal changes and fattening. Birds eat as much as they can and accumulate extra fat under their skins. They get restless and may gather in flocks on the coast at suitable points before setting off, usually at night. Suitable weather conditions are important during migration. Headwinds, rain and storms can cause flocks of birds to land on the nearest available spot. Hence remote islands such as Pedro Cays, and exposed headlands like Luana/Font Hill and Morant Point, are good places to look for migrants during the September–October and March–April migration periods, when many birds are moving through the region. It is always a good idea to check suitable habitats for birds after the passage of a storm or frontal system at those times of year.

Birds do not necessarily follow the same tracks on their southward and northward migrations. The numbers and species-compositions of migrating flocks differ between the two seasons. For example, Rose-breasted Grosbeak is usually seen in Jamaica only on its northward migration. Also there are usually more birds involved in the autumn migration, because of the additional young of the year.

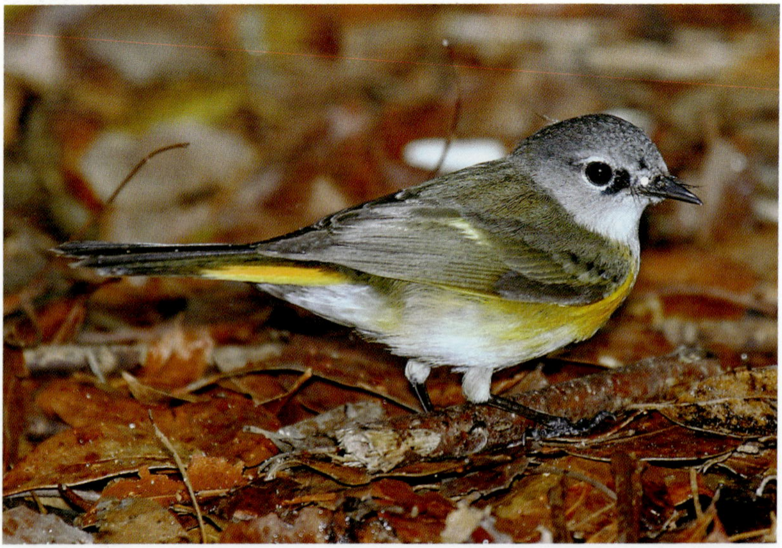

An American Redstart – a common wintering migrant on Jamaica. This is an immature male.

One of the best ways to study bird migration employs ringing (or banding). Birds are permanently marked with standard, individually numbered bands or rings. For North American migrants, information is stored by the US Fish and Wildlife Service, and if a band is recovered they will provide information about where and when the bird was banded. Although recoveries away from the banding site are rare (about one in ten thousand) this technique has provided a lot of useful information. For example, band recoveries have indicated that many migrant warblers that winter in Jamaica breed in New England, while the Blue-winged Teal that visit Jamaica in the winter mostly breed in the Great Lakes area.

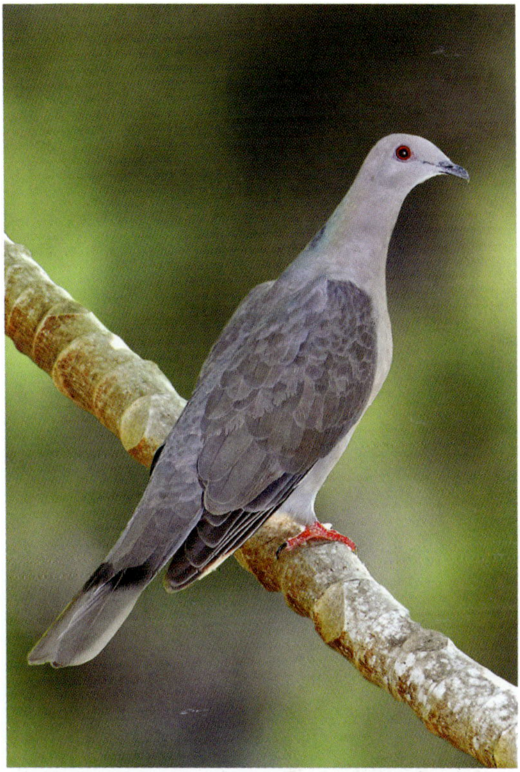

Adult Ring-tailed Pigeon. Port Royal Mountains, Jamaica, November.

Austral Migration

Austral migrants are species that breed in Jamaica but spend the winter further south, returning north for the summer to breed. Ten species of austral migrants occur in Jamaica, including Black-whiskered Vireo, Antillean Nighthawk and Gray Kingbird, as well as several seabirds. Black-whiskered Vireos winter in Brazil. Antillean Nighthawks are thought to winter in South America. The wintering grounds of the Gray Kingbirds from Jamaica are not known, but populations from the Lesser Antilles winter in South America. Gray Kingbirds vanish from Cuba in the winter, as they do from Jamaica, but in Hispaniola they are present year-round, although numbers decline in winter.

ALTITUDINAL MOVEMENTS

Although Jamaica is small, seasonal changes in the abundance of resident birds are noticeable. There are two types of internal migration and both appear to involve changes in altitude. Post-breeding dispersal is the movement of young birds from the nesting area once they have fledged. Several species of pigeons, mostly White-crowned Pigeons and White-winged Doves, breed in the hills in May–August. The young migrate to dry forests on the coast, especially Portland Bight in September,to feed on seasonally abundant berries and fruits such as burnwood. By chance this movement coincides with the pigeon hunting season. By the end of September they have moved away – or been shot. Hunters believe that some of the increase in numbers in September is due to inter-island migration from Hispaniola and Cuba but this has not been proved.

The second type of altitudinal migration involves movement down-slope from the mountains in winter. Species like the Rufous-throated Solitaire that breed on the higher slopes in summer are absent from mid-level forests (e.g. around Mandeville) in summer. As soon as the weather gets cool in the mountains they move down, becoming rare on the breeding grounds and common at mid-levels. Greater Antillean Elaenias probably do the same thing, but they are almost undetectable on their wintering grounds. Similarly some Plain Pigeons spend October–December in the coastal forests of Portland Bight, where Thatch Palm berries are seasonally abundant. In the summer they breed in the mountains, for example in Cockpit Country.

For some species movement patterns are less clear. The movements of Ruddy Quail Doves are not documented. They are rare everywhere for most of the year but suddenly become abundant at mid-levels during the breeding season (March–April).

DAILY MOVEMENTS

Many species have regular daily movements from their roosting grounds to their feeding grounds. They include non-colonial land birds such as parrots, which are best seen at dawn and dusk as they move out from the Cockpit Country to feed in the surrounding agricultural areas, as well as colonial species such as Cattle Egrets. Nocturnal ducks, like West Indian Whistling Ducks and Masked Ducks, reverse the process, roosting during the day and leaving for the feeding grounds at dusk, usually choosing the time when mosquitoes start to bite.

BREEDING SEASONS

In the absence of the constraints of sharply defined seasons, breeding seasons in the tropics tend to be flexible. Many species, such as Bananaquits, can breed at almost any time of year. In Bananaquits, breeding is initiated by rain. For most terrestrial species the peak breeding season is April–May.

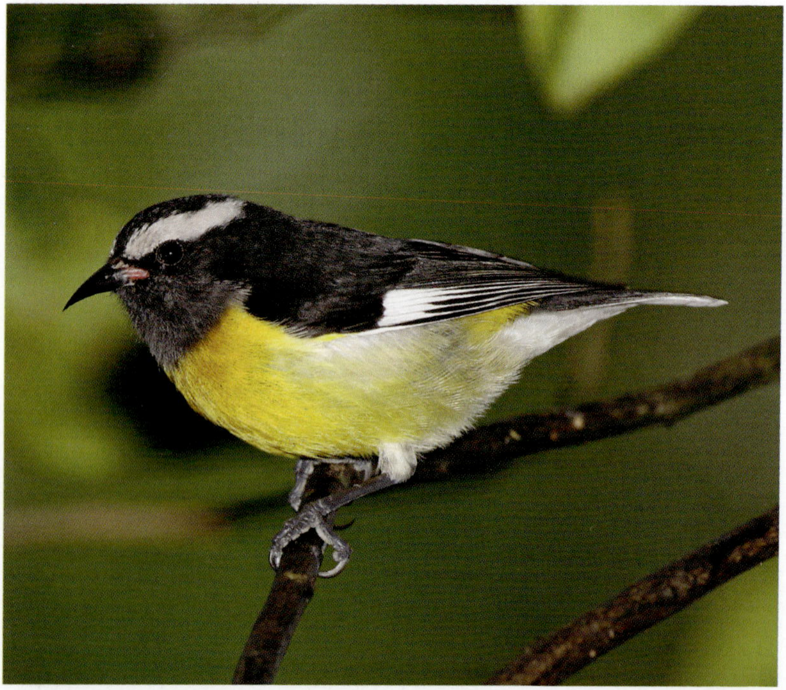

An adult Bananaquit. Rocklands, Jamaica, November.

MOULTS AND PLUMAGES

Birds depend on their feathers for flight, camouflage, waterproofing, temperature regulation and sexual signalling. Thus maintaining their plumage in good condition is very important. Regular preening helps but birds replace their feathers at least twice a year. This usually includes a full moult when all feathers are replaced, and a partial moult. Birds usually moult into breeding plumage before the breeding season and have a second, less extensive moult after breeding. It takes a lot of energy to moult feathers and most birds are less active during moult. Depending on the species, a complete moult may take a few weeks to several years. Many (Bananaquit for example) fledge with a distinctive juvenile plumage, often characterised by loose feathers and dull colours, which may be very different from the adult plumage. Juvenile flight and back feathers are often pale-tipped or broadly tipped brown. In Jamaica this is particularly obvious in the thrushes, flycatchers and doves. The post-juvenile moult, a month or longer after fledging depending on the species, produces the adult plumage type in most passerines but in non-passerines there may be an intermediate, immature plumage which persists for a year or longer. Some passerines do have a distinguishable immature plumage: for example male Orangequits and Greater Antillean Bullfinches do not achieve adult plumage until late in their second year. In addition, many species exhibit two forms of adult plumage, breeding and non-breeding: egrets are a good example.

CONSERVATION OF JAMAICA'S AVIFAUNA

Status of Jamaican birds

Perhaps surprisingly, given the vulnerability of species on islands and the high number of endemic species, most Jamaican bird species appear to be surviving fairly well. Only six species in Jamaica are considered to be globally threatened or endangered. Including the two extinct or near-extinct species, these are the West Indian Whistling Duck, Ring-tailed Pigeon, Plain Pigeon and Black-billed Parrot. In addition, one endemic subspecies – the Golden Swallow – has not been reliably reported since the mid-1980s, and the endemic race of Uniform Crake is similarly probably extinct. Most other endemic species are relatively common and widespread, mainly because they are flexible in their habitat use. Indeed two of the birds that belong to endemic genera – the Red-billed Streamertail and the Orangequit – are among the most catholic in their distributions, occurring from sea level to the mountain peaks.

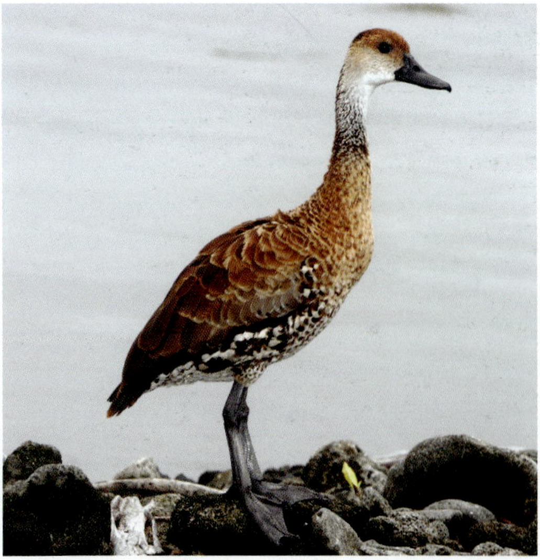

West Indian Whistling Duck is one of several globally threatened species that occur on Jamaica.

The relative abundance of many endemic species should not be a cause for complacency. Species that could become vulnerable in the future are those that depend on wetlands and old-growth forest. This includes the Jamaican Blackbird and Ring-tailed Pigeon, which are rare and restricted to mature montane and wet limestone forests. Their habitats are being destroyed for coffee plantations, mining and illegal cultivation. The Ring-tail is also affected by illegal hunting. Another species that needs to be monitored is the Black-billed Streamertail, as

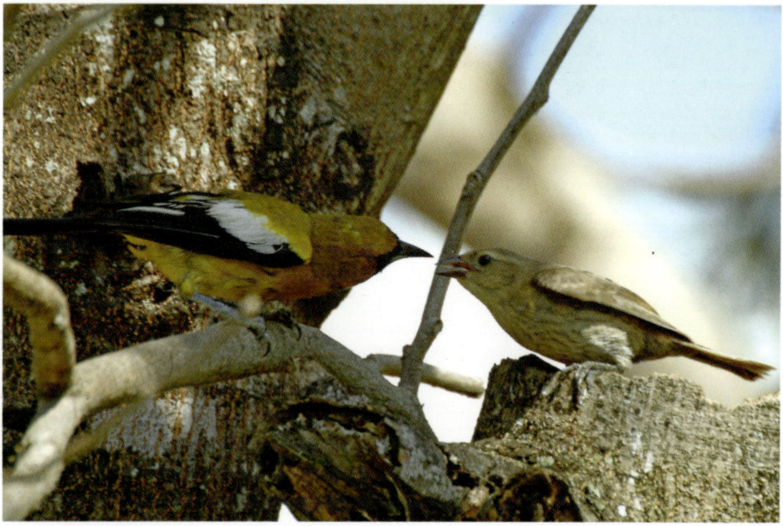

A Jamaican Oriole feeds a Shiny Cowbird chick. This introduced brood-parasite may in time become a significant threat to several Jamaican endemics.

there have been suggestions that Red-bills may be moving into its habitats; these habitats may also have been damaged by the two recent hurricanes (Ivan in 2003 and Dean in 2007). Hurricane Ivan in 1988 blew Black-billed Streamertails from Portland to Kingston.

The importance of wintering and stop-over habitats to migratory species is increasingly being recognized. Many migratory species in Jamaica use mainly disturbed woodlands and thus are not likely to be short of habitat. A few, such as Bicknell's Thrush, depend on old-growth forest and are very rare and hard to see.

Threats

Threats to Jamaican birds include loss and degradation of habitats through human activities, such as conversion of land for housing, tourism, industry and agriculture, and from hurricanes and climate change. Predation and competition with introduced species, especially the Shiny Cowbird, a brood-parasite that was first detected in Jamaica in the early 1990s, is also a problem. Hunting for sport, the pet trade and subsistence is governed by laws that could be better enforced and is also a threat, as is disease. Seabirds on islands are particularly vulnerable to introduced cats, rats and dogs. The presence of just one cat was enough to extirpate Brown Noddies from Big Half Moon Cay in Portland Bight in 2002–03.

Legal protection

Almost all birds and their eggs are fully protected in Jamaica under the Wild Life Protection Act (1974), which prohibits hunting, harassment, capture and possession of 'the whole or any part' of a protected bird. Excluded from this provision are specified game birds in a shooting season and all introduced birds, even if they are established in the wild. Shooting season dates (if any), bag limits and other provisions are declared annually by the Minister. The Wild Life Protection Act is administered by the Natural Resources Conservation Authority. Other relevant legislation includes the Endangered Species Act, which controls trade in native species.

Habitat protection

Over the last 25 years several protected areas have been declared including the Blue and John Crow

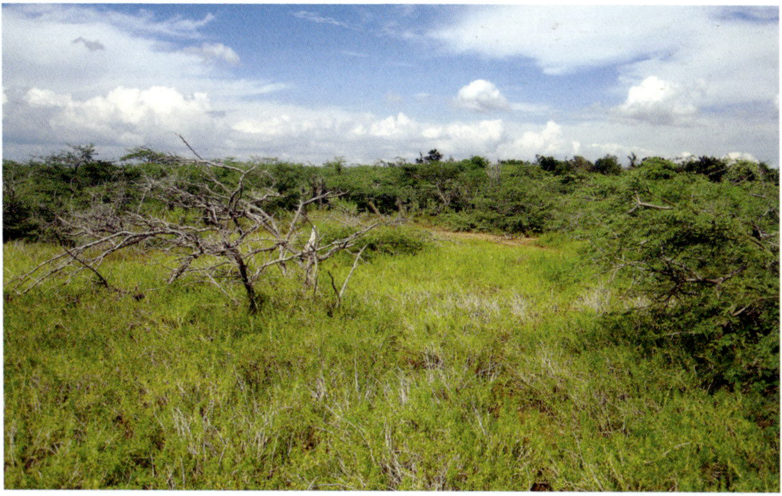

Thorn scrub at the Portland Bight Protected Area.

Mountains National Park, Portland Bight Protected Area, Port Royal Protected Area, Coral Gardens Protected Area and Negril Environmental Protection Area. Many other areas of importance for birds have some protection as Forest Reserves. The creation of a more effective system of national parks is urgently needed to preserve Jamaica's rich heritage of birds and other species.

Other conservation measures

Public education on the ecological and economic value of birds, law enforcement, revision and consolidation of environmental legislation, the improvement of the government agencies responsible for the environment and research are also needed to ensure the future of Jamaica's bird populations.

HISTORY OF ORNITHOLOGY IN JAMAICA

Jamaica's birds were first described in detail by Hans Sloane after he visited Jamaica in 1687–9 (Sloane 1707). His collections eventually formed the basis for London's Natural History Museum. In 1844–46, Philip Henry Gosse came from England to explore Jamaica's natural history. He made extensive collections and wrote and illustrated the first books devoted to the birds of Jamaica (Gosse 1847, 1849). Theories of island biogeography were tested by David Lack, based on detailed study of Jamaican birds in the 1970s (Lack 1976)

The Gosse Bird Club was founded in 1963 to stimulate the study and conservation of Jamaican birds. In 1998 the Gosse Bird Club became a partner designate of Birdlife International and the name of the Club was changed to BirdLife Jamaica. Although this arrangement ended in 2003, BirdLife Jamaica continues to operate. Notes and short articles about birds are published in the *Broadsheet*, which is sent to members twice a year. New members, enquiries and notes and articles for the *Broadsheet* are always welcomed. Meanwhile, BirdLife International continues its work in Jamaica, including the identification of Important Bird Areas.

Since the 1980s Jamaica has been the focus for a series of important studies of migrant warblers on their wintering grounds. These studies of American Redstarts, Black-throated Blue Warblers and Ovenbirds have demonstrated sexual segregation of warblers on the wintering grounds, the importance of food availability in defining habitat quality, and the importance of habitat quality on the wintering grounds in determining subsequent nesting success. Ongoing studies have demonstrated the importance of birds in controlling pests in coffee plantations.

BIRDING IN JAMAICA

WHERE TO WATCH BIRDS

Fortunately for birders, many of Jamaica's endemic species are common and widespread and can be seen in almost any wooded site, roadside or garden. Others are more restricted in range and are best seen at only a few places. The prime birding sites are listed in alphabetical order below. Numbers refer to Figure 6.

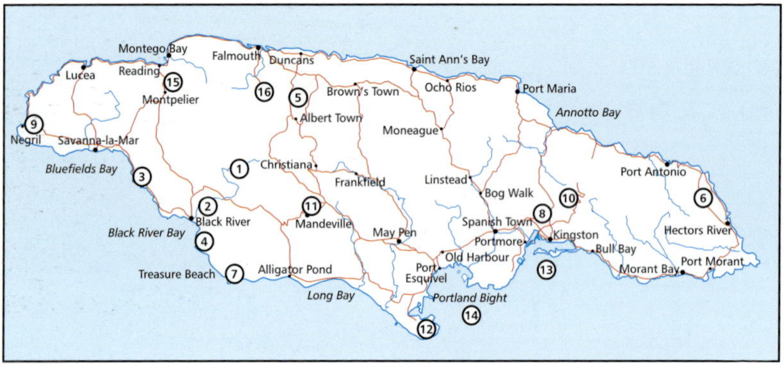

1. Black River Upper Morass
2. Black River Lower Morass
3. Bluefields
4. Parottee Pond
5. Barbecue Bottom
6. Ecclesdown Road
7. Great Pedro Ponds
8. Hope Gardens
9. Negril Royal Palm Reserve
10. Newcastle/Hardwar Gap
11. Marshall's Pen
12. Portland Ridge
13. Port Royal Cays
14. Portland Bight
15. Rocklands Bird Sanctuary
16. Windsor

Figure 6. Map of birding areas.

Black River Upper and Lower Morasses

The largest wetland in Jamaica. It includes:

Elim Pools (1) (in the Black River Upper Morass). A good place to look for ducks, herons and rails. West Indian Whistling-Duck, Caribbean Coot, Least Bittern, Limpkin and Black-crowned Night Heron nest there.

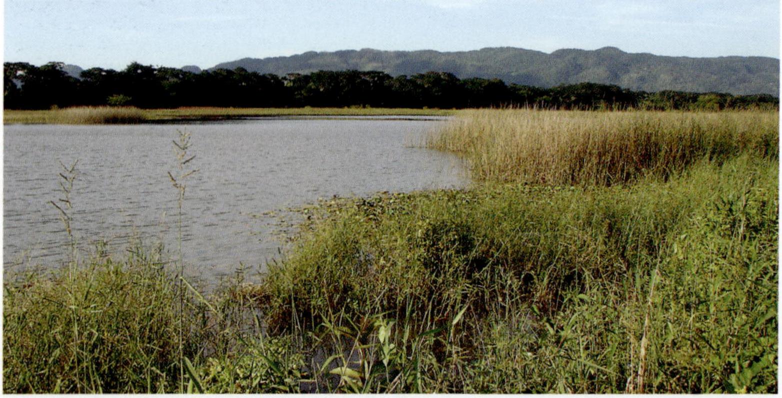

Elim Pools – prime habitat for West Indian Whistling Duck.

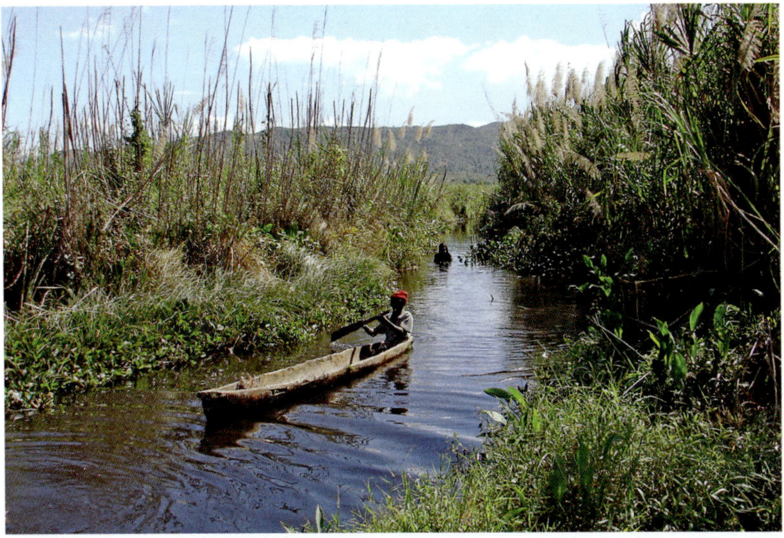

Black River Lower Morass – a shrimp fisherman in a dugout canoe.

Black River Lower Morass (2). Has several scenic rivers that provide good views of crocodiles and common waterbirds such as jacanas. Boat tours leave from Black River. For better views of uncommon birds such as Yellow-breasted Crake private tours to the Middle Quarters River can be arranged.

Parottee Ponds (4): One of the best places in Jamaica to observe waders in the winter months, when water levels are not too high.

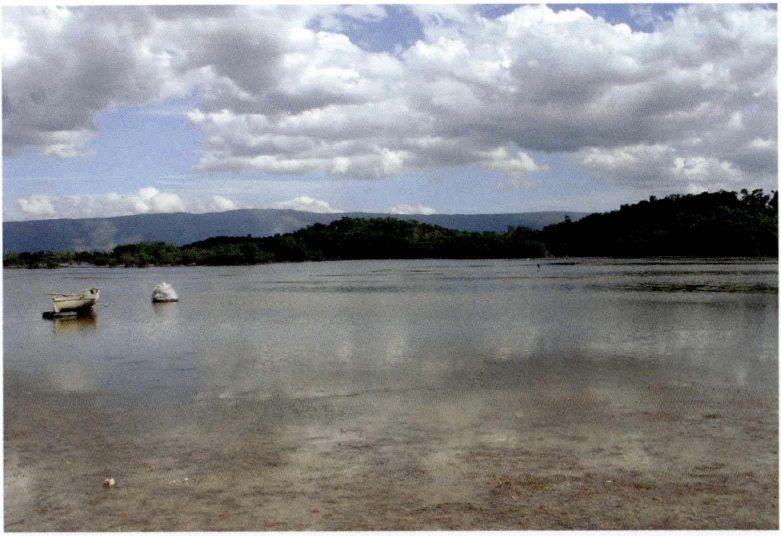

Parottee Ponds – one of the most important sites in Jamaica for migrant ducks, shorebirds and terns.

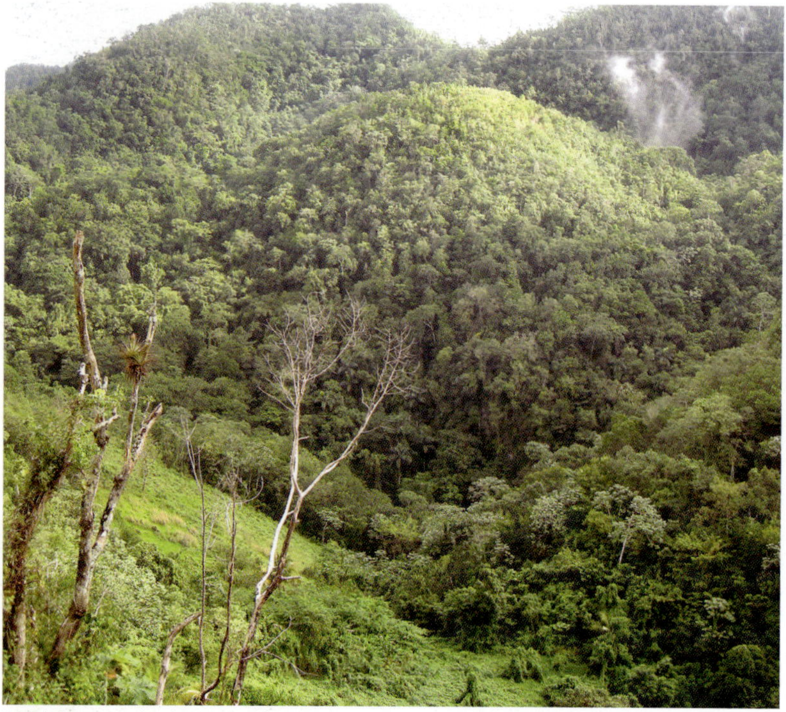

Cockpit Country – a view from Burnt Hill Road.

Cockpit Country (5 & 16)

The wet limestone forest and karst landscape provide a spectacular habitat for many endemic species, including Black-billed and Yellow-billed Parrots and Jamaican Crow. Good access points include Windsor on the north and the north/south road from Clark's Town to Albert Town, known as the Burnt Hill or Barbecue Bottom Road.

Ecclesdown Road, Portland (6)

The scenic inland road from Manchioneal to Reach in eastern Portland passes through the foothills of the John Crow Mountains and is a good place to see all the endemic species, especially Black-billed Streamertail.

Great Pedro Pond, near Treasure Beach, St Elizabeth (7)

The large brackish coastal pond and other small seasonal ponds in the area are excellent for waders and ducks, and for migrants in winter.

Hectors River, Portland (6)

White-tailed Tropicbirds nest in the cliffs near Happy Grove School and can be seen from December to May.

Hope Gardens, off Old Hope Road, Kingston (8)

The grounds of the public gardens and zoo provide habitat for hummingbirds, Yellow-billed Parrots and ducks. Migrant warblers are common in the winter.

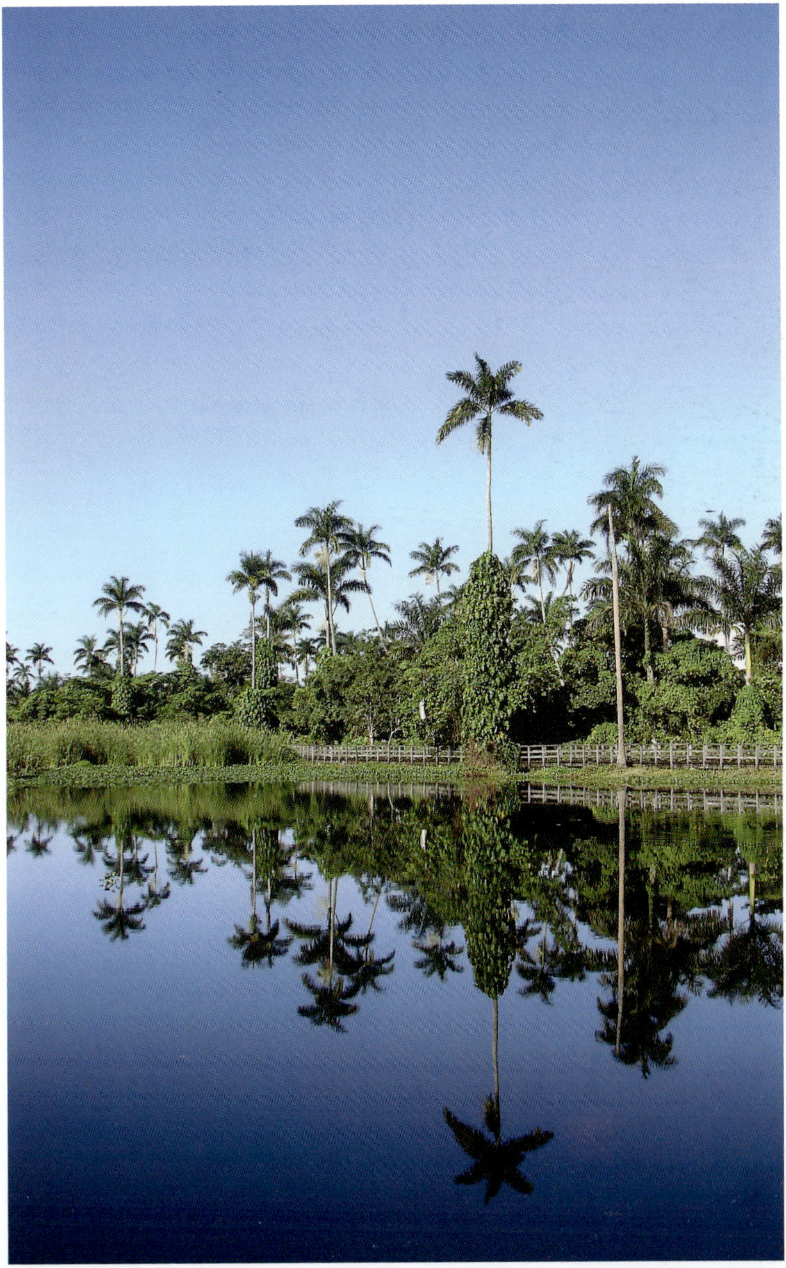

Swamp forest at Negril Royal Palm Reserve.

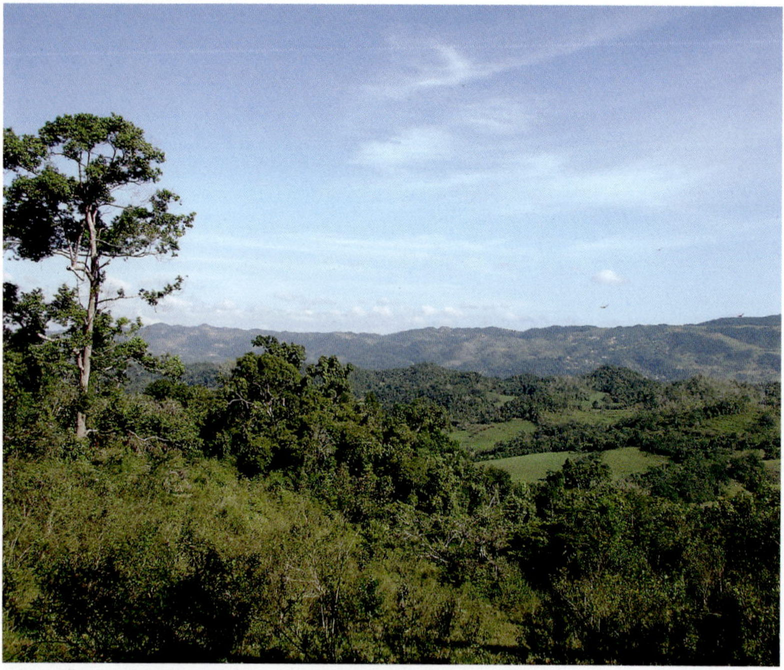

A Red-tailed Hawk soars high above the woodland and pasture of Marshall's Pen.

Negril Royal Palm Reserve, near Sheffield, Westmoreland (9)

A wetland boardwalk that offers good views of West Indian Whistling-Duck, Limpkin and several other wetland and forest birds. Yellow-breasted Crake is a possibility.

Newcastle to Hardwar Gap, St Andrew-Portland (10)

This scenic drive through mature montane forests provides good opportunities to see many endemic birds, including Jamaican Blackbird, Crested Quail Dove and Blue Mountain Vireo. There are nature trails and picnic areas in the Holywell Forest Reserve but the best strategy is to look for birds along the road.

Marshall's Pen, near Mandeville, Manchester (11)

Marshall's Pen is a private cattle property and nature reserve with extensive tracts of mid-level limestone forest. Twenty-three of Jamaica's endemic species (including the Jamaican Owl) are easy to see on the property. Birders and tours are admitted by prior arrangement only.

Portland Ridge, Clarendon (12)

At the most southerly point in Jamaica dry limestone scrub and adjacent mangroves provide a good place to see Bahama Mockingbird, Caribbean Dove and Stolid Flycatcher. Wetland species including Clapper Rail are also common.

Rocklands Bird Sanctuary, near Anchovy, St James (15)

A private bird feeding station where Red-billed Streamertails, Jamaican Mangos and Black-faced Grassquits come to the hand to be fed. Open from 3.00 p.m. to 5.00 p.m. daily. An entrance fee is charged. The surrounding property is ruinate woodland and many of Jamaica's endemic species and migrants can be observed from the road. It is possible to arrange for a local guide to this area.

INFORMATION FOR VISITING BIRDERS

Planning your trip

There are many good trip reports available on the web. For personal advice contact Ann Sutton (asutton@cwjamaica.com).

Clothing Lightweight, light-coloured trousers and long-sleeved shirts are the best attire for birdwatchers. A hat is essential – preferably broad-brimmed to reduce the risk of burning ears and neck. In the mountains a light jacket or sweater will be needed, because early mornings and foggy afternoons can be chilly and damp. Tucking trousers into long waterboots (Wellingtons) or socks is one of the best ways to reduce the risk of being attacked by ticks when birding in pastures or other areas frequented by cattle. It can be showery at almost any time of the year. A light waterproof bag or day pack will help to protect binoculars, cameras and books.

Electricity Supply Most parts of Jamaica have mains electricity. The supply is 110V and 50 cycles. Sockets are the same as those used in the United States.

Water All public water supplies are treated and tap-water is safe to drink.

Food There are small eating places and restaurants throughout Jamaica offering an excellent variety of cooked meals and local snacks, such as patties and bun-and-cheese. Generally there are few places to buy food or water in the best birding spots, so taking snacks or a picnic is advised.

Transportation None of the best birding sites are easy to reach by public transport. If planning to tour the island to see birds it is essential to have a vehicle at your disposal. Cars, four-wheel-drive vehicles and buses can be rented. During the high season (mid-December to April) rental cars are in great demand and it is advisable to book in advance with a reputable company – and even then you may be disappointed. Cars are driven on the left. The alternative is to hire a taxi and driver – but this can be very expensive.

Security and crime Like most places in the world, Jamaica is suffering from an increase in crime. Fortunately the localities usually visited by birders are remote and uninhabited. Cautious birders are unlikely to have any problems. However it is essential to be aware of the risks and not take chances. Birders are advised not to wear expensive jewellery or watches and never to leave expensive equipment or purses in clear view on the seats of cars. Potentially tempting items should be locked out of view. Unattended vehicles should always be locked. Visitors should not give lifts to strangers.

Other hazards Jamaica has no venomous snakes and very few dangerous insects. There are no chiggers. However, you may get the impression that mosquitoes and sandflies pick on newcomers. They tend to bite at dawn and dusk – repellent applied at these times will reduce suffering. Sandflies are a problem at dawn and dusk on beaches where there are freshwater inflows.

Ticks are common in pastures and paths used by cattle, especially in the driest months of January–April. The smallest are smaller than the size of a pin-head. Tucking long trousers into long boots, spraying legs, arms and waists generously with the most powerful repellent you can tolerate, avoiding brushing against vegetation at the sides of paths, avoiding sitting down in pastures, and constantly checking for signs of ticks on clothes are ways to reduce the chances of a serious attack. If found, they should be brushed off immediately, and infested clothing should be changed and not worn again until it has been washed. For visits to prime tick country (e.g. Cockpit Country) during the tick season, a spare pair of trousers could be carried so that you can change if necessary. Beating your legs with a leafy stem of pimento leaves (which often grow in tick-prone areas) also helps. The best way to get ticks off the body when you get back to base is to ask a friend to remove them with a pair of tweezers! For the less athletic, antihistamine creams applied liberally to itchy spots will suffocate ticks while reducing itching. All this said, ticks in Jamaica are not known to carry any diseases transmissible to humans.

Biting Ants Piles of loose soil on lawns and slopes may indicate the presence of biting ants. Inadvertently standing on an ants' nest will result in a swift and devastating ant attack. The best first line of defence on discovering that one has made this unfortunate mistake is to stamp your feet vigorously, while brushing off the ants.

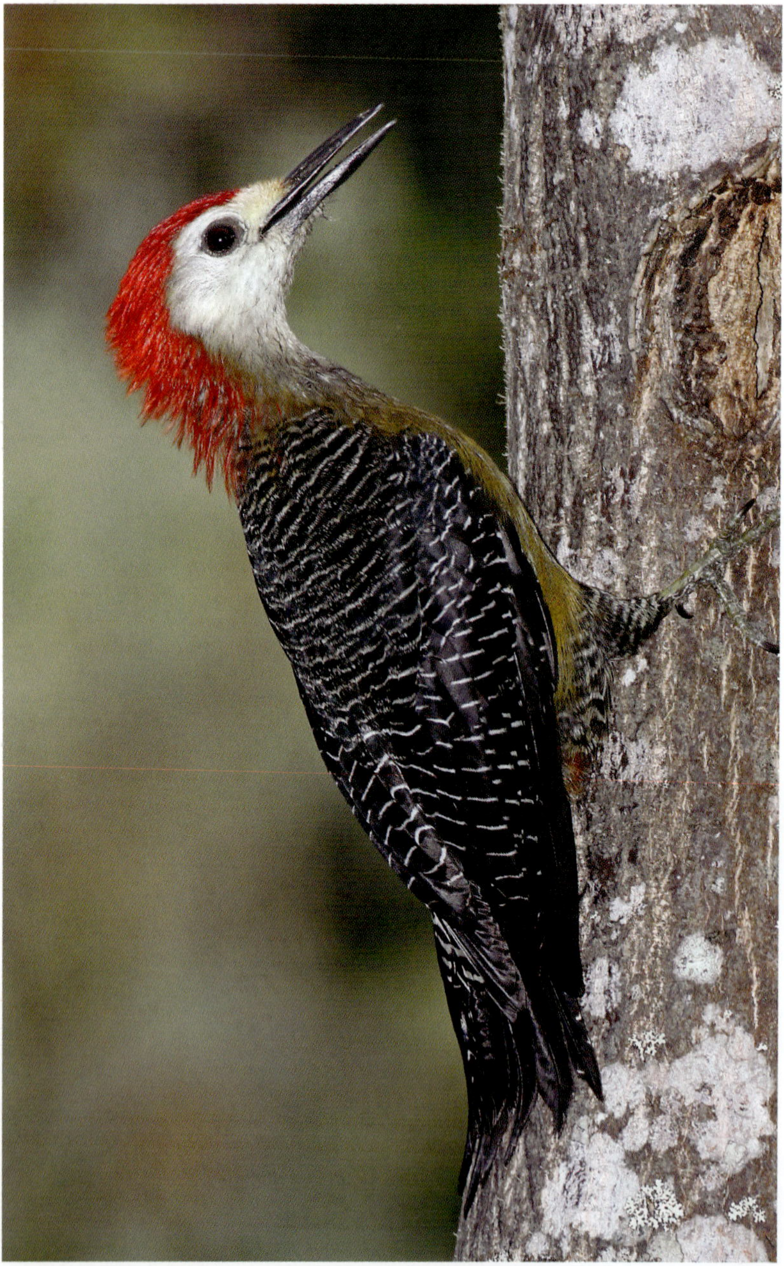

Adult male Jamaican Woodpecker. Port Royal Mountains, Jamaica, February.

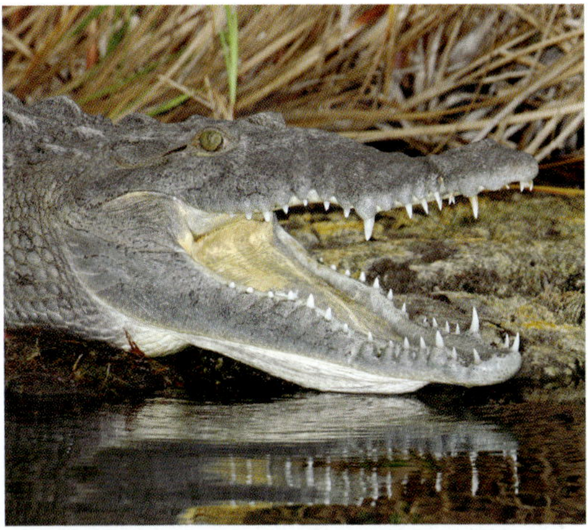

American Crocodile *Crocodylus acutus* at the Black River Lower Morass.

Crocodiles On remote south coast beaches there is a small chance of disturbing basking crocodiles. In normal circumstances crocodiles will go to great lengths to avoid people but you should be aware of their presence as females guarding their nests could be dangerous.

Noxious plants There are several trees and creepers which can cause allergic reactions similar to that of Poison Ivy. The worst, Maiden Plum (also known as Poison Sumac), is common in limestone forests. For some people contact with any part of the tree causes serious burn-like, itchy blisters that can be spread by scratching. The sap is the most potent part but some people are even affected by water dripping off the plants. Several vines and herbaceous plants are collectively known as cow-itch. Their effects are irritating but temporary. The irritants of some types of cow-itch are carried on small hairs. Rubbing dry earth on the affected area will quickly relieve the irritation. It is wise to avoid contact with these plants by keeping to paths and tracks.

Submitting records

Please report unusual sightings to Ann Sutton (asutton@cwjamaica.com) or to BirdLife Jamaica (birdlife@yahoo.com). BirdLife also welcomes articles and notes on Jamaican avifauna and its conservation.

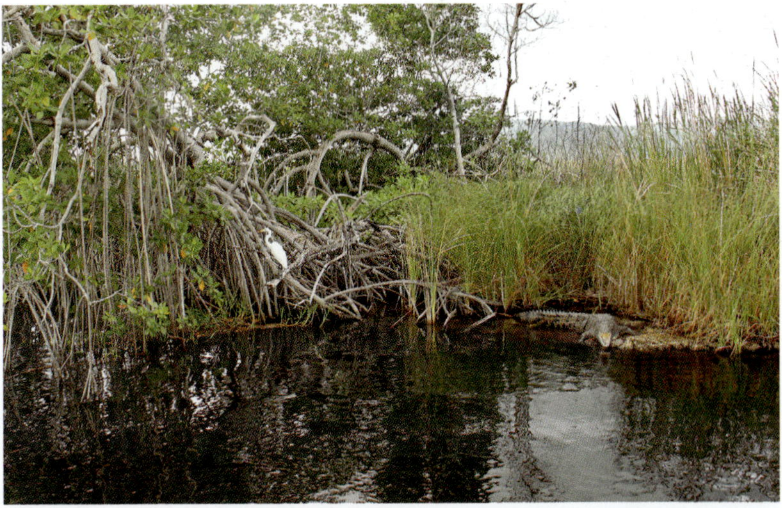

A basking croc on the banks of the Black River Lower Morass.

HOW TO USE THIS BOOK

This book includes descriptions of all the bird species that have been reliably reported from Jamaica in the last 50 years. As far as possible, these are represented by photographs and descriptions based on field observations and observations of birds in the hand made by the authors.

NOMENCLATURE AND TAXONOMY

The classification, nomenclature and systematic arrangement used in this book follow *A Complete Checklist of Birds of the World* (Clements 2007), although the waterfowl and gamebirds are placed at the start of the list in accordance with widespread recent practice. There are many local Jamaican bird names, some of which are used for different birds in different parts of the island; therefore only the most widespread and commonly used local names have been listed.

PHOTOGRAPHS AND DESCRIPTIONS

As far as possible, photographs are included of all adult and immature birds likely to be seen in Jamaica in a typical year. With a few exceptions (detailed on page 4), all of the photographs were taken by Yves-Jacques Rey-Millet and are of wild birds in their natural habitats. Wherever possible these are of birds in the wild in Jamaica but some photographs were of necessity taken in other places. In such cases great care was taken to ensure that the photographs show the appropriate races. Where this was not possible the location and races are indicated in the captions. Photographs of typical habitats are included in the introduction. Species that are now probably extinct on Jamaica are detailed in Appendix 1 on page 275. Vagrants are listed in Appendix 2 on page 277.

SPECIES ACCOUNTS

Taxonomy The taxonomic status of the species is identified as monotypic (consisting of only one taxonomic race) or polytypic (consisting of more than one taxonomic race). For polytypic species, the number of races listed by Clements (2007) is shown in brackets and the name of the race that occurs in Jamaica is given where this is known.

Description L refers to the body length (from the top of the crown to the tip of the tail). Wherever possible this was measured on live birds in the hand. Where no local measurements were available the measurements were taken from Raffaele *et al.* (2003). Separate descriptions of the male, female and juvenile are provided when necessary. Descriptions are based on the authors' observations of wild birds in the field and in the hand.

Similar species Brief descriptions are given of the major field-marks used to distinguish species that might be confused with the species being described.

Voice The voices of all birds are described, based on field observations and Robert Sutton's comprehensive collection of tapes – a lifetime's collection. Many of these songs are available on *Bird Songs in Jamaica*, a set of CDs by George Reynard and Robert Sutton available from the Library of Natural Sounds, Cornell Laboratory of Ornithology (http://birds.cornell.org) (see Appendix 7 on page 297) .

Habitat and behaviour The main habitats used by the species are listed. Aspects of behaviour which are useful in distinguishing or finding a species are described (e.g. colonial and feeding behaviour, flight patterns). Dates, descriptions of nests and clutch size are given (where known) based on field descriptions and published accounts for Jamaica. There are many gaps in the data on nesting for many species in Jamaica, therefore this section is incomplete in many cases. Observers are invited to send their own records and observations to Ann Haynes-Sutton (asutton@cwjamaica.com).

Range The geographical distribution of the species is described, including a summary of breeding and wintering grounds. Endemic species or subspecies are indicated. Global conservation status (based on the BirdLife International/IUCN Red List) is given for threatened species only. The following terms are used:

Critically Endangered	species is facing an extremely high risk of extinction in the wild.
Endangered	species is facing a very high risk of extinction in the wild.
Vulnerable	species is facing a high risk of extinction in the wild.
Near Threatened	species does not qualify for 'Critically Endangered', 'Endangered' or 'Vulnerable' status now, but is close to qualifying for or is likely to qualify for a 'threatened' category in the near future.

Please see http://www.iucnredlist.org/static/categories_criteria_3_1 for further details.

Status in Jamaica In this section the abundance and distribution on Jamaica for each species is summarised. The following terms are used:

Vagrant	occurs by chance, very few (usually one or two) records since 1960.
Rare	very unlikely to be seen even in preferred habitat.
Uncommon	occurs but unlikely to be seen in preferred habitat.
Fairly common	usually seen in preferred habitat.
Local	seen regularly in some examples of preferred habitat.
Common	very likely to be seen singly or in small numbers in its preferred habitat.
Abundant	always easy to see in large numbers in its preferred habitat.
Resident	present year round (unless qualified by winter or summer) and breeds.
Non-breeding resident	present year round but does not breed.
Visitor	migrant species, does not breed.
Summer	April/May to September/October.
Winter	September/October to April/May.
Introduced	non-native species, established in the wild.

Maps Range maps are included for all species. These maps are not intended to show a comprehensive record of all the places that each species has ever been seen; rather, they are just to give a general indication of general patterns of distribution. Although there are many records of the distributions of birds in Jamaica, they have never been compiled into maps. And since birders tend to go to the same places, the records tend to represent the places to which birders go, rather than the actual distribution of birds.

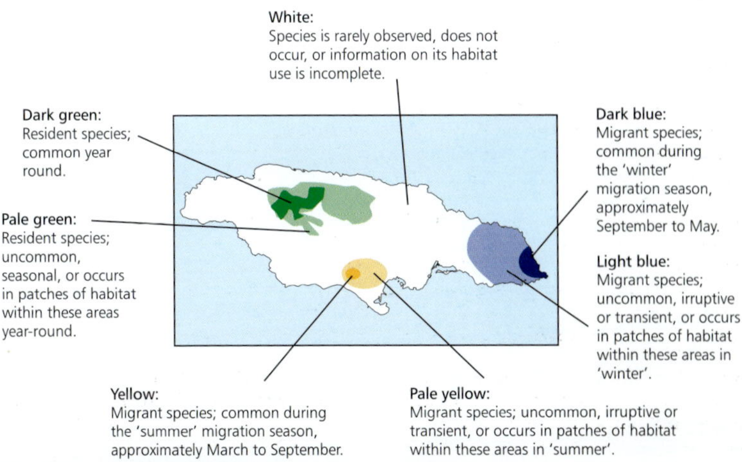

BIRD TOPOGRAPHY

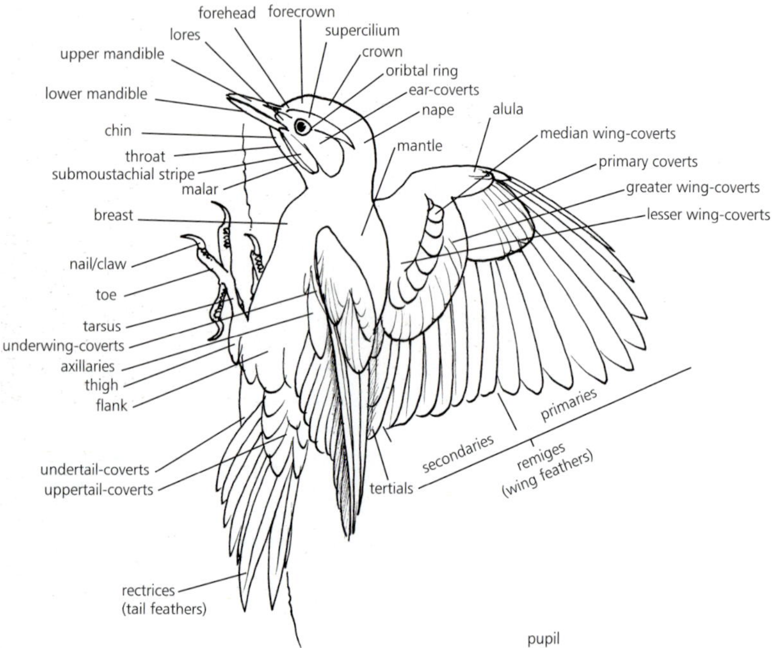

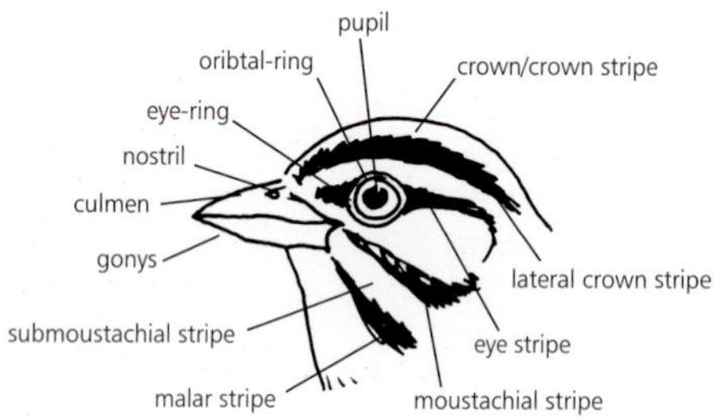

Fulvous Whistling Duck
Dendrocygna bicolor

Taxonomy Monotypic.
Description L 44–51cm (17–20in). A large, pale buffy, goose-like duck with dark back and wings. In flight wings are all-dark, contrasting with buffy underparts, white uppertail-coverts and white band on tail. Bill and legs dark.
Habitat and behaviour Wooded wetlands and river banks. Often seen with West Indian Whistling Ducks.
Similar species West Indian Whistling Duck is darker and browner with white spots on sides. Black-bellied Whistling Duck has orange bill, black belly and light patch on wings.

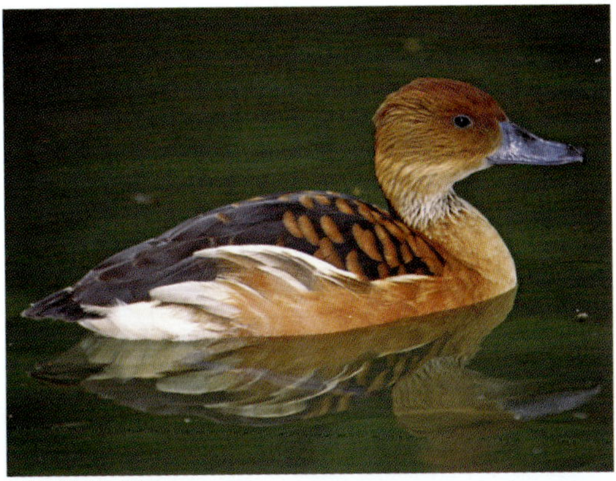

Adult. Florida, February.

Range Southern United States, Central and South America, East Africa, southern Asia. Seems to be extending its breeding range in the West Indies, including Cuba, Hispaniola and Puerto Rico.
Status Rare winter visitor.

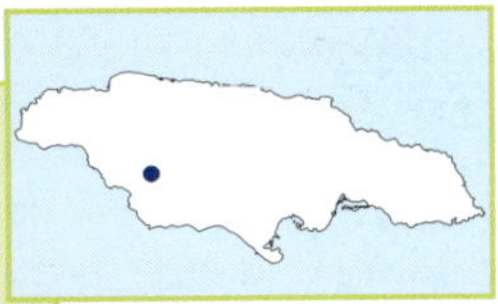

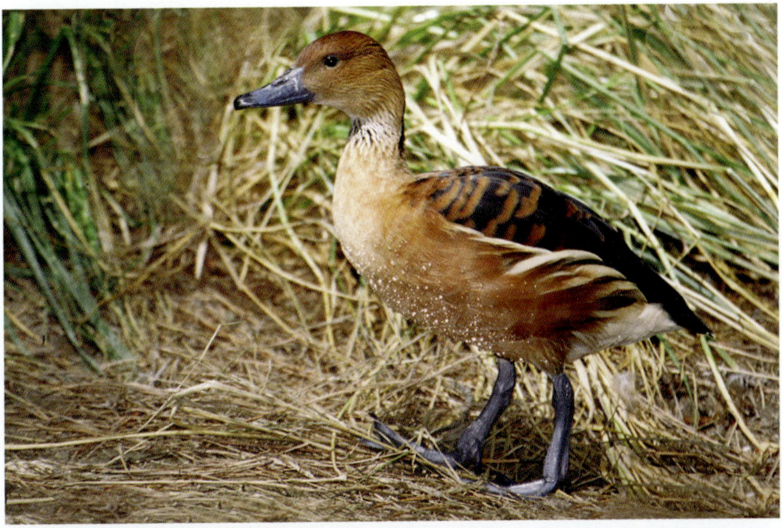

Adult. Note pale head and dark legs. Florida, February.

West Indian Whistling Duck
Dendrocygna arborea

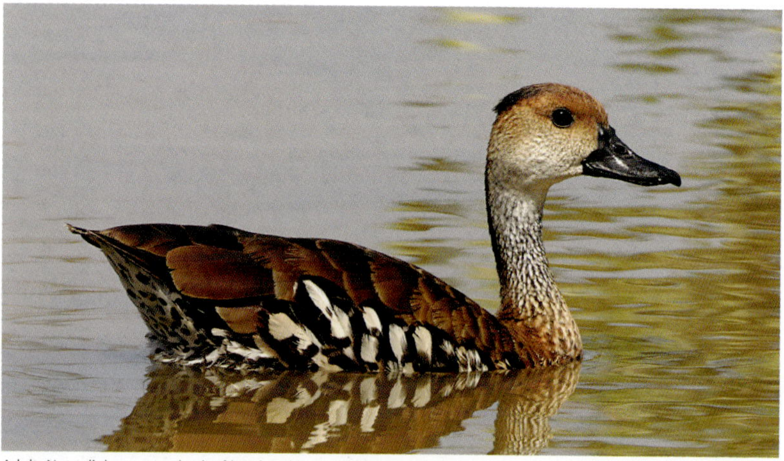

Adult. Note slight crest on back of head and large white spots on flanks. Cayman Islands, February.

Local name Whistler.
Taxonomy Monotypic. West Indian endemic.
Description L 48–56cm (19–22in). A large, upright, goose-like brown duck with long neck and legs. Conspicuous black and white spotting on flanks. Head slightly crested. Crown and centre of hind-neck black. Upper face rufous. Lower face and foreneck white. Base of neck speckled with black and white. Upper breast rufous. Lower breast pale buff, finely streaked with black speckles, becoming broader on white abdomen. Flanks black with distinctive tear-drop shaped white spots. Back and wings brown. Legs long and dark grey. Tail black. In flight has distinctive, arched profile – head is held lower than body, and legs trail behind tail. Light bases of wing feathers resemble wing-patches. Immature is less distinctly marked.
Similar species Fulvous Whistling Duck has rufous underparts and lacks black-and-white speckling on flanks; in flight shows white uppertail-coverts, all-dark wings. Black-bellied Whistling Duck has a black belly and conspicuous white patches on wings.
Voice In flight and from roosts gives a high-pitched, haunting, five-note descending whistle

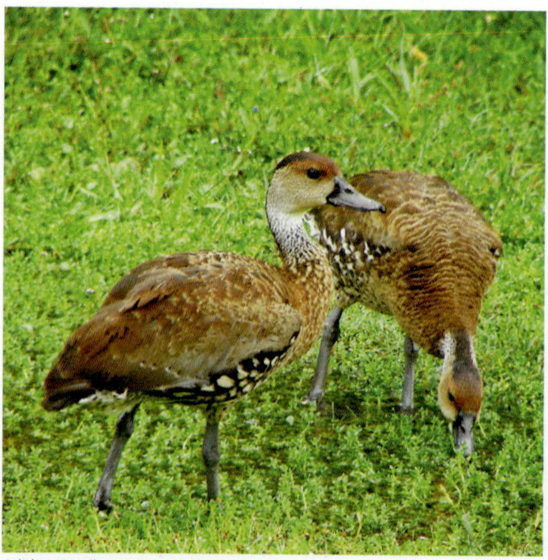

Adults. Negril, Jamaica, September.

pi-pi-pee-pee-pee-eh.
Habitat and behaviour Usually nocturnal. Occasionally seen in daytime flying low, standing or swimming in swamps or rivers, or loafing or roosting in or under trees. Generally flies to feeding grounds at dusk, but sometimes during the day. Feeds usually while walking, foraging with head in shallow water or grazing, in marshes, fields and mudflats on seeds, grasses and fruits and some small invertebrates. Roosts (generally during the day) on the ground in sheltered spots such as reed beds, grassy marshes or among mangrove roots, or perches in trees close to water. Family groups of up to 14 live together. Swims but does not dive. Nests year-round, depending on rainfall; nest placed on the ground in reeds or mangrove roots or above ground in bromeliads or tree holes in trees. Lays up to 14 white eggs.

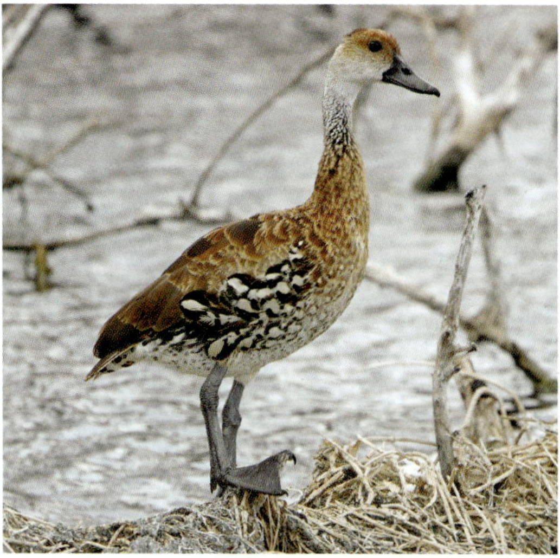
Adult. Note erect stance, dark legs. Cayman Islands, February.

Range Bahamas, Turks and Caicos, Greater Antilles and northern Lesser Antilles as far south as Antigua and Barbuda and Martinique.
Status in Jamaica Uncommon resident. The Jamaican population is the largest outside Cuba. This duck is threatened throughout its range because of hunting and habitat loss. Fortunately it recovers quickly once protected and populations are increasing in some islands as a result. Global conservation status: Vulnerable.

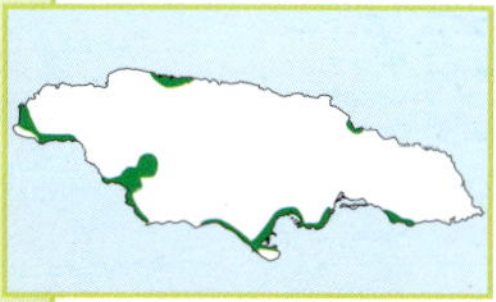

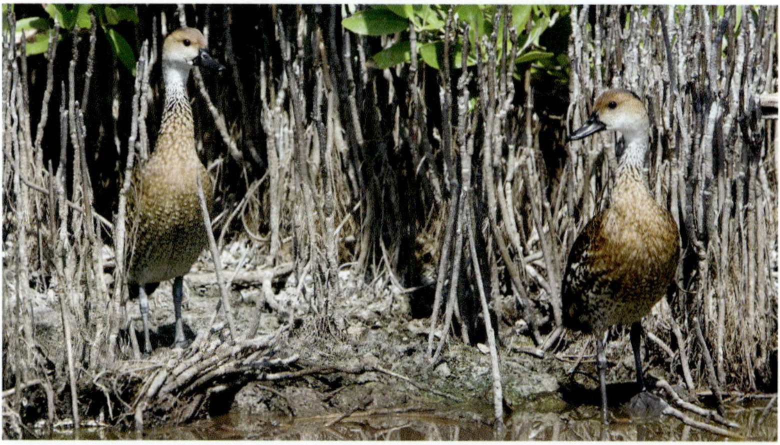
Adults may roost among mangrove roots during the day. Cayman Islands, February.

Black-bellied Whistling Duck
Dendrocygna autumnalis

Taxonomy Monotypic.
Description L 46–53cm (18–21in) Adult is a large goose-like duck with orange bill and legs, and a black belly. Head is grey. Eye ring white. Neck brown. Folded wings brown with large pale patch. Legs pink. In flight upperwing shows a conspicuous white line.
Similar species West Indian Whistling Duck is darker and browner with white spots on sides.
Habitat and behaviour Usually seen in wetlands with West Indian Whistling Duck.

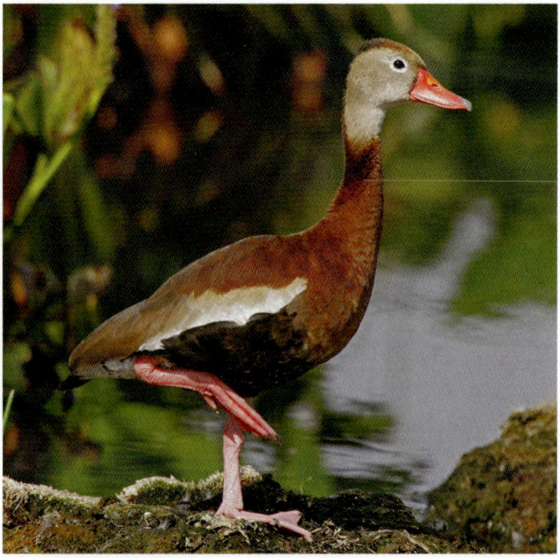

Adult. Note orange bill and legs. Florida, February.

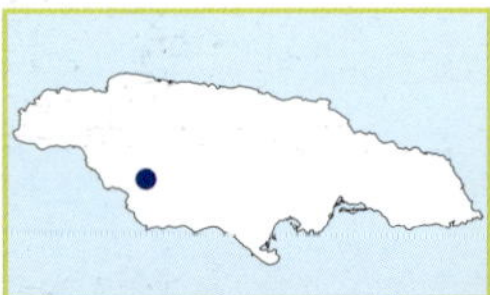

Range Texas to Argentina. Rare breeding bird in the West Indies (Cuba and Puerto Rico).
Status in Jamaica Very rare winter visitor or vagrant.

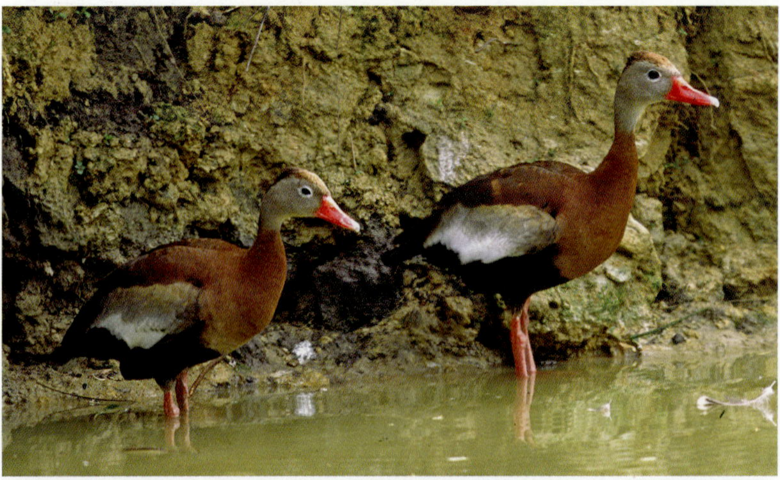

Adults. Florida, February.

Wood Duck
Aix sponsa

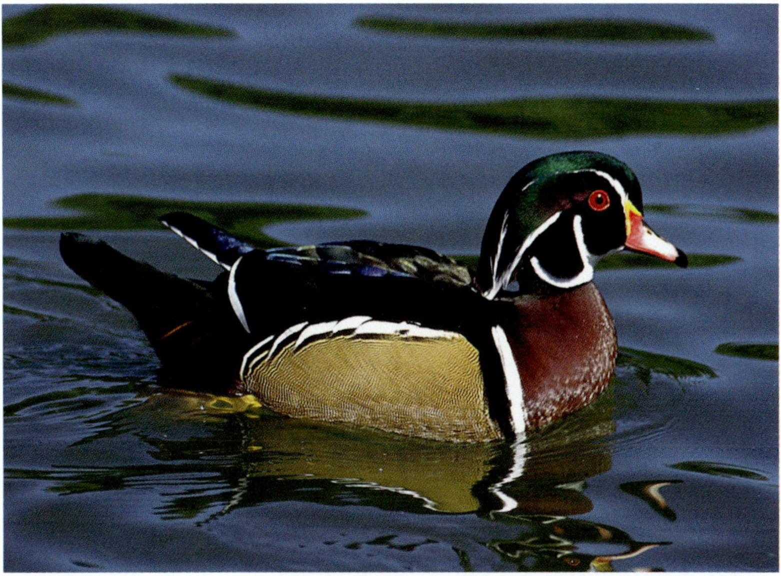

Adult male. January.

Taxonomy Monotypic.
Description L 47–54cm (18–21in). Medium-sized surface-feeding duck with a crested head. Male has a unique green, white and purple pattern on head and crest. Female is brownish, with a wide white eye ring that tapers to the rear.
Similar species None.
Habitat and behaviour Freshwater wetlands.

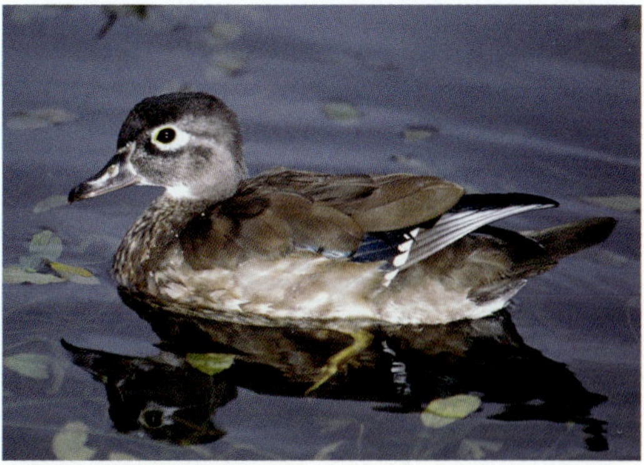

Adult female. Note eye-ring. May.

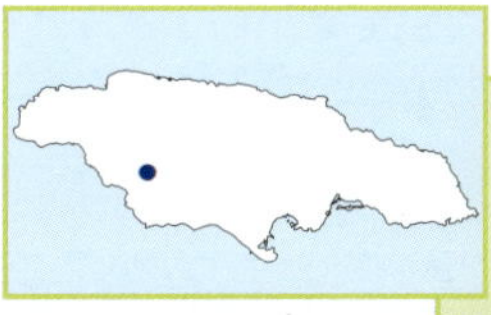

Range Breeds in eastern and western North America and Cuba. Winters south to Mexico.
Status Uncommon winter visitor or vagrant on freshwater wetlands, such as the Black River Morass.

American Wigeon
Anas americana

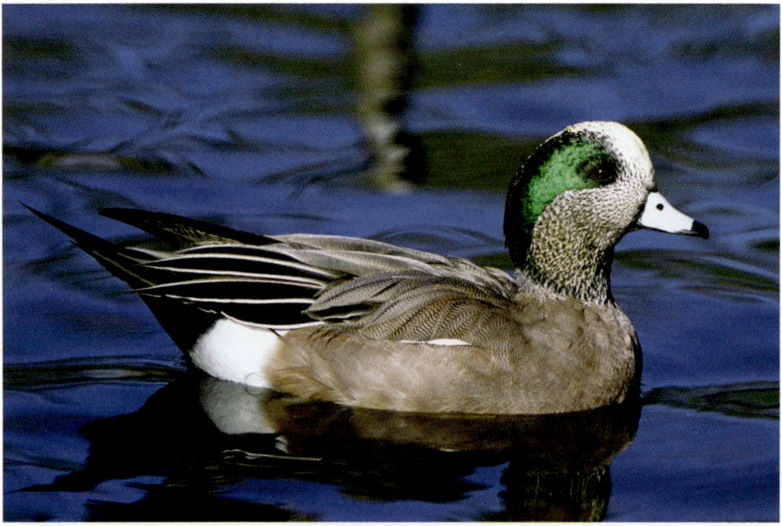

Adult male. Note white crown, small bill with dark tip. January.

Local name Baldpate
Taxonomy Monotypic.
Description L 46–58cm (18–23in). Large surface-feeding duck with a long, broad, light-coloured bill with a dark border. In flight distinguished by white belly, green speculum and white patch on forewing. Male in breeding plumage is brown with white crown, and iridescent green patch behind eye. Female and male in eclipse are reddish brown with greyish brown heads, darker around eyes.
Similar species None
Voice Usually silent in Jamaica.
Habitat and behaviour Fresh and saltwater ponds. Usually seen in mixed flocks with Blue-winged Teal and Northern Shoveler. Feeds in shallow water in fresh, brackish or hypersaline ponds and dams.

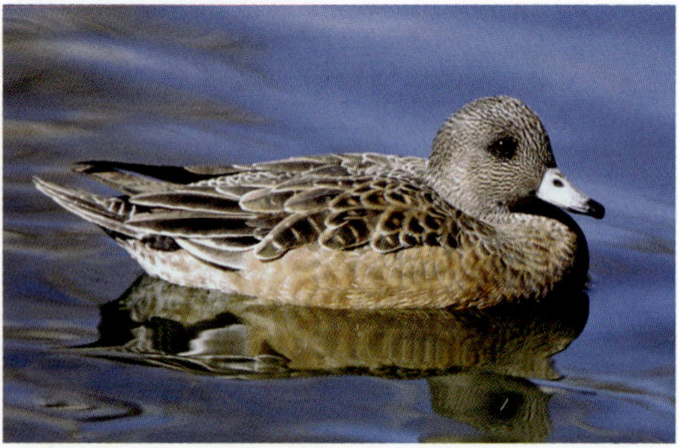

Adult female. Note head is greyer than flanks. January.

Range Breeds in North America. Winters south to northern South America via the West Indies and Central America.
Status in Jamaica Fairly common winter visitor and passage migrant (November–March).

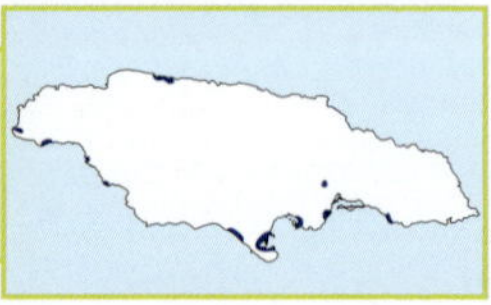

Green-winged Teal
Anas carolinensis

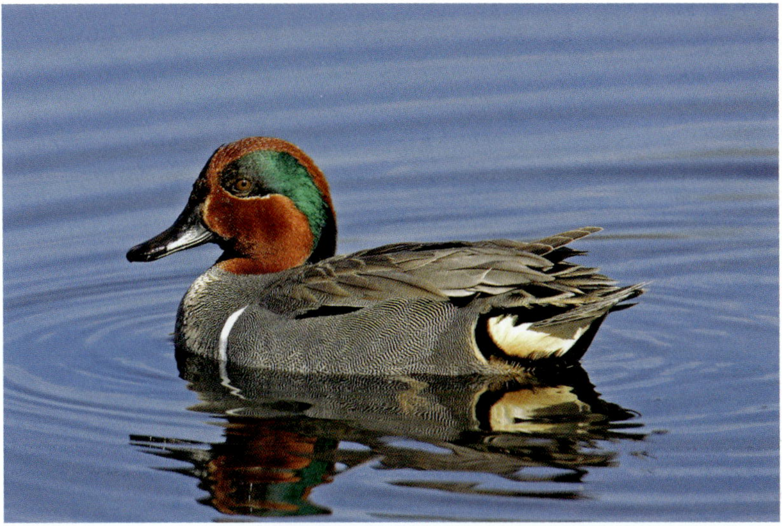

Adult male. Note white bar on flanks. Florida, February.

Taxonomy Monotypic.
Description L 33–39cm (13–15in). Surface feeding, small brown duck with green speculum. Adult male in breeding plumage has rich chestnut head. Mask (from eye to nape) iridescent green. Breast buffy, divided from grey side by vertical white stripe. Speculum bright green, edged brown above, white below. Female, juvenile and male in eclipse are brown.
Similar species Female resembles female Blue-winged Teal but lacks loral spot and has brown forewing.
Habitat and behaviour Wetlands, usually freshwater.

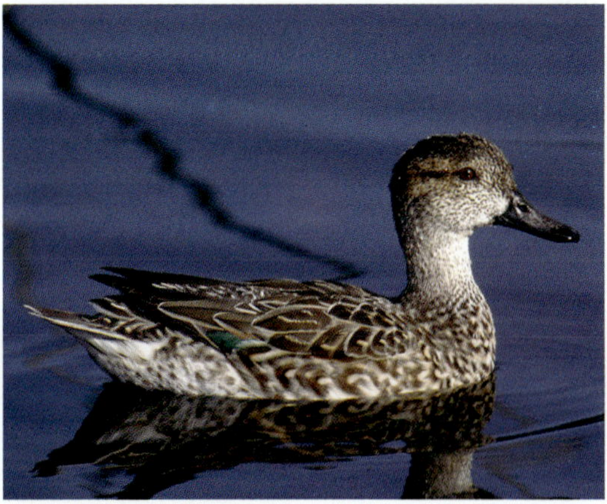

Adult female. Florida, February,

Range Breeds North America. Winters south to central Mexico.
Status in Jamaica Very rare winter visitor.

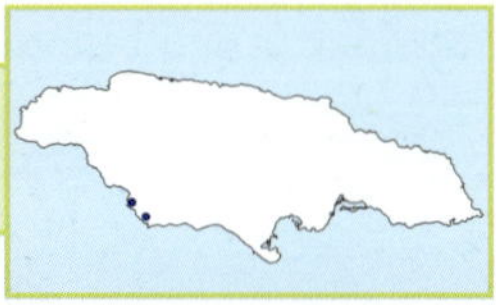

Blue-winged Teal

Anas discors

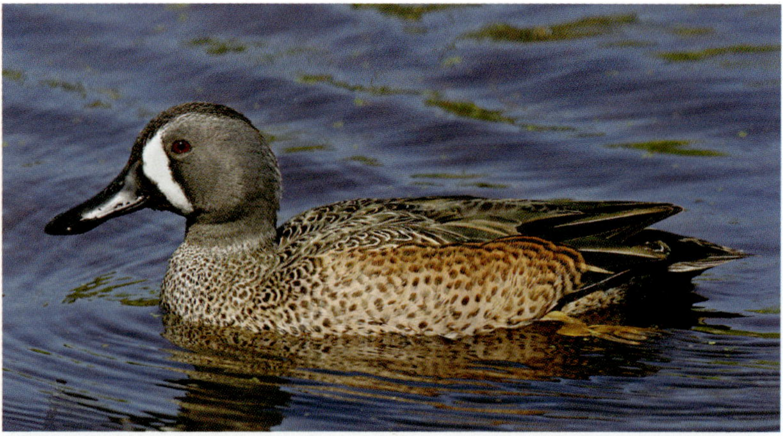

Adult male breeding. Florida, February.

Local name Duck-and-Teal.
Taxonomy Monotypic.
Description L 38–40cm (15–16in). mall brown dabbling duck with mottled body. In flight shows bright blue patch on forewing. Adult male in breeding plumage has a conspicuous, rear-facing, white crescent on the face and a green speculum edged with white and black. Males can be seen in this plumage throughout the winter. Adult female, juvenile and male in eclipse have a dark eye line and small white spot on the lores instead of the white crescent.
Similar species Female and juvenile Green-winged Teal lack blue on forewing. Those of Cinnamon Teal have cinnamon wash on belly. Northern Shoveler has very large bill.
Voice Male high-pitched peeping notes. Female very soft quacks.
Habitat and behaviour Fresh, brackish and saline ponds; also reservoirs and rivers. Feeds by dabbling on the surface or grazes in grass or pond margins. Often roosts on pond margins during the day. Sometimes seen in rafts at sea, usually inshore. Pairs are formed on the wintering grounds.

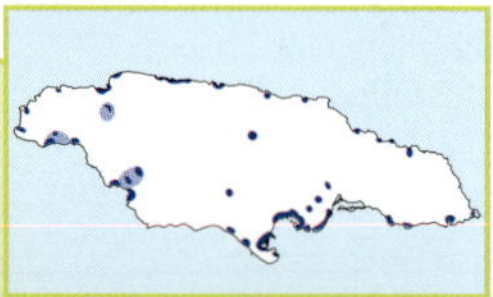

Range Breeds in the prairies and winters south as far as Argentina. Ringing recoveries suggest that Jamaican populations originate mainly from the Great Lakes area.
Status in Jamaica Common winter resident and passage migrant (August–May). The most abundant and widespread duck in Jamaica during the winter.

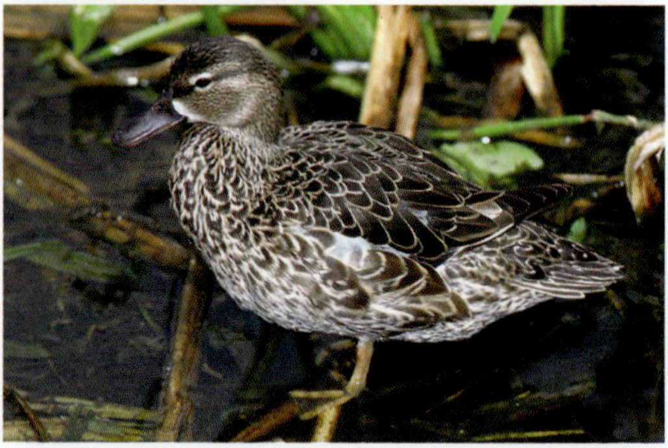

Adult female. Note broken white eye-ring, white at base of bill. Florida, Jan.

Cinnamon Teal
Anas cyanoptera

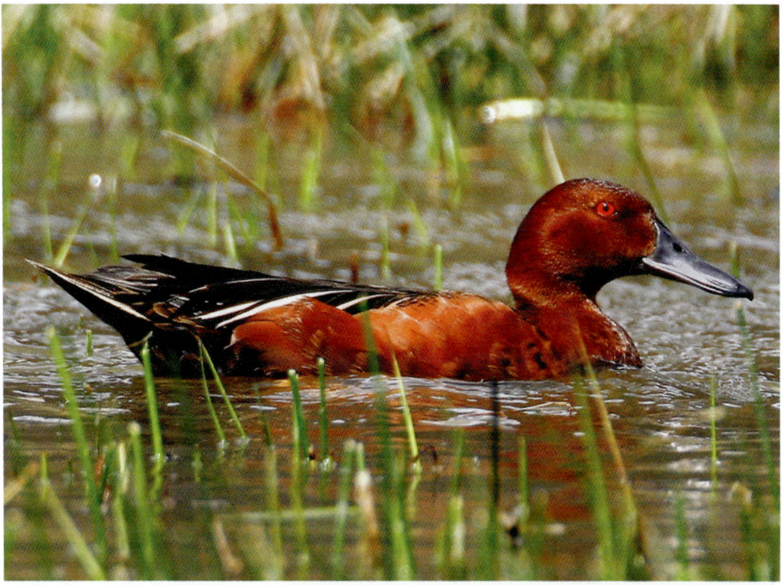

Adult male. Note length of bill. November.

Taxonomy Polytypic (5). Race is *A. c. septentrionalium*.
Description L 38–40cm (15–16in). Small size, brown body. In flight shows blue patch on forewing. Adult male (rarely seen in Jamaica) has bright cinnamon body. Iris is red. Adult female, juvenile and male in eclipse are brown and mottled, with cinnamon wash on underparts.
Similar species Blue-winged Teal has white on face and breast.
Habitat and behaviour Shallow, fresh or brackish ponds.

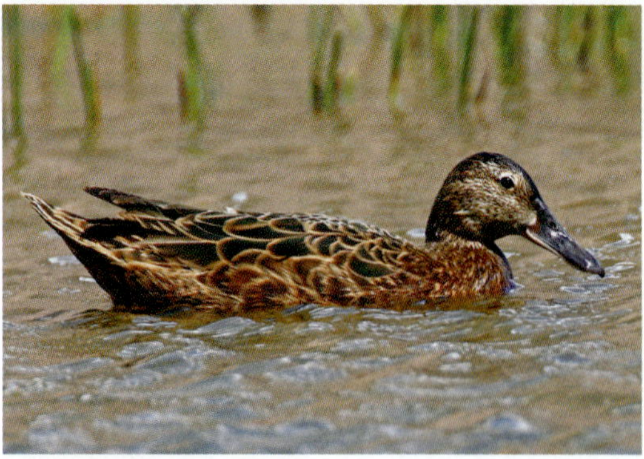

Adult female. November.

Range Breeds in the prairies of North America and winters south to northern South America.
Status Rare winter visitor.

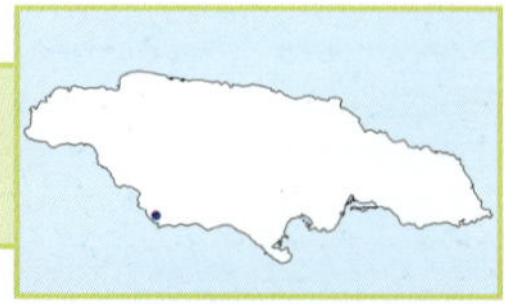

Northern Shoveler

Anas clypeata

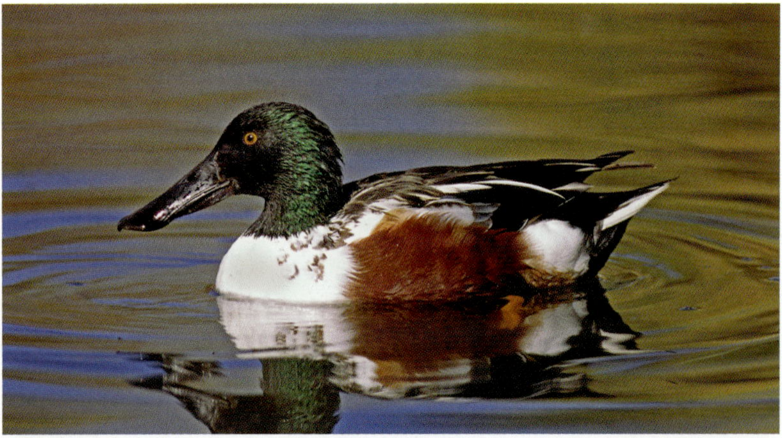

Adult male. Note large bill, white breast and rufous flanks. February.

Local name Spoonbill.
Taxonomy Monotypic.
Description L 45–50cm (18–20in). About a third larger than Blue-winged Teal, with a very large flattened bill. In flight shows blue patch on forewing, green speculum is edged with white. Adult male (common throughout Jamaica in winter) has distinctive contrast between green head, white breast and bright cinnamon underparts and flanks, white saddle markings on back seen in flight. Adult female, juvenile and male in eclipse are brown, with head less mottled and greyer than body.
Similar species Blue-winged Teal has similar green speculum and blue on wing but noticeably smaller, especially the bill.
Voice Usually silent in Jamaica.
Habitat and behaviour Freshwater, brackish and saline water bodies. Feeds at the surface by swimming with the bill in the water, filtering out small invertebrates, or dabbles to graze on submerged plants. Often seen in mixed groups with teal.

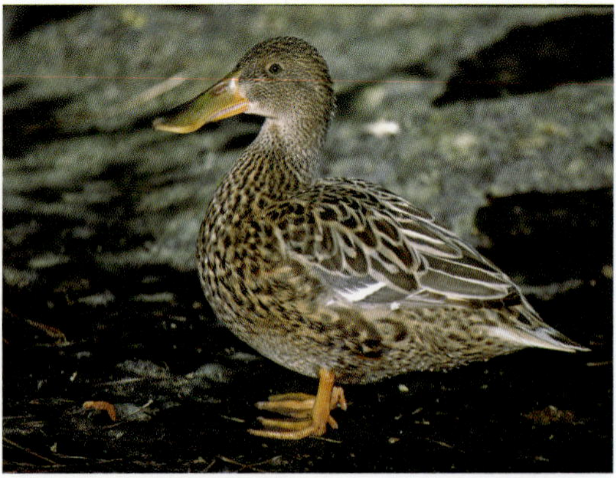

Adult female. Arizona, January.

Range Breeds in North America, wintering south to the West Indies and northern South America.
Status in Jamaica Common winter visitor (October–April).

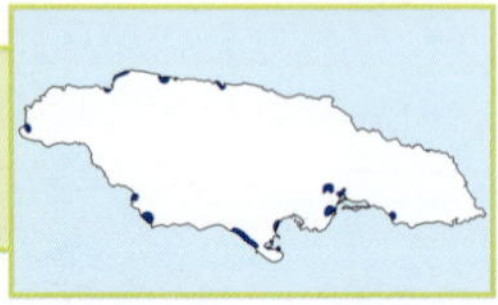

Ring-necked Duck
Aythya collaris

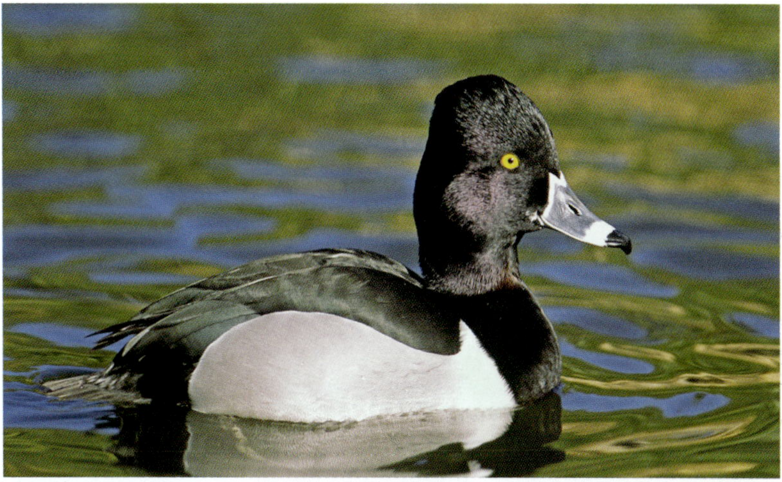

Adult male. Note peaked head, white ring and margins to bill. February.

Taxonomy Monotypic.
Description L 39–46cm (15–18in). Medium-sized diving duck with white ring on the bill. Peaked head. Light feathers surround base of bill. In flight secondaries grey. Adult male is black and white with black head, neck and back; grey sides, white vertical bar in front of wing, white underparts. Adult female is brown and white with white eye ring, head and neck brown, underparts white, flanks mottled brown. Male in eclipse resembles female but has a yellow iris.
Similar species Male Lesser Scaup lacks ring on bill, has grey wings, white primaries, lacks thin white line around base of bill. Female Lesser Scaup has broad white band around base of bill. Other similar species (such as Greater Scaup) are very rare in Jamaica.
Voice Usually silent in Jamaica.

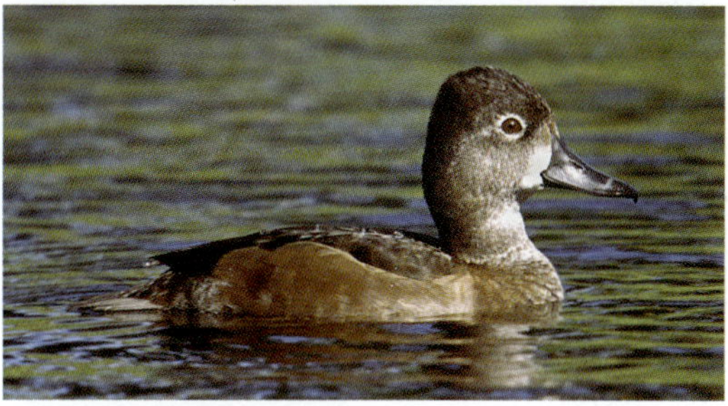

Adult female. Note white at base of bill. Dominican Republic, April.

Habitat and behaviour Freshwater wetlands, rivers and canals. Not usually seen with teal because it prefers deeper waters. Feeds by diving for roots, seeds, insects and molluscs. Runs over water to take off.

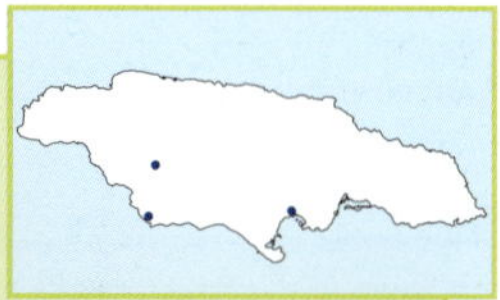

Range Breeds in northern North America and winters south of Panama. Non-breeders present year-round in Cuba. Occurs in Jamaica mainly on migration but a few individuals over-winter.
Status in Jamaica Uncommon winter visitor. Numbers seem to be increasing.

Lesser Scaup
Aythya affinis

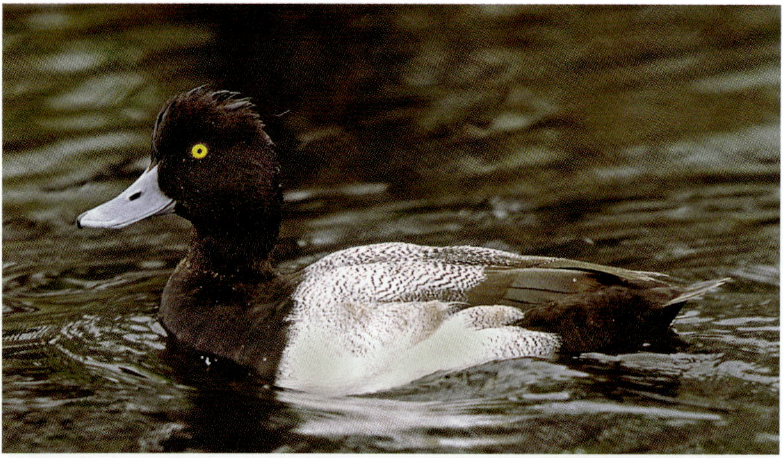
Adult male. Note peaked head. Florida, February.

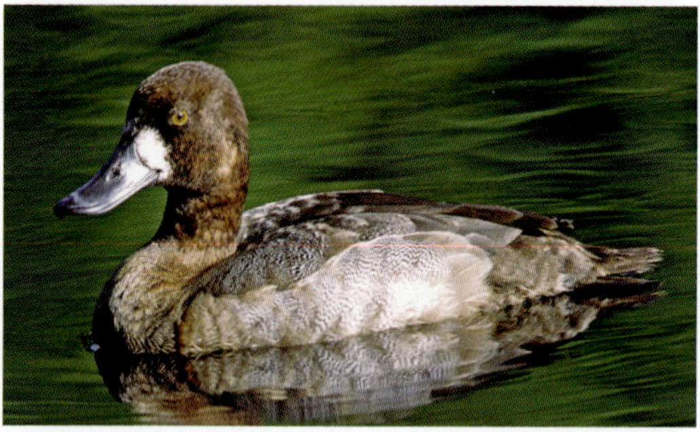
Adult female. Note brown head and neck, light base to bill. Cayman Islands, Feb.

Local name Black Duck, Black Head.
Taxonomy Monotypic.
Description L 31cm (12in). Diving duck with rounded peak at rear of crown. In flight wings show contrast between white secondaries and dark primaries. Adult male is black, white and grey with distinctive light greyish flanks and back. Black head with purple sheen (only visible in good light). Iris yellow. Female brown, with lighter brown flanks and back. Feathers around bill base are white.
Similar species Ring-necked Duck has a conspicuous white ring near the bill-tip. Greater Scaup (very rarely recorded in Jamaica) has a more rounded head with a green sheen rather than purple, and the white wing-bar extends to the primaries.
Voice Usually silent in Jamaica.
Habitat and behaviour Open water, including fresh and brackish ponds, sheltered bays and reservoirs. Rafts found at edge of ponds, or in the middle of reservoirs, sometimes among teal.

Range Breeds in North America. Winters in northern South America via Central America and the West Indies. Non-breeding birds present year-round in Bahamas and Cuba.
Status in Jamaica Locally common winter visitor and passage migrant (October–May).

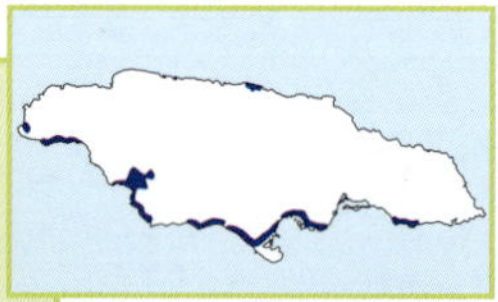

Masked Duck
Nomonyx dominica

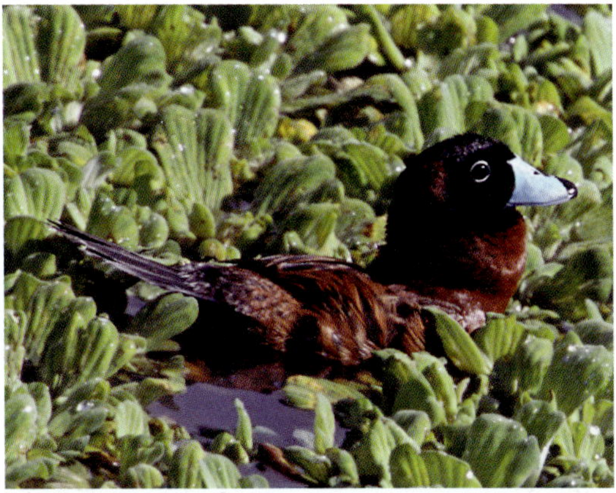
Adult male. Note blue bill and black head. Puerto Rico, April.

Local name Squat Duck
Taxonomy Monotypic.
Description L 25cm (10in). A chunky, brown, stiff-tailed duck. In flight (and when displaying) both sexes show a white patch in wing. Breeding male reddish-brown with black facial mask and very bright blue bill. Female, non-breeding male and juvenile are buffy and mottled, with dark crown and two horizontal dark stripes across face, one through eye and the other below; face of female tends to be cinnamon coloured. Bill dark. Ducklings buffy brown, with two dark stripes on face.
Voice A soft, dove-like cooing and a descending *du-du-du-du-du*.

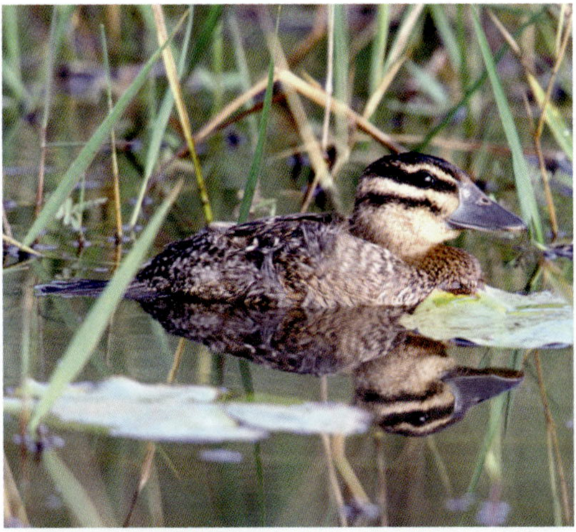
Juvenile or non-breeding female. Puerto Rico, April.

Habitat and behaviour
Fresh, brackish or saline ponds with dense vegetation around their margins or islands in the centre. Generally very secretive and therefore often overlooked. During the day may hide with only the head out of water, among dense vegetation such as water lilies; the undersides of the leaves match the male's breeding plumage. Swims with tail erect and fan-like, or trailing along the top of water. Sinks rather than flies when disturbed. Flies from roosting site at nightfall to feeding grounds, returning just before dawn. Feeds by diving. Nests in the winter (also year-round depending on conditions) in dense marsh vegetation.

Range Southern United States to Argentina, Brazil, the Bahamas and West Indies.
Status in Jamaica Locally common resident. More common following wet years.

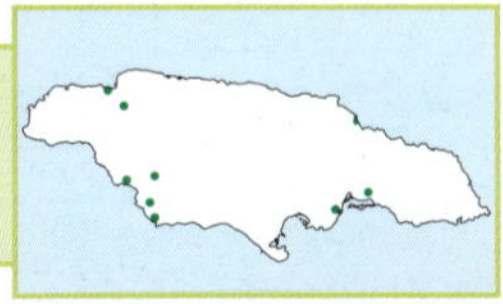

Ruddy Duck
Oxyura jamaicensis

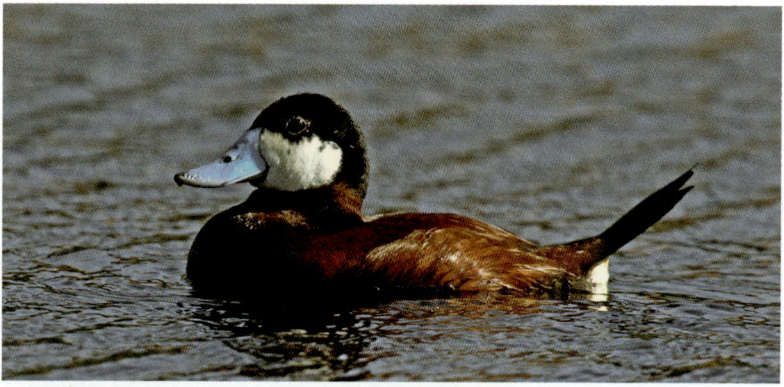

Adult male breeding. Note blue bill and white cheek. Dominican Republic, April.

Local names Diving Teal, Red Diver
Taxonomy Polytypic (2). Race in Jamaica is *O. j. jamaicensis.*
Description L 28cm (11in). A small, squat, stiff-tailed duck. Breeding male is conspicuous among rafts of other ducks, coots and grebes because of its bright rufous body and sky-blue bill. Head dark with large, white cheek patches. Tail may be held erect while swimming. Non-breeding male dull grey-brown, with white cheek patch and grey bill. Female and juvenile are both brown with one dark stripe across the pale brown cheek. Back brown, underparts paler. Ducklings dark brown.
Similar species Masked Duck male lacks white cheek patch while female and juvenile have a single stripe across lower face.
Voice a fast *kick-ik-ik-ik-ik.*
Habitat and behaviour Open stretches of fresh water (inland or on the coast) with vegetation at edges. Keeps tail on top of water or elevated above back while swimming. Ducklings dive expertly. Nests in thick rushes, mainly February–May but possibly year-round depending on conditions.

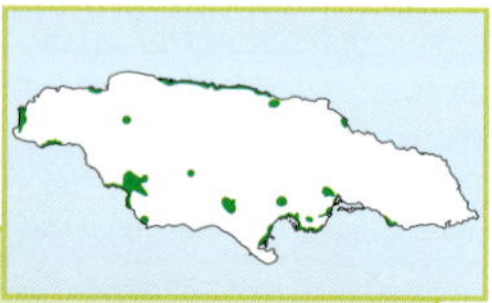

Range *O. j. jamaicensis* breeds from southwest Canada to Mexico and West Indies. Rare in Lesser Antilles; also breeds in Colombia. Northern populations migrate south in winter. Southern populations are resident.
Status in Jamaica Locally common resident.

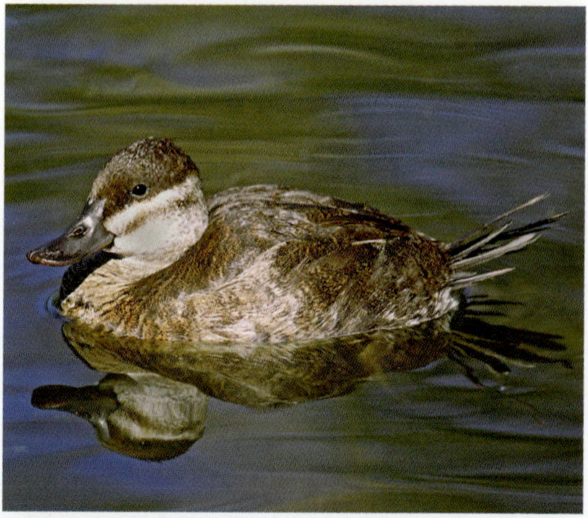

Juvenile. Note single indistinct line on face. January.

Least Grebe
Tachybaptus dominicus

Local name Diving Dapper.
Taxonomy Polytypic (4). Race in Jamaica is *T. d. dominicus.*
Description L 24cm (9in). Very small dark grebe with yellow iris and white fluffy undertail-coverts. Bill straight and thin. In breeding plumage head, throat and back are blackish. Iris bright yellow. In non-breeding plumage the throat is white. Nestling has black-and-white stripes and a red patch on crown. Juvenile streaked with black and white, iris dull yellow.
Similar species Pied-billed Grebe is larger, paler and has thick bi-coloured bill. Iris dark.
Voice Rapid, raspy *tetetetete*… Single *keeek* when alarmed. Downy chicks are very vocal, peeping incessantly and loudly, only pausing to swallow food.

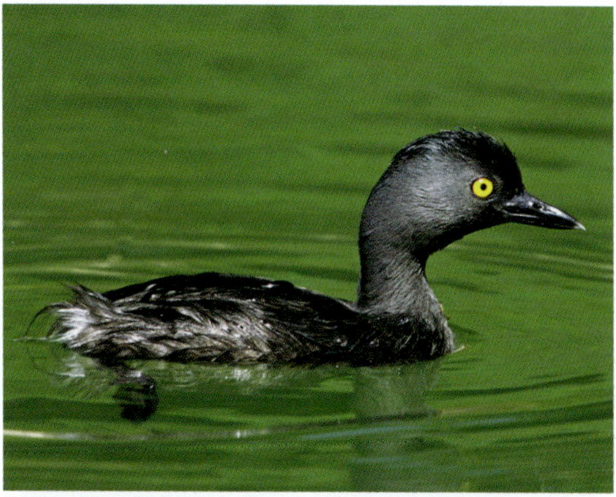

Adult breeding. Note bright yellow iris and black face. Jamaica, April.

Habitat and behaviour Fresh-water ponds, cattle ponds and reservoirs, and temporary bodies of fresh water. When alarmed, usually dives or sinks, staying submerged for long periods, emerging far from point of submersion. Rarely flies during the day; takes off by pattering across the water noisily. Often displays by rising up and flapping wings. Feeds on aquatic invertebrates and small fish. Nests year-round. A platform is made of rotting pond vegetation, leaves, sticks and mud, usually attached to emergent plants. Lays two or more white eggs at intervals of a day or more. The parents cover the eggs with wet vegetation on leaving the nest so the eggs get stained brownish. Chicks often ride on their parents' backs (secured by the wings), even during dives. A pair may raise two broods per year.

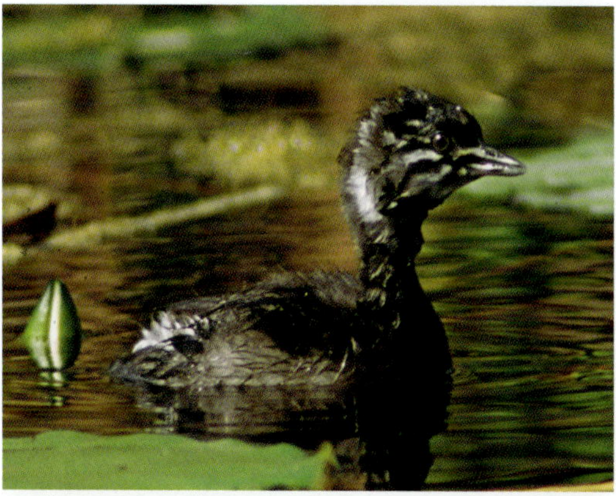

Juvenile. Note stripes. Cuba, May.

Range *T. d. dominicus* is resident and breeds throughout the Greater Antilles, Bahamas, Virgin Islands and Cozumel Island (Mexico). Other races occur in Central and South America.
Status in Jamaica Common resident.

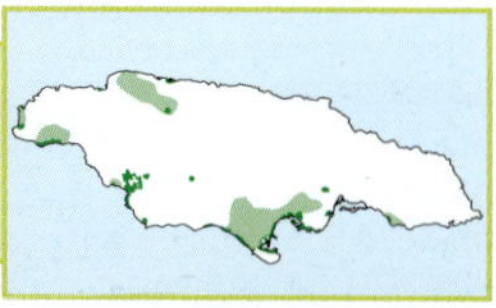

Pied-billed Grebe
Podylimbus podiceps

Taxonomy Polytypic (3). *P. p. antillarum* is endemic to the West Indies.
Description L 30–38cm (12–15in). Breeding adult (plumage seen year-round) has a black throat and a black mark across the middle of the thick, conical bill. Non-breeding adult has a white throat and unmarked bill. Nestling and juvenile streaked black-and-white with red on head, the streaks on the body disappearing as the bird ages until only the head is streaked.

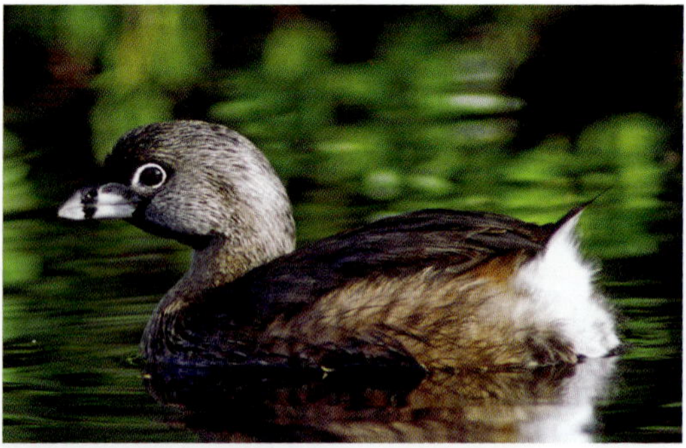
Adult breeding. Note dark band on heavy bill. Cayman Islands, May.

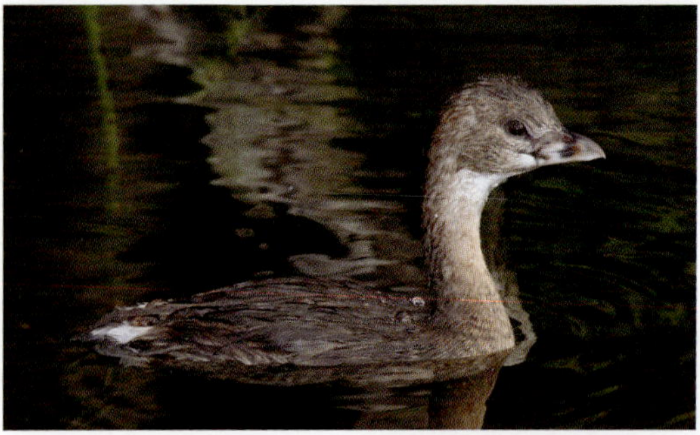
Adult non-breeding. Pale coloration. Cayman Islands, May.

Range *P. p. antillarum* occurs in Greater Antilles and Bahamas. Other races in North and South America.
Status in Jamaica Common. Population greatly increased by North American visitors (of race *P. p. podiceps*) in winter and especially during migration in March–April.

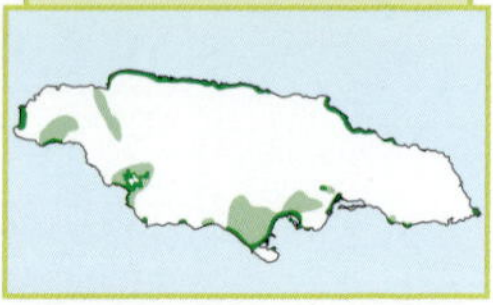

Similar species Least Grebe is smaller, has a yellow iris and a thinner bill.
Voice A loud *cow-cow-cow-cow* ... Downy chicks peep incessantly.
Habitat and behaviour Ponds, rivers and reservoirs, especially larger lowland water bodies. When alarmed it may sink until only the head is visible, submerge completely, dive, or patter across the water in a struggle to take-off. Feeds on aquatic invertebrates. Nests year-round. A floating platform made of pond vegetation is attached to emergent plants near the bank. Lays two or more eggs. The parents cover the eggs with wet vegetation when they leave the nest.

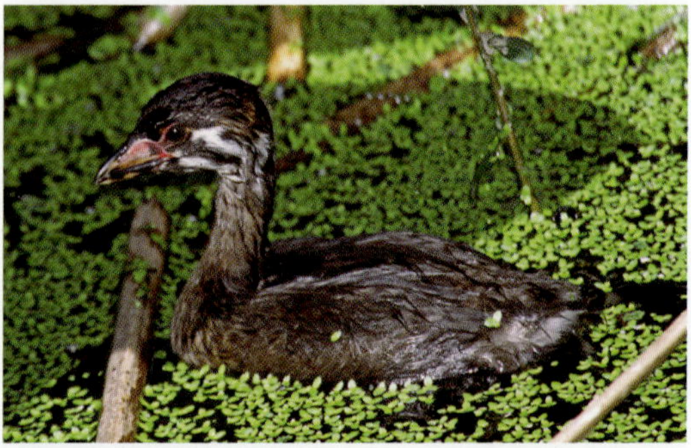
Juvenile. Florida, May.

Audubon's Shearwater
Puffinus lherminieri

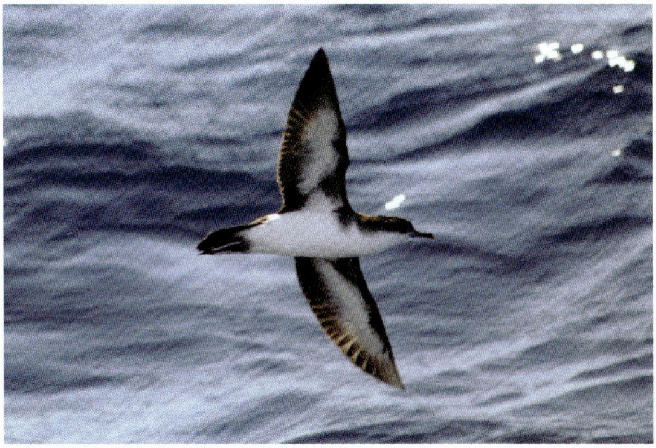

Adult. 90 miles off the coast of Massachussets, August.

Taxonomy Polytypic (9). Race in region is *P. l. lherminieri.*
Description L 28cm (11in). Adult blackish-brown above and white below. Dark crown extends to below eye. Upperwing blackish-brown, darkest on primaries. Underwing mainly white with contrasting dark flight-feathers. Dark undertail-coverts. Relatively long-tailed and short-winged for a shearwater. Nestling is grey.
Similar species None in region.
Habitat and behaviour Pelagic outside the nesting season, mainly off the southern Atlantic coast of North America near the Carolinas. Flies with fluttery wingbeats and short glides close to water surface. Feeds on small fish, squid and crustaceans at the water surface, or makes shallow dives. Nests colonially in rock crevices or under dense vegetation. Nesting has not been confirmed for Jamaica but presence of adults ashore at night on the Morant Cays suggests there may be a small colony. Lays one egg.

Range *P. h. lherminieri* breeds in West Indies and Bahamas. Other races pantropical.
Status in Jamaica Scarce but probably overlooked.

White-tailed Tropicbird
Phaethon lepturus

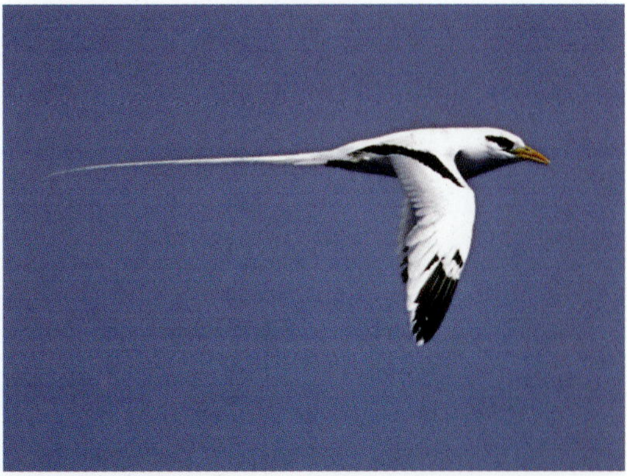

Adult. Note long tail and dark stripe on wing. Cayman Islands, April.

Local name Bo'sun Bird.
Taxonomy Polytypic (6). *P. l. catesbyi* is a Caribbean endemic race.
Description L (including streamers) 60–80cm (24–26in). Adult is a white, tern-like seabird with a broad black bar on the inner upperwing and black outer primaries. Wedge-shaped tail with two central tail feathers elongated into streamers (one or both of which might be missing). Black line through the eye. Orange bill. Immature lacks the streamers, is barred on head

and back and has a yellow bill.
Similar species Royal Tern has a forked tail and no black markings above. Red-billed Tropicbird has not been reported from Jamaica but has red bill, black barring on back and lacks broad black stripes on inner wings.
Voice A tern-like *keek-keek*.
Habitat and behaviour At sea most of the year, only coming to land to nest. Fishes by diving into the sea, often near the shore. Nests colonially (mainly January-March) in crevices in coastal cliffs, mainly on the north-east coast. Usually seen early morning and late evening (when nest change-overs occur) or feeding at sea.

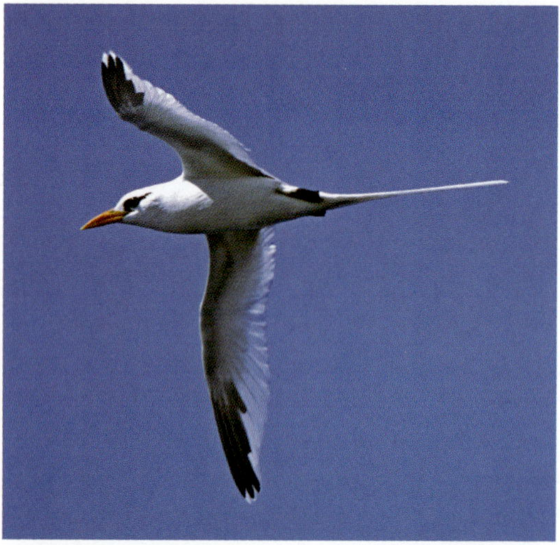

Adult. Cayman Islands, April.

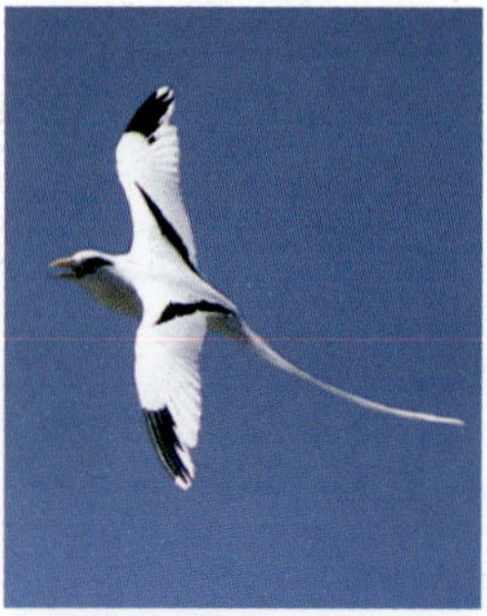

Range *P. l. catesbyi* West Indies and Bahamas. Other races are pantropical.
Status in Jamaica Locally common winter–spring resident, from October or November to June. Considered threatened in the West Indies.

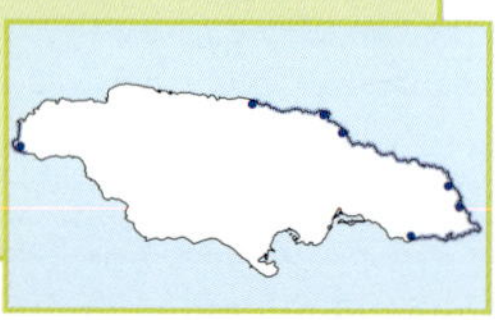

Left: Adult. Note pure white back. Cayman Islands, April.

Below: Adult. Note dark wing tips and yellow bill. Cayman Islands, April.

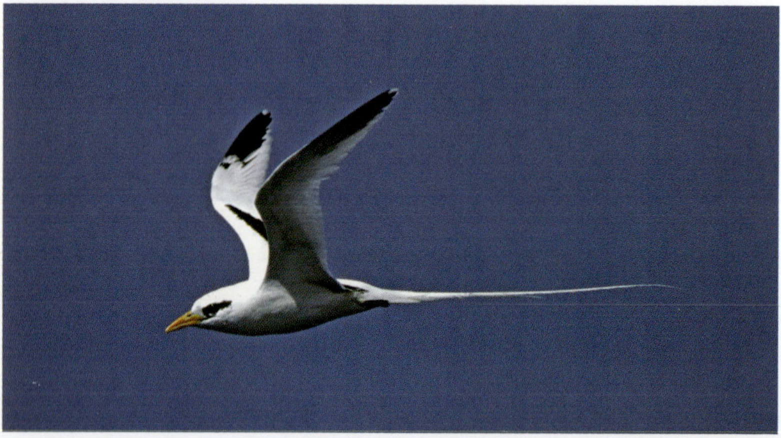

Brown Pelican
Pelecanus occidentalis

Local name Old Joe.
Taxonomy Polytypic (5). P. *o. occidentalis* is a West Indian endemic race.
Description L 100–137cm (39–54in) Breeding adult has yellow and white crown, chestnut-coloured hind-neck, white fore-neck and black belly. Irises white. Wings and back pale grey. Non-breeding adult has white hind-neck. Immature has brown head and neck, whitish underparts and brown irises. Nestlings are initially bare, later developing white down.
Similar species American White Pelican is white with dark primaries. It does not dive.
Voice Usually silent but chicks in the nest are noisy, with groaning and screaming.
Habitat and behaviour Harbours, coastal waters, reservoirs, ponds, and marshy areas. Seen perching at the water's edge on trees, at sea on buoys, or gliding close to the water surface. Flies in lines, each bird following the leader's flaps and glides. Feeds on fish by plunge-diving or scooping from the surface. At fish-farms takes fish weighing up to 0.5kg (1lb). Nests colonially (December–July) in mangroves on Refuge Cay (Port Royal Cays). Lays 2–3 white eggs.

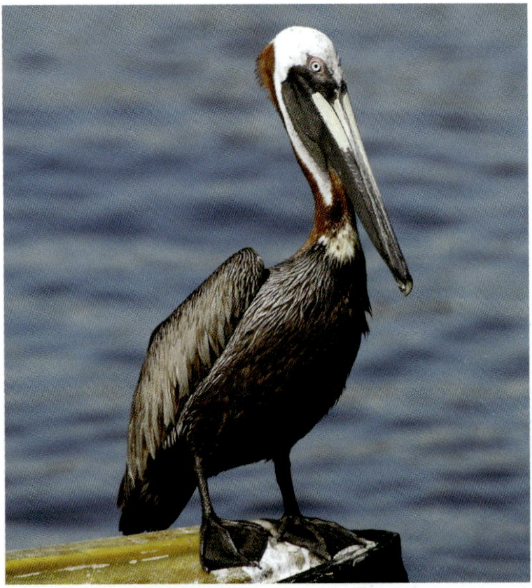

Adult breeding. Port Royal, Jamaica, March

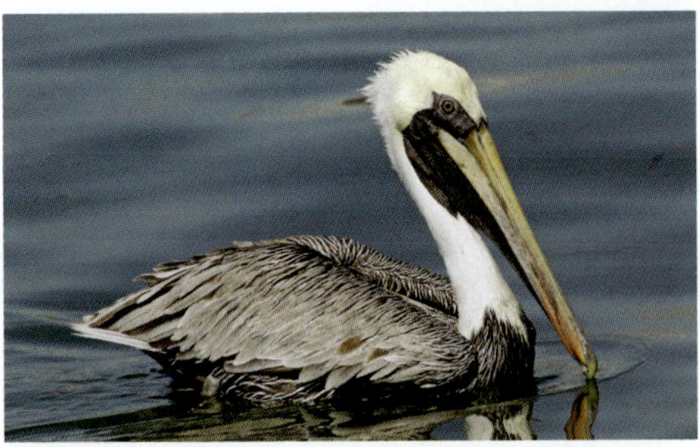

Adult non-breeding. Note white neck. Port Royal, Jamaica. March.

Range *P. o. occidentalis* occurs in the West Indies. Other races are found in coastal North, Central and South America.
Status Common resident. Residents are probably joined by migrants from North America in winter.

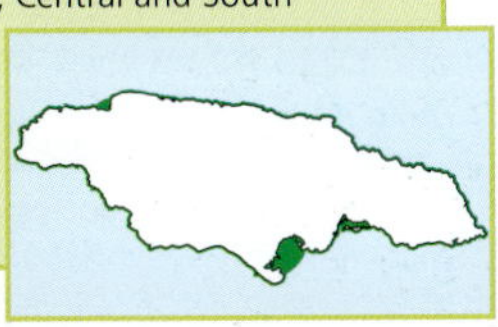

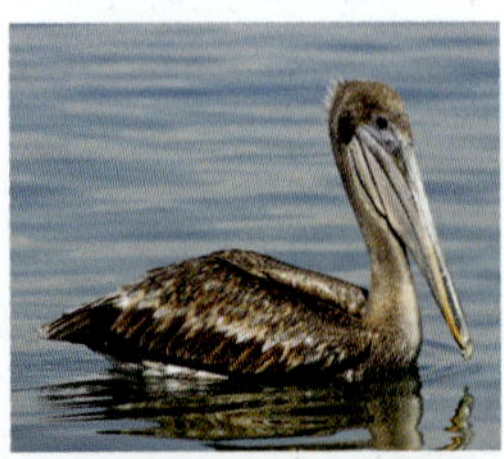

Juvenile. Note greyness. Jamaica, Mar.

Masked Booby
Sula dactylatra

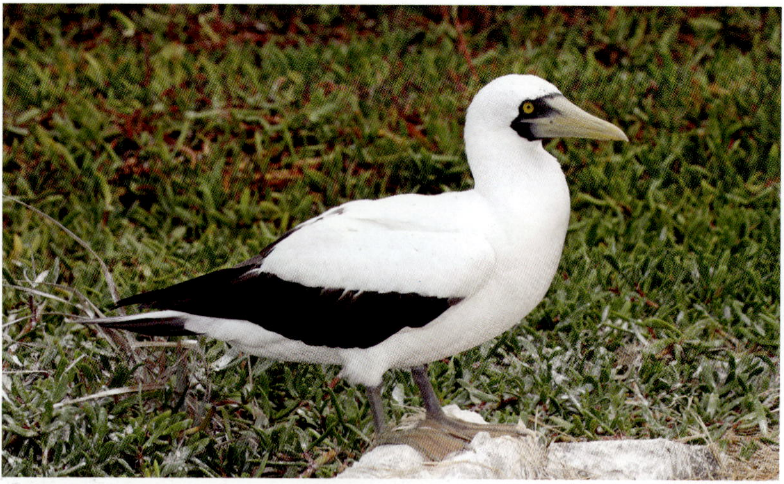

Adult. Note yellow bill with dark base. Pedro Cays, Jamaica, March.

Taxonomy Polytypic (4). Race in Jamaica is *S. d. dactylatra*.
Description L 74–86cm (29–34in). Adult white, with black flight feathers and tail. Bill yellow. Nestling white. Immature brown on back and head, with white collar, rump and underparts.
Similar species Brown Booby has a dark brown head and neck and lacks white collar. White phase Red-footed Booby has white head and tail.
Voice At nest female honks loudly. Male makes series of high-pitched whistles.
Habitat and behaviour Seen year-round on the Pedro Cays. Feeds by plunge-diving, entering the water almost vertically with very little splash. At the nest, pairs exhibit highly ritualised displays. Nests colonially, year-round (mainly August–November) on hard-packed sand with little or no nesting material.

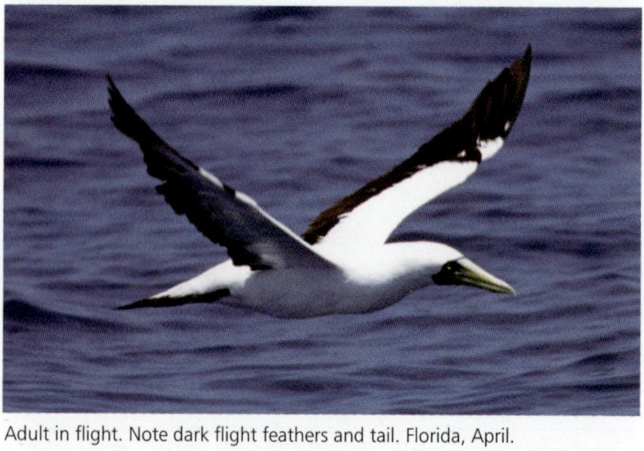

Adult in flight. Note dark flight feathers and tail. Florida, April.

Range *S. d. dactylatra* is resident on West Indian and southwest Atlantic islands. There are only about four known nesting colonies in the West Indies and the Pedro Cays colony is one of the largest. Other races are pantropical.
Status in Jamaica Very local resident. One of the rarest boobies in the Caribbean, but more than 200 pairs nest on Middle Cay and South West Cay on the Pedro Bank.

Red-footed Booby
Sula sula

Taxonomy Polytypic (3). Race that occurs in Jamaica *S. s. sula*.
Description L 71cm (28in). Small booby with greyish bill and bright red feet. Dark morph adult brown with all-white tail. Light morph adult white with black primaries and secondaries. Immature dull brown with yellowish-grey feet.
Similar species *Dark morph*: adult Brown Booby has darker head and neck and white tail. Juvenile Brown Booby is darker, especially on head and neck. *Light morph* (very rarely reported in Jamaica) resembles Masked Booby but has white tail and red feet.
Habitat and behaviour Flies bouncily. Feeds by plunge-diving steeply.

Range *S. s. sula* breeds on small, usually uninhabited cays from West Indies (not Jamaica) to Brazil. Other races pantropical.
Status in Jamaica Rare visitor or vagrant (mostly immature birds).

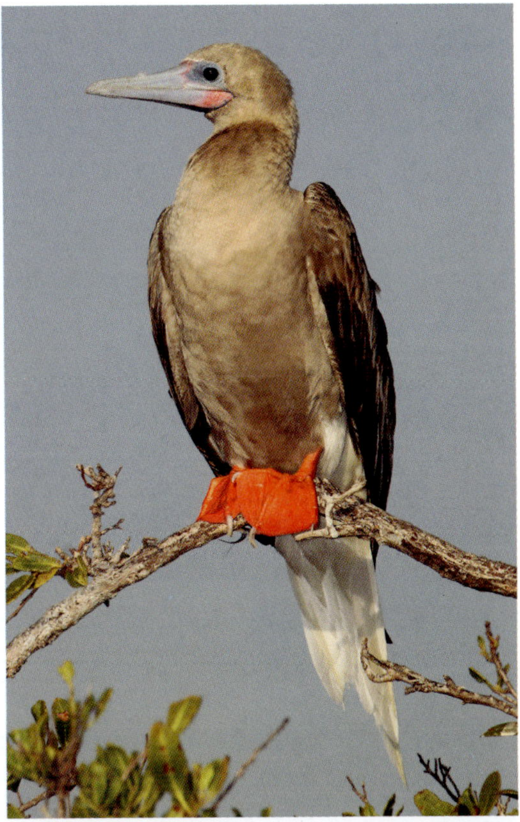

Adult dark morph. Red feet, pink-based greyish bill. Cayman Islands, March.

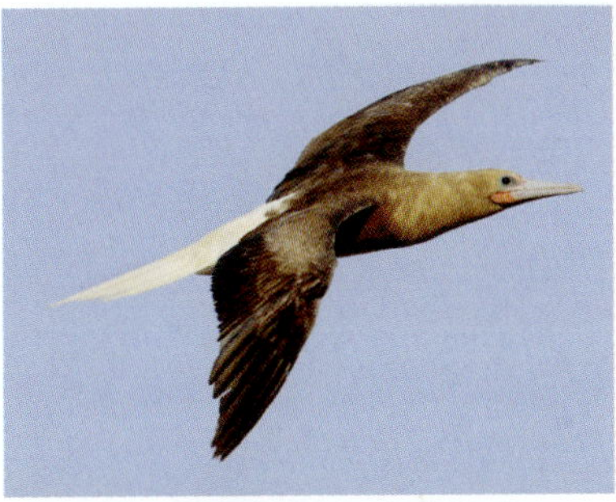

Adult dark morph in flight. Cayman Islands, March.

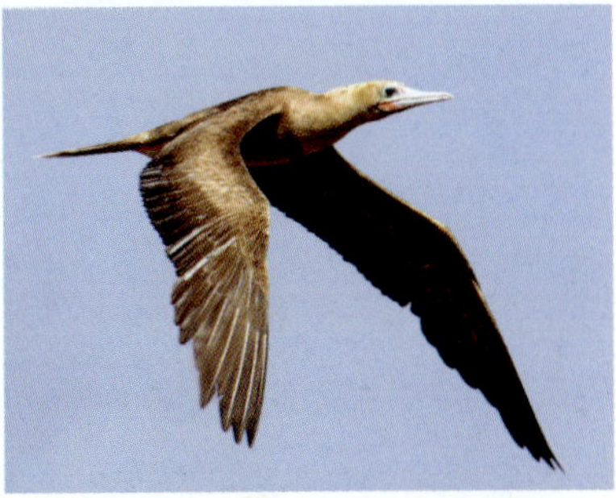

Juvenile dark morph in flight. Cayman Islands, March.

Brown Booby
Sula leucogaster

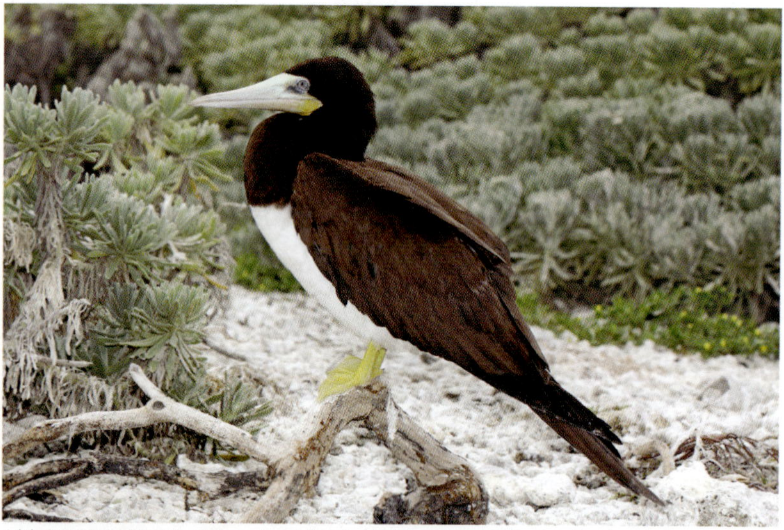

Adult. Note sharp delineation between brown breast and white abdomen. Pedro Cays, Jamaica, March.

Taxonomy Polytypic (4). Race in Jamaica is *S. l. leucogaster.*
Description L 71–76cm (28–30in). Adult has chocolate-brown head, upperparts and flight feathers. The brown breast is sharply delineated from the white abdomen. Bill mostly yellow or grey in adults. Legs and feet yellow. Immature (first year) is dusky brown above and white, mottled brown below, with dark bill and yellowish feet and legs.
Similar species Immature Masked Booby has white collar. Brown phase Red-footed Boobies are much paler.
Voice At the nest immature and female honk, males have high whistle.
Habitat and behaviour Pelagic outside the breeding season. Flies with neck outstretched, frequently gliding. Feeds by plunge-diving steeply or from a shallow angle. Immature is regularly seen from the mainland. Nests colonially (September–November) on the ground among plants on Pedro Cays.

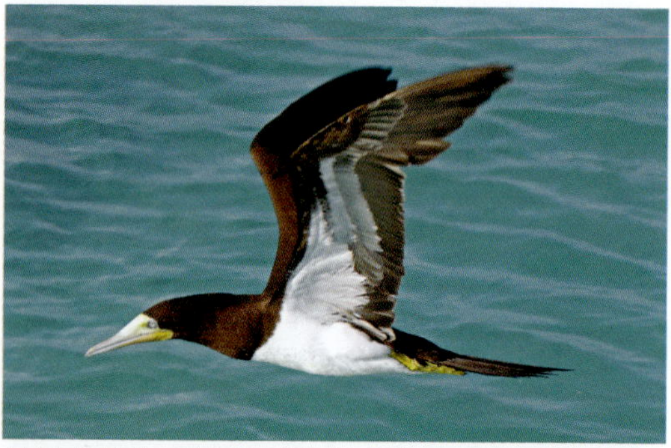

Adult in flight. Note all dark head, back and tail. Cayman Islands, January.

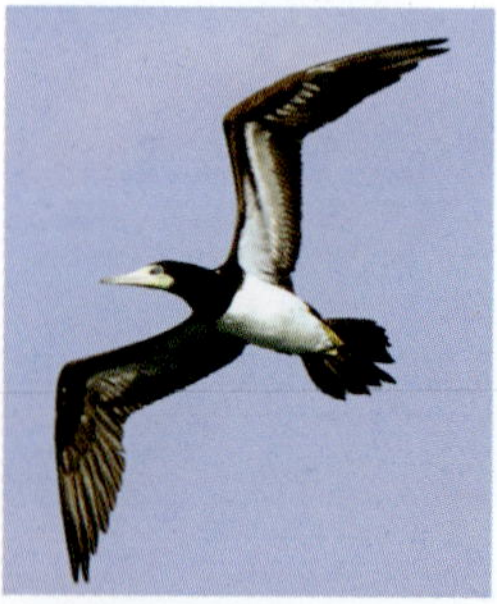

Adult. Cayman Islands, January.

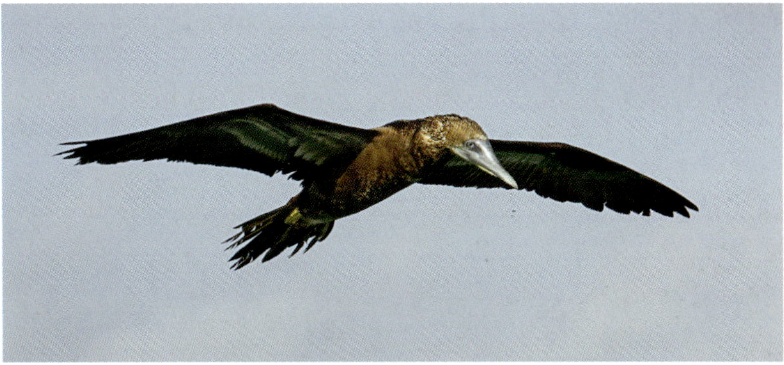

Juvenile in flight. Note dark bill and abdomen. Cayman Islands, January.

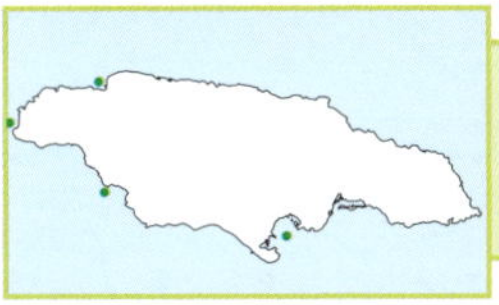

Range *S. l. leucogaster* breeds on West Indian and Atlantic islands. Other races are pantropical.
Status Local resident, breeds on the Pedro Cays and occasionally seen in coastal waters (Portland Bight, Montego Bay and Negril).

Double-crested Cormorant
Phalacrocorax auritus

Taxonomy Polytypic (4). *P. a. floridanus* is the most likely race to occur in Jamaica.
Description L 84cm (33in). A large dark bird with a long neck. The wide orange pouch on the throat is rarely visible. Usually identified by its profile as it stands erect on pond margins, or swims with just head exposed, or flies with kinked neck. Immature brown.
Similar species Neotropic Cormorant (vagrant to Jamaica) is smaller and has a yellowish gular region with a white border.
Habitat and behaviour Seen perched on the margins of sewage ponds or swimming (for example, on Montego Bay). Dives to feed on fish, submerging for extended periods and emerging far from the place where it originally dived.

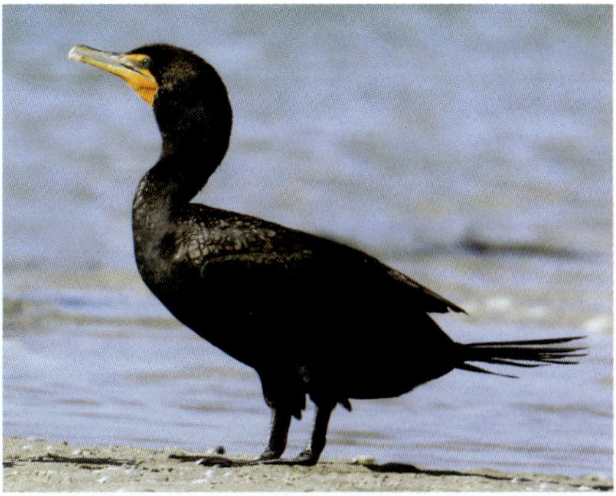

Adult. Note the kinked neck. Florida, February.

Range Breeds in North America, Cuba and the northern Bahamas. Winters south to Belize.
Status Rare winter visitor.

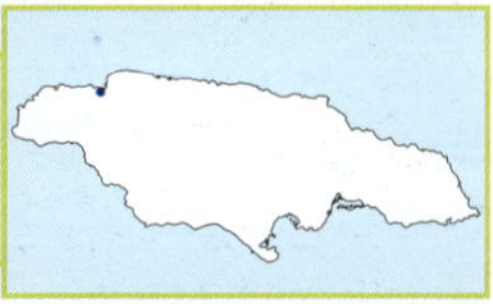

Magnificent Frigatebird
Fregata magnificens

Local name Man-O'-War Bird, Scissor's Tail, Gwaloo.
Taxonomy Monotypic.
Description L 89–114cm (35–45in). Large black seabird with characteristic bent-winged silhouette in flight. Tail deeply forked. Adult male all-black with a long, hook-tipped grey bill and a large, red, balloon-like gular pouch which is inflated in courtship. Adult female black with white across the breast. Immature brownish-black with white head and underparts.
Similar species None
Voice Usually silent, except at the nest where breeding males inflate their red pouches and sit on bushes or the nest, clattering their bills and making a fast, low, clicking sound until a female is attracted.
Habitat and behaviour Coastal harbours and beaches. Glides effortlessly with minimal flapping over fresh or salt water, or perches on trees and shrubs near the water. Flocks often soar on thermals over fishing beaches or harbours. Feeds by harassing gulls, terns and pelicans to steal fish from them. Also swoops over the surface to scavenge waste fish discarded by fishermen. Sometimes catches flying fish or picks floating refuse from the surface of the sea (but never lands on it, because its feathers are not water resistant). Also preys on tern chicks. Nests colonially (August–March) on coastal shrubs on cays, or on mangrove trees on the mainland. The nest is a platform constructed of twigs and sticks (often stolen from nearby nests) topped by a white layer of smoothed faeces. Female feeds chick for a year and only breeds in alternate years.

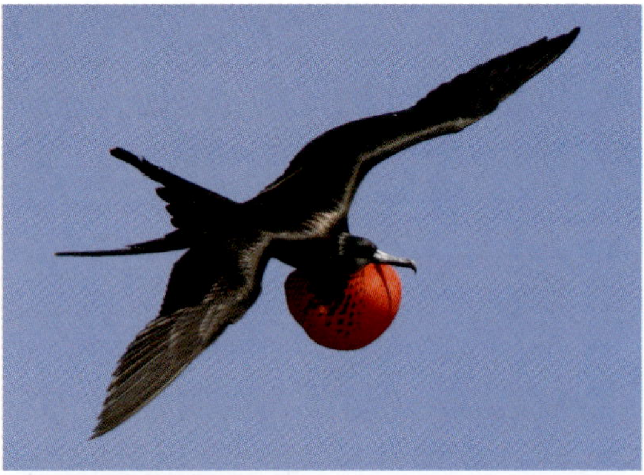

Adult male with inflated pouch. All flying birds Cayman Islands, November.

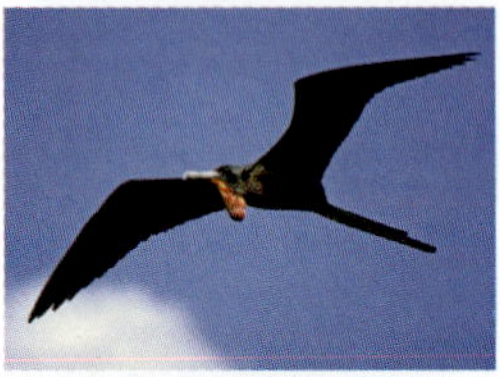

Adult male with deflated pouch.

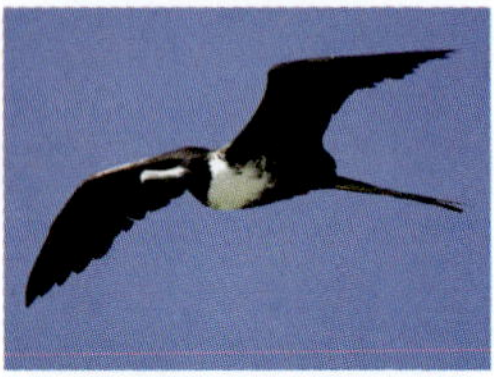

Adult female.

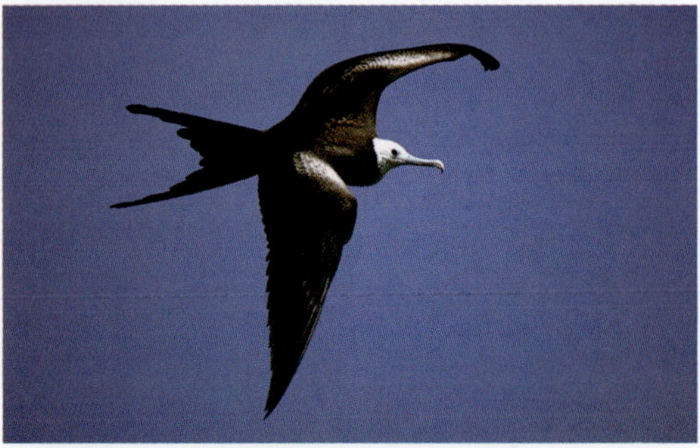

Juvenile in flight. Cayman Islands, November.

Range Tropical western Atlantic and East Pacific.
Status in Jamaica Locally common resident.

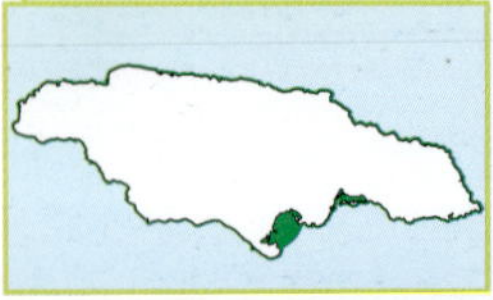

Great Blue Heron
Ardea herodias

Local name Blue Gaulin, Morgan
Taxonomy Polytypic (5). Race in Jamaica *A. h. occidentalis*.
Description L 96–137cm (38–54in). Adult is a large grey-blue heron with black forehead and crown (with a white centre) and black underparts. Long grey neck. Eye and large, thick, straight bill yellow, legs mostly brownish. Wings and back brownish-grey. *White morph* all white with yellow bill and legs; has not been recently reported from Jamaica. Juvenile has whole crown black, lower mandible yellow contrasting with dark upper mandible.
Similar species None. All other non-white herons and egrets are much smaller.
Voice A nasal *cro-aaark*.
Habitat and behaviour Saline or freshwater ponds and sometimes by the sea. Usually stands quietly and alone in a pond. Sometimes seen perching on top of mangroves. Feeds mainly on small fish and invertebrates.

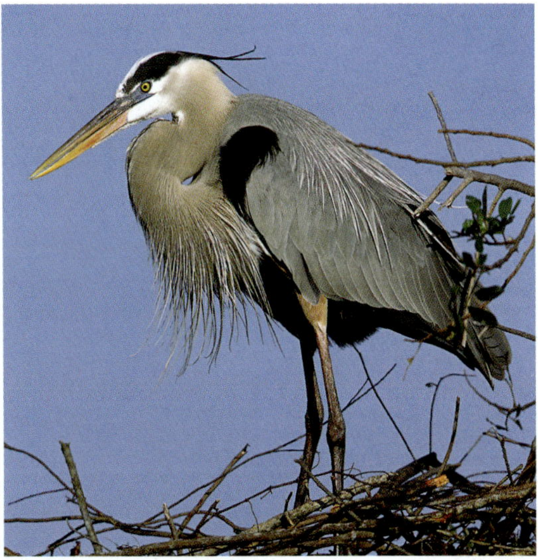

Adult with black plumes. Florida, January.

Range *A. h. occidentalis* in southern Florida, West Indies (breeding in Cuba, rarely in Hispaniola) and islands off Venezuela. Other races breed North America, winter in West Indies and Central America south.
Status in Jamaica Common winter visitor (August–May). A few non-breeders spend the summer. Previously bred in Jamaica but has apparently been extirpated as a breeding species.

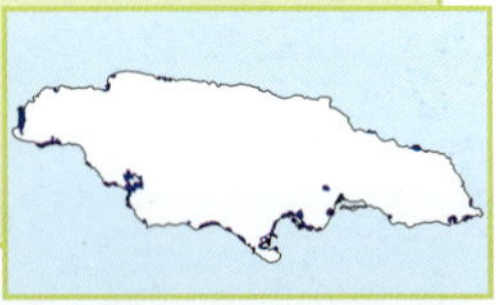

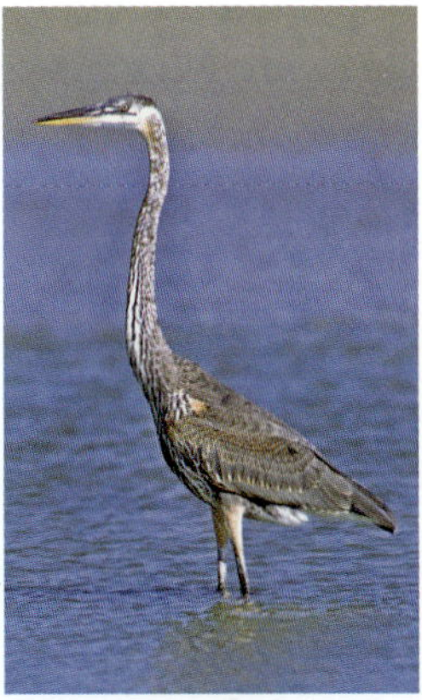

Juvenile. Florida, February.

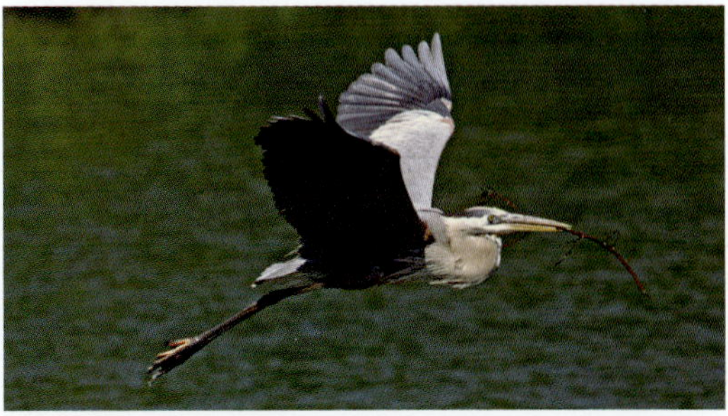

Adult in flight. Note contrasting dark flight feathers. Florida, January.

Great Egret
Ardea alba

Local name Crane.
Taxonomy Polytypic (4). Race in Jamaica is *A. a. egretta*.
Description L 94–104cm (37–41in). Adult and juvenile are large white egrets with black legs and yellow bills. Irises yellow. Very long neck often held in an 'S' shape in flight.
Similar species Cattle Egret is much smaller, has a relatively short neck and is not usually seen standing in water. Snowy Egret is much smaller and has golden yellow feet.
Voice A snoring *cro-aak*.
Habitat and behaviour Generally seen stalking through shallow waters or on the open grassy banks of rivers, wetlands and fish-farms. Sometimes in shallow water between the reef and the shore. Nests colonially (March–June) in mangroves, often with various other herons and egrets. The nest is a large platform of mangrove sticks. Lays 2–3 blue-green eggs.

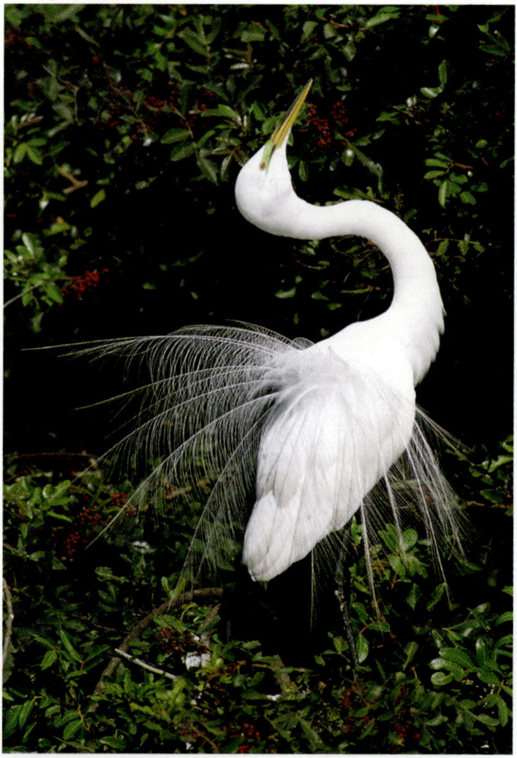

Adult breeding with long plumes and greyish lores. Florida, February.

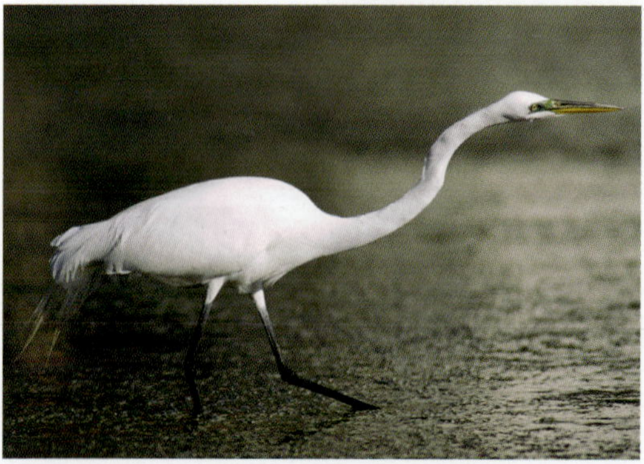

Adult foraging. Note yellow bill and dark legs. Florida, February.

Range *C. a. egretta* occurs in West Indies and North, Central and South America. Other races occur worldwide.
Status in Jamaica Very common in wetlands. Numbers are increased by migrants in winter.

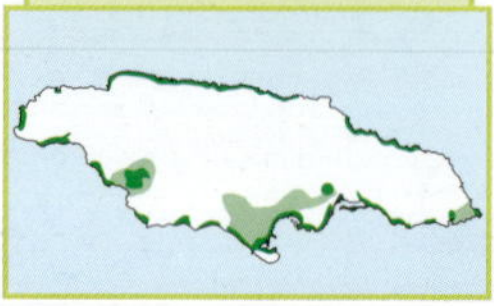

Reddish Egret
Egretta rufescens

Taxonomy Polytypic (2). Race in Jamaica is *E. r. rufescens.*

Description L 70–80cm (26–32in). Similar in size to Tricolored Heron, larger than Snowy Egret or juvenile Little Blue Heron. *White morph* white with a heavy bicoloured bill, with pinkish base and black tip. Legs blue. Iris pale yellow. Juvenile white morph is white with all-dark bill and dark legs, gradually changes to adult plumage in second year. *Rufous morph* greyish with reddish-brown head and neck. Juvenile rufous morph head and face washed light cinnamon, neck rufous. Wings grey with some black and white feathers, wing-coverts edged white. Lower underparts grey. Feet and legs black. Intermediate juveniles can show a mixture of white and rufous features.

Similar species *White morph*: juvenile Little Blue Heron has grey-blue bill tipped with black. Snowy Egret has all dark bill and yellow feet. Tricolored Heron has vertical white stripe on neck. *Rufous morph*: adult Little Blue Heron has grey-blue bill tipped with black.

Voice Usually silent in Jamaica.

Habitat and behaviour Saline ponds and marshy areas. Unlike other herons and egrets practices canopy feeding: characteristically pursues fish very actively, prancing around in the shallows with wings spread.

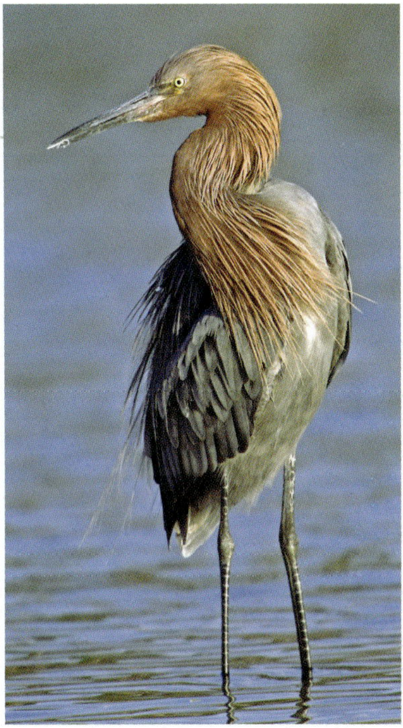

Adult rufous. Florida, February.

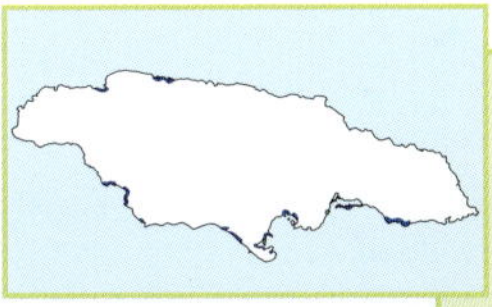

Range *E. r. rufescens* breeds elsewhere in the Greater Antilles (wintering in northern South America) but there are no recent reports of breeding in Jamaica. Other race occurs in coastal southwest North America and Central America.

Status in Jamaica Uncommon winter visitor (October–March). White morph more common than rufous morph.

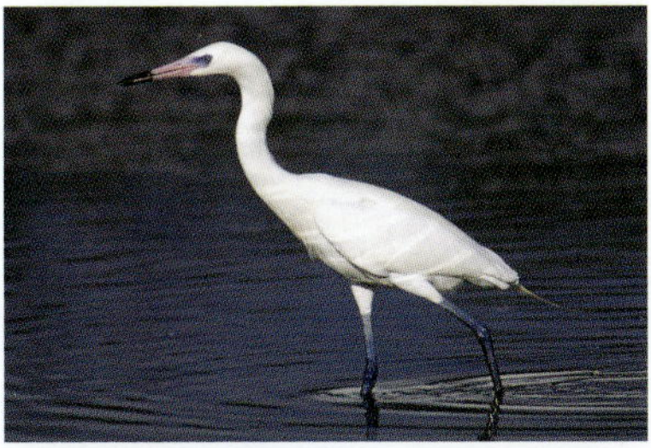

Adult white. Bahamas, April.

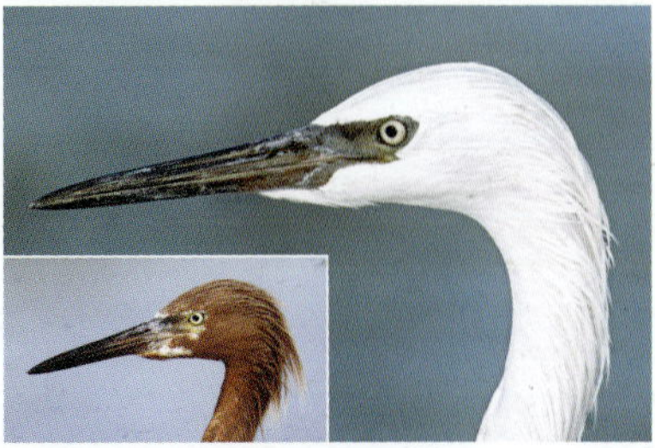

Juvenile rufous and white morphs.

Tricolored Heron
Egretta tricolor

Taxonomy Polytypic (2). *E. t. ruficollis* is a Caribbean endemic race.
Description L 60–70cm (24–28in). Medium-sized, grey, white and brown heron with a long neck and a very long, dark-tipped, yellowish bill. Upper mandible darker than lower. Adult has dark grey head, neck, wings and upper breast strongly contrasting with white belly and undertail-coverts. A distinctive rufous and grey chain-like pattern runs down the white foreneck. Legs and feet greenish-yellow. In breeding plumage has white plumes at back of head, lores turn blue, bill becomes bicoloured. Juvenile brown, with rufous hindneck and wing-coverts. Foreneck pattern similar to adult. Legs greenish yellow.
Similar species Little Blue Heron is all-dark.
Voice A harsh *croaak*.

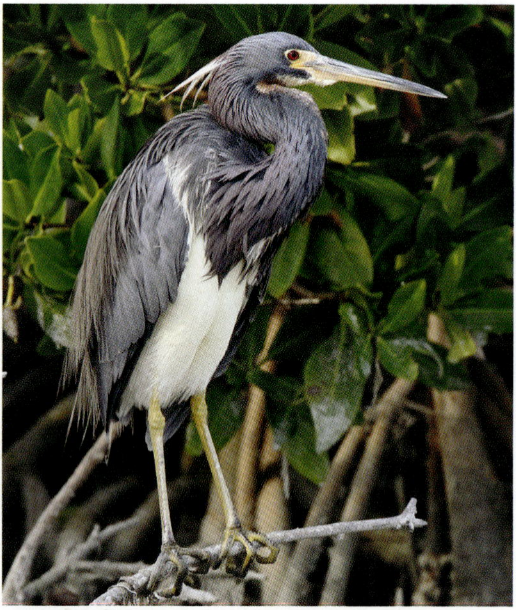

Adult breeding with plumes and bicoloured bill. Florida, April.

Habitat and behaviour Mangrove ponds and the seashore. Forages very actively. Feeds mainly on fish. Nests colonially (April–June) with other egrets and herons. Lays three bluish eggs on a loose platform of twigs.

Range *E. t. ruficollis* breeds Bahamas and Greater Antilles. Other race inhabits the continental Americas.
Status Fairly common resident. Local populations increased by migrants in winter.

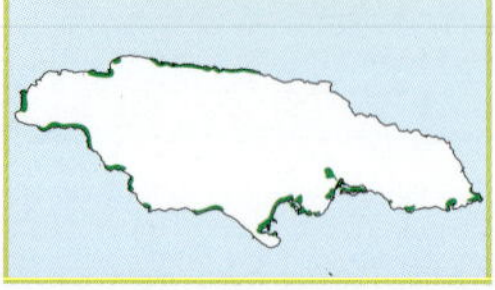

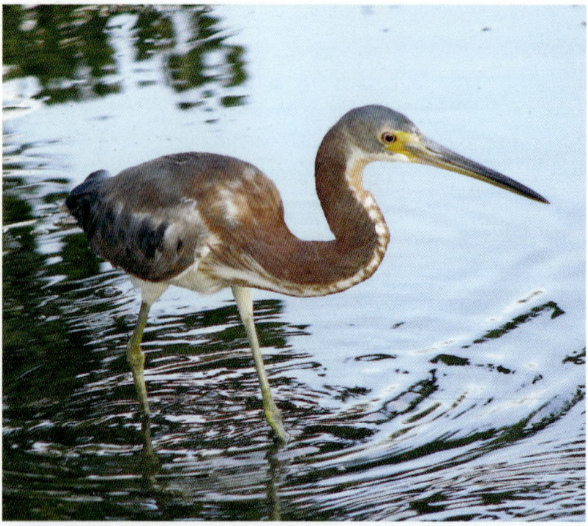

Immature. Yallah's Salt Pond, Jamaica, June.

Little Blue Heron
Egretta caerulea

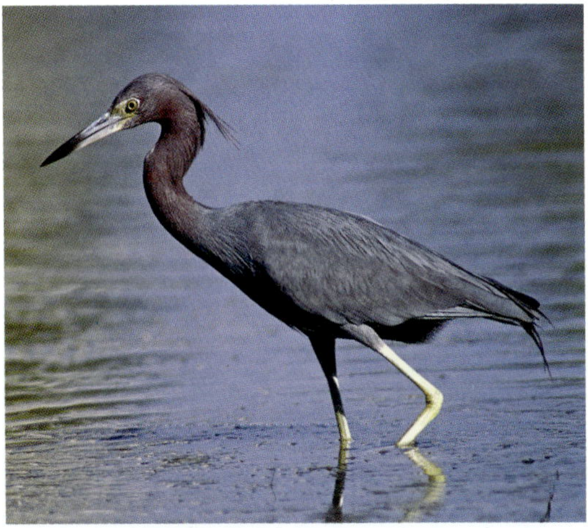

Adult breeding. Note plumes and dark tip to bluish bill. Florida, March.

Local name Blue Gaulin.
Taxonomy Monotypic.
Description L 56–74cm (22–29in). Small greyish-blue heron with dark-tipped bluish bill and greenish legs and feet. Juvenile white, developing dark feathers on the body and wings as it matures producing a characteristic patchy or calico pattern.
Similar species Snowy Egret resembles juvenile but has golden feet. White morph Reddish Egret is larger and has pinkish base to bill. Dark morph Reddish Egret has rufous neck. See also Little Egret (in Appendix 2).
Voice Usually silent.
Habitat and behaviour Fresh- or saltwater ponds and ditches, and shallow back-reef lagoons. Usually seen singly, standing in water. Feeds on invertebrates (particularly crabs) and small fish. Nests colonially (March–June) often with Cattle Egrets. Nest a loose platform of twigs. Lays 2–3 bluish-green eggs.

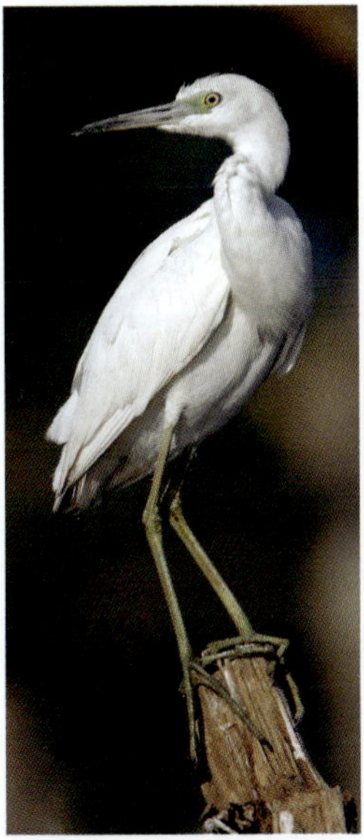

Juvenile. Costa Rica, March.

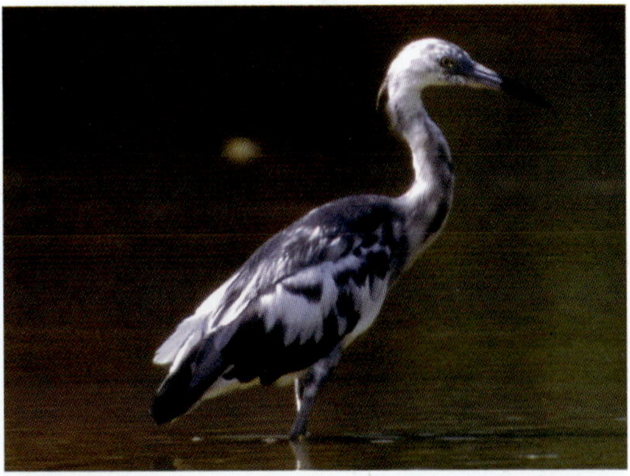

Juvenile moulting to adult plumage. Florida, March.

Range West Indies and the Americas.
Status in Jamaica Fairly common resident. Local populations increased by migrants in winter.

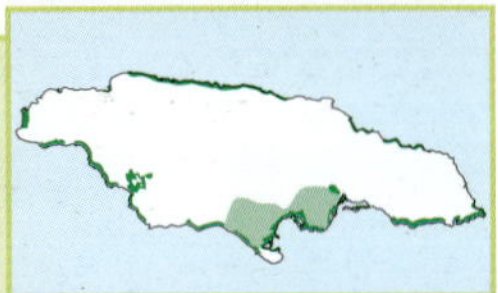

Snowy Egret
Egretta thula

Local names White Gaulin, Golden Slippers.
Taxonomy Polytypic (2). Race in Jamaica is *E. t. brewsteri*.
Description L 51–66cm (20–26in). Small, slender, white heron with black legs and contrasting bright yellow feet. Bill black. In breeding plumage yellow lores and feet turn bright orange, legs turn black, and head and back feathers develop long, lacy plumes. Juvenile is white with legs dark in front and greenish yellow behind, feet are yellow.
Similar species Other common white herons and egrets lack golden yellow feet.
Voice A harsh *aaark*.
Habitat and behaviour Shallow salt- and freshwater ponds, bays or marshes. When feeding often shuffles feet, probably using the yellow toes as lures. Nests colonially (March–June) in mangroves usually in mixed colonies with Cattle Egrets. 2–3 bluish-green eggs.

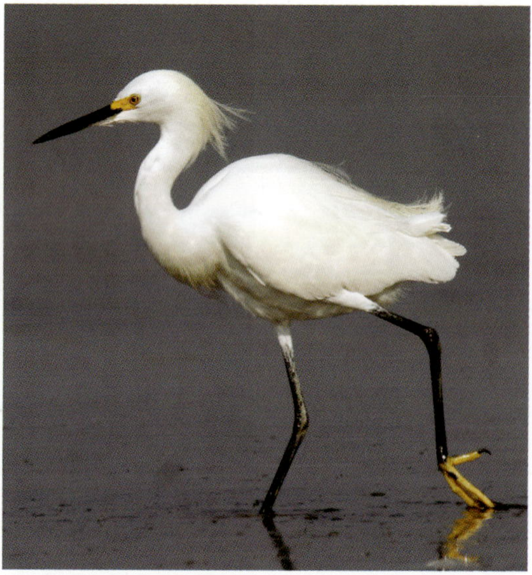

Adult breeding. Note contrast between legs and feet. Florida, March.

Range *E. t. brewsteri* in eastern North America, South America and the West Indies. Nominate in western Americas.
Status Common resident. Numbers increased by migrants in winter.

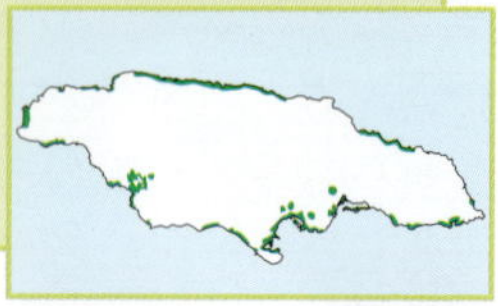

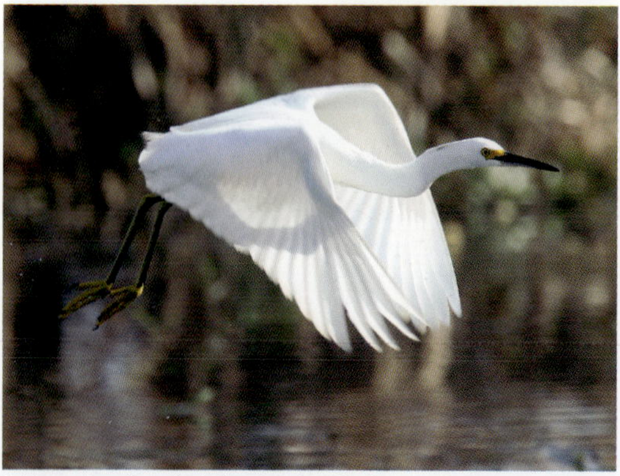

Adult non-breeding. Black River, Jamaica, March.

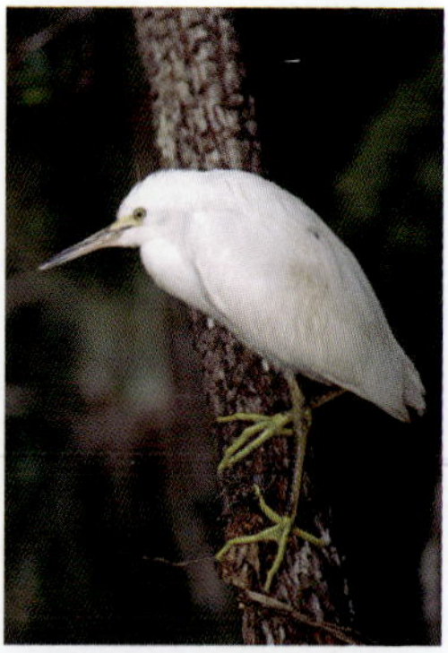

Juvenile. Cayman Islands, January.

Cattle Egret
Bubulcus ibis

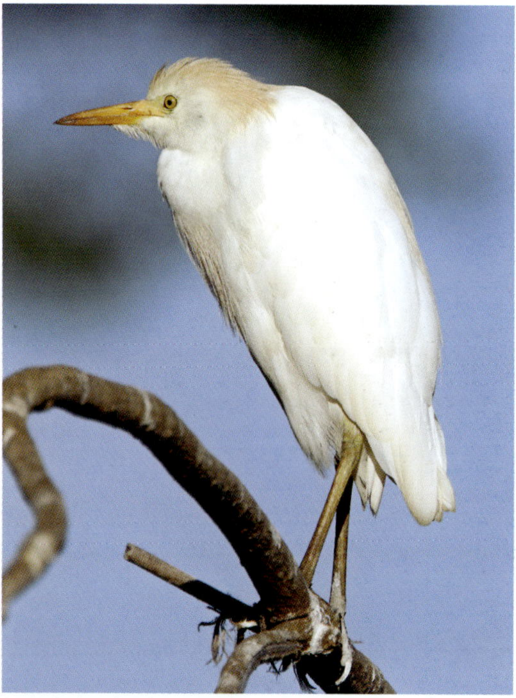

Developing breeding plumage with orange on head/breast. Costa Rica, April.

Local name Ticks Bird, White Gaulin.
Taxonomy Polytypic (2). Race in Jamaica *B. i. ibis.*
Description L 43–56cm (17–22in). Small, stocky white heron with a short neck. In non-breeding plumage white with a yellow bill and dull greenish or yellow legs. In breeding plumage crown becomes rufous, a rufous patch develops on the back and breast and bill and legs become reddish orange. Juvenile white with black bill and legs.
Similar species Great Egret nearly twice as tall with longer neck, much longer dark legs, usually seen close to water. Snowy Egret more slender with black legs, yellow feet.
Voice A cackling *kercuk-oo, kercuk-oo*, usually in flight. Grunts and squawks at the nesting colony.
Habitat and behaviour Generally seen in disturbed habitats, such as cattle pastures, cultivation. Does not usually feed in water. May forage on the ground for insects stirred up by browsing animals or mowing machines. Nests colonially (mainly April–July, but also other months) often with other herons. Colonies are usually on trees in or by rivers, ponds, or lagoons. Most colonies occupy the same site year-round but some have regular seasonal movements.

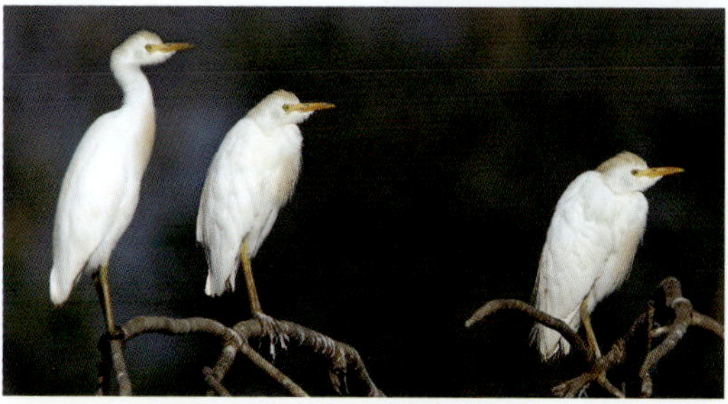

Non-breeding adults. Costa Rica, April.

Range *B. i. Ibis* Spain to Iran, North and South America. Spread to the West Indies in the 1950s. Other race in eastern Asia.
Status in Jamaica Very common resident (first recorded in Jamaica in 1953). Probably increasing.

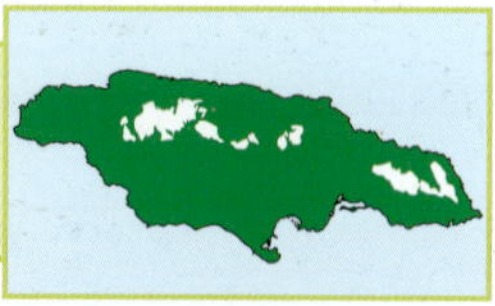

Green Heron
Butorides virescens

Local name Green Gaulin.
Taxonomy Polytypic (4). Race in Jamaica *B. v. virescens.*
Description L 40–48cm (16–19in). A small, squat, dark heron. Crown greenish, relatively short neck chestnut on rear and sides. Back iridescent green and mauve. Upper mandible black, lower greenish. Lores bare, green. Iris pale yellow. Wings dark brown. Legs greenish-yellow, becoming orange when breeding. Juvenile has heavily streaked underparts and browner wings.
Similar species Least Bittern is more rufous, has buffy wing patches and is unlikely to be seen in the open.
Voice When flushed utters a loud *kwow* or otherwise *cuk-cuk-cuk-ow*. Also growls from nest.

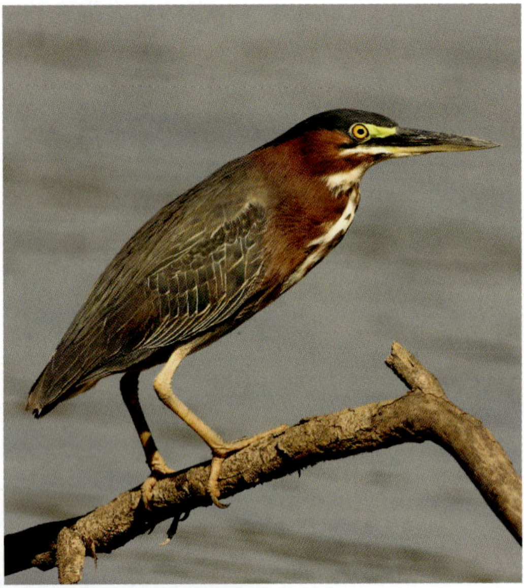

Adult. Note the orange legs. Cayman Islands, March.

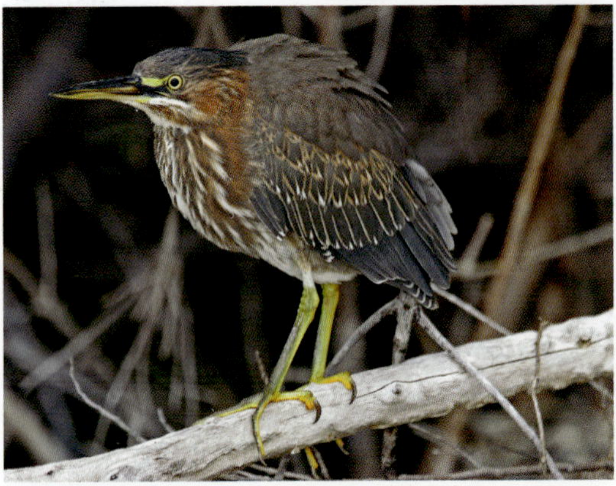

Juvenile, with spotted wings and streaky breast. Cayman Islands, February.

Habitat and behaviour
Usually solitary, seen on edges of ponds, dams, streams and rivers and in mangroves. Island-wide near fresh or salt water in open areas or in dense vegetation. Feeds on fish and aquatic invertebrates by lunging from a crouched position, but may also stalk prey such as lizards. Nests in summer, solitarily or in mixed colonies with other herons and egrets, in trees or bushes. Nests are loose platforms of twigs. Eggs are greenish-blue.

Range *B. v. virescens* is found in eastern North America, Central America and West Indies. Other races in Bahamas and California.
Status in Jamaica Very common resident.

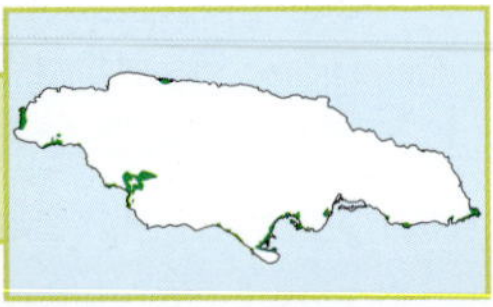

Black-crowned Night-heron
Nycticorax nycticorax

Local name Quok
Taxonomy Polytypic (4). Race in Jamaica is *N. n. hoactli.*
Description L 53–66cm. (21–26in). Large, chunky, short-necked night-heron with a black crown; has white head-plume in the breeding season. Back black, wings and tail grey, underparts white. Irises red, legs yellow. Feet only partly extend beyond tail in flight. Juvenile brown with heavily streaked back and underparts and large white spots on wing-coverts. Legs greenish, bill brown with yellowish lower mandible. Takes three years to mature. During this time the streaking decreases, and the head and bill darken, though they still remain brownish.
Similar species Immature Yellow-crowned Night-heron is very similar but has thicker, mostly black bill and smaller white spots on wing-coverts. Feet extend completely beyond tail in flight.
Voice *Quok*. Immature in nest *tuk-tuk-tuk-tuk* like a thirsty chicken.

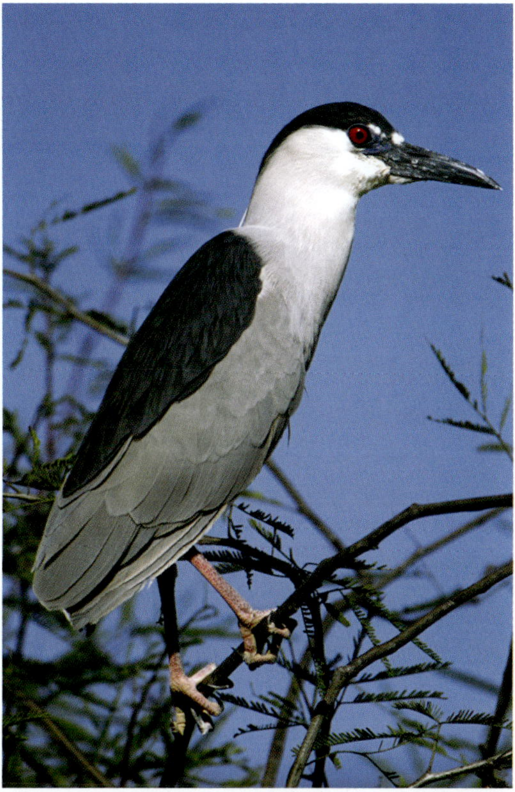

Adult with black head and crown. January.

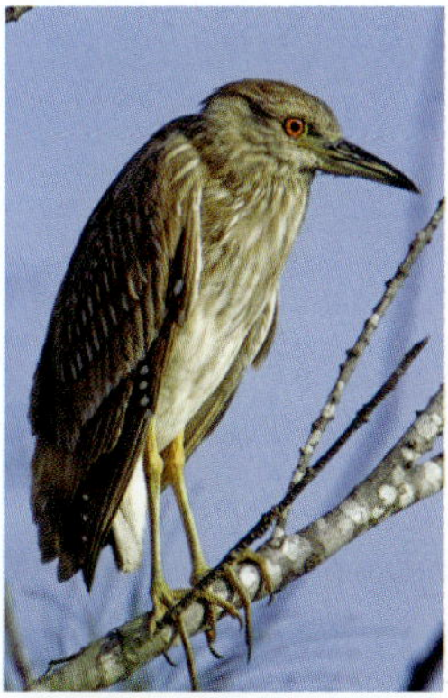

Juvenile. Bahamas, April.

Habitat and behaviour Usually found near freshwater or brackish ponds and rivers. Perches during the day in shrubs and trees close to ponds or rivers. Feeds mainly at dusk and dawn, mainly on crustaceans and other aquatic invertebrates. Nests in small colonies (April–August) in trees (as at Royal Palms, for example), or in rushes.

Range *N .n. hoactli* occurs from Canada to Argentina. Other races occur worldwide, except Australia.
Status in Jamaica Locally common resident.

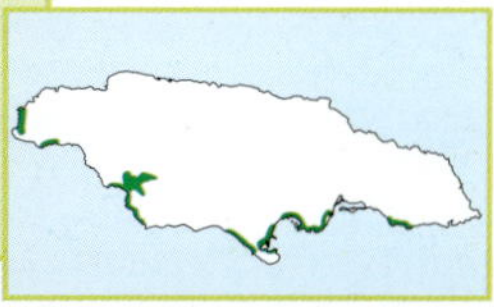

Yellow-crowned Night-heron
Nyctanassa violacea

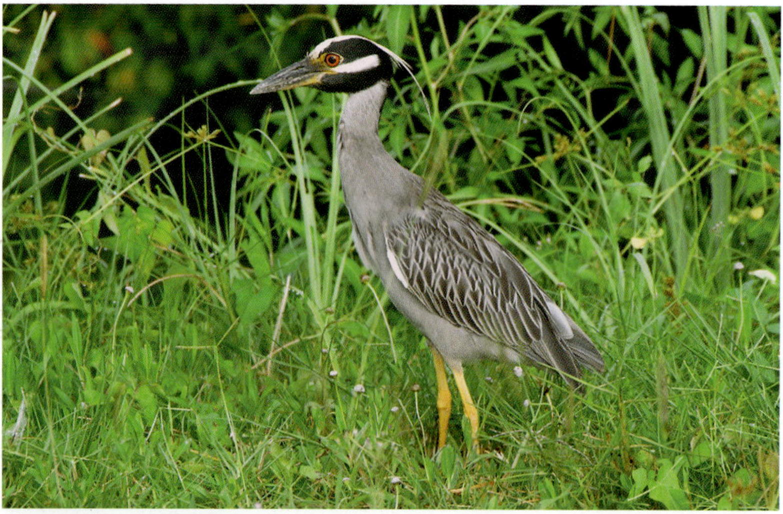

Adult. Note white on cheek. Montego Bay, Jamaica, June.

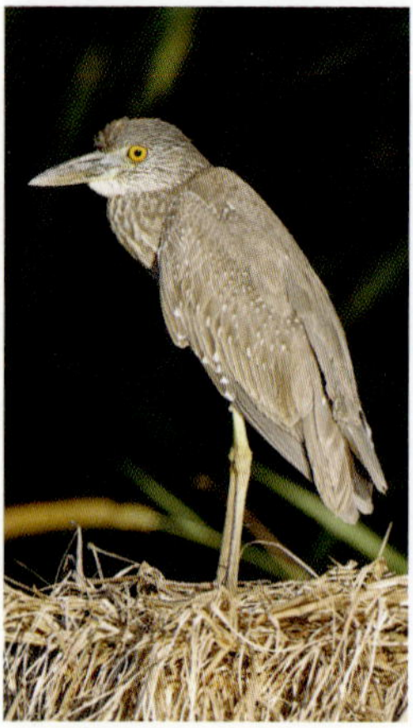

Juvenile. Costa Rica, March.

Local name Quok, Crab Catcher.
Taxonomy Polytypic (6). Race in Jamaica is *N. v. bancrofti.*
Description L 51cm (20in). Head black with white stripe below and behind eye, buffy-white crown with long white plume. Wings and underparts light grey. Iris rufous. Bill heavy and black, legs yellow. Feet extend beyond tail in flight. Juvenile has streaked head and underparts, back and wing-coverts brown tipped with white. Bill dark, legs greenish.
Similar species Immature Black-crowned Night-heron is similar but stockier, more heavily spotted white and has a partly yellow bill; its legs are shorter and in flight only part of its feet extend beyond tail. It is not found on beaches.
Voice *Quaaark*.
Habitat and behaviour Beaches, mangroves, golf links. Has an upright gait. More diurnal than Black-crowned Night-heron. Feeds mainly on crabs at dawn and dusk on the ground and in relatively open areas close to the sea. During the day it perches in trees. Nests in small colonies or singly (April–August), sometimes far from water in dense shrubs or undergrowth. Nest is a loose platform of twigs. Lays 2–4 bluish-green eggs.

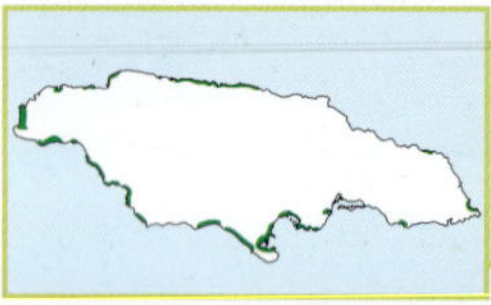

Range *N. v. bancrofti* occurs throughout the West Indies and Central America to California. Other races occur in North and South America and on the Galápagos islands.
Status in Jamaica Locally common resident.

Least Bittern
Ixobrychus exilis

Taxonomy Polytypic (4). Race in Jamaica is *L. e. exilis*.
Description L 28–36cm (11–14in). Very small. Male has a black crown and back, rufous wings with buffy wing-coverts and dark flight feathers. Neck buff, faintly streaked white. Female brown on back. Immature resembles female but paler mottled brown, boldly streaked on breast and back.
Similar species Green Heron is larger with dark wings lacking buffy wing-patch.
Voice A harsh *cyak-cyak-cyak-cyak-cyak*… and soft, fast *uh-uh-uh* …
Habitat and behaviour Mangroves and marshes, reed-beds beside ponds and rivers. Shy and inconspicuous. Sits motionless, camouflaging itself by raising bill skywards so that it looks like a reed, or flutters briefly between perches. Extends the neck slowly until within reach of fish or aquatic insect prey before making a sudden lunge. Nests (May–July) in sedges or mangroves, just above water line.

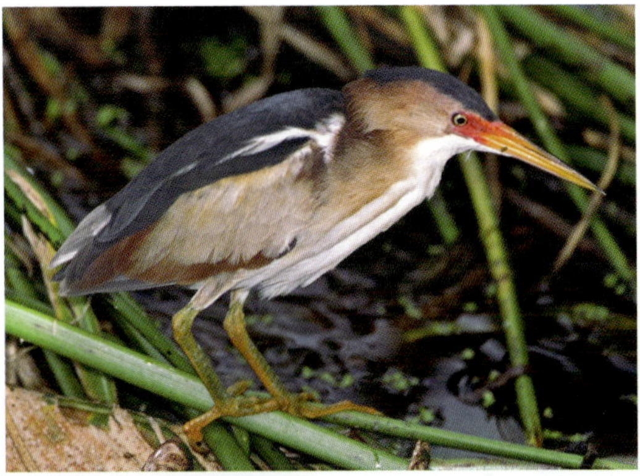
Adult male with black back and crown. Florida, April.

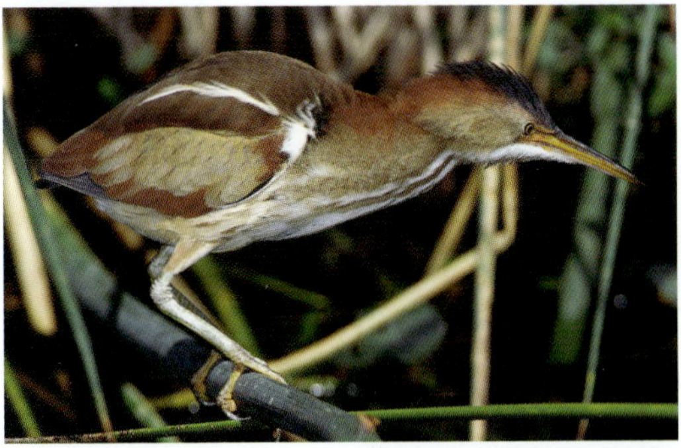
Adult female. Note brown back. Florida, April.

Range *L. e. exilis* West Indies and North and Central America. Other races in Central and South America. **Status in Jamaica** Locally common resident. North American migrants increase local populations in winter.

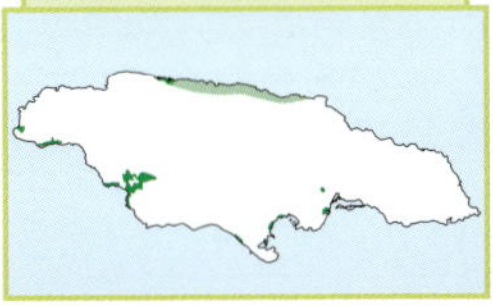

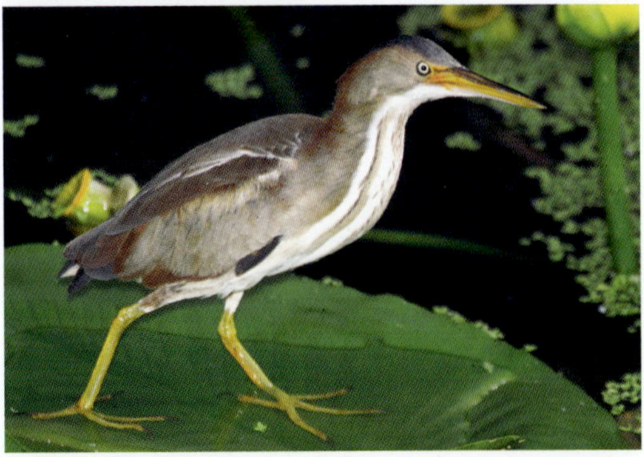
Juvenile. Note greyish back. Florida, April.

White Ibis
Eudocimus albus

Local name Curlew
Taxonomy Monotypic.
Description L 56cm (22in). Large white heron-like bird with long, decurved bill and red legs, which get brighter in the breeding season. Bill is thick and orange-coloured. Black outer wing-tips are clearly visible in flight. Juvenile variably brown, dark on the back and lighter on head and neck, with white underparts, orange bill and grey legs. Downy young are dark grey with some white on the throat and head.
Similar species Juvenile Glossy Ibis is uniformly dark, bill grey.
Voice Alarm call is *hunk, hunk, hunk*.
Habitat and behaviour Near water, especially mangroves, and grassy margins of fish ponds. Usually seen in small groups. Feeds by probing in mud for small invertebrates. Flies with quick wing-beats followed by short glides. Nests colonially (April–May), often with night-herons, herons and egrets. Nest is a deep cup built of sticks in the fork of mangrove branches or in the foliage at the end of a branch.

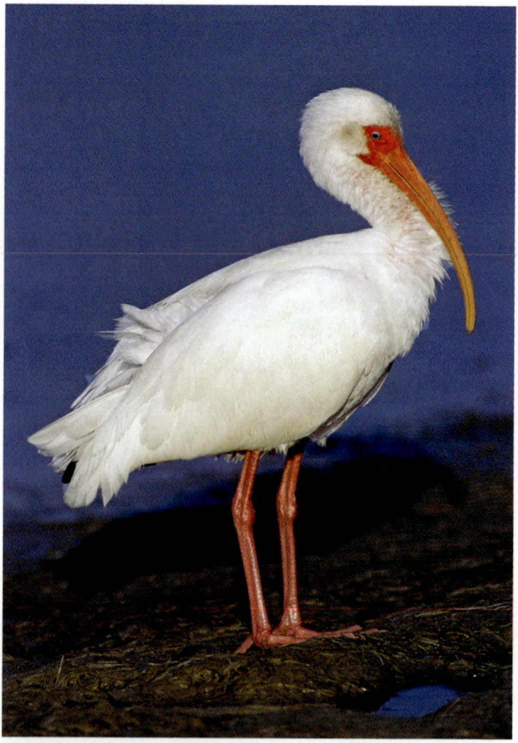
Adult. Note red legs and large orange bill. Florida, May.

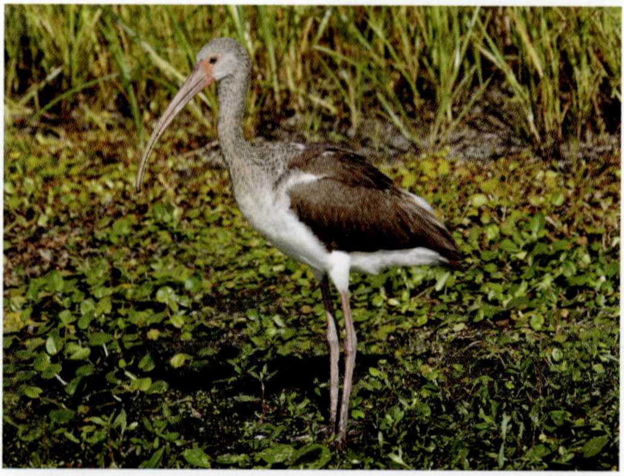
Juvenile; grey with dull legs. Florida, April.

Range Southern United States to Brazil.
Status in Jamaica Locally common resident.

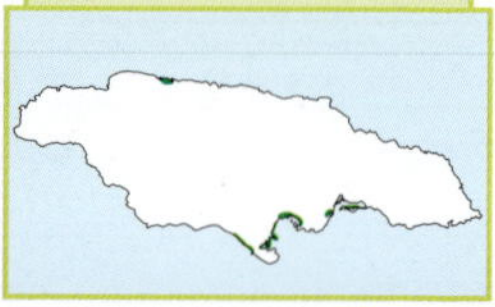

Glossy Ibis
Plegadis falcinellus

Local name Curlew
Taxonomy Monotypic.
Description L 48cm (19in). Dark, heron-like bird with decurved bill. Wing-coverts dark green. Legs dark in non-breeding adults, red in breeding adults. Immature dark brownish, with pale brown irises and dark bill and legs.
Similar species Juvenile White Ibis has a white belly and orange bill.
Voice *ka-onk, ka-onk*.
Habitat and behaviour Muddy areas, fish ponds and wetlands. Feeds on invertebrates by probing in shallow water, mud or soil. Nests colonially with White Ibises, herons or egrets (April-May) in mangroves. Eggs bluish-green, sparsely spotted with brown.

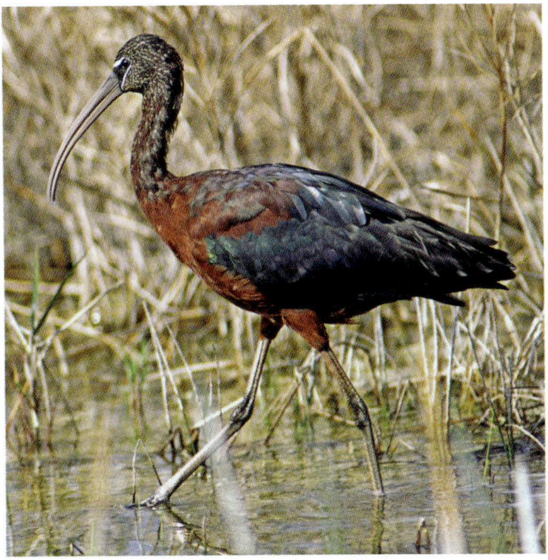

Adult breeding. Note white lines on face. Florida, February.

Range Worldwide.
Status in Jamaica Common resident and winter visitor. Populations are increasing in Jamaica because aquaculture has increased suitable habitat.

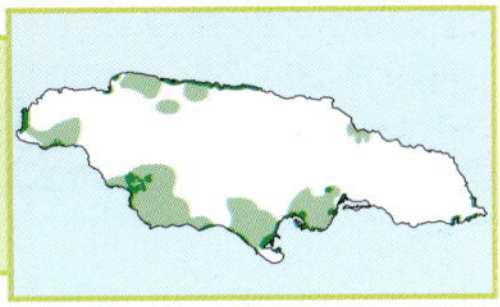

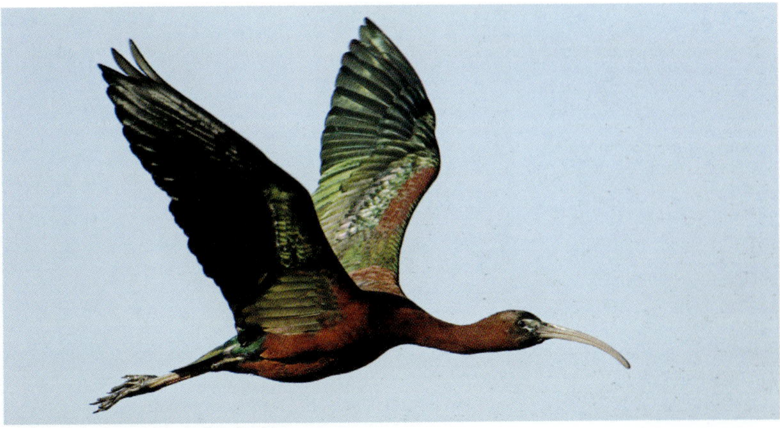

Adult in flight. Note the arched profile. May.

Greater Flamingo
Phoenicopterus ruber

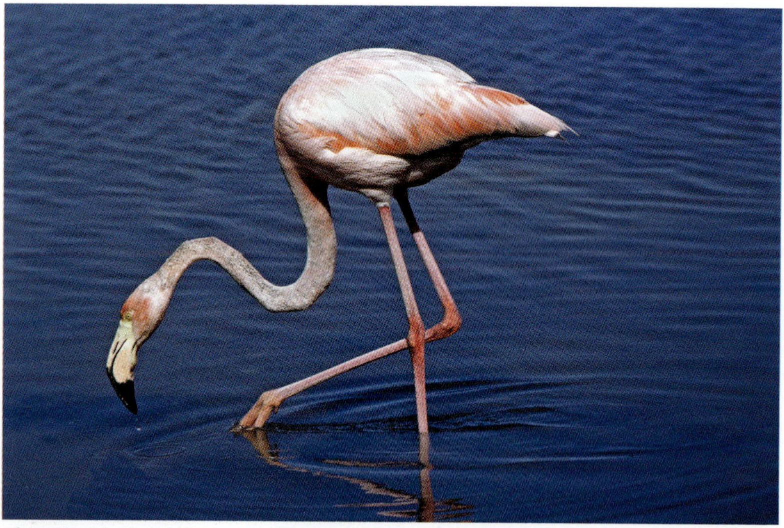

Adult. Cuba, February.

Local name Fillymingo.
Taxonomy Polytypic (2). Race in region is *P. r. ruber*.
Description L 107cm (42in). Unmistakable, large, salmon-pink, long-legged, long-necked bird with a thick, decurved bill. Adult has tricoloured bill, black at the tip, pink in the middle and white at the base. Immature paler; yellowish instead of pink in the bill.
Similar species None.
Habitat and behaviour Usually seen singly on fringing reefs or in saline ponds. Feeds by turning the bill upside-down and sifting small aquatic organisms from the water.

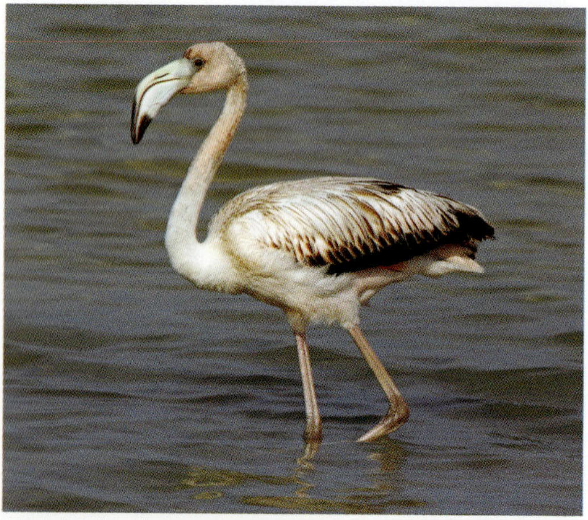

Subadult. Cuba, February.

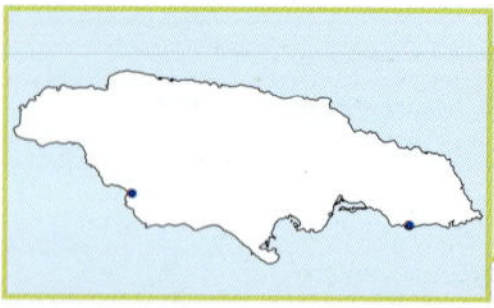

Range *P. r. ruber* a common resident on Cuba, Hispaniola, some of the Bahamas and Caicos, Caribbean coasts of Central and South America. Other race Africa and southern Europe to India.
Status in Jamaica Uncommon visitor, rarely seen in groups.

Turkey Vulture
Cathartes aura

Local name John Crow
Taxonomy Polytypic (4). Race on Jamaica is *C. a. aura.*
Description L 64–81cm (25–32in). Large, blackish-brown bird with small, red, unfeathered head, and horn-coloured, hooked bill. Underwing two-toned with blackish coverts and grey flight feathers. Partial albinos occur occasionally. Downy chicks white. Juvenile has dark head.
Similar species Red-tailed Hawk is sometimes seen soaring with Turkey Vultures but is typically brownish above and pale below, with reddish tail.
Voice Generally silent, but may give a hoarse hiss (usually at nest).
Habitat and behaviour Usually seen soaring singly or in flocks on thermals in the sky. In flight soars with wings outstretched in a stiff, upward V-shape, often rocking from side-to-side. Rarely flaps. Depends on thermals for soaring flight so does not take to the wing in the morning until the air gets warmed by the sun. After rain, sometimes perches on a tree with wings extended to dry. Often aggressively mobbed by kingbirds in June–July. Feeds on carrion which it can locate by smell. Often seen at carcasses on roadsides. Nests (January–May) on the ground at the base of large trees or under overhanging rocks. Sometimes seen in flocks of more than 100 (near sea cliffs, for example).

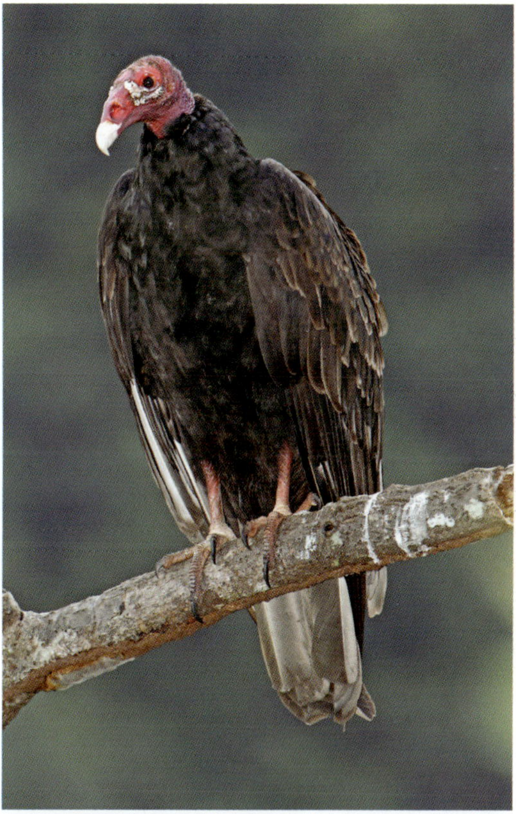

Adult. Cockpit Country, Jamaica, November.

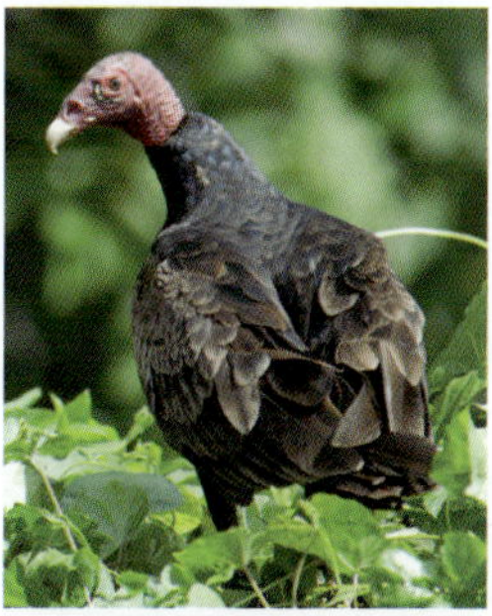

Adult. Bog Walk Gorge, Jamaica, Mar.

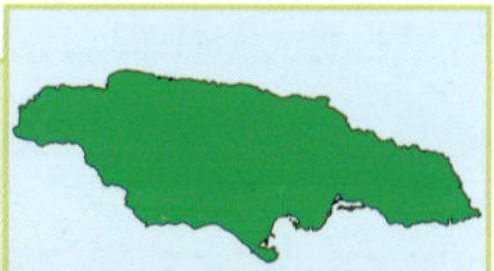

Range *C. a. aura* occurs Greater Antilles, western North America to Costa Rica and some islands of the Bahamas. Other races in eastern North America, Central and South America.
Status in Jamaica Very common resident.

Osprey
Pandion haliaetus

Local name Fish Hawk, Fish Eagle.
Taxonomy Polytypic. On Jamaica mostly North American race *P. h. carolinensis*. Also (rarely) Caribbean race *P. h. ridgwayi*.
Description L 56cm (22in). Adult *carolinensis* has dark brown upperparts and white underparts, with a dark-speckled breast-band. Head white with a dark line through the eye that continues down the sides of the neck to the back. Tail is barred. Juvenile has pale tips to upperpart feathers. In flight bent-looking wings make an 'M' shaped silhouette, black carpal patches distinctive. Caribbean race *ridgwayi* (rarely observed in Jamaica) is paler, especially on head and breast and with a limited dark line through the eye.
Similar species In flight Turkey Vultures and Red-tailed Hawks do not have angled wings.
Voice Sharp, shrill *cheep*!
Habitat and behaviour Rivers, fish-farms and open water in wetlands, fresh and brackish ponds, pools and rivers, mainly near to or on the coast. In spring and autumn passage migrants may be seen flying high over the centre of the island. Feeds by hovering and then plunging feet-first into water to seize fish near the surface. Prey is taken to a perch to eat.

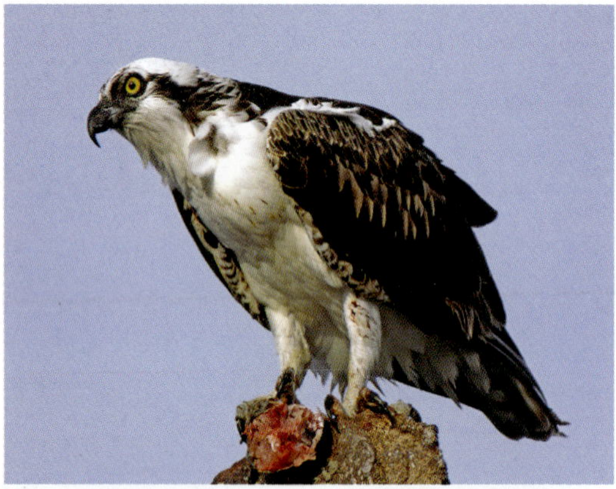
Adult *P. h. carolinensis* with broad dark line on face. Florida, February.

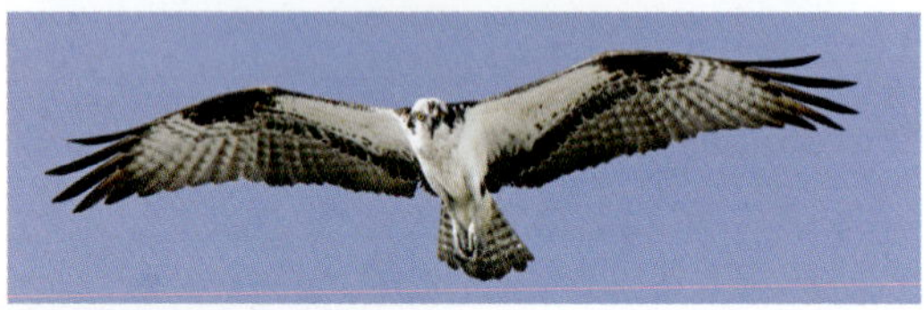
Adult *P. h. carolinensis* in flight; note dark carpal patches. Florida, February.

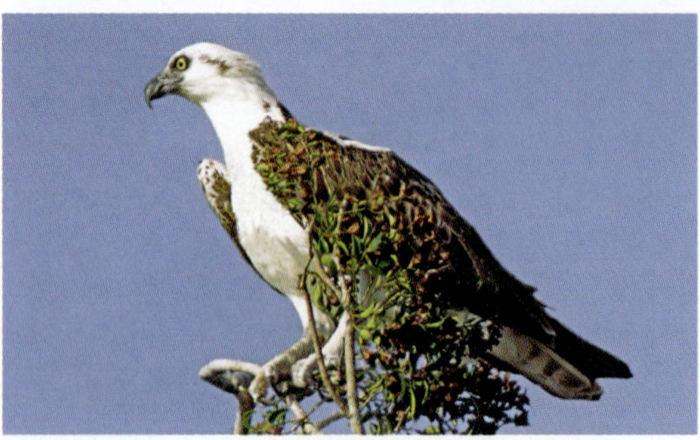
Adult *P. h. ridgwayi*. Note the paleness. Cuba, April.

Range *P. h. carolinensis* breeds in North America and migrates south to central South America and the West Indies. *P. h. ridgwayi* breeds in Cuba and Bahamas but does not breed in Jamaica. Other subspecies are found worldwide.
Status in Jamaica Fairly common winter visitor; mainly September–May, though non-breeders are present year-round.

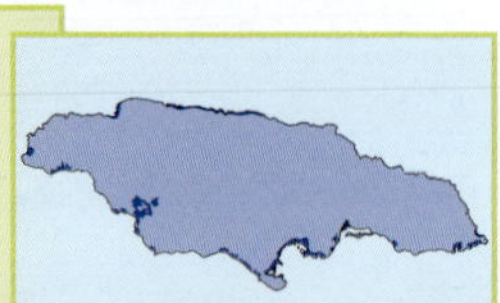

Swallow-tailed Kite
Elanoides forficatus

Taxonomy Polytypic (2). Race in Jamaica is *E. f. forficatus.*
Description L 50–64cm (20–25in). White head and body contrast with black flight feathers and deeply-forked long tail. Wings sharply pointed.
Similar species None.
Habitat and behaviour Most likely to be seen on migration soaring high over the island, singly or in small groups.

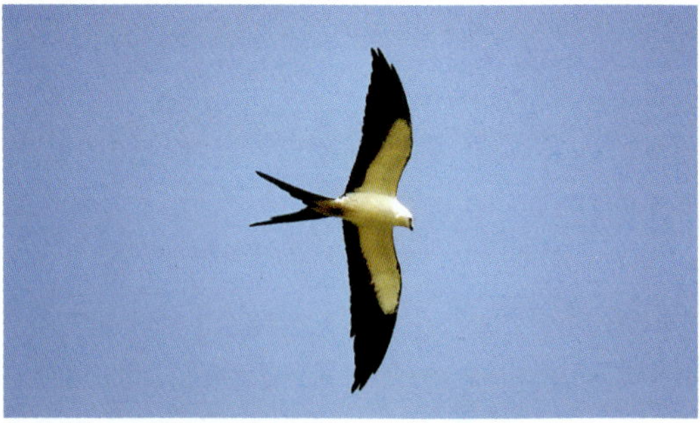

Adult. Florida, December.

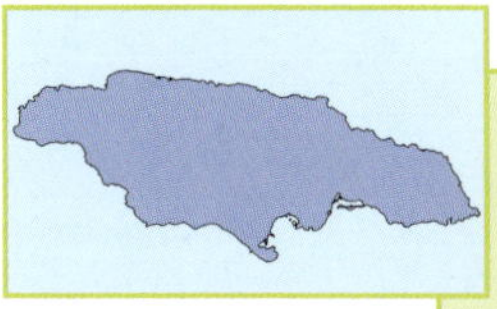

Range Breeds in southeast United States and Central America. Winters south to South America.
Status in Jamaica Rarely observed but satellite tracking shows that it may be a regular passage migrant over Jamaica, more likely in autumn (August–October) than spring (February–June).

Red-tailed Hawk
Buteo jamaicensis

Local name Chicken Hawk
Taxonomy Polytypic (14). *B. j. jamaicensis* is a West Indian endemic race.
Description L 46cm (18in). Large hawk with rounded wings and reddish tail finely tipped with black then white. Head, back and wings are mottled dark brown, underparts are buffy-white with mottled dark bar across the belly. Immature streaked on the underparts, tail brownish and finely barred.
Similar species No similar species are present in the summer but rare vagrants can occur in the winter, when all hawks should be carefully checked.
Voice The high explosive *chi-uuuuu* descending at the

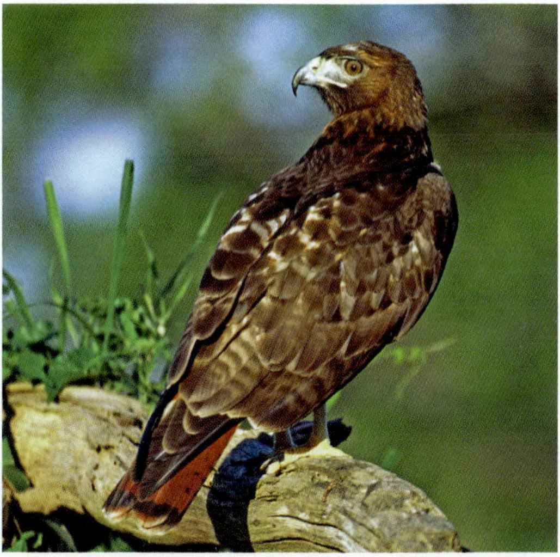

Adult with red tail. Dominican Republic, April.

end is often the first indication of its presence.

Habitat and behaviour Usually seen soaring over wooded areas, or open areas close to woods. Sometimes mobbed by Grey Kingbirds, which may even land on its back. Feeds on small birds and occasionally on chickens from farms, becoming a pest. Often soars with Turkey Vultures. Nest (March–May) is made of sticks, including green leafy branches, in tall trees at forest edges.

Range *B .j. jamaicensis* in Greater Antilles (bar Cuba) and northern Lesser Antilles. Other races throughout the Americas.

Status Common but possibly declining due to human persecution.

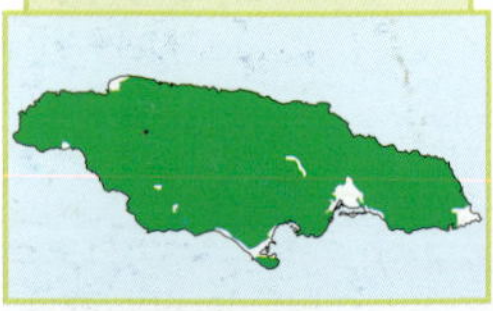

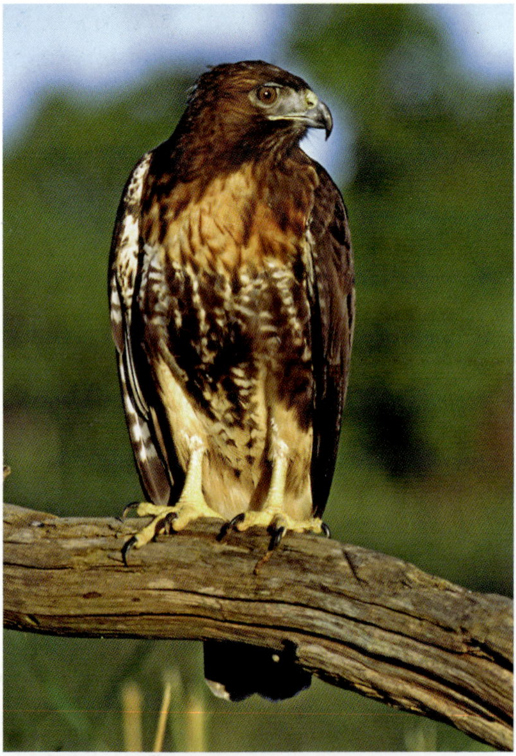

Adult. Dominican Republic, April.

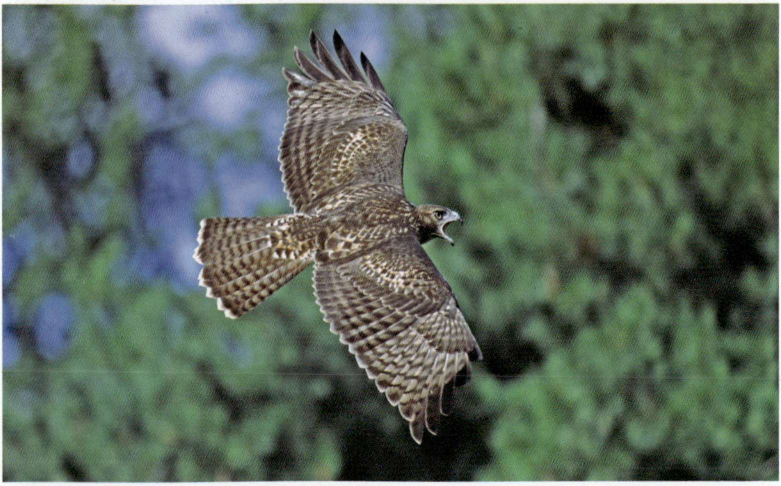

Immature. Note the fine barring. Dominican Republic, May.

Merlin
Falco columbarius

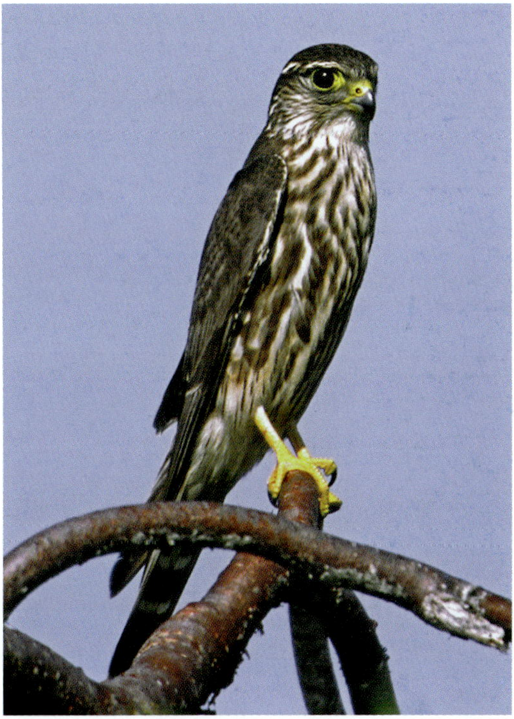

Adult female. Note grayish barring on upper breast. Cayman Islands, May.

Local name Pigeon Hawk, Bird Hawk.
Taxonomy Polytypic (9). Race in Jamaica is *P. c. columbarius.*
Description L 24–31cm (10–12in). Male has head, back and upperwings plain blue-grey. Face has indistinct dark moustachial stripe. Faint buff supercilium. Bill dark grey, with a lighter cere. Throat is white. Underparts white, heavily streaked. Dark grey tail is finely barred with white (2–5 bars), with a broader black band at the end. Feet and legs are bright yellow. Female and juvenile similar but have browner head and back.
Similar species American Kestrel has much rufous and two facial stripes. Peregrine much larger, with prominent black moustachial stripes.
Voice Silent.
Habitat and behaviour Usually seen near coastal ponds, and sometimes further inland when migrating. Captures dragonflies by flying over ponds, grabbing them, and returning to the roost to eat them, discarding the wings. Also eats beetles, lizards and small passerines.

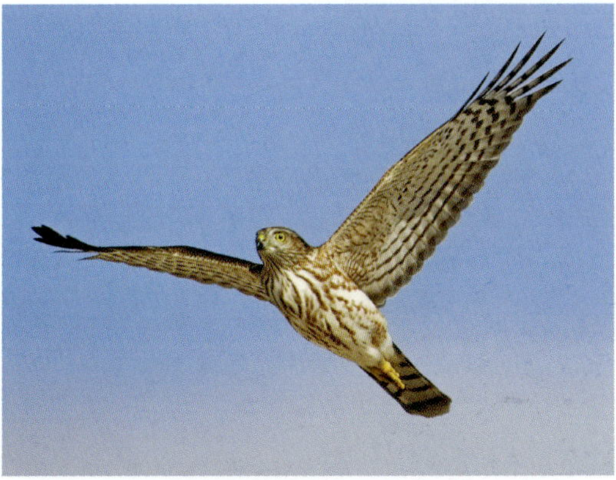

Juvenile. Note brown barring. October.

Range *F. c. columbarius* North America. Winters south to West Indies and northern South America.
Status in Jamaica Common winter visitor (September–May).

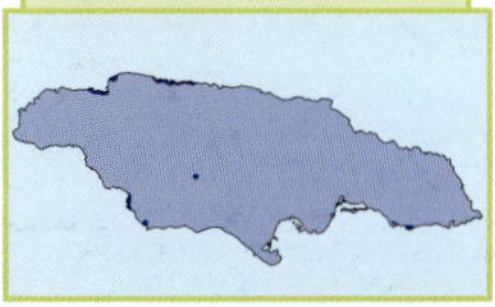

American Kestrel
Falco sparverius

Local name Sparrow Hawk, Killy-Killy.
Taxonomy Polytypic (17). A relatively recent colonist to Jamaica. The races involved may include *F. s. sparveroides* and *F. s. dominicensis* and their hybrids, but this has not been conclusively established.
Description L 22–31cm (9–12in). A small falcon with a distinctively marked head: face white with two vertical black streaks on each cheek. Back and tail reddish brown. Crown is grey, ending in a grey triangle at the nape. Two white patches with dark grey centres at the back of the head look like the eyes of an owl when seen from behind. Male has dark blue wings and a plain rufous back and tail, the latter tipped black with a white margin. Female has dark grey barring on back, wings and tail. White and rufous morphs occur in Jamaica; the rufous morph is less common. *White morph* underparts are pure white or with variable amounts of short, vertical streaks that tend to be concentrated on the upper breast. *Rufous morph* male and female have completely rufous breasts contrasting with the white lower belly and undertail-coverts. Juvenile of either morph has white underparts heavily streaked rufous.
Similar species Merlin has a dark tail and dark-streaked underparts, and lacks rufous. Peregrine Falcon is much larger, with prominent black moustachial stripes.
Voice A high-pitched, rapid

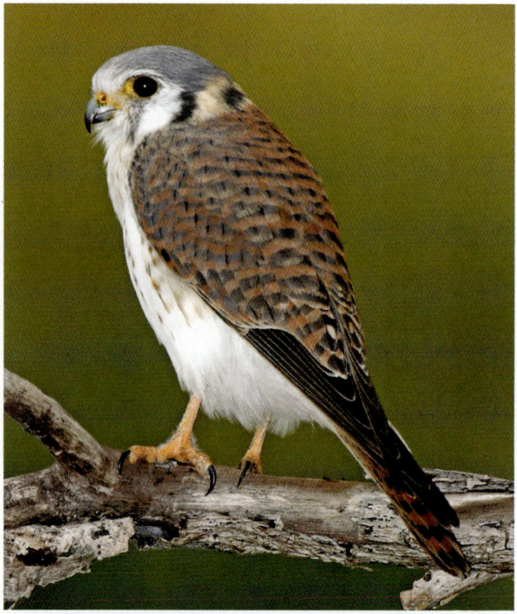
Adult female with finely barred back. Hope Gardens, Jamaica, November.

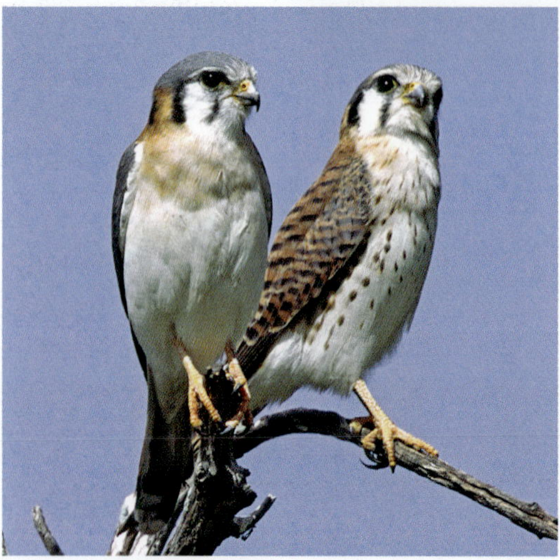
Adult male (left), adult female (right) white morph. Hope Gardens, November.

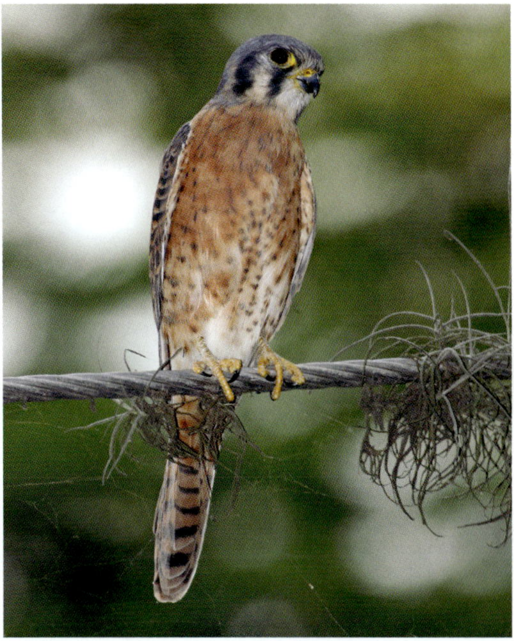

Adult female rufous morph. Hope Gardens, Jamaica, November.

killy-killy-killy or *yip-yip*.

Habitat and behaviour
Found in cleared areas island-wide but more frequent at low and middle elevations. Hovers over commons or canefields. Sometimes dives with wing-tips bent back. Often seen perched at roadsides on power lines or posts. Swoops down to capture prey from the ground or bushes. Feeds mainly on lizards and insects but also (rarely) takes birds such as Cave Swallow, Orangequit, Bananaquit and even hummingbirds. Has been seen killing birds up to the size of a Zenaida Dove that are too heavy for it to carry away. Nests (March-April) in hollows on broken-off tops of coconut palms, in or under the bases of palm fronds, or in old woodpecker holes.

Range *F. s. dominicensis* Jamaica, Hispaniola and adjacent islands. *F. s. sparveroides* Jamaica, Bahamas, Cuba and the Isle of Pines. Other races occur on other West Indian islands and in North, Central and South America.
Status Common and widespread. Was very rare until the 1940s, since when it appears to have greatly increased its range and population, possibly following habitat changes and colonisation from Cuba and Hispaniola.

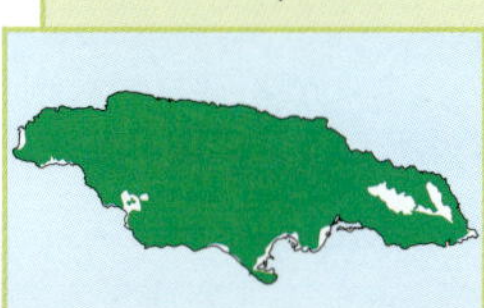

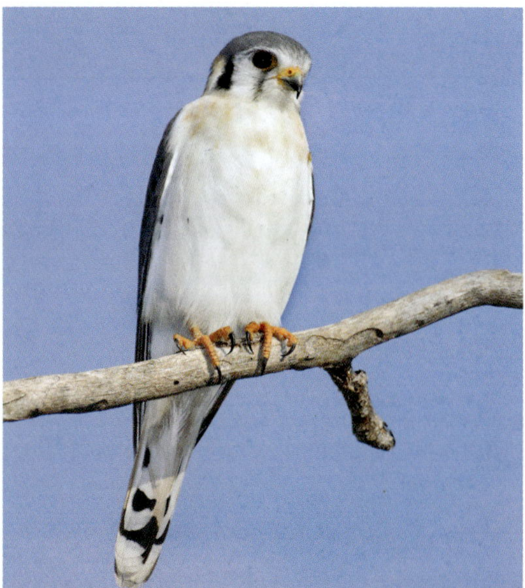

Adult male white morph. Hope Gardens, Jamaica, November.

Peregrine Falcon
Falco peregrinus

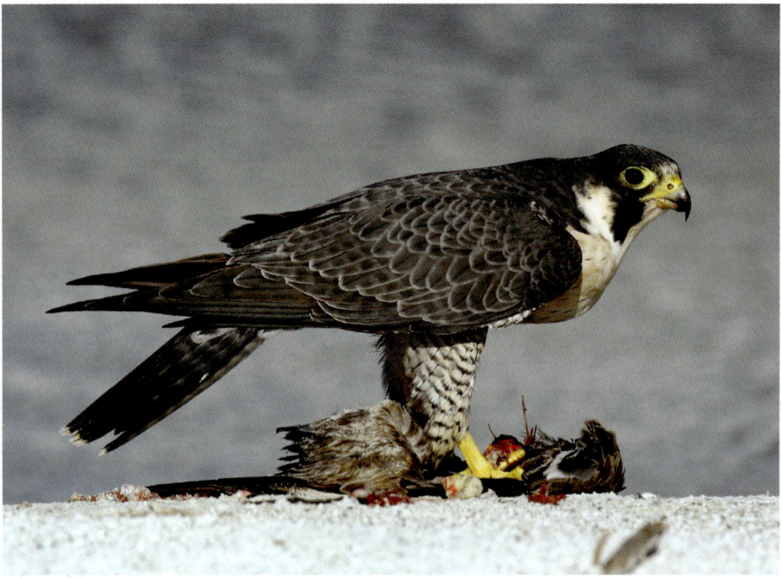

Adult. Florida, February.

Local name Duck Hawk.
Taxonomy Polytypic (18).
Description L 38–50cm (15–20in). Adult has dark grey head with prominent black moustachial stripes contrasting with white neck-sides. Dark blue-grey above, paler below. Underparts and tail are barred. Legs yellow. Juvenile brown with heavily streaked underparts.
Similar species Merlin is much smaller.
Voice Usually silent in Jamaica.
Habitat and behaviour Usually seen near saline mudflats on the mainland or on the offshore cays. Feeds on birds (such as adult terns and chicks on the Pedro Cays) and ducks (on the mainland). Nesting is very rarely reported.

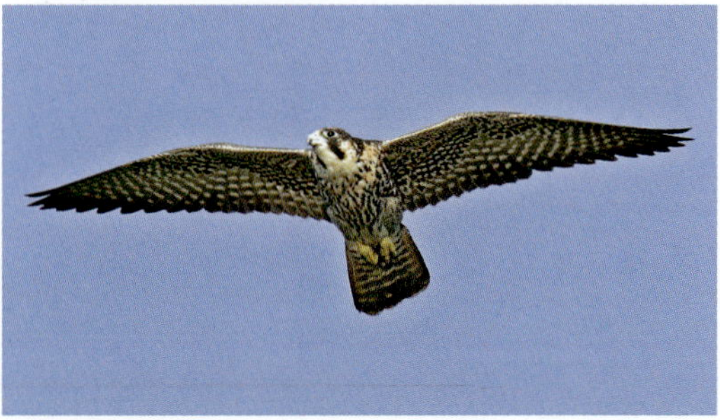

Immature in flight. Note barring. Costa Rica, February.

Range Worldwide. North American populations migrate to South America via Central America and the West Indies.
Status Uncommon winter visitor (November-March) and rare year-round resident.

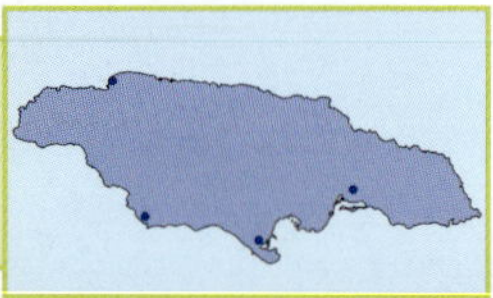

Limpkin
Aramus guarauna

Local name Clucking Hen
Taxonomy Polytypic (4). Race on Jamaica is *A. g. pictus*.
Description L 56cm (22in). Large, long-legged, long-necked ibis-like waterbird. Brown head and underparts are heavily streaked with white. Back and sides brown with distinct white spots. Wings brown. Slightly decurved yellow bill has a darker tip. Legs long and grey-black.
Similar species None
Voice Very distinctive series of wailing calls: *krau-krau-krau kreow*, also shrill screams and clucking. Often calls from tree tops.
Habitat and behaviour Usually first detected by call. May be seen roosting in trees or walking (with characteristic limp and tail jerks) on the ground in swamps, wet grasslands, fields near rivers, close to the coast or inland on suitable river banks. In flight resembles an ibis but bill is shorter. Never found in flocks. Feeds by probing in soft muddy soil, mainly on water-snails, but also on frogs, crabs, crayfish and worms. Nest (April–June) is a platform of dried marsh grasses usually built close to the ground among grasses, in bushes or sometimes in trees.

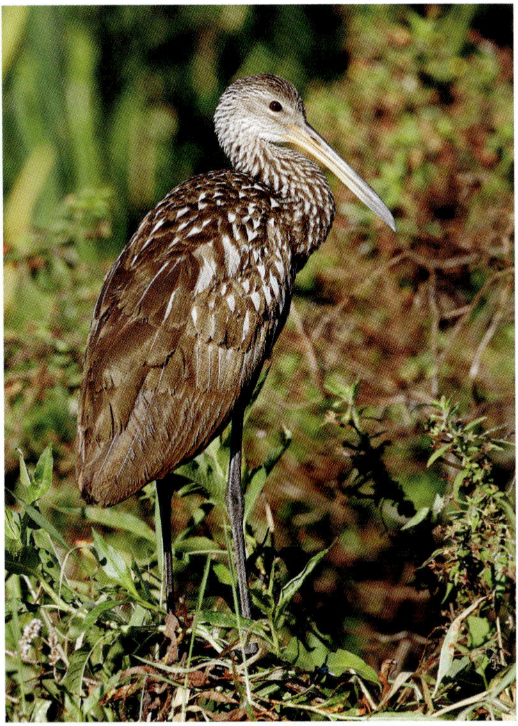

Adult. Florida, December.

Range *A. g. pictus* Cuba, Isle of Youth, Jamaica and Florida. Other races in Hispaniola, Puerto Rico, some Bahaman islands, Central and South America.
Status in Jamaica Locally common resident. May be declining in numbers due to habitat loss.

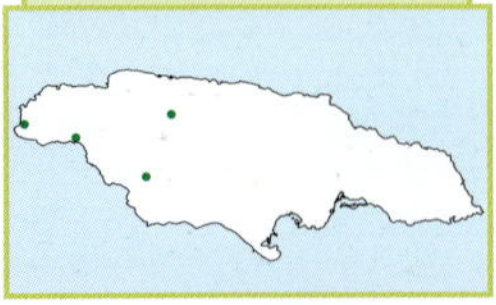

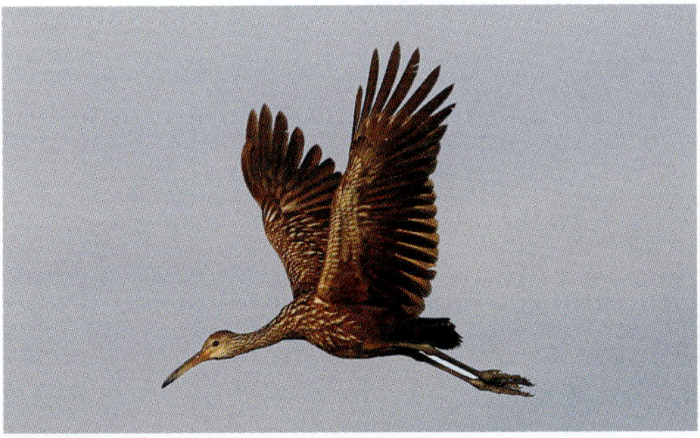

Adult. Florida, December.

Black Rail
Laterallus jamaicensis

Taxonomy Polytypic (4). Race in Jamaica is *L. j. jamaicensis.*
Description L 10–15cm (4–6in). A tiny, dark and secretive bird. Adult is black with white spots. Iris red, nape rufous.
Similar species Black downy chicks of coots and gallinules lack spots and red irises.
Voice *Kiki-du* repeated. Much more frequently heard (especially at night and after rain) than seen.
Habitat and behaviour Sedges and grasses in wetlands. Very secretive. Tends to scuttle into vegetation when disturbed. Probably breeds in remote wetlands but not recently confirmed.

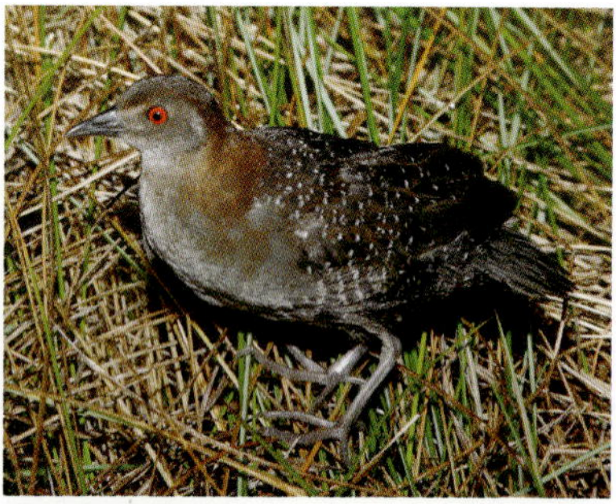

Adult female. April.

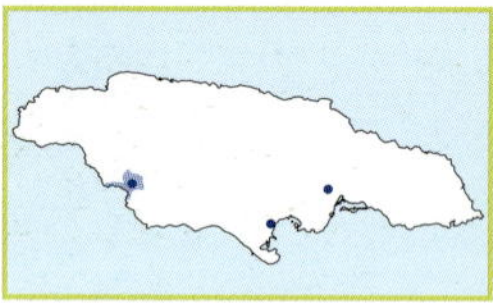

Range *L. j. jamaicensis* in eastern United States, Jamaica and Cuba. Other races in the Americas. Rare throughout range. Global conservation status: Near Threatened.
Status in Jamaica Very rare resident and winter visitor.

Clapper Rail
Rallus longirostris

Local name Mangrove Hen, Mangrove Chicken.
Taxonomy Polytypic (17). Race in Jamaica probably *R. l. caribaeus.*
Description L 31–41cm (12–16in). A chicken-sized, grey-brown and black bird with a long, slightly decurved bill. Sides barred black and white. Back dark grey with lighter streaks.
Similar species No similar species occur in the mangroves.
Voice A harsh chatter. When one bird calls others in the area may respond in a cascade.
Habitat and behaviour Usually seen running on mangrove roots or feeding on the edge of mudflats on invertebrates. Nest (April–June) of sticks placed on the ground among mangrove roots. Lays 5–9 creamy-coloured spotted eggs.

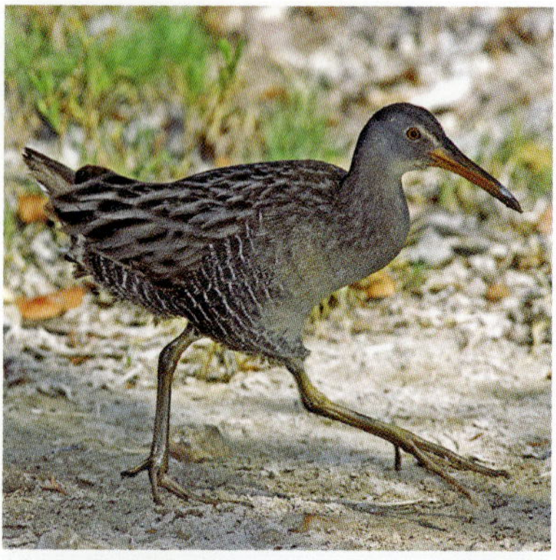

Adult. Cuba, April.

Range *R. l. caribaeus* is a West Indian race, locally common in Greater Antilles and Virgin Islands. Other races occur in Bahamas and mainland Americas.
Status in Jamaica Fairly common resident.

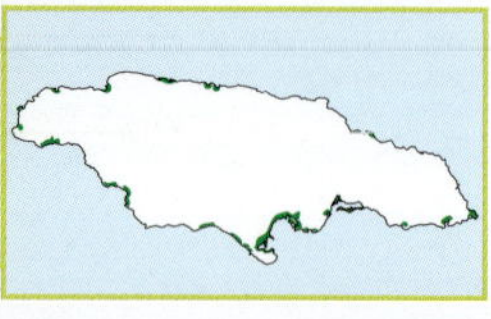

Sora
Porzana carolina

Taxonomy Monotypic.
Description L 17–25cm (7–10in). Small, greyish with dark head and back, short yellow bill, black face and throat (less distinct in females). Flanks have vertical black-and-white stripes. Upper breast grey, abdomen white. Legs yellow.
Similar species None.
Voice *Keet*. Also a descending whistle.
Habitat and behaviour Fresh or saline swamps and ponds.

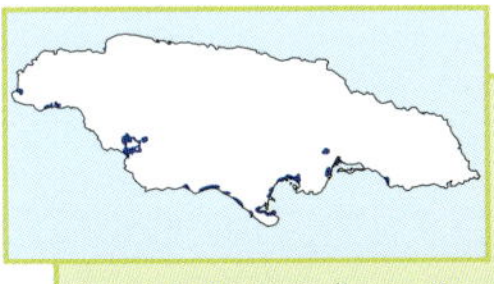

Range Breeds in North America. Winters south through West Indies to South America.
Status Uncommon but regular winter visitor and passage migrant (October–April).

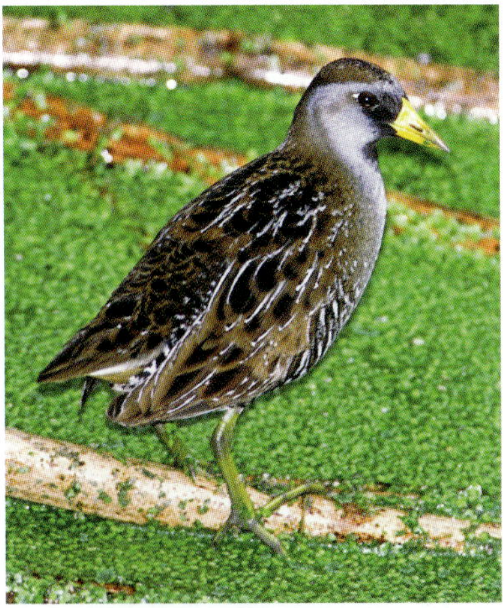

Adult, Note short yellow bill. Florida, January.

Yellow-breasted Crake
Porzana flaviventer

Local name Twopenny Chick.
Taxonomy Polytypic (6). *P. f. gossii* is a Caribbean endemic.
Description L 14cm (6'). A tiny yellowish crake. Crown black. Bill dark olive, supercilium white, eye stripe black. Iris red. Sides of head and neck yellowish grey. Back yellow-brown flecked white. Flanks broadly barred black. Underparts white with yellow wash on breast. Feet yellow. Juvenile browner than adult.
Similar species None.
Voice A prolonged high-pitched ascending *peeeeeeep*.
Habitat and behaviour Pond edges, wet meadows. Difficult to flush; flies feebly with legs dangling. Nests (April–June) in sedge; four spotted eggs.

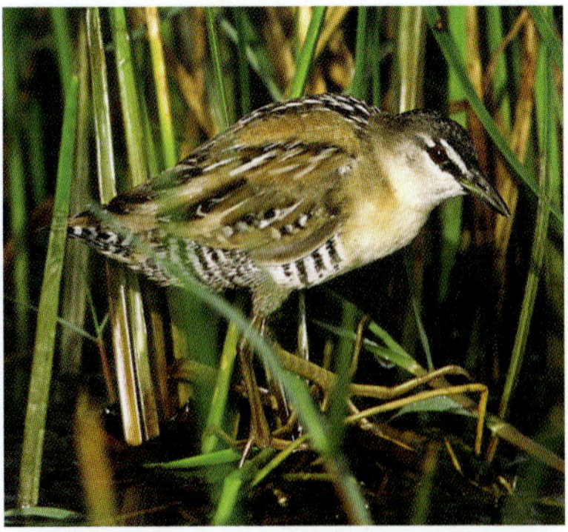

Adult *P. f. hendersoni*. Note pale breast. Puerto Rico, April.

Range *P. f. gossii* Jamaica and Cuba. Other races in Hispaniola, Puerto Rico, Central and South America.
Status in Jamaica Rare resident and very hard to find. May be declining due to loss of habitat.

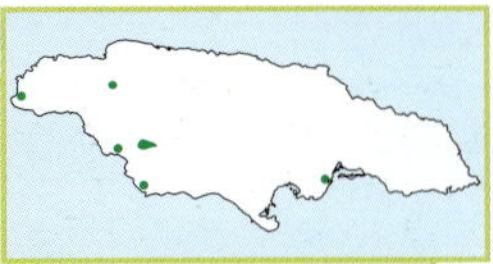

Spotted Rail
Pardirallus maculatus

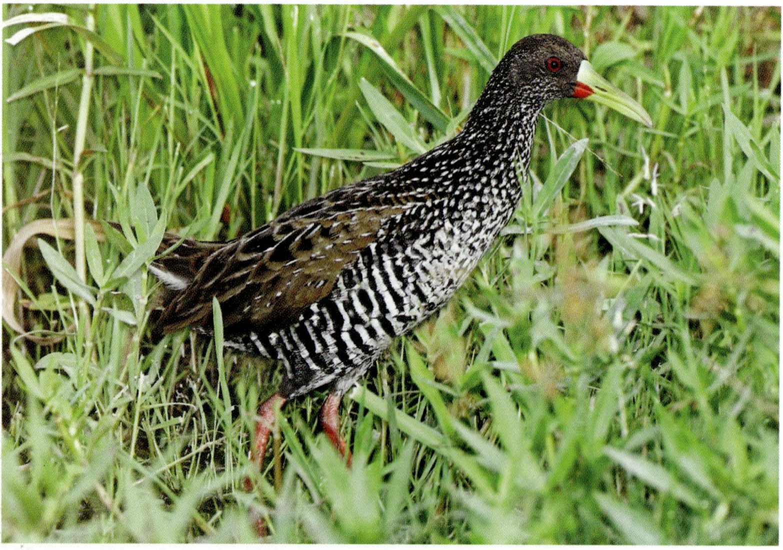

Adult. Note red legs and red spot on bill. Puerto Rico, April.

Taxonomy Polytypic (2). *P. m. maculatus* is endemic to the Greater Antilles.
Description L 28cm (11in). Medium-sized rail with red legs. Bill green with a bright red spot at base of the lower mandible. Head and neck very dark brown, finely spotted with white. Back and wing-coverts dark, edged with lighter brown. Wings black with light brown edges and a few white spots. Lower underparts to vent greyish brown, undertail-coverts cinnamon. Flanks and front of thighs dark brown, barred with white. Juvenile browner.
Similar species No other rail likely to be seen in Jamaica has bright red legs.
Voice *Tuk-tuk-tuk-tuk*, faster toward the end.

Habitat and behaviour
Secretive and generally nocturnal. Glimpsed at freshwater pond margins in dense rushes and grasses. Flies short distances with legs trailing.

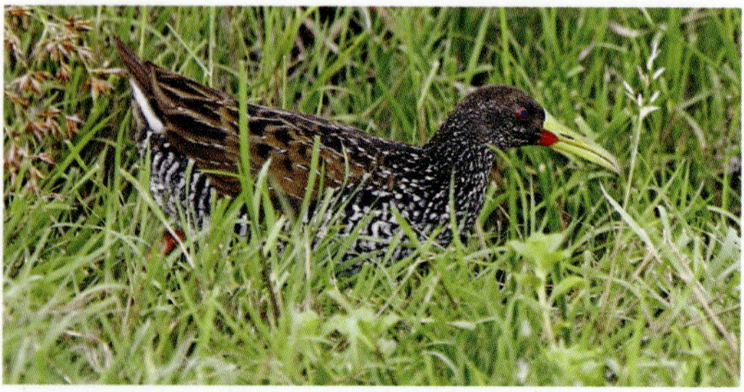

Adult. Puerto Rico, April.

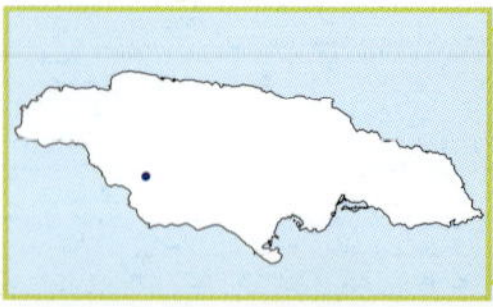

Range *P. m. maculatus* is resident in Cuba and probably Dominican Republic and Jamaica. Other races in Mexico, Central and South America (including Trinidad and Tobago).
Status in Jamaica Very rare. Current breeding status unknown but small resident population may be boosted by winter visitors.

American Purple Gallinule
Porphyrio martinica

Taxonomy Monotypic.
Description L 27–36cm (11–14in). A colourful waterbird with iridescent bluish-purple head, neck and underparts. Bill red with yellow tip and pale blue shield on forehead. Upperparts iridescent turquoise-green. Undertail-coverts white. Legs and feet yellow, toes very long. Juvenile has brown head and buffy underparts, greenish bill and white throat.
Similar species Adult is unmistakeable. Juvenile resembles juvenile Common Moorhen.
Voice Adult gives a high-pitched *pink!* Immature birds give a loud whistle.
Habitat and behaviour Usually seen clambering in dense vegetation on the margins of freshwater ponds and rivers with emergent vegetation. Jerks tail while walking. Feeds on seeds and invertebrates among dense grasses or on lily pads. Nest (April–July) is a loose platform of grasses among pond vegetation; generally attached to aquatic vegetation. Eggs (3–12) are spotted.

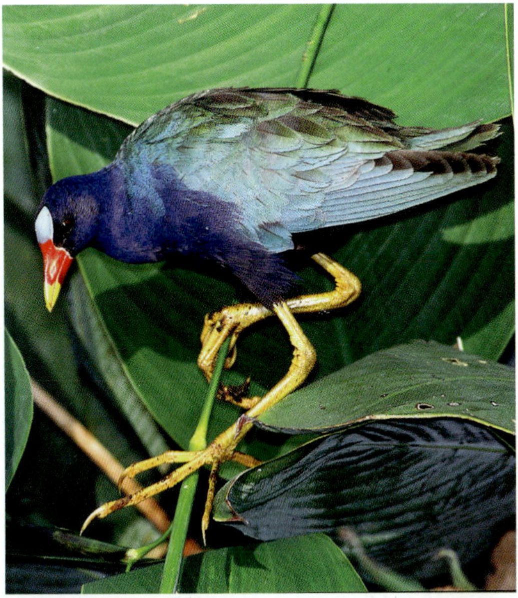
Adult. Florida, April.

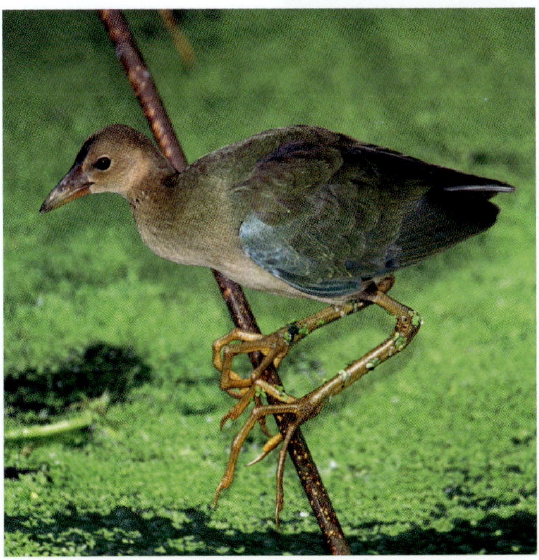
Juvenile. Note blue-tinged wings. Florida, April.

Range North, Central and South America and Greater Antilles.
Status Fairly common resident.

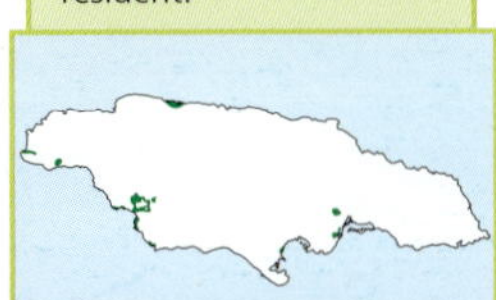

Common Moorhen
Gallinula chloropus

Local names Water Hen, Red Seal Coot.
Alternative name Common Gallinule.
Taxonomy Polytypic (12). *G. c. cerceris* is an endemic West Indian race.
Description L 27–35cm (11–14in). Dark grey-black with white stripe on flanks. White undertail-coverts have black median stripe. Breeding adult has red bill tipped with greenish-yellow, and a red frontal shield. Non-breeding adult has a dark bill. Juvenile has brownish wings, grey-brown body and bill, throat paler. Downy chicks mostly black with red bills and a bare patch behind eye.

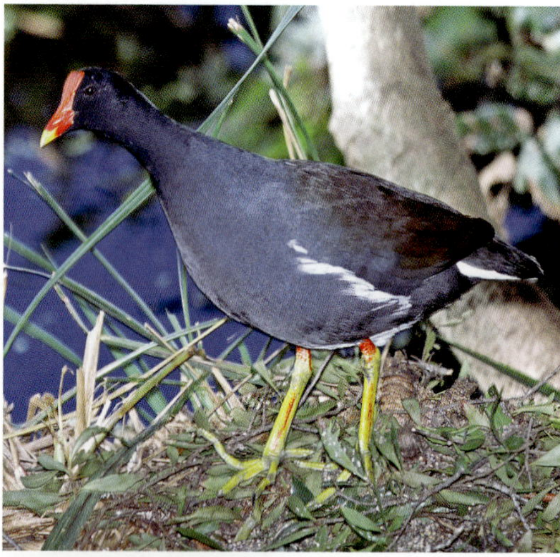
Adult breeding with red shield. Cuba, April.

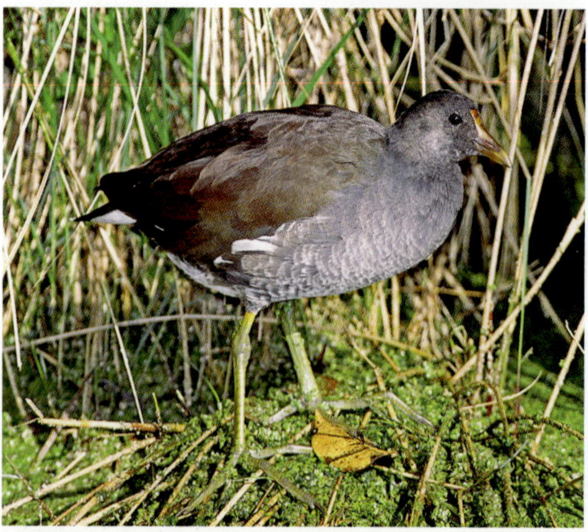
Juvenile. Note the brown wings. Cuba, April.

Similar species Juvenile American Purple Gallinule lacks the white flank line. Juvenile coots are grey with whitish bills.
Voice A variety of harsh, quarrelsome high-pitched notes.
Habitat and behaviour Fresh and brackish ponds, lowland rivers and their margins. May be seen walking on banks. Often seen swimming in open water, pumping its head back and forth. If disturbed, may take off, pattering noisily across water. Feeds on a wide range of aquatic plants and invertebrates. Nest (year-round) is a large platform made of vegetation lined with grass, on bank or floating. Lays 3–9 spotted eggs.

Range *G. c. cerceris* on the West Indies. North American migrants (race not determined) increase local populations in winter. Other races occur worldwide.
Status Very common resident.

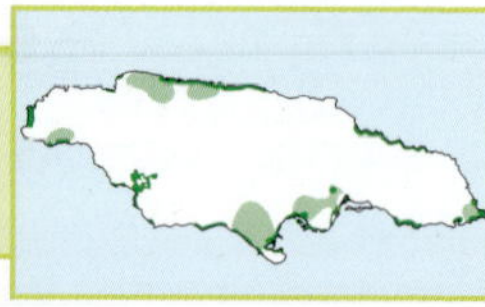

American Coot
Fulica americana

Local name White Seal Coot
Taxonomy Polytypic (2). Race in Jamaica is *F. a. americana.*
Description L 31–40cm (12–16in). A large, greyish-black waterbird, darker on the head, undertail-coverts white. Bill white, usually with a dark spot near the tip; white frontal shield has a reddish-brown knob at the top. Legs and feet yellowish with lobed toes. Juvenile is paler and greyish, with a yellowish bill. Downy chicks (often seen swimming with adults) are mostly black with red bills. When they are first hatched, the neck feathers have red or buffy tips. Later the chicks become grey with white on the throat, front of neck and underside.
Similar species Caribbean Coot has a larger white shield and lacks the brown shield knob. Juveniles indistinguishable.

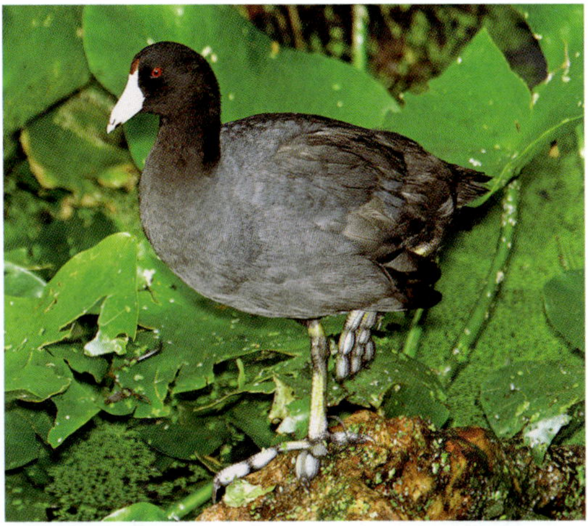

Adult. Florida, May.

Voice A series of low rasping variations of *ca-cuk-cuk-cuk-cow-cows*.
Habitat and behaviour Often seen in groups or singly on fresh and brackish ponds. Pumps head when swimming. When disturbed, patters across water to take off. Large groups often seen rafting together on open water or walking on banks. Feeds on invertebrates, roots and leaves, foraging from the surface like a duck, or diving. Nest (year-round) is a large mound of fresh green grasses and plants on shallow mud, or floating and attached to emergent plants.

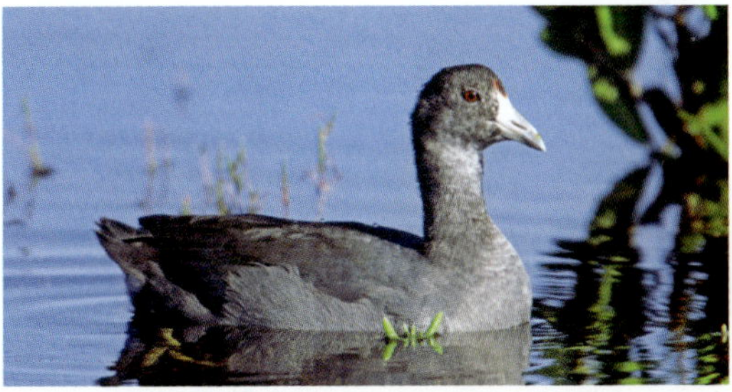

Juvenile. Note yellowish bill. Florida, May.

Range *F. a. americana* breeds in North and Central America and the Bahamas, Greater Antilles and northern Lesser Antilles. Winters south to northern South America. Other race in western South America.
Status in Jamaica Common resident. Local populations greatly increased by migrants in winter.

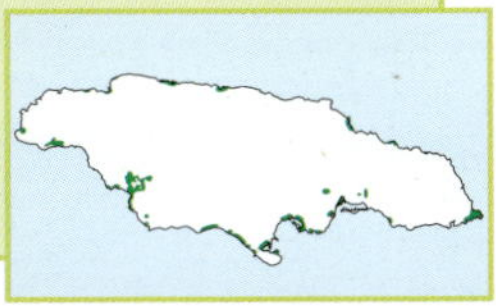

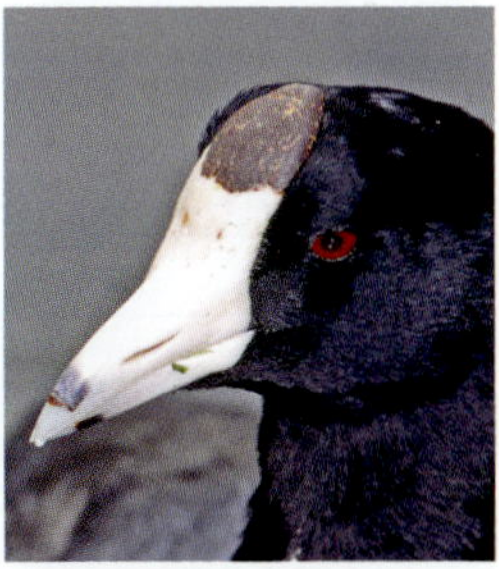

Adult head. Florida, May.

Caribbean Coot
Fulica caribaea

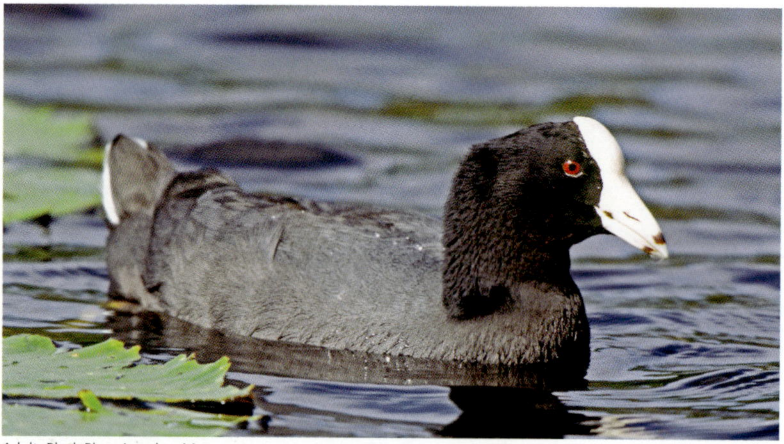

Adult. Black River, Jamaica, May.

Local name White Seal Coot
Taxonomy Monotypic.
Description L 31–36cm (12–14in). A large greyish-black waterbird, darker on the head. Has a prominent white shield that may be tinged with yellow, which extends above the crown at the top and so is visible from behind. Bill is white with a dark spot near the tip. Undertail-coverts are white. Juveniles and downy chicks are very similar to those of American Coot.
Similar species American Coot is very similar but generally has a smaller white shield, with a brown knob at the top. Juveniles are indistinguishable.
Voice Similar to American Coot: *c-cuk*.
Habitat and behaviour Ponds and rivers. Usually observed swimming in open stretches of water or walking along the bank. When disturbed, patters across water to take off. Often seen in rafts with groups of American Coots. Feeds on invertebrates, roots and leaves, often taken from the surface like a dabbling duck, though will also dive. Nests (mainly April–June) are usually floating platforms of vegetation, attached to emergent plants. Rarely hybridises with American Coot.

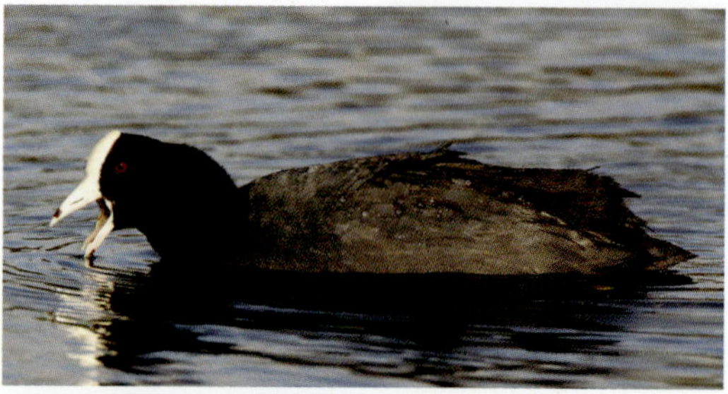

Adult threat display. Dominican Republic, March.

Range West Indies (mainly Greater Antilles) and northern South America. Rare and probably declining throughout its range. Global conservation status: Vulnerable or Near Threatened.
Status in Jamaica. Uncommon resident.

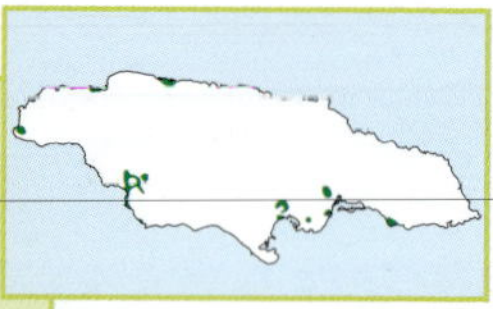

Northern Jacana
Jacana spinosa

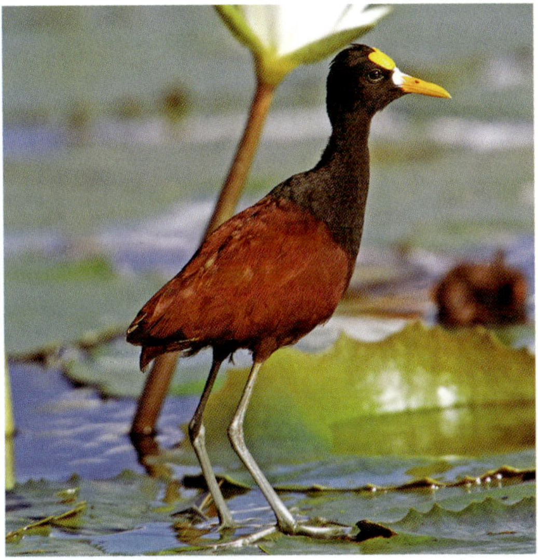

Adult. Hispaniola, May.

Local names River Chink, Pond Coot, Jesus Bird.
Taxonomy Polytypic (3). *J. s. violacea* is endemic to the Greater Antilles.
Description L 18cm (7in). A bright reddish-brown plover-like bird with black head and neck and bright yellow bill and shield. Flight feathers bright yellow. Wings have a prominent carpal spur. Legs longish and grey, toes grey and very long. Male smaller than female. Juvenile has pale brown back and head, pale yellow bill and white underparts.
Similar species None.
Voice A noisy cackling.
Habitat and behaviour Usually seen in fresh and brackish ponds and rivers, walking on floating vegetation such as water lilies, where the very long toes help to distribute weight and prevent sinking. Feeds on plant material and on aquatic invertebrates. Flies with dangling legs close to the water surface. Wings are characteristically elevated after landing and during territorial and mating displays, showing the yellow flight-feathers. Nest (April–August) is a small floating platform among dense aquatic vegetation, roughly built and shallow. Two brown eggs are laid.

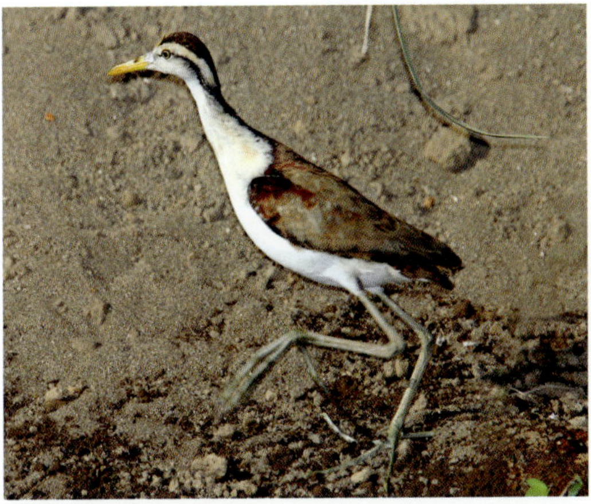

Juvenile. Domincan Republic, May.

Range *J. s. violacea* occurs in Jamaica, Cuba and the Isle of Pines, and Hispaniola. Other races in Central and South America.
Status in Jamaica Common resident.

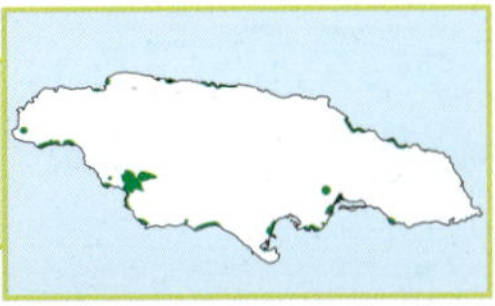

Black-necked Stilt
Himantopus mexicanus

Local name Cap'n Lewis.
Taxonomy Polytypic (2). Race in Jamaica is *H. m. mexicanus.*
Description L 33–39cm (13–15in). Distinctive black-and-white shorebird with very long, spindly red legs. Bill is long, straight, very thin and black. Breeding male black-and-white. Female and non-breeding male are brownish and white. Nestlings white below, buffy with dark spots above.
Similar species None.
Voice Noisy and often first detected by call, which is a characteristic loud and prolonged yipping.
Habitat and behaviour Fresh or saline ponds (including sewage ponds and salt pans), shallow coastal wetlands. Nest (March–June) is a shallow scrape, lined with grass or other vegetation, just above the water level on sandy beaches or pond margins. Lays 3–4 buffy eggs with fine dark spots. Downy chicks are mobile on hatching. When disturbed near the nest the adult may perform a 'broken leg' display to distract predators.

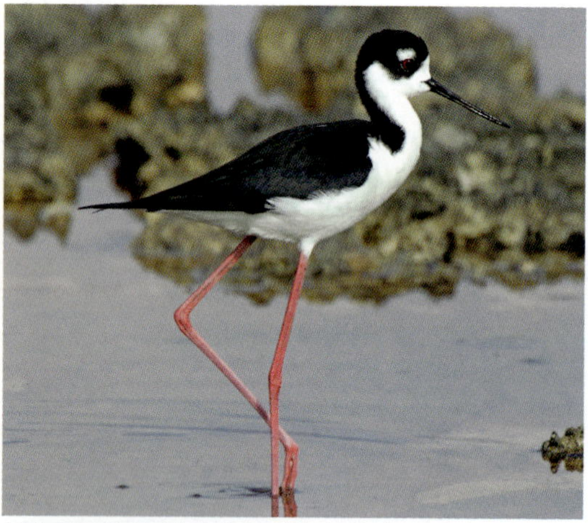
Adult male. Note black back. Cayman Islands, March.

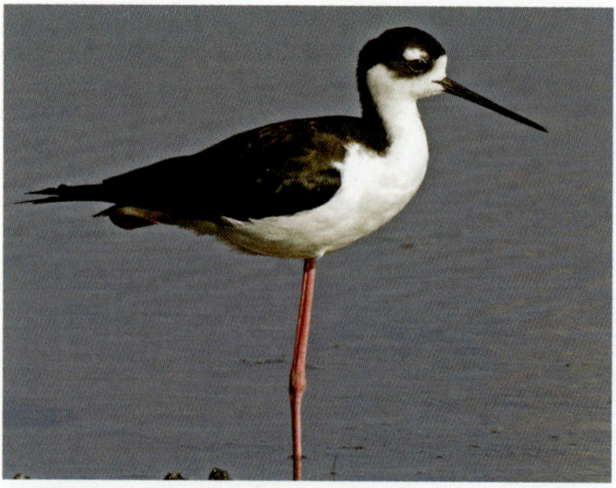
Adult female. Note brown back. Cayman Islands, March.

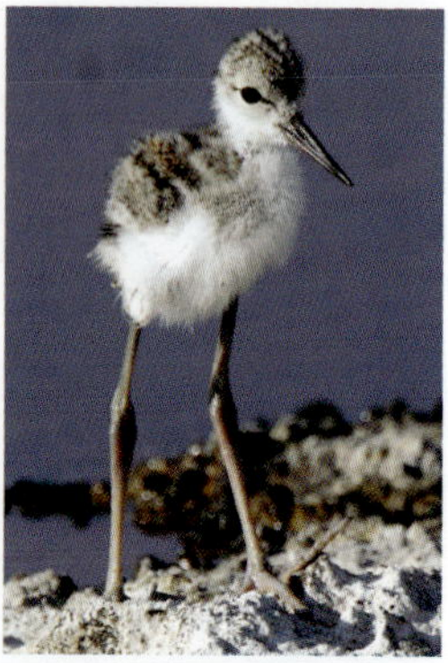
Nestling. Cayman Islands, March.

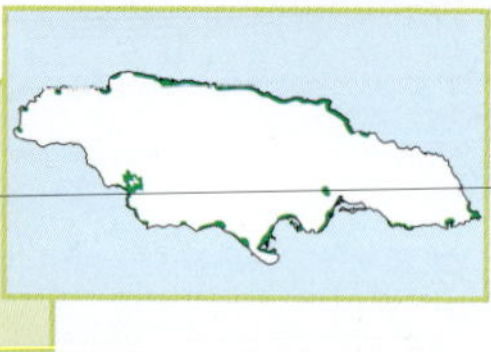

Range *H. m. mexicanus* in Bahamas, Greater Antilles and northern Lesser Antilles and throughout North and South America. Other race in Hawaiian Islands.
Status in Jamaica Common resident. Local populations probably increased by migrants from North America in winter.

Grey Plover
Pluvialis squatarola

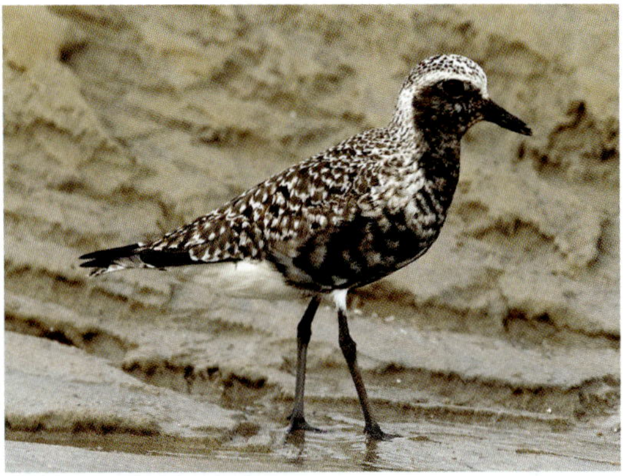

Adult in breeding plumage. August.

Local name Lapwing.
Alternative name Black-bellied Plover.
Taxonomy Monotypic.
Description L 24–26cm (10–11in). Stocky grey plover, with large black eyes, heavy black bill and conspicuous black 'wing-pits' (axillaries). Juveniles browner. When adults first arrive in August they are in breeding plumage, with black underparts and silvery-grey and black spangled upperparts. By November the black has turned to white with fine streaking on throat. White wing-bar. Tail barred, legs grey. In March, black areas begin to appear on underparts and some moult into full breeding plumage before departing in early May.
Similar species American Golden Plover is smaller and lacks black 'wing-pit'. Sandpipers have longer bills.
Voice A clear whistle *tlee-oo-ee*.
Habitat and behaviour Sandy margins of shallow salt ponds and beaches. Flocks often found feeding with other plovers and sandpipers.

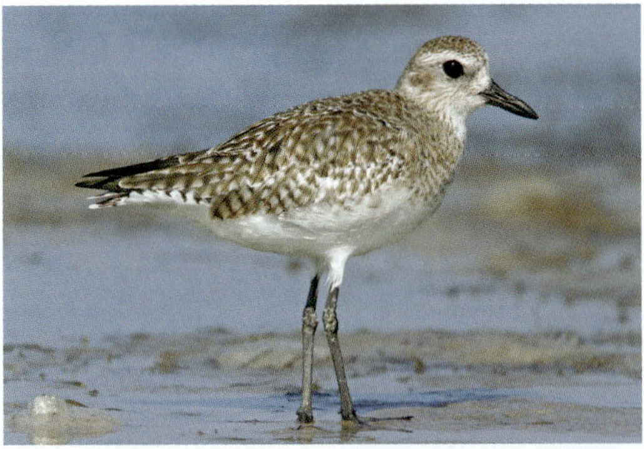

Adult non-breeding. Florida, February.

Range Breeds in the Arctic tundra and winters south, virtually worldwide.
Status in Jamaica Fairly common winter visitor. A few non-breeders spend the summer in Jamaica.

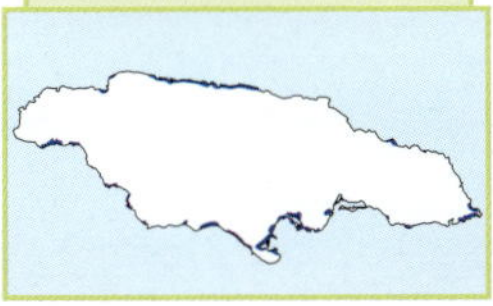

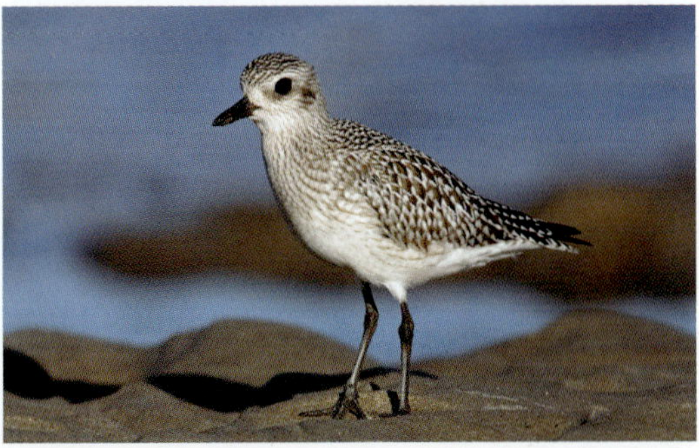

Juvenile. Note whitish 'spangled' appearance of upperparts. October.

Semipalmated Plover
Charadrius semipalmatus

Taxonomy Monotypic.
Description L 17–19cm (7–8in). A small shorebird, brown above, white below, with a small, stubby bill, one dark band across breast and a complete white collar around base of neck. White or buffy line over eye and bill. White wing-bar. Legs are light orange. Breeding male has breast-band, eye patch and band across crown black; female duller. Bill usually bicoloured with variable amounts of yellow at base, dark at tip but may be all dark (in non-breeding and immature plumages). Breast-band and head markings brownish in non-breeding adults. Juvenile resembles non-breeding adult, legs yellowish.
Similar species Wilson's Plover is larger and has longer and heavier bill.
Voice A plaintive *chewee*.
Habitat and behaviour Sandy beaches and mudflats. Runs swiftly along the shore, sometimes 'freezing' in mid-step.

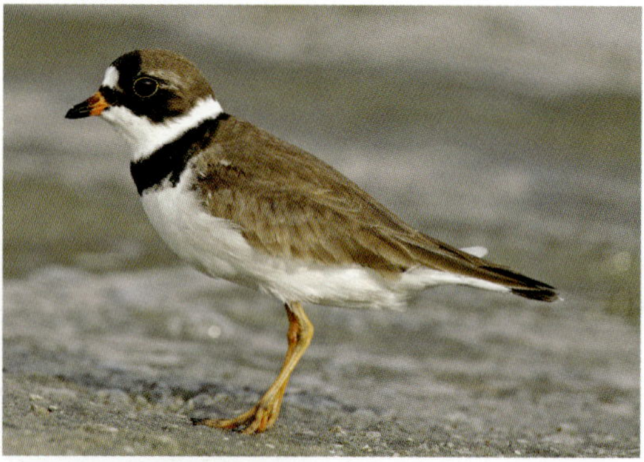

Adult breeding plumage. Note bill and single breast stripe. Florida, May.

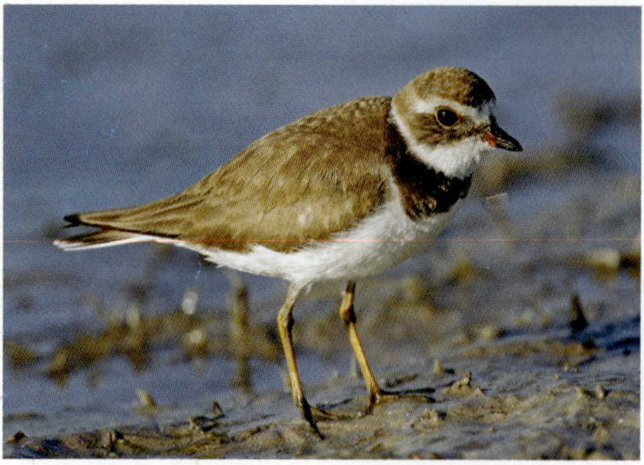

Adult non-breeding. Florida, February.

Range Breeds in North America. Winters south through Mexico and the West Indies to central South America.
Status in Jamaica Fairly common winter visitor (August–March).

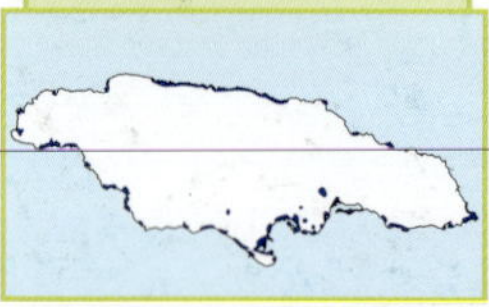

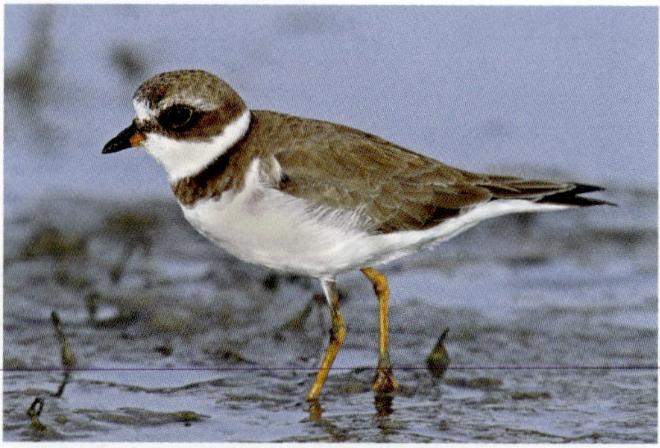

Juvenile. Note brown breast band. Florida, February.

Wilson's Plover
Charadrius wilsonia

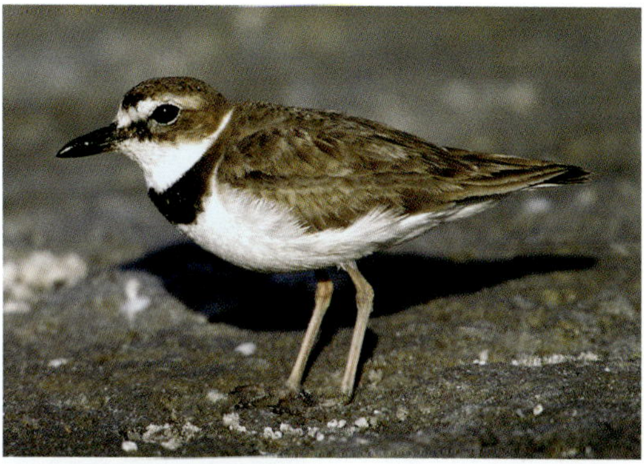

Adult male breeding. Note heavy black bill, pinkish legs. Florida, May.

Taxonomy Polytypic (3). Race in Jamaica is *C. w. wilsonia*.
Description L 16–20cm (6–8in). A medium-sized shorebird with a thick black bill. Crown and back dark brown. White below with a broad breast-band, black in breeding male, brown in non-breeding male, female and juvenile. Legs dull pinkish.
Similar species Semipalmated Plover, a winter migrant, has a short thin bill, yellow legs and a different call. Killdeer is larger, has two bands on breast and is generally found on grassland.
Voice A high *wheet*. Also a grating rattle when disturbed.
Habitat and behaviour Usually seen running along sandy beach, mudflat or pond margin near the coast. Occasionally stops abruptly and freezes. Feeds mainly on crustaceans, also insects and small snails. Nests (mainly April–July) on sandy beaches, in depressions in the sand such as hoof-marks left by horses or cattle, sometimes lined with a few pieces of shell. Lays two spotted eggs.

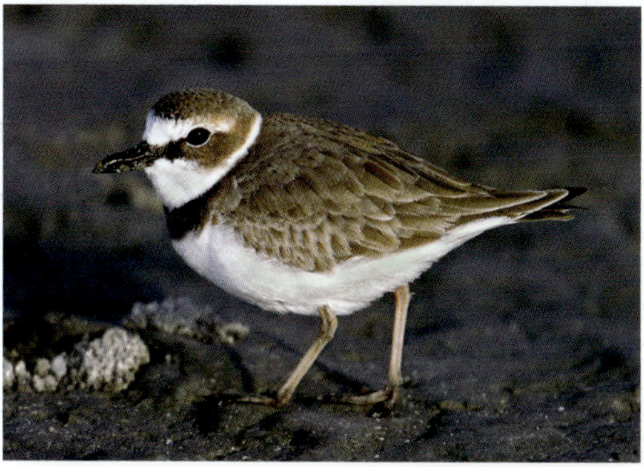

Adult non-breeding. Florida, February.

Range *C. w. wilsonia* in West Indies and coastal United States, wintering south to Brazil. Other races in California and South America.
Status in Jamaica Fairly common resident.

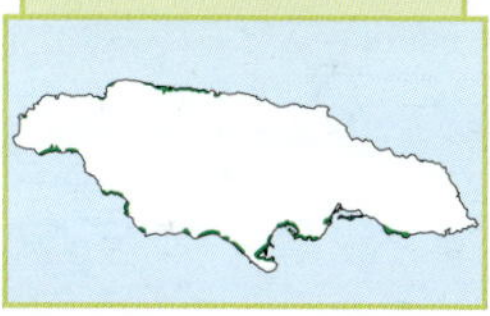

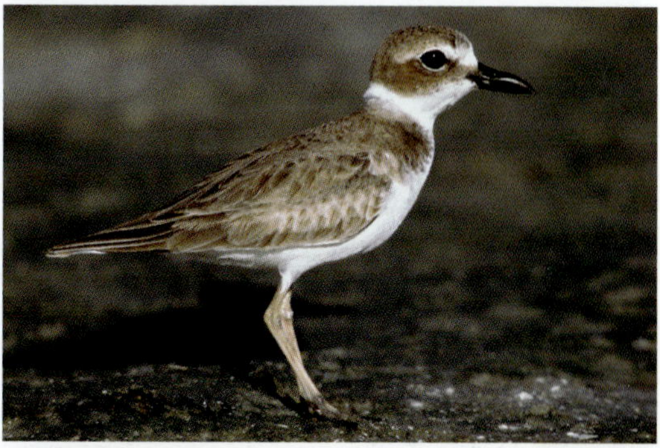

Juvenile. Florida, February.

Killdeer
Charadrius vociferus

Local names Tilderee, Tell-Tale, Killdee.
Taxonomy Polytypic (3). *C .v. ternominatus* is endemic to the Greater Antilles and Bahamas.
Description L 20–28cm (8–11in). A long-tailed, banded plover, distinctive with its two breast-bands. Brown above, fringed rufous when non-breeding, white below.
Similar species No other plovers in the region have double breast-bands.
Voice A characteristic shrill *killdeeeer*, and a rising *deee*.
Habitat and behaviour Wet grassy places, not necessarily beside water. Nests (May–July) are shallow scrapes, sometimes lined with stones, often found in the open, on grassy edges of ponds and reservoirs, also on fields and golf courses. Lays 3–4 olive eggs, with more dark splotches on broad end. Downy chicks are well camouflaged, grey above, with dark line through eye, and white below. They are mobile almost immediately after hatching. Adults may do a 'broken wing' display to distract predators and thus protect nest or chicks.

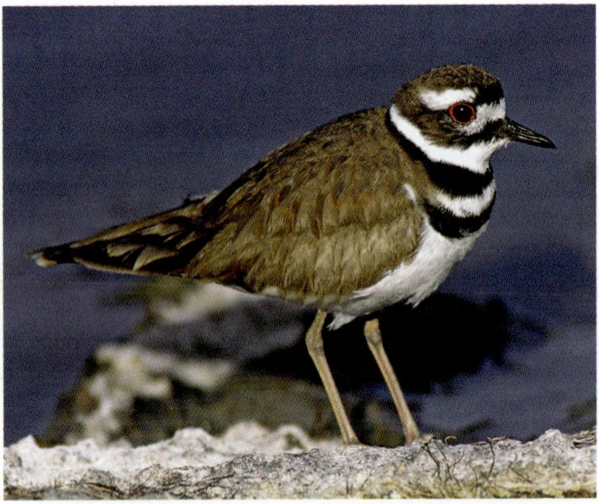

Adult breeding. Note double breast band. Bahamas, April.

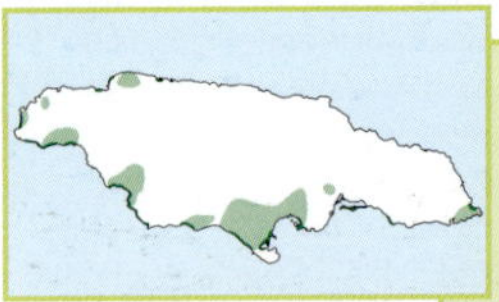

Range *C .v. ternominatus* in Greater Antilles and Bahamas. Other races in North America, wintering to northern South America via Bahamas and Greater Antilles; also in South America.
Status in Jamaica Common resident. North American migrants increase local populations in winter.

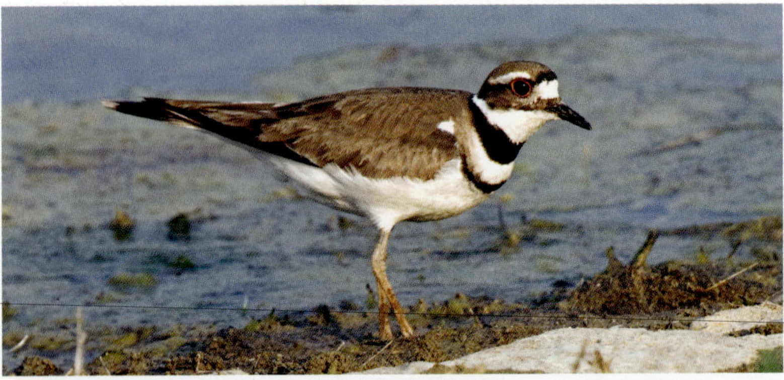

Juvenile. Note the yellow eye-ring – otherwise very similar to the adult. May.

Wilson's Snipe
Gallinago delicata

Taxonomy Monotypic.
Description L 23cm (9in). A plump shorebird with a distinctive brown, buff and white-striped head, and a very long, dark bill, lighter at base. Legs short. Dark brown overall, heavily streaked black and rufous. Four distinct longitudinal buffy stripes along back.
Similar species None in region.
Voice Usually silent, but utters a harsh note when flushed.
Habitat and behaviour Grassy wet pastures and pond edges. Secretive. Crouches in grass then flushes, flying away in a zig-zag manner.

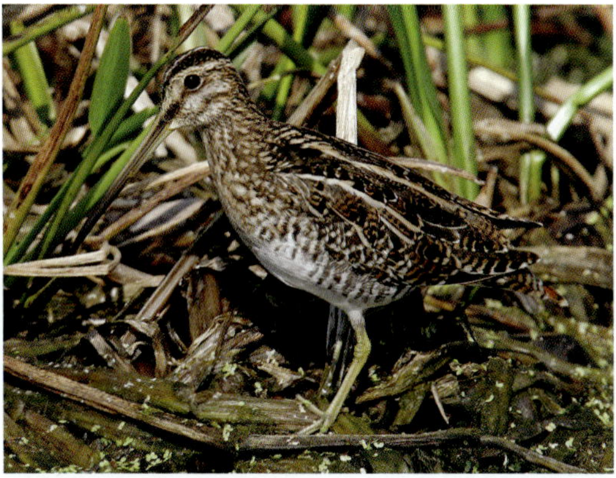

Adult. Note long bill. Florida, February.

Range North America. Breeds in the northern United States and Canada and migrates through the West Indies and Central America to southern South America.
Status in Jamaica Fairly common winter visitor (October–March).

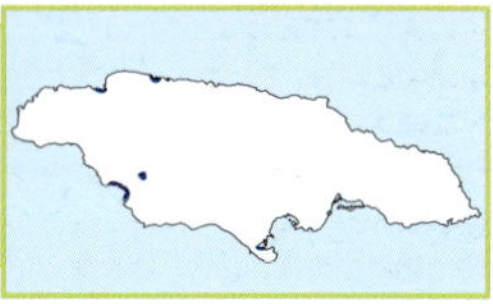

Short-billed Dowitcher
Limnodromus griseus

Taxonomy Polytypic (3). Race on Jamaica undetermined.
Description L 25–29cm (10–12in). Fairly large grey-brown shorebird with a long, straight, dark bill (lighter at base) and greenish legs. White 'V' on back. Juvenile browner above with rufous feather edging.
Similar species Long-billed Dowitcher is scarcer in Jamaica; best distinguished by call.
Voice A sharp, rapid *tu-tu-tu*.
Habitat and behaviour Lagoons and mudflats close to the coast. Usually seen in small flocks. Perpendicular 'sewing machine-like' feeding pattern distinguishes dowitchers from other shorebirds of similar size.

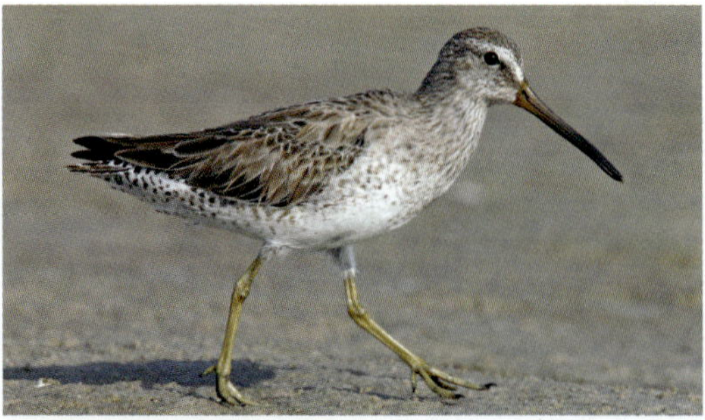

Adult. Grey with barred flanks. Florida, March.

Range Alaska and Canada, wintering south to Brazil
Status in Jamaica Common winter visitor (August–May).

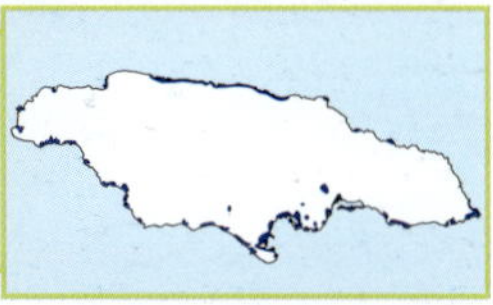

Long-billed Dowitcher
Limnodromus scolopaceus

Taxonomy Monotypic.
Description L 25–29cm (10–12in). A fairly large, dark grey shorebird with white 'V' on back. Bill straight, very long and dark. Legs greenish. Dark cap and eye-stripe contrast with white supercilium. Juvenile browner above with rufous feather edging.
Similar species Very difficult to distinguish from Short-billed Dowitcher in winter plumage. The latter is more common, and smaller with a shorter bill. Best separated by voice.
Voice A high-pitched single *keek*, sometimes repeated.
Habitat and behaviour Swamps, more likely in fresh water than Short-billed Dowitcher; feeds in shallow water with characteristic 'sewing machine-like' action.

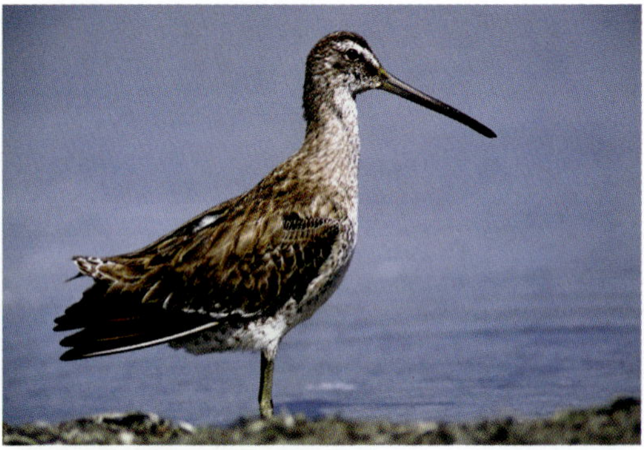

Adult in non-breeding plumage. Florida, March.

Range Breeds in northern N America and Eurasia. Migrants from N America winter in C America.
Status Uncommon winter passage migrant (August–May).

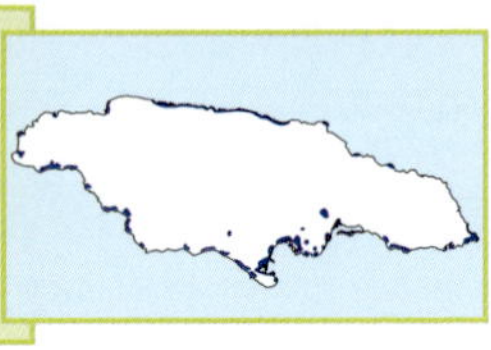

Whimbrel
Numenius phaeopus

Taxonomy Polytypic (4). Race in Jamaica is *N. p. hudsonicus.*
Description L 38–46cm (15–18in). Large brown shorebird with a long decurved bill and a dark crown with central white crown stripe. Legs grey.
Similar species None in region.
Voice Usually silent. When flushed a long, trilling *tu-tu-tu-tu-tu-tu-tu*.
Habitat and behaviour Wetlands and shallow coastal ponds. Usually seen singly, in marshes.

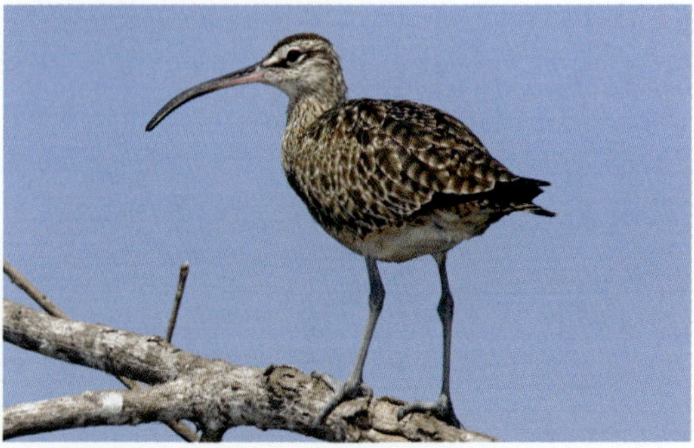

Adult. Note decurved bill. Costa Rica, March.

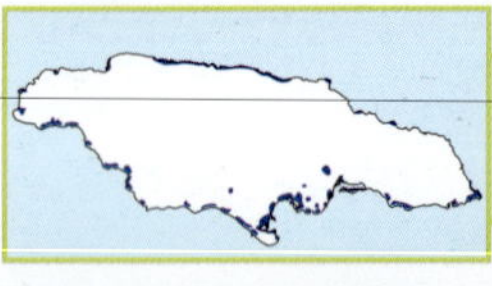

Range *N. p. hudsonicus* breeds in the Arctic, winters S. America.
Status in Jamaica Uncommon but regular winter visitor and passage migrant (mostly December–February).

Spotted Sandpiper
Actitis macularius

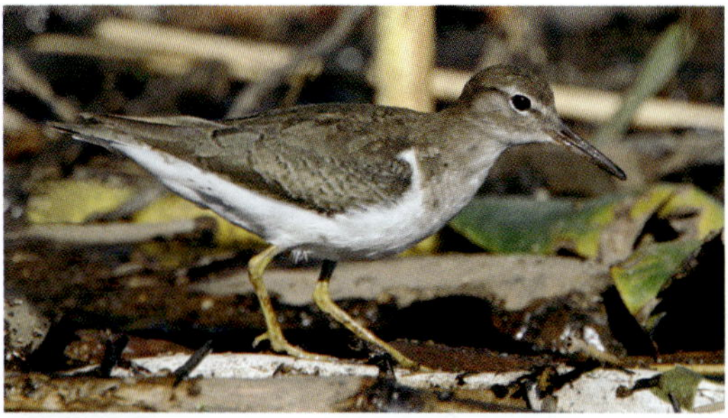
Adult non-breeding. Costa Rica, February.

Taxonomy Monotypic.
Description L 16–20cm (6–8in). A small sandpiper, brown above, light below, legs yellowish. Adult in breeding plumage has dark spots on white underparts. Bill orange with dark tip. Underpart spots lost by Nov., regained in March. Adult non-breeding grey above, white below; bill dark
Similar species Solitary Sandpiper larger, browner, with longer legs and clearer eye-ring.
Voice A soft *peet* or *peet-weet*.
Habitat and behaviour Sandy or muddy edges beaches, rivers or ponds. Constantly 'teeters' when on the ground. Flies with rapid, stiff wing-beats.

Range Breeds in North America and winters south via the West Indies to southern South America.
Status in Jamaica Common winter visitor, year-round non-breeding resident and passage migrant (July–March).

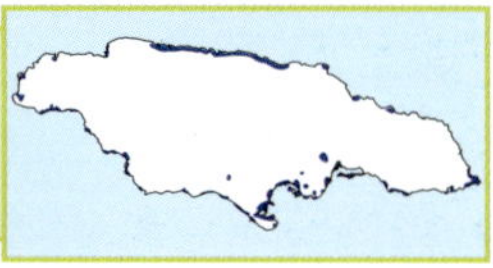

Solitary Sandpiper
Tringa solitaria

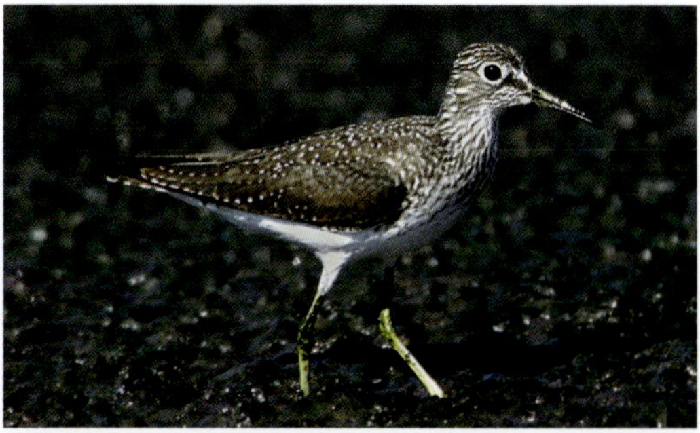
Adult. Note prominent white eye-ring. Florida, February.

Taxonomy Polytypic (2). Race on Jamaica is *T. s. solitaria.*
Description L 18–23cm (7–9in). Medium-sized shorebird with dark brown, finely spotted upperparts, head and neck, whitish underparts. Eye-ring white and prominent. Long, thin, straight bill, dark with greenish base. Legs dull greenish. In flight shows uniformly dark wings above and below and dark rump and tail centre.
Similar species Both yellowlegs have yellow legs.
Habitat and behaviour Mudflats, muddy margins of ponds and wet areas across the island. Often feeds alone.

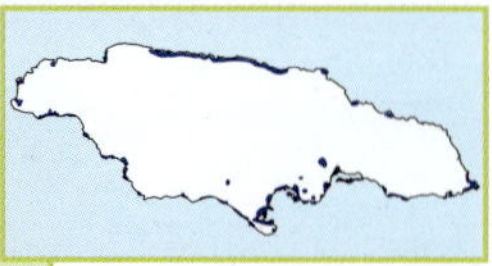

Range *T. s. solitaria* breeds western North America; other race northeast North America. Both winter in South America.
Status Rare winter visitor and passage migrant (August–May).

Greater Yellowlegs
Tringa melanoleuca

Taxonomy Monotypic.
Description L 28–33cm (11–13in). Large shorebird with brown speckled back, white underparts and yellow legs. White rump obvious in flight. Slightly upturned bill is about 1.5 times longer than head length. Juvenile is more spotted above.
Similar species Lesser Yellowlegs is smaller, with a relatively shorter and more slender bill.
Voice A loud, liquid, three or four note *tchew-tchew-tchew*, often repeated. Sounds a little like *yell-ow-legs*.
Habitat and behaviour Swamps and saline ponds. Usually seen in flocks with Lesser Yellowlegs and other shorebirds. Feeds on a variety of aquatic invertebrates.

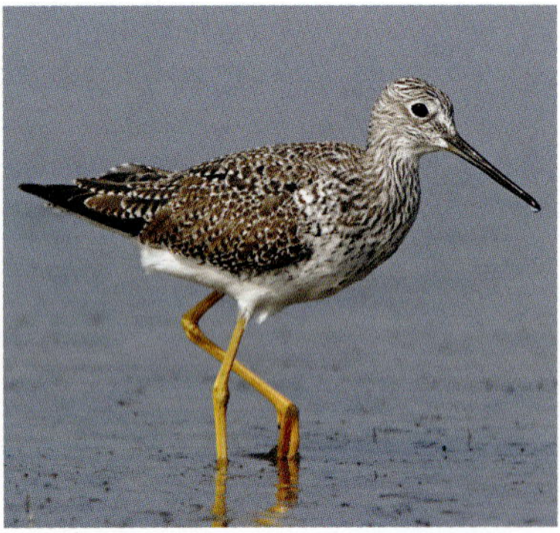
Adult. Bill is 1.5 x width of head. Florida, March.

Range Breeds in North America. Winters south to southern South America.
Status in Jamaica Common winter resident and passage migrant (mainly on southbound migration).

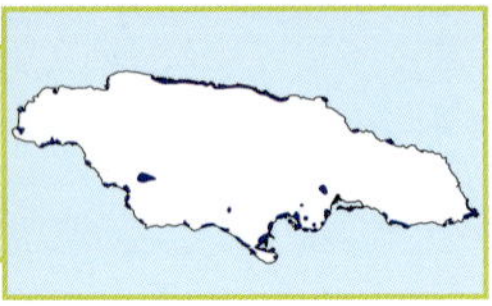

Lesser Yellowlegs
Tringa flavipes

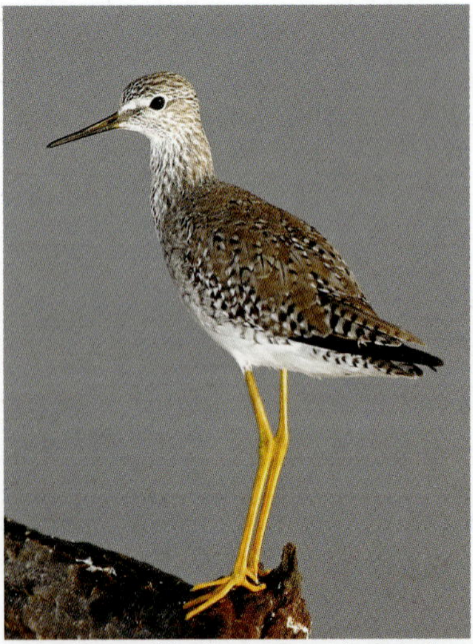
Adult. Bill is 1 x width of head. Cayman Islands, April.

Taxonomy Monotypic.
Description L 22–25cm (9–10in). Medium-sized shorebird with yellow legs. Upperparts dark grey. Neck and breast streaked. White rump obvious in flight. Thin, straight bill about equal to head length. Juvenile more spotted above.
Similar species Greater Yellowlegs larger with a longer, stouter bill and a longer call.
Voice One to three (usually two) *tew* notes.
Habitat and behaviour Occurs in mixed-species flocks in salt ponds and swamps. Feeds on invertebrates by probing in shallow water.

Range Breeds in North America. Winters south to South America.
Status in Jamaica Fairly common passage migrant and winter visitor.

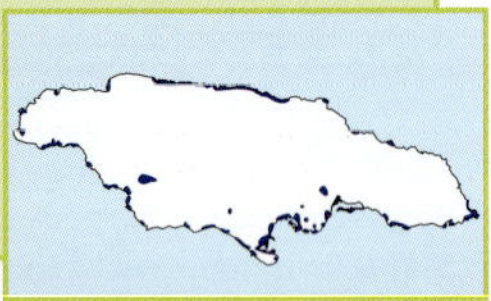

Willet

Tringa semipalmatus

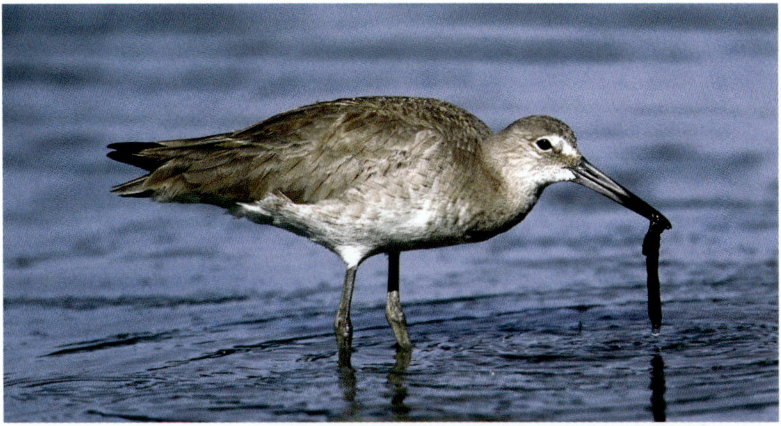

Adult. Note grey legs and bill. Florida, March.

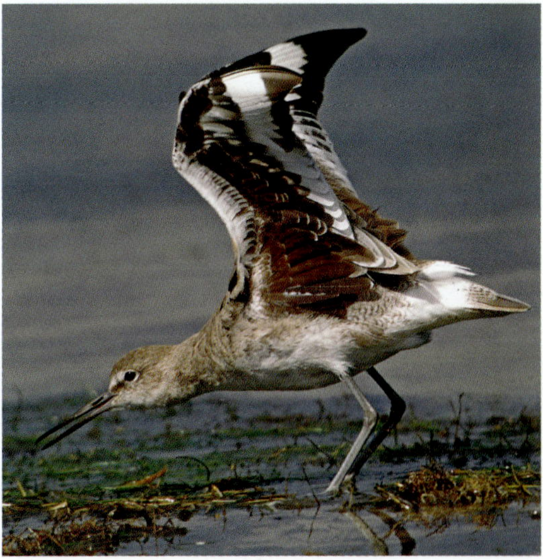

Adult. Note black-and-white underwing. Florida, March.

Taxonomy Polytypic (2). Race in Jamaica is *T. s. semipalmatus*.
Description L 33–41cm (13–16in). Large shorebird, greyish above, white below with brownish head and upperparts. Legs grey, as is thickish straight bill. In flight striking black-and-white W pattern on wings.
Similar species None.
Voice Distinctive *will-will-willet* (in flight) or *kip-kip* (when alarmed).
Habitat and behaviour Shallow mangrove ponds and mudflats. Black-and-white pattern on underwing is obvious when wings are raised, and in flight. Feeds mainly on small invertebrates. Nest (April– July) is a slight depression in the sand. Lays four spotted eggs.

Range *T. s. semipalmatus* breeds on east coast of North America, also Greater Antilles and northern Lesser Antilles. North American migrants winter to South America via the West Indies.
Status in Jamaica Locally common resident.

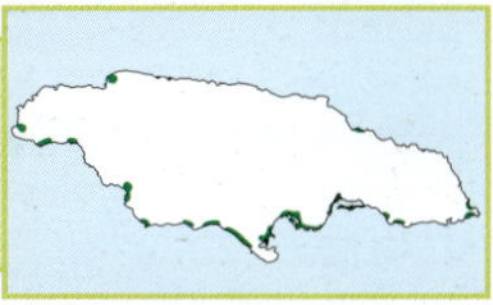

Ruddy Turnstone
Arenaria interpres

Taxonomy Polytypic (2). *A. i. morinella*.
Description L 18–26cm (7–10in). Chunky shorebird, brown above, white below, with shortish, chisel-shaped black bill and bright orangey-red legs. Adult birds arrive in August in breeding plumage, with black-and-white heads, black bibs and black-and-chestnut backs. By November they have moulted into winter plumage; brown upperparts, brown bibs, white underparts and red legs. Juvenile resembles winter adult but

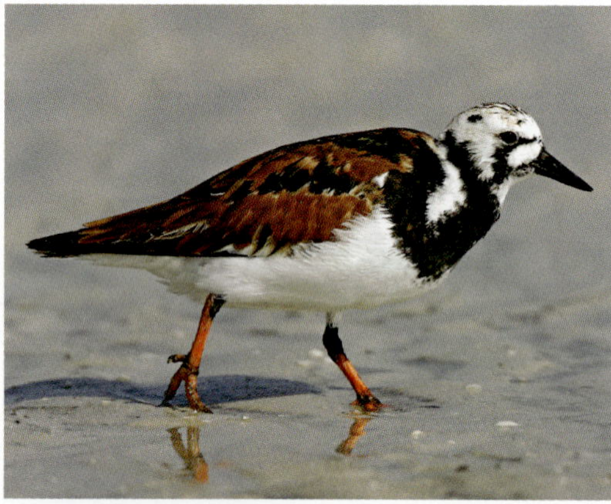

Adult female in breeding plumage. Florida, May.

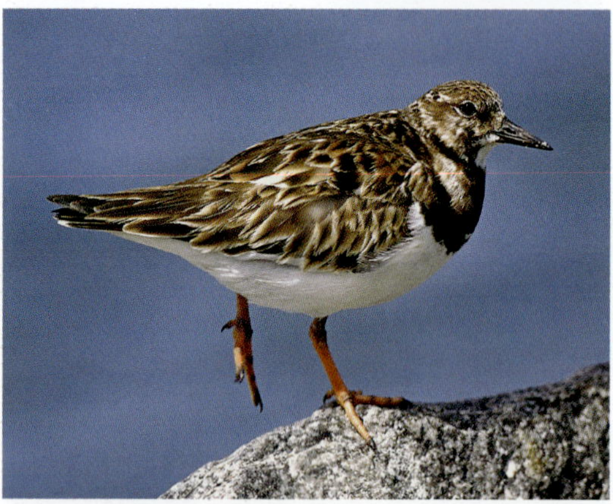

Adult non-breeding. Bahamas, April.

has a scaly back – all feathers have rusty fringes. In flight black-and-white pattern on wings is distinctive.
Similar species No other shorebird has red-orange legs and distinctive brown, black and white plumage.
Voice Usually silent in Jamaica.
Habitat and behaviour Usually seen in loose groups on beaches, cays and rocky coastlines. Feeds on invertebrates by turning over detritus on the strandline. Also eats tern eggs.

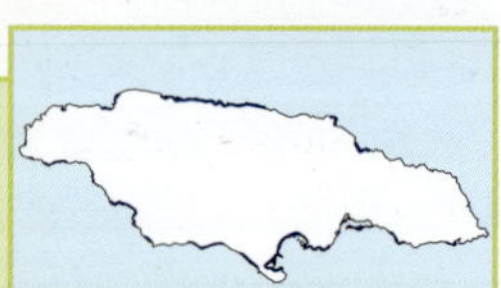

Range *A. i. morinella* breeds in the Arctic, wintering south to Chile via the West Indies. Other race in Palearctic and western North America.
Status in Jamaica Locally common winter visitor (August–May). A few non-breeding adults are present year-round, especially on the Morant and Pedro Cays.

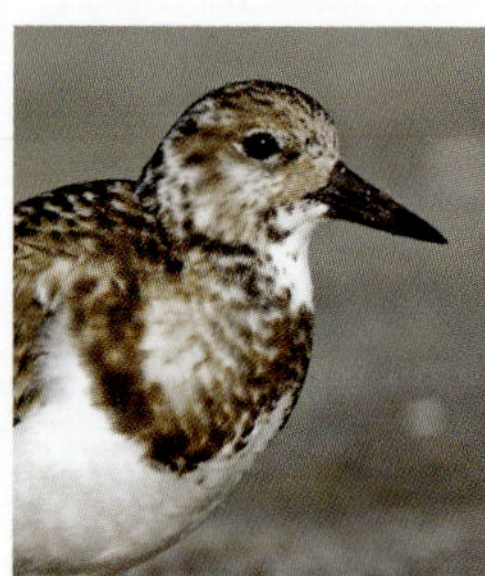

Juvenile. Florida, February.

Red Knot
Calidris canutus

Taxonomy Polytypic (6). Race in Jamaica is *C. c. rufa.*
Description L 23–25cm (8–9in). Chunky sandpiper with a relatively short black bill. Legs are short and dull greenish. Non-breeding adult grey above, with a pale line above the eye. Breast grey, belly white. In flight the rump is grey and lightly barred. Back, tail and wings pale grey.
Similar species Care is needed to distinguish this species in its plain non-breeding plumage from other wintering shorebirds. Least, Spotted and Western Sandpipers are smaller and less chunky. Similarly-sized Stilt Sandpiper has downcurved bill. The yellowlegs are larger and have yellow legs.

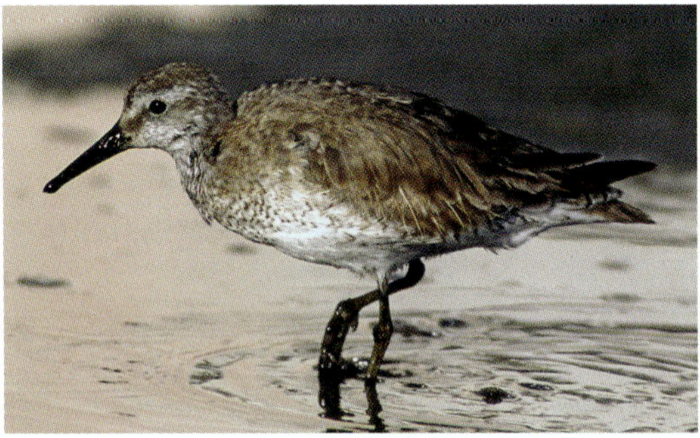

Adult non-breeding. Note chunky outline, relatively short bill. Florida, Feb.

Range *C. c. rufa* breeds in the Arctic and winters south to South America.
Status in Jamaica Rare winter visitor or vagrant.

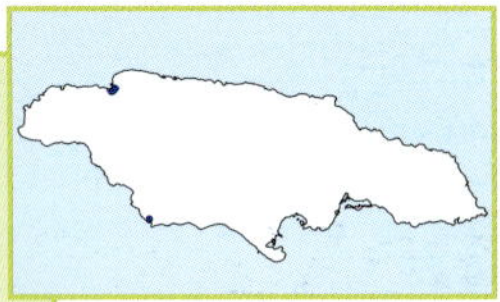

Sanderling
Calidris alba

Taxonomy Monotypic.
Description L 17–22cm (7–9in). Small, plump and rounded in silhouette, with short black bill and legs. Pale grey above, underparts and face white. Dark mark on bend of wing sometimes visible. Lacks hind toe. Wing-bar conspicuous in flight. In winter, the palest shorebird in Jamaica. In breeding plumage (rare in Jamaica) head and breast are reddish brown with black spots.
Similar species Non-breeding Red Knot is darker, stouter and lacks dark carpal mark.
Voice Silent in winter.
Habitat and behaviour Surf line of beaches; often seen in flocks running up and down, following the surf.

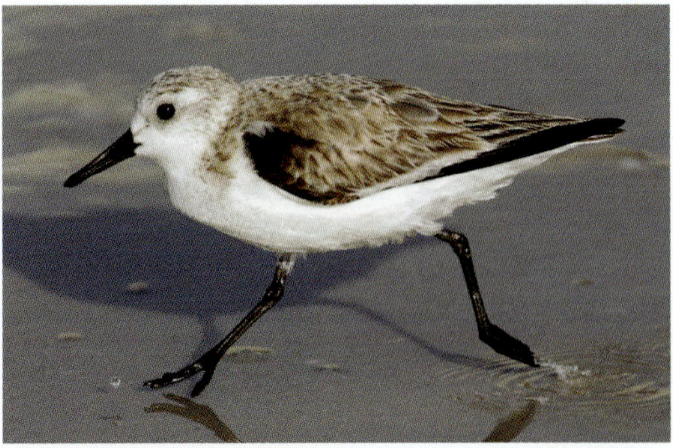

Adult non-breeding. Florida, February.

Range Breeds in the Arctic; almost worldwide in winter.
Status in Jamaica Locally common winter visitor and passage migrant (Oct–March).

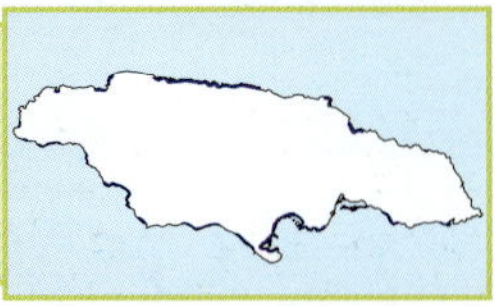

Semipalmated Sandpiper
Calidris pusilla

Taxonomy Monotypic.
Description L 13–15cm (5–6in). A small shorebird with straight black bill and short black legs. Feet have partial webbing between middle and outer toes. Wing-tip is level with or just beyond tail-tip. Adult in winter plumage has a brown-tinged head and back with a faint breast band. Underparts white. Juvenile browner on back.
Similar species Western has slightly downcurved bill and different call. Least Sandpiper is smaller, has greenish-yellow legs and a browner back.
Voice *Chit* or *cheh*.
Habitat and behaviour In flocks on mudflats, beaches, swamps. Flocks turn and weave in highly coordinated waves.

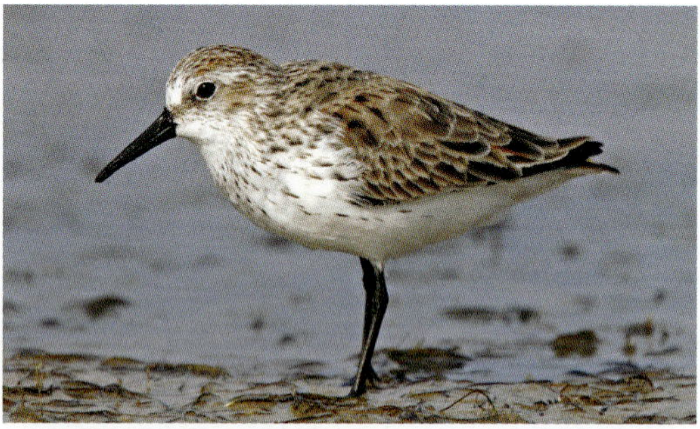

Adult in winter plumage. Note black bill and legs. Florida, March.

Range Breeds in the Arctic, wintering to northern South America via the West Indies.
Status in Jamaica Common winter visitor and passage migrant (August–March).

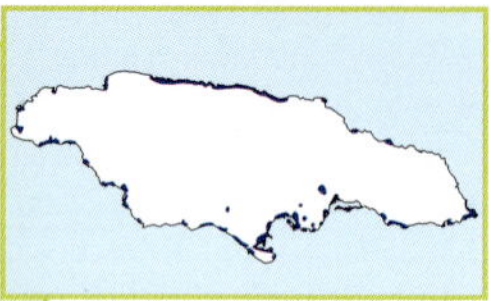

Western Sandpiper
Calidris mauri

Taxonomy Monotypic.
Description L 14–17cm (5–7in). A small shorebird with a long, black bill. Legs short and black. Various plumages seen in Jamaica. Non-breeding brownish-grey above, white below. Breeding has brownish back contrasting with rufous crown, cheeks and scapulars; black chevrons on flanks may be retained in winter. Juvenile has rufous scapulars.
Similar species Semipalmated in winter has shorter bill, blunt at tip. Best separated by call.
Voice A thin *jeeet* or *cheep*.
Habitat and behaviour Flocks occur in swamps and shallows island-wide. Often seen with other sandpipers on mudflats and beaches.

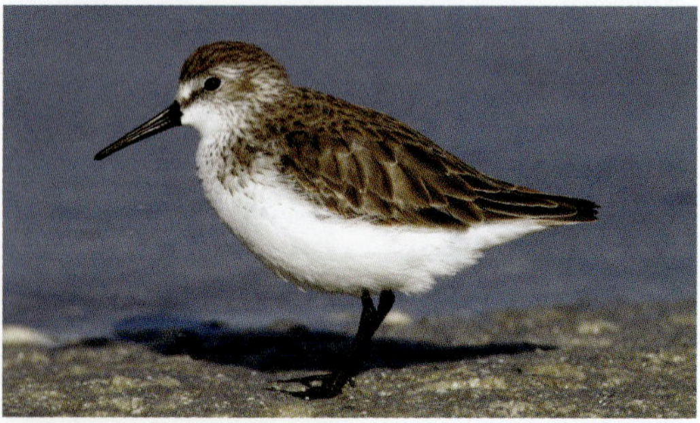

Adult non-breeding. Florida, February.

Range Breeds in Alaska, wintering North America and West Indies, south to Peru.
Status in Jamaica Common winter visitor and passage migrant (August–March).

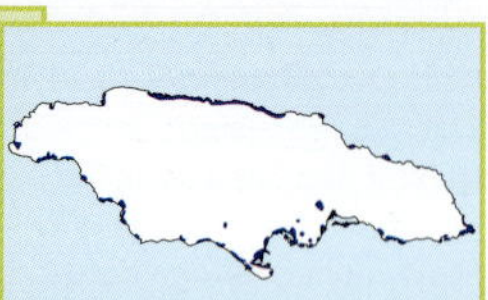

Least Sandpiper
Calidris minutilla

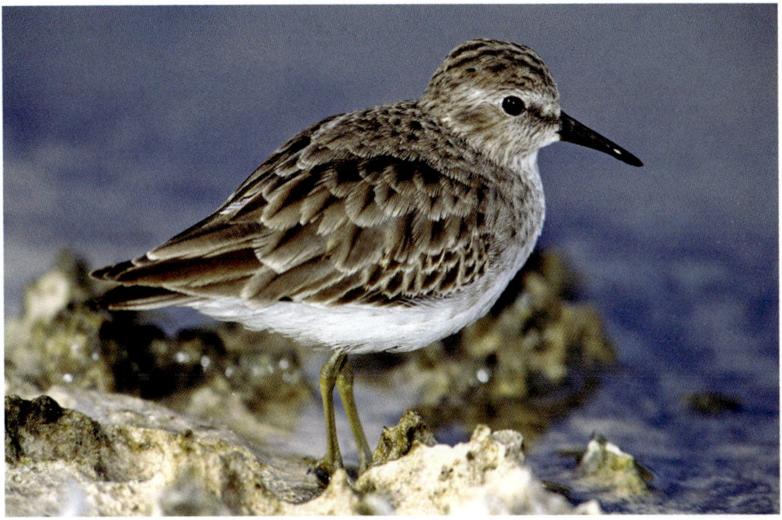

Adult non-breeding. Note greenish-yellow legs. Cayman Islands, March.

Taxonomy Monotypic.
Description L 12–15cm (5–7in). A very small sandpiper with yellow or greenish legs (sometimes stained darker by mud). Toes not webbed. Dark brownish above (more rufous when breeding) with black feather-centres. Underparts are white. Bill is black, thin, and slightly decurved. Tail projects very slightly beyond folded wings. Juvenile has rufous feather-edging above.
Similar species Semipalmated and Western Sandpipers are slightly larger, paler and have black legs. Pectoral Sandpiper is larger with sharp distinction between breast and belly.
Voice A thin *kee-eet*.
Habitat and behaviour Shallow pools on beaches, saline or brackish mudflats and wetlands. Like other sandpipers, flocks fly in unison like a single entity.

Range Northern North America, wintering to South America via the West Indies.
Status in Jamaica Locally common wnter visitor (August–May).

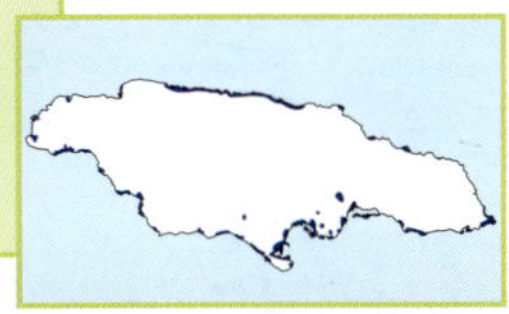

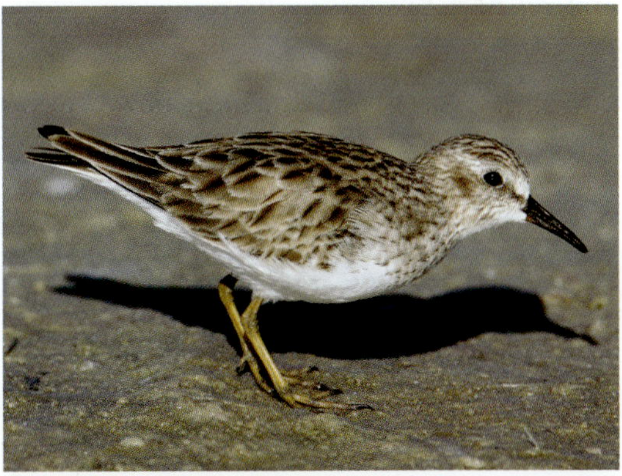

Adult non-breeding. Slightly decurved bill and brownish breast. Florida, Feb.

White-rumped Sandpiper
Calidris fuscicollis

Taxonomy Monotypic.
Description L 15–18cm (6–8in). Medium-sized with dark back and white rump. When standing, wings extend beyond tail. Legs black. Bill black with a slight droop at the end. In flight white uppertail-coverts are diagnostic.
Similar species Semipalmated and Western Sandpipers are smaller and lack white uppertail-coverts.
Voice Usually silent. Rarely gives an insect-like *jeet*.
Habitat and behaviour Reedy edges of freshwater ponds and wetlands island-wide. Not seen in flocks in winter but often among other sandpipers and easily overlooked.

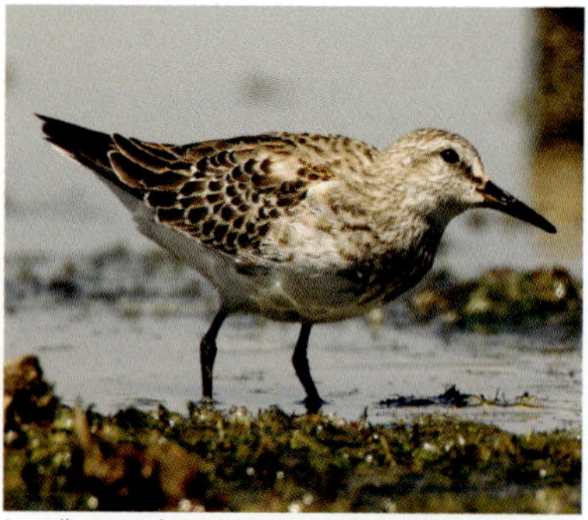

Juvenile. Note that wings extend beyond the tail. Nov.

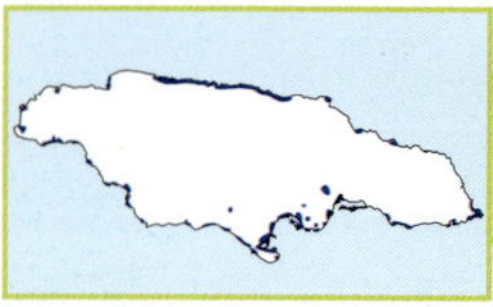

Range Breeds in northern North America and winters throughout South America via Central America, West Indies.
Status in Jamaica Uncommon winter passage migrant.

Pectoral Sandpiper
Calidris melanotos

Taxonomy Monotypic.
Description L 19–24cm (7–10in). Medium-sized with dark brown upperparts. Legs dull yellow. Neck and breast dark brownish and finely streaked black, sharply delineated from plain white belly. Rump and uppertail-coverts dark with white edges. Bill has slight droop; brown-black, darker at tip.
Similar species Yellowlegs lack sharply defined underpart contrast. Least Sandpiper is smaller.
Voice A throaty *pruk-pruk*.
Habitat and behaviour Shallow fresh or brackish coastal wetlands. May adopt an upright stance when alarmed.

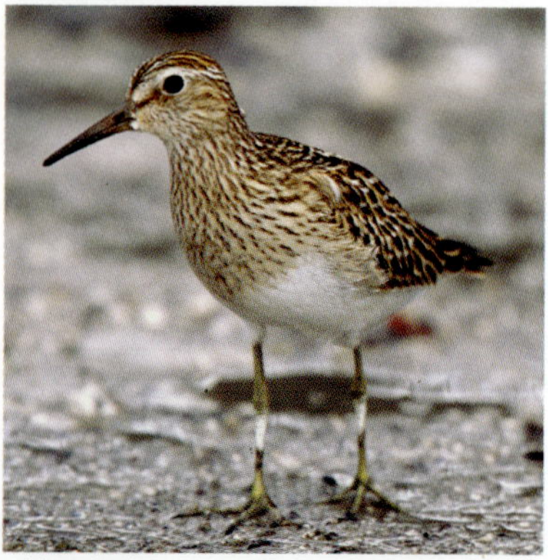

Adult. Note sharp delineation between breast and belly. Bahamas, March.

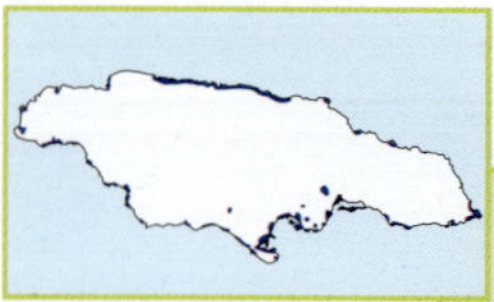

Range Breeds in the Arctic; winters southern South America.
Status in Jamaica Uncommon autumn passage migrant (mostly August–November).

Dunlin
Calidris alpina

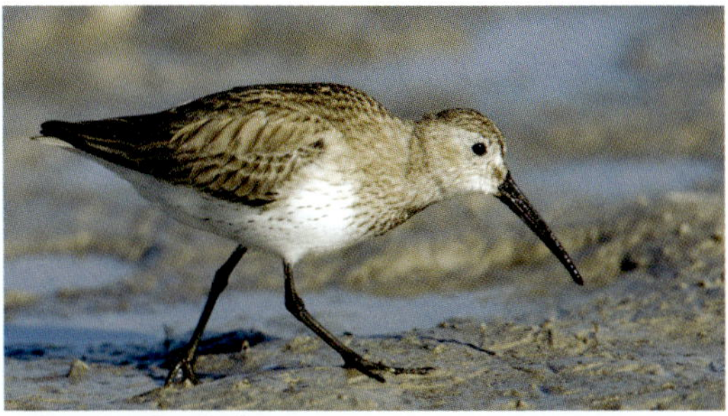

Adult non-breeding. Florida, February.

Taxonomy Polytypic (9). Race in Jamaica is *C. a. hudsonia.*
Description L 16–22cm (6–9in). Medium-sized shorebird with decurved black bill. Legs black. Non-breeding brown-grey on back. Underparts mostly white but brownish wash to head and breast. Breeding adult streaky with black belly patch; upperparts rufous with black feather centres. Juvenile brownish, with white or rufous feather edges above.
Similar species Non-breeding Red Knot larger with shorter, straight bill and greenish legs. Western and Semipalmated Sandpipers smaller, shorter-billed; White-rumped has white rump.
Voice Silent in Jamaica.
Habitat Coastal saline ponds and mudflats.

Range *C. a. hudsonia* breeds in northern North America. Winters south to Mexico. Recorded from Bahamas, Greater Antilles, northern Lesser Antilles and Barbados.
Status in Jamaica Rare winter visitor or passage migrant.

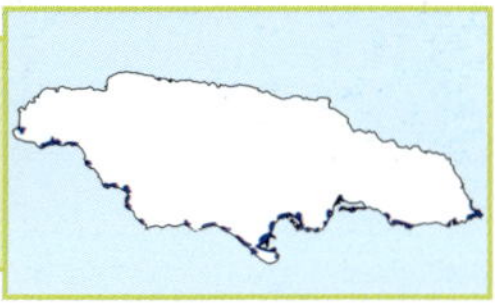

Stilt Sandpiper
Calidris himantopus

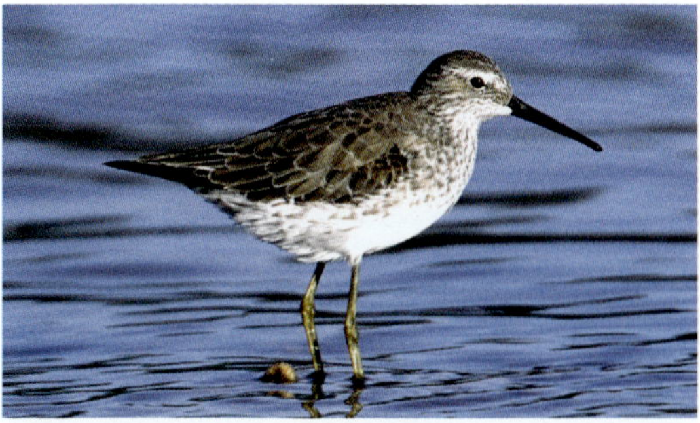

Adult non-breeding. Dominican Republic, March.

Taxonomy Monotypic.
Description L 19–23cm (7–9in). Medium-sized, erect and slender. Greyish above, white below with fine grey streaks. Distinct white supercilium. Bill long, dark, slight droop at tip. Legs long, greenish. Juvenile browner above with pale feather edging.
Similar species Yellowlegs have yellow legs and long, straight bills. Dowitchers longer-billed.
Voice A quiet 'tew' in flight.
Habitat Freshwater wetlands.

Range Breeds Alaska and Canada. Winters south to central South America via West Indies and Central America.
Status in Jamaica Uncommon winter visitor (November–May)

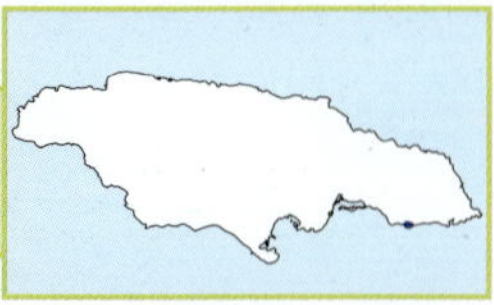

Ring-billed Gull
Larus delawarensis

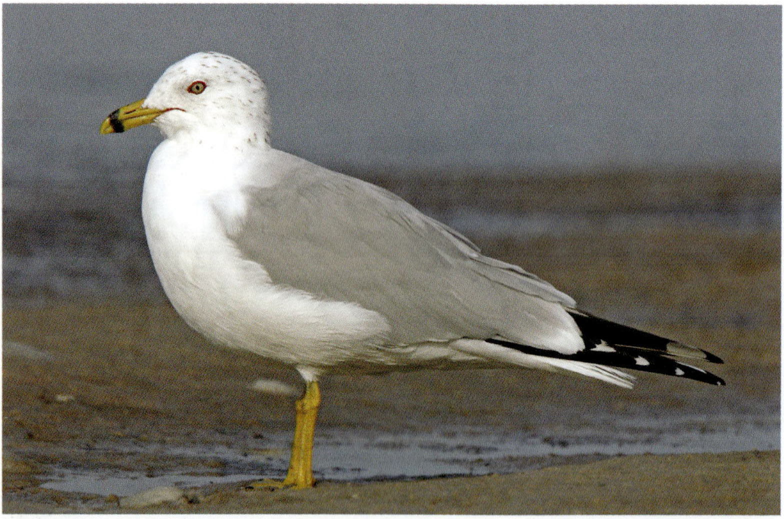

Adult breeding. Note red orbital skin. Florida, March.

Taxonomy Monotypic.
Description L 46–51cm (18–20in). Adult white with pale grey mantle and upperwings and black wing-tips. Bill yellow with black band near tip. First-winter has contrasting dark brown outer primaries, dark secondaries and brownish forewing-coverts.
Similar species American Herring Gull is larger; adult has yellow bill with red spot on lower mandible, first-winter is uniformly brownish.
Voice Similar to American Herring Gull but higher-pitched. Rarely heard in Jamaica.
Habitat and behaviour Often seen on mudflats with other gulls and terns.

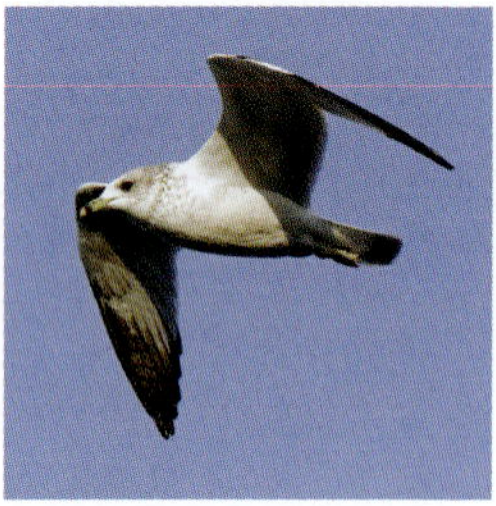

Adult non-breeding. Florida, Dec.

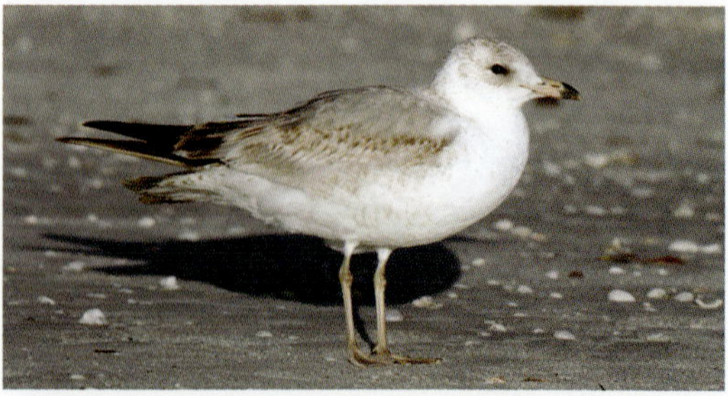

First-winter. Florida, February.

Range Breeds northern North America. Migrants winter in the Bahamas, Greater Antilles and Central America.
Status in Jamaica Rare winter visitor or passage migrant.

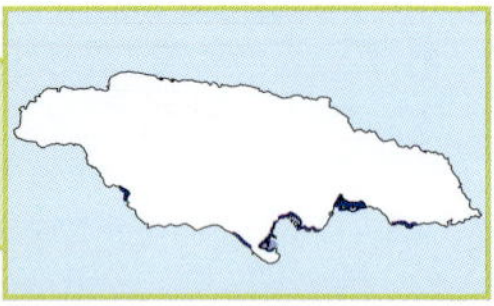

American Herring Gull
Larus smithsonianus

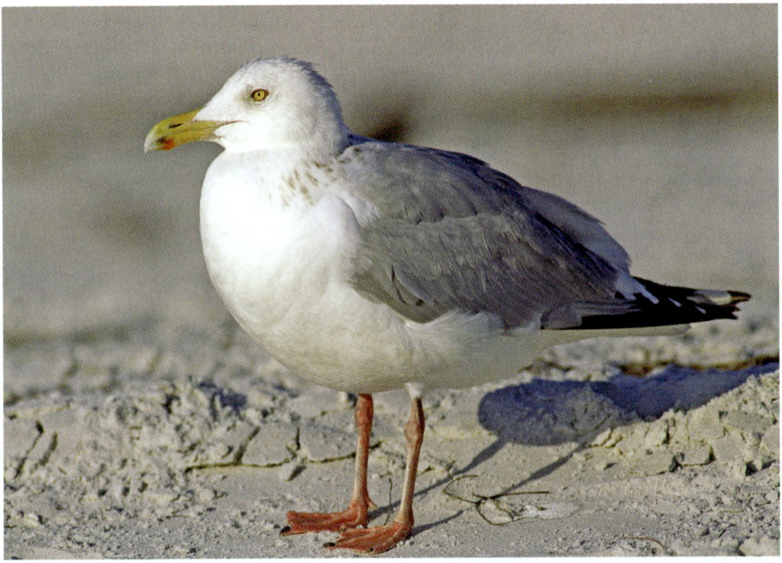

Adult non-breeding. Florida, March.

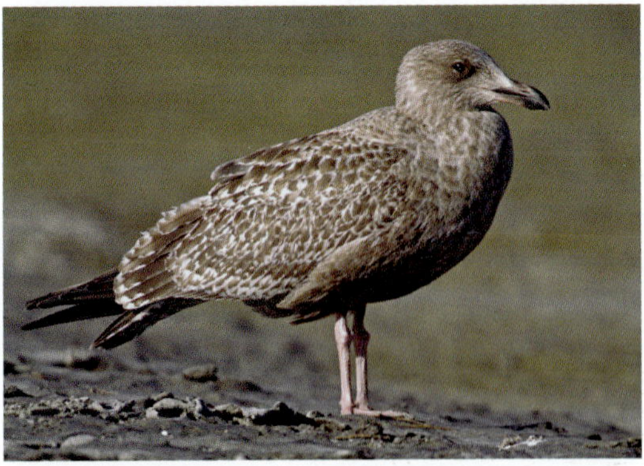

First-winter. Florida, March.

Taxonomy Monotypic.
Description L 56–66cm (22–26in). Large. Non-breeding adult white with grey mantle and upperwing, wing-tips black with white spots. Head and neck often streaked brown. Bill is yellow with a red spot on lower mandible close to the tip. Legs and feet are pinkish. Juvenile and immatures mainly brown and speckled, bill black, with pink base.
Similar species Ring-billed Gull is smaller; adult has dark ring mark on bill, immatures show more contrast.
Voice A bugle-like *kowk*, often repeated. Rarely heard in Jamaica.
Habitat and behaviour Usually seen on mudflats or flying over Kingston Harbour.

Range Breeds in northern North America and winters south to Mexico.
Status Rare winter visitor or passage migrant.

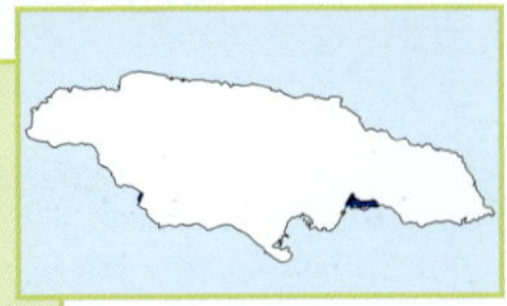

Laughing Gull
Larus atricilla

Taxonomy Polytypic (2). Race in Jamaica is *L. a. atricilla*.
Description L 39–46cm (15–18in). Breeding adult has a blackish hood, broken white eye ring. White overall with mantle and upperwings dark grey and black wing-tips. Legs dark. Non-breeding adult has a white head with variable amounts of grey smudges behind the eyes. Bill black. Nestling brown with darker streaks on head and spots on back. Juvenile (August–November) has greyish brown upperparts including head and neck. Tail white with broad black terminal band. Underparts white. It takes three years to reach maturity and there are several intermediate plumages. In the first winter grey develops on back. In the first summer the dark head starts to develop. By the second winter bird resembles adult but with traces of the tail-band and blacker wing-tips.
Similar species First-winter Ring-billed Gull is much paler than first-winter Laughing Gull and has pink legs and bill-base. First-winter Herring Gull is larger, uniformly brownish and has pink legs.
Voice A harsh *cheeer-ah* or (when breeding) a rising *he-he-he-heee* – the source of its name.
Habitat and behaviour Commonly seen in winter on fishing beaches, harbours and offshore cays. Feeds omnivorously. Voracious predator of tern eggs. Nests colonially (April–July) on Morant and Pedro Cays. Nest is a cup made of grasses, built on the ground or on low vegetation. Lays 2–4 dark brown, dark-speckled eggs.

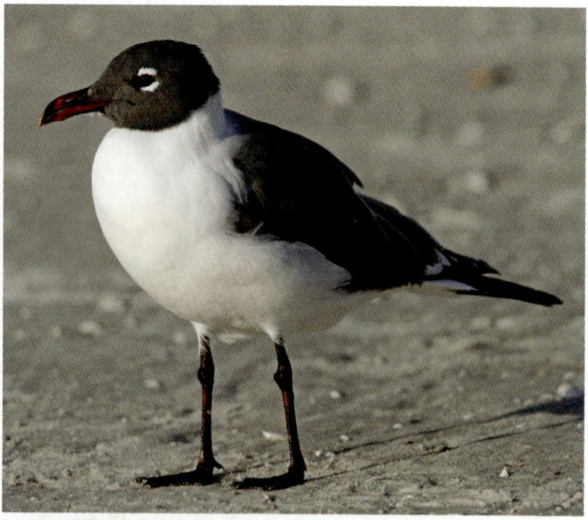

Adult breeding. Bahamas, April.

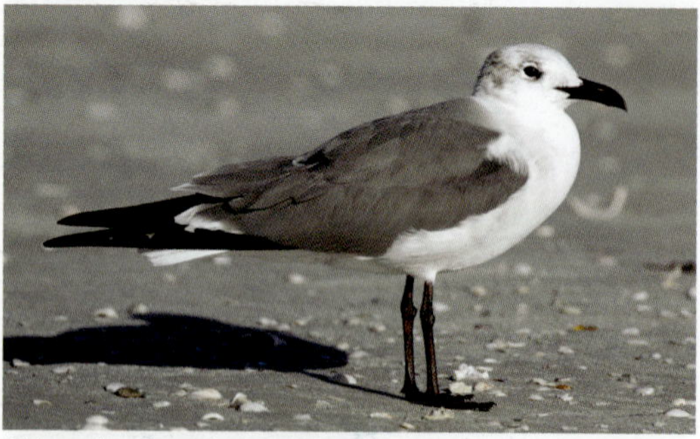

Adult non-breeding. Bahamas, February

Range Coasts of North America, West Indies and northern South America.
Status in Jamaica The only gull that is likely to be seen in summer. *L. a. atricilla* breeds in the West Indies and winters to northern Brazil. Fairly common resident in Jamaica. Another race breeds in the US; migrants may increase local populations in winter. Jamaica population increasing.

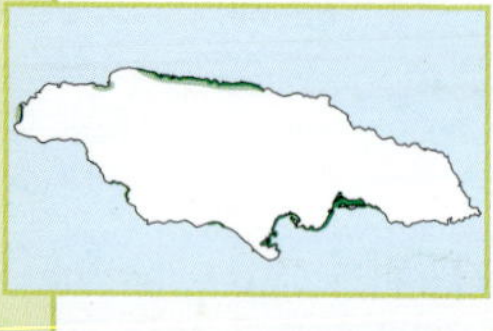

Brown Noddy
Anous stolidus

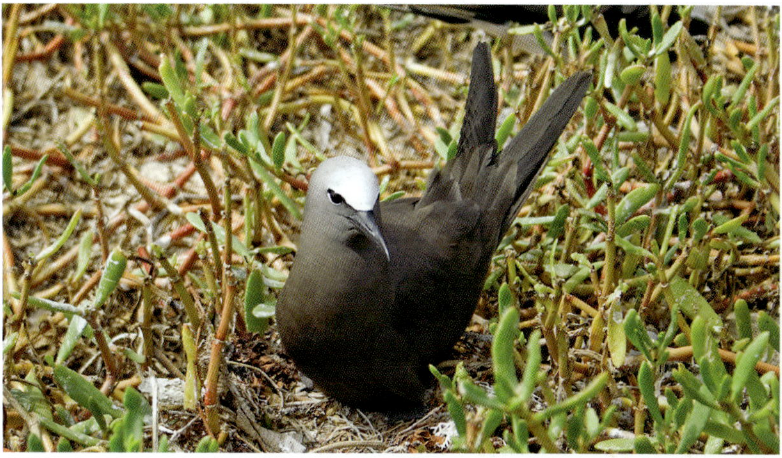
Adult breeding. Morant Cay, Jamaica, May.

Local name Sea Pigeon.
Taxonomy Polytypic (5). Race in Jamaica is *A. s. stolidus.*
Description L 38cm (15in). Medium-sized dark brown tern with a white crown and white crescent below eye. Tail is rounded and relatively short compared to other terns. Downy chick occurs in two colour phases – white or sooty grey. Juvenile is all brown, with no white on head.
Similar species Black Noddy (very rare vagrant) is darker and more slender.
Voice A growling *karrrrk* or dry *kak*. Also growls.
Habitat and behaviour Feeds close to shore by plunging on small fish and squid close to shore (often nearer to shore than Sooty Terns); may occur in mixed feeding flocks. Nests on small cays. Mating display includes head-nodding, from which it gets its name. Nests colonially (May–August). Nest is a pile of sticks, stones, seaweed and pieces of plastic, on tree branches, on the top of shrubs or on the ground, varying from a few items to large heaps. Lays one buffy, brown-speckled egg (rarely two). Incubation takes about 28 days.

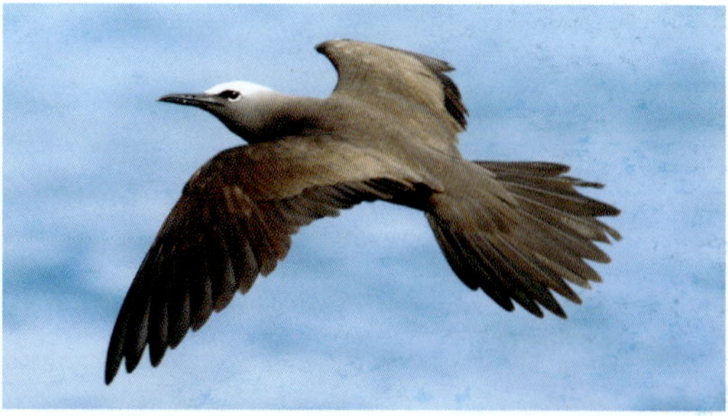
Adult. Florida, April.

Range *A. s. stolidus* nests on islands in the West Indies and tropical South Atlantic; other races pantropical.
Status in Jamaica Common summer resident on breeding colonies at Portland Bight Cays, Morant and Pedro Cays or at sea.

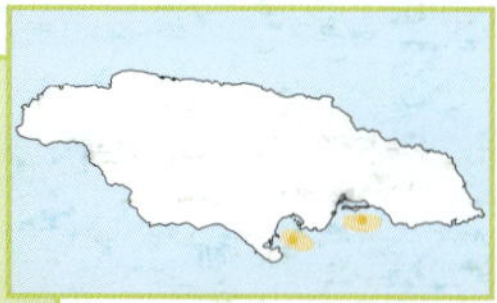

Sooty Tern
Onychoprion fuscatus

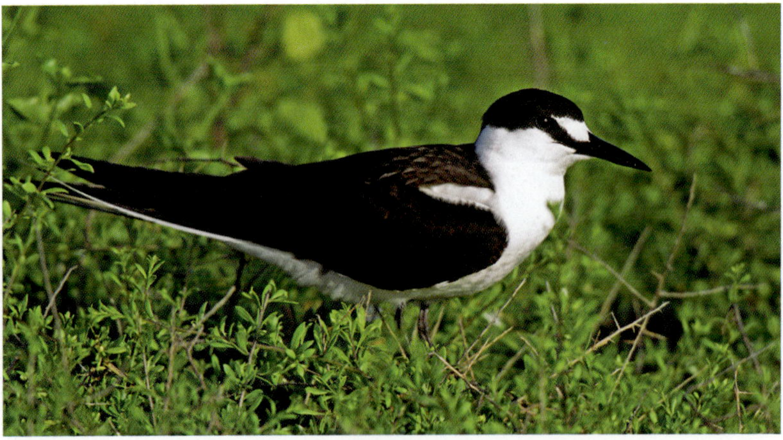

Adult breeding. Note white forehead does not extend beyond the eye Dominican Republic, May.

Taxonomy Polytypic (8). Race in Jamaica is *O. f. fuscatus.*
Description L 36–41cm (14–16in). Medium-sized; black above, white below. White on forehead does not extend beyond eye. Long black wings. No contrast between black head and back. Tail has narrow white edge. Primaries all-dark below. Bill and legs black. Juvenile dark brown, feathers broadly tipped with tan. Immature has dark underparts, head and back. Wings and back finely spotted white.
Similar species Bridled Tern has very different call; white on forehead extends beyond eye, nape white, tail shorter, wings shorter and narrower with more white below; spread tail has more white on margins.
Voice A nasal *wideawake* or *wacky-wack*. Breeding colonies are very noisy, day and night.
Habitat and behaviour Pelagic outside breeding season; breeds on offshore cays. Feeds in flocks. Does not plunge dive but swoops down to scoop prey from the surface. Feathers are not waterproof so has to stay on the wing continuously while away from land. Nest (April–August) is a shallow scrape in the sand beside or under vegetation, on Pedro and Morant Cays. Lays one pointed egg (very rarely two), buffy with brown and maroon speckles.

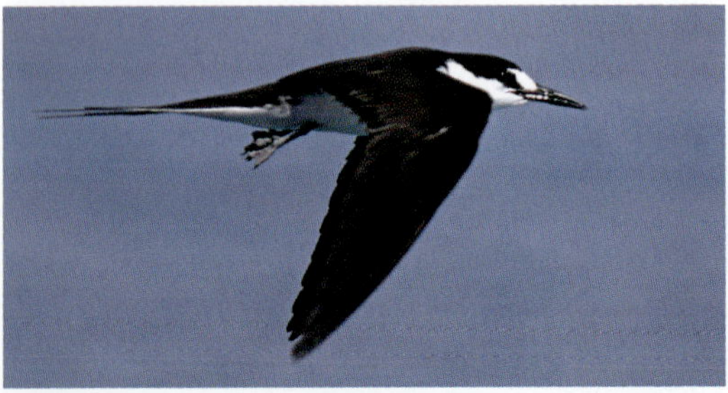

Adult breeding. Dominican Republic, May.

Range *O. f. fuscatus* breeds from the Carolinas to Mexico, including cays throughout Caribbean. Immatures follow currents east to islands in the Gulf of Guinea. Winters south to Brazil.
Status in Jamaica Locally very common summer resident on breeding grounds on Morant and Pedro Cays or at sea. Populations have declined drastically in recent years.

Bridled Tern
Onychoprion anaethetus

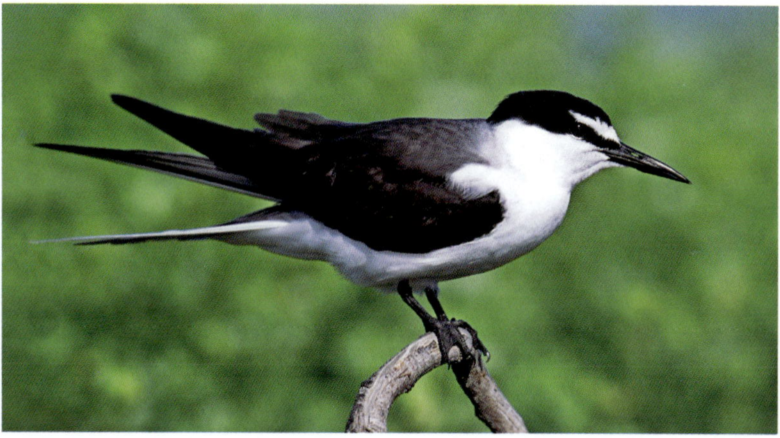

Adult breeding. Note narrow white forehead extends beyond the eye. Dominican Republic, May.

Taxonomy Polytypic (6). Race in Jamaica is *O. a. recognitus.*
Description L 30–32cm (12–14in). Adult in breeding plumage is a medium-sized tern with dull, dark grey-brown upperparts, black crown contrasts with grey back, white hind-neck white. Narrow white forehead patch extends back beyond eye and appears V-shaped from the front. Outer tail feathers appear as long white streamers. Bill and legs black. Downy chick is dark grey. Juvenile dark with white spots. Immature has white head, dark grey back with white spots.
Similar species In Sooty Tern the white forehead patch does not extend behind eye, black crown and nape do not contrast with back. At sea, Bridled Terns may perch on floating debris but Sooty Terns never do.
Voice A high-pitched *erk* – like the squeaky bark of a small puppy.

Habitat and behaviour Never seen near land except during breeding season. Forages offshore, often in mixed flocks with other terns, hovering then plunging for small fish from the surface. Nests in small groups in rocky crevices and under vegetation on inshore and offshore cays (Port Royal Cays, Portland Bight Cays, Morant and Pedro Cays).

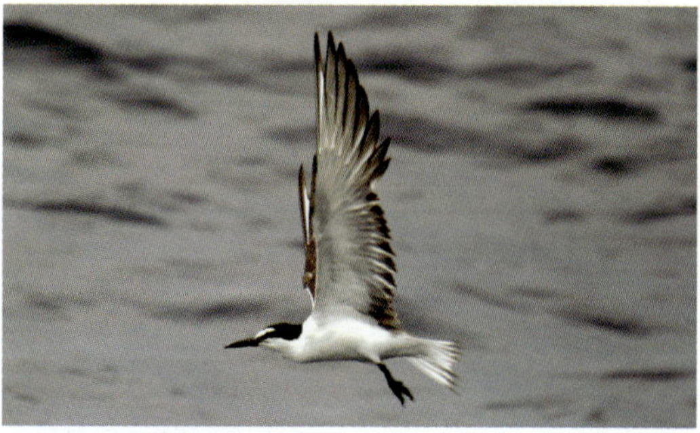

Adult breeding. August.

Range *O. a. recognitus* West Indies, Belize and islands off Venezuela. Other races pantropical.
Status in Jamaica Uncommon and local summer resident.

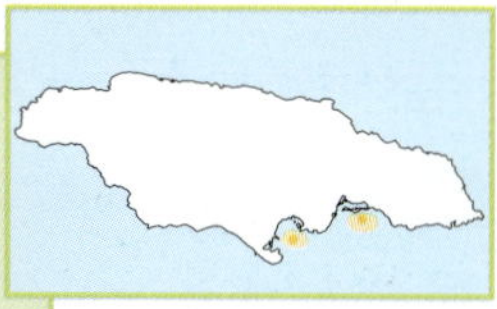

Least Tern
Sternula antillarum

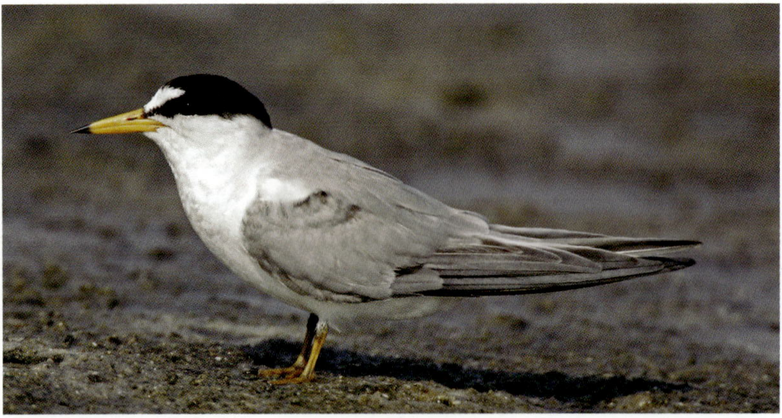

Adult breeding. Note yellow bill with dark tip, and yellow legs. Florida, May.

Taxonomy Polytypic (3). Race in Jamaica is *S. a. antillarum*.
Description L 21–23cm (8–9in). Tiny tern (the smallest in the region) with grey back, white underparts and a deeply forked tail. Wings grey with black outer primaries. Breeding adult has black crown that contrasts with V-shaped white mark above bill and grey upperparts. Bill orange with a small amount of black on tip. Legs orange. Non-breeding adult has black bill and white on forehead extends on to crown. Nestling mottled buffy above, white below and has pinkish feet and bill. Juvenile has mottled crown, faint barring on upperparts, dark carpal patch and dark eye-patch; bill dark, legs orange.
Similar species Juvenile Black Tern slightly larger with a squarer tail, dark grey upperparts and a dark spot on upperside of breast.
Voice A high-pitched *cheereep* and a short *kip, kip*.

Habitat and behaviour
Sandy islands in salt ponds or beaches. Hovers and plunges to feed on fish in shallow mangrove ponds or at sea. Nests colonially or singly (April–August) in a scrape in the open sand, often close to the waterline. The only tern that nests on mainland Jamaica. Lays two spotted eggs.

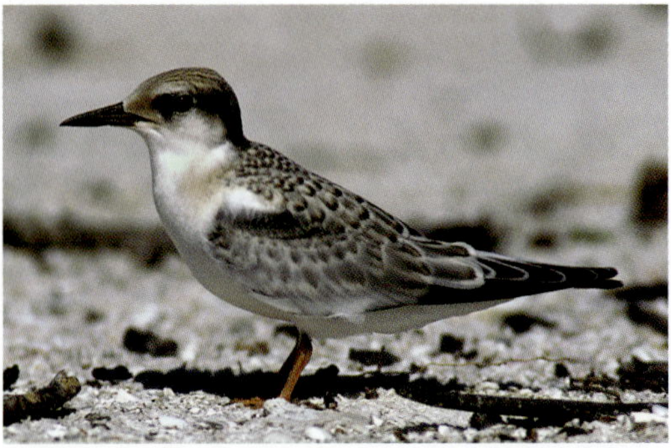

Immature. Florida, June.

Range *S. a. antillarum* breeds Central America and Caribbean. Winters Central America southwards.
Status in Jamaica Common summer resident.

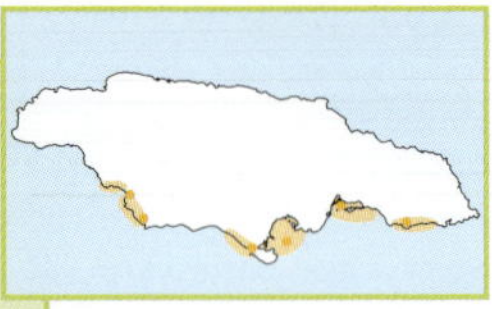

Gull-billed Tern
Gelochelidon nilotica

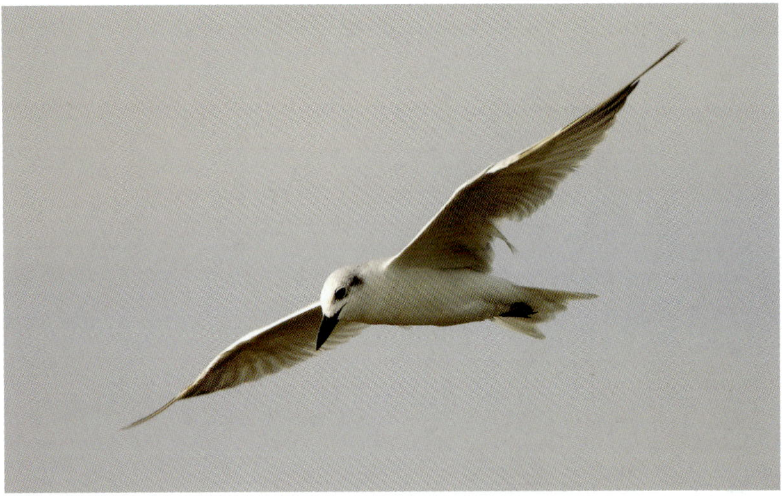

Adult non-breeding, July.

Taxonomy Polytypic (6). Race in Jamaica is *G. n. arenea*.
Description L 33–38cm (13–15in). A chunky white-bodied tern with thick, short, black bill, longish black legs and a short forked tail. Wings grey. Usually seen as a non-breeding adult with white head and a dark smudge behind eye. Breeding adult has black cap.
Similar species No other tern has a similar bill.
Voice A nasal *ker-weck*. Rarely heard in Jamaica.
Habitat and behaviour Usually seen foraging close to the shore (including offshore cays) or resting quietly in wetlands on saline mudflats.

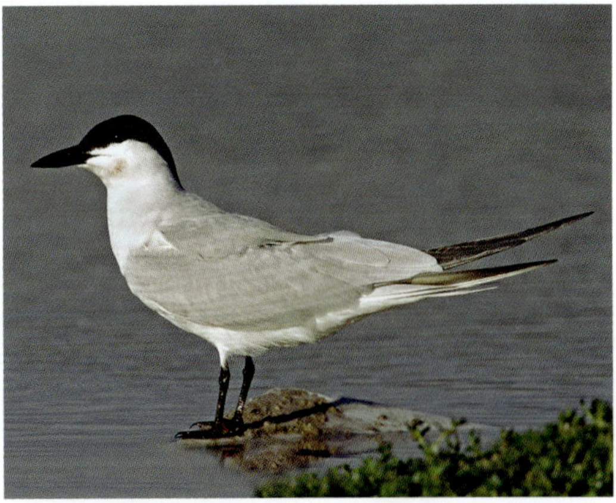

Adult breeding. This plumage is seen only rarely in Jamaica; usually in non-breeding plumage. Bahamas, May.

Range *G. n. arenea* breeds in North America, Bahamas and Virgin Islands. Winters through Central America and West Indies to South America. Other races almost worldwide.
Status in Jamaica Uncommon winter passage migrant.

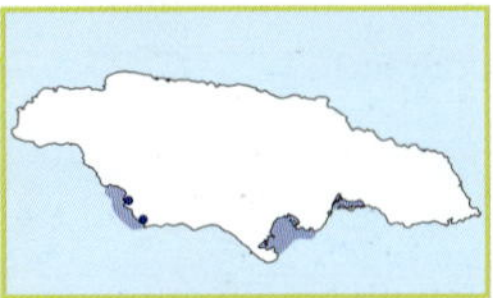

Caspian Tern
Hydroprogne caspia

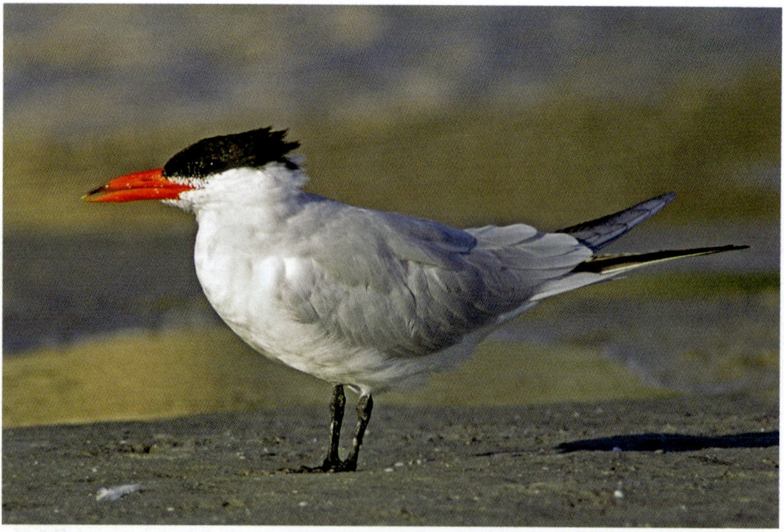

Adult. Note heavy, bright red bill. Florida, April.

Taxonomy Monotypic.
Description L 51cm. (20in). A large, chunky, very white tern, with a long, thick, heavy, dark-tipped red bill, and black legs. Adult non-breeding has streaky forehead. Immature has a barred tail and yellow legs.
Similar species Royal Tern is smaller and has a yellower bill lacking dark tip.
Voice Usually silent in Jamaica.
Habitat and behaviour Saline or brackish wetlands. Usually seen standing alone or with Royal Terns, perching on fence posts or in pairs in swamps.

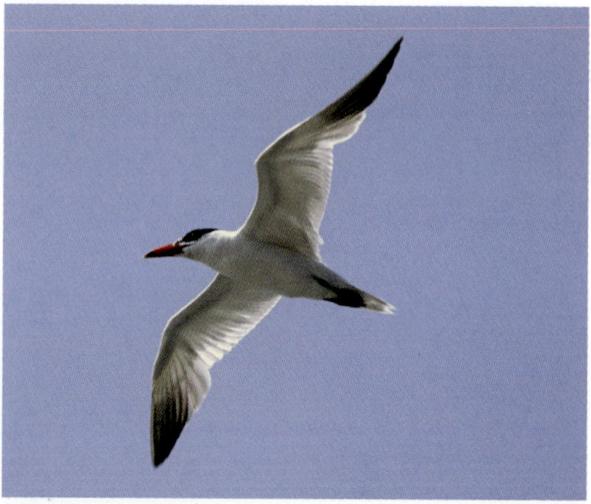

Adult breeding. May.

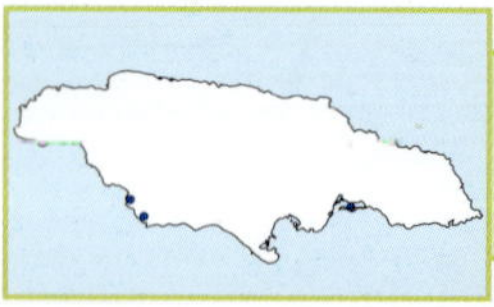

Range Worldwide. Nearctic birds breed North America, winter southern North America and West Indies.
Status in Jamaica Occasional winter passage migrant.

Black Tern
Chlidonias niger

Taxonomy Polytypic (2). Race in Jamaica is *C. n. surinamensis.*
Description L 23cm (9in). Adult is usually seen in non-breeding plumage. It is a small white tern, with black on the back of the head and shoulder, contrasting with grey wings, back and tail. Bill dark, legs reddish. Tail notched. Juvenile resembles non-breeding adult but mantle, back and upperwing-coverts darker.
Similar species Least Tern similar in size but much paler above and tail is more deeply forked.
Voice *Kik, keek*.
Habitat and behaviour Usually seen flying over fresh or brackish water. Distinctive, fluttering flight style as food (aquatic invertebrates) is picked from the surface of the water.

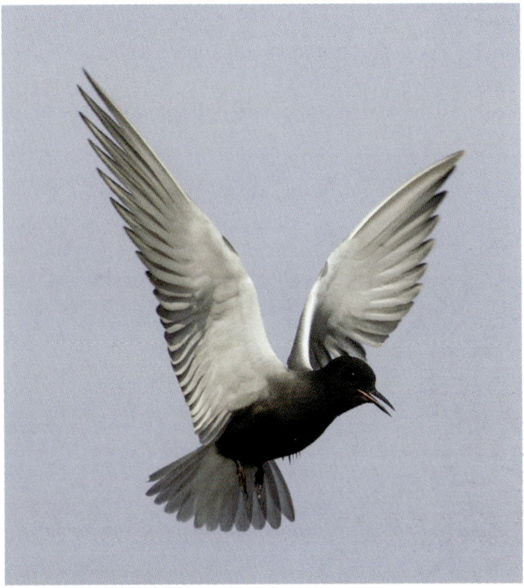

Adult in breeding plumage. May.

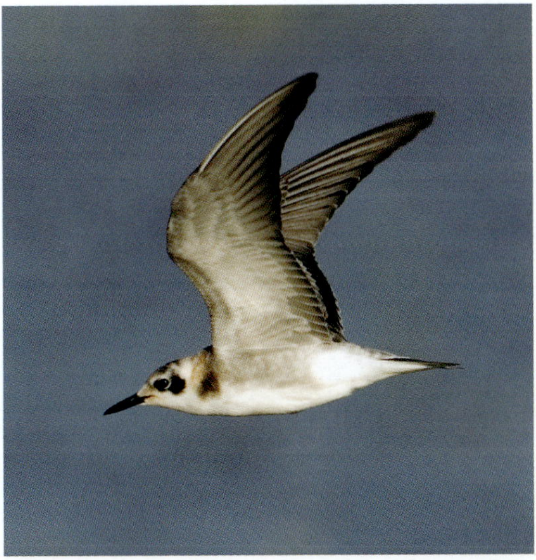

Juvenile. August.

Range *C. n. surimamesis* breeds in North America, wintering from Panama south to Chile via the West Indies. Other race occurs in the Old World.
Status in Jamaica Uncommon summer visitor and passage migrant (June–November).

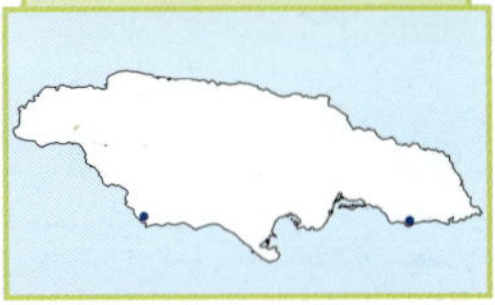

Roseate Tern
Sterna dougallii

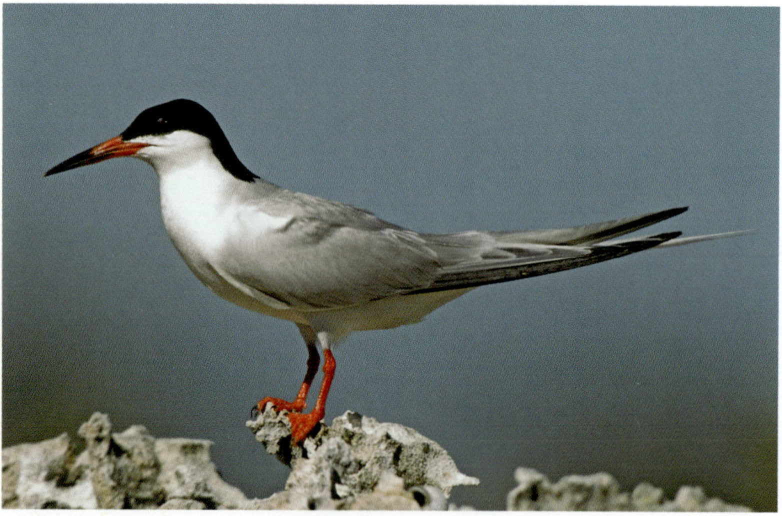

Adult breeding. Note red and black bill, red legs and long streamers. Bahamas, May.

Taxonomy Polytypic (5). Race in Jamaica is *S. d. dougallii*.
Description L 33–41cm (13–16in). Breeding adult is a slender white tern with black on crown, extending to the nape. The bill is long, red at the base, with a black tip. Legs red. Tail is white and deeply forked. When standing, the outer tail feathers extend markedly beyond the folded wing. Non-breeding adult has a white forehead and black bill but is rarely seen in Jamaica. Nestling has grey and speckled upperparts, bluish or pinkish-grey bill with a darker tip, and grey legs. Juvenile has black cap and blackish feather edging above, giving a scaly appearance.
Similar species Roseate Terns in the West Indies have much more red on the bill than those in North America. This contributes to misidentification as Common Tern. Common Tern is darker on the back and the wingtips have more black underneath than in Roseate Tern. Common Tern flies with shallower wingbeats than Roseate. Common Tern is longer-winged and shorter-tailed, so when perched the tail feathers do not protrude beyond wing tips. Sandwich Tern is larger and has a black bill with a yellow tip.
Voice Often first detected by its call, a sharp but rough *chi-vick*, mainly given in flight.
Habitat and behaviour Pelagic except during the nesting season. Feeds on fish by plunge-diving into deep 'blue water.' Nest (June–July) is a scrape in the open sand or on rocks, sometimes next to vegetation. Lays one or two olive-spotted eggs. Frequently changes breeding colony sites between years.

Range *S. d. dougallii* breeds coastal United States to West Indies, the Azores, Europe and Africa. Other races worldwide. Global conservation status: Near Threatened.
Status in Jamaica Uncommon summer breeding species. One of the scarcest breeding terns in the West Indies, with a declining population. Caribbean population listed as Threatened by the United States Fish and Wildlife Service.

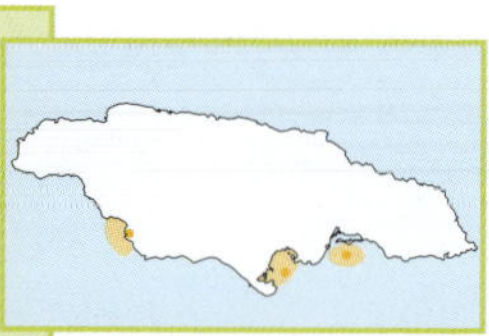

Common Tern

Sterna hirundo

Taxonomy Polytypic (4). Race in Jamaica is *S. h. hirundo*.
Description L 33-40cm (13–16in). Breeding adult (on Jamaica during spring migration) is a medium-sized white tern, with black crown and nape, orange-red bill and legs. When standing, the tail does not project behind the folded wings. Non-breeding adult and juvenile (present on Jamaica during autumn migration) have white forehead, black legs and bill. Juvenile has dark outer primaries and a dark carpal bar.
Similar species Roseate Tern is paler, with shorter wings, longer tail streamers and less contrast between back and rump.

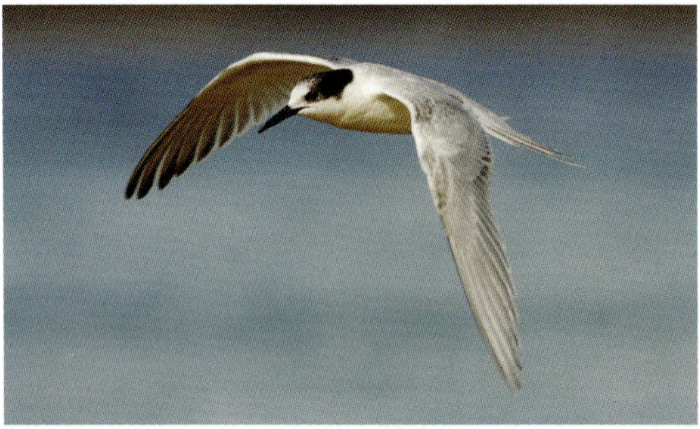

Adult non-breeding. Dominican Republic, October.

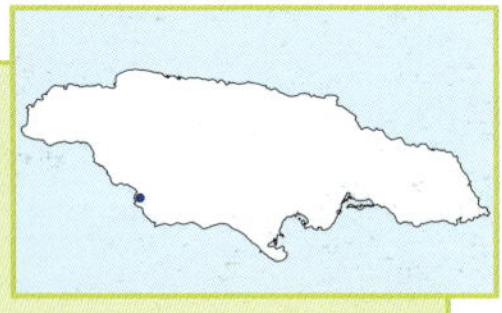

Range *S. h. hirundo* breeds in North America, rarely Cuba, also West Africa, Western Palearctic. Winters south. Other races in Asia.
Status in Jamaica Very rare passage migrant or vagrant.

Forster's Tern

Sterna forsteri

Taxonomy Monotypic.
Description Seen in non-breeding plumage on Jamaica, where it is a very scarce winter visitor. A white tern, pale grey above, with a distinctive dark mask around the eye and on the ear coverts. Nape white.
Similar species Non-breeding Roseate and Common Terns have black napes.

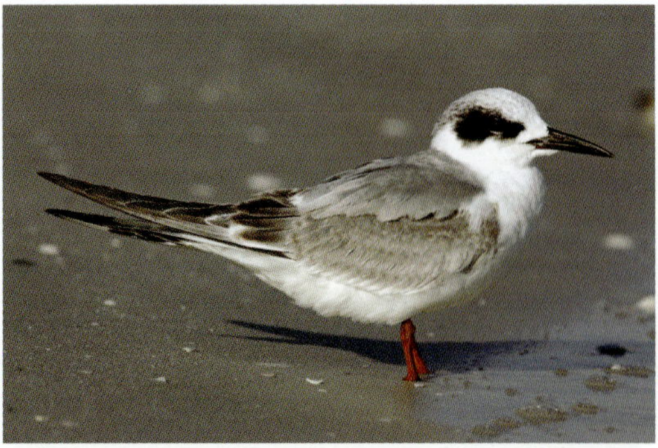

Adult non-breeding. Note black mask, black bill and red legs. Florida, Feb.

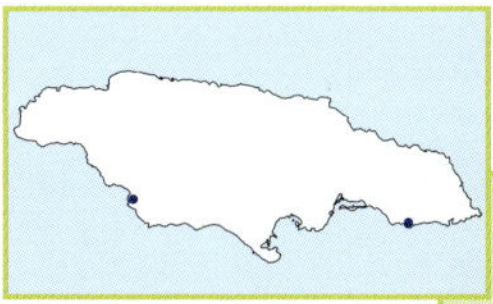

Range Breeds inland in North America. Winters south to Costa Rica and Panama.
Status Very rare winter passage migrant.

Royal Tern
Thalasseus maximus

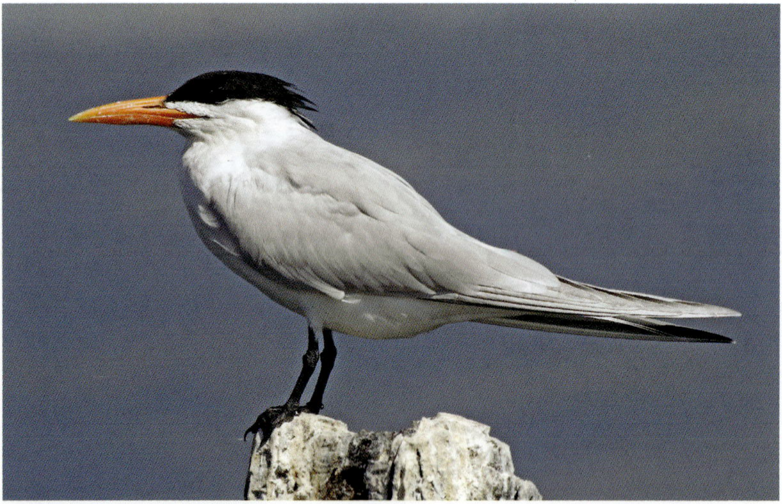

Adult breeding. Port Royal, Jamaica, March.

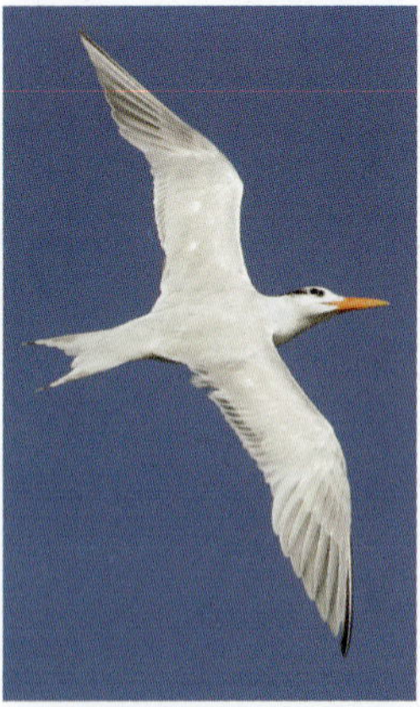

Adult non-breeding. Florida, February.

Taxonomy Polytypic (2). Race on Jamaica is *T. m. maximus*.
Description L 46cm (18in). A large, stocky, white tern with pale grey upperparts, dark wing tips and black crest, fairly thick yellow-orange bill and black legs. Adult breeding has a black cap and bright orange yellow bill. Adult non-breeding has a white forehead. Nestling is pale white or buffy (sometimes spotted), with yellowish or grey bill and legs. Juvenile has dark grey carpal bar, secondaries and outerwing, bill pale yellow.
Similar species Caspian Tern is much larger and has a cherry-red, dark-tipped bill.
Voice A nasal *zeheit* or *chireep*.
Habitat and behaviour Coastal beaches and saline or brackish ponds. Often seen perched on posts in brackish water or flying singly along the coast. Feeds close to shore by plunge-diving. Nests (April–July) on Morant and Pedro Cays, in small, very dense colonies. Sometimes nests with Sandwich Terns. Nest is a shallow scrape in open sand, usually without any lining. Lays one creamy-white spotted egg.

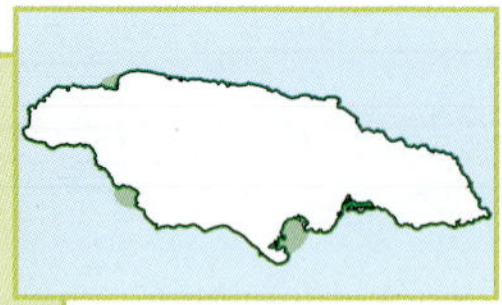

Range *T. m. maximus* occurs West Indies, southern North America, wintering south to Argentina. Ringing recoveries indicate that most winter visitors to Jamaica originate in the Carolinas. Other race is in West Africa.
Status in Jamaica Common resident. The most likely seabird to be seen from the mainland year-round.

Sandwich Tern
Thalasseus sandvicensis

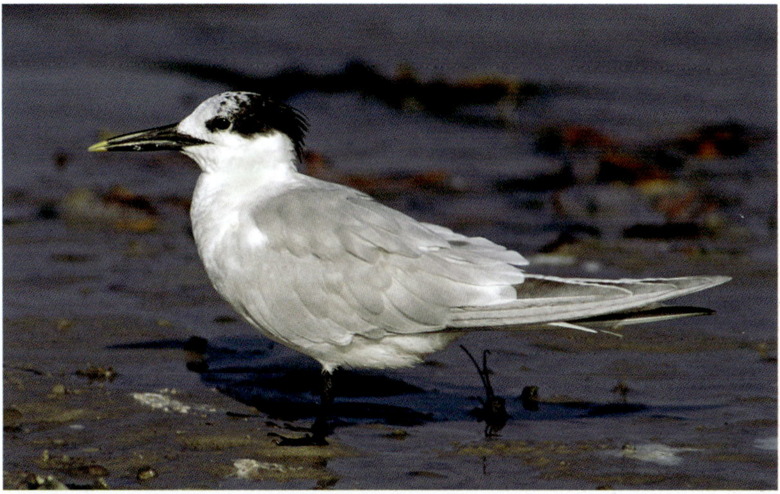

Adult non-breeding. Note yellow tip to bill. Florida, February.

Taxonomy Polytypic (3). Race in Jamaica is *T. s. acuflavidus*.
Description L 38cm (15in). Medium-sized white tern with a yellow tip to black bill. Legs black. Upperparts pale grey. Tail forked. When standing the wings do not extend beyond the tail. In flight wing-tips look paler from below than from above. Adult breeding (May–July) has black cap, which is gradually lost as the season progresses. Adult non-breeding (plumage usually seen in Jamaica) has white forehead; only back of head is black. First-winter has darker flight feathers and black tail-tip. Nestling buffy with grey legs and feet.
Similar species None.
Voice Rough two-note *ker-rick*.
Habitat and behaviour Generally seen by salt ponds. Feeds in mixed flocks, hovering then plunge-diving for small fish. Nests colonially (June-July). Nest is shallow scrape on shelly sand close to lagoon. May be lined with plant material.

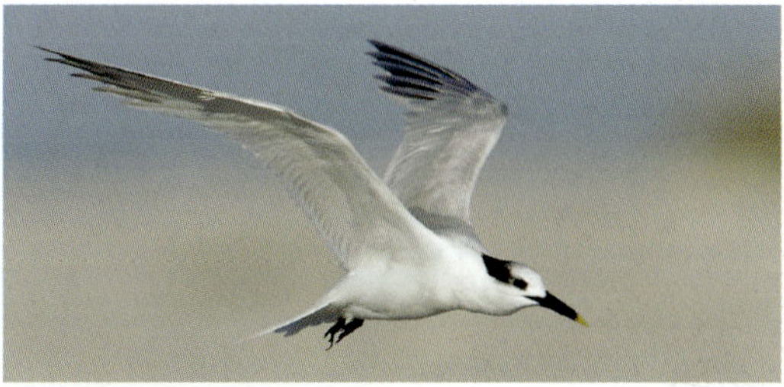

Adult non-breeding in flight. Florida, November.

Range *T. s. acuflavidus* North America to northern South America. Other races Western Palearctic, South America. North American birds winter from West Indies south. Jamaican birds include some ringed as nestlings in North Carolina and Virginia.
Status in Jamaica Common winter visitor (October–March) on mainland coasts. Occasional summer breeding species on Pedro Cays.

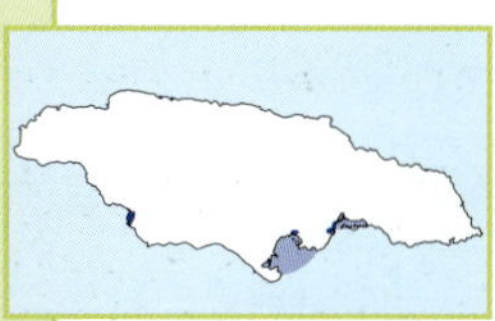

Black Skimmer
Rynchops niger

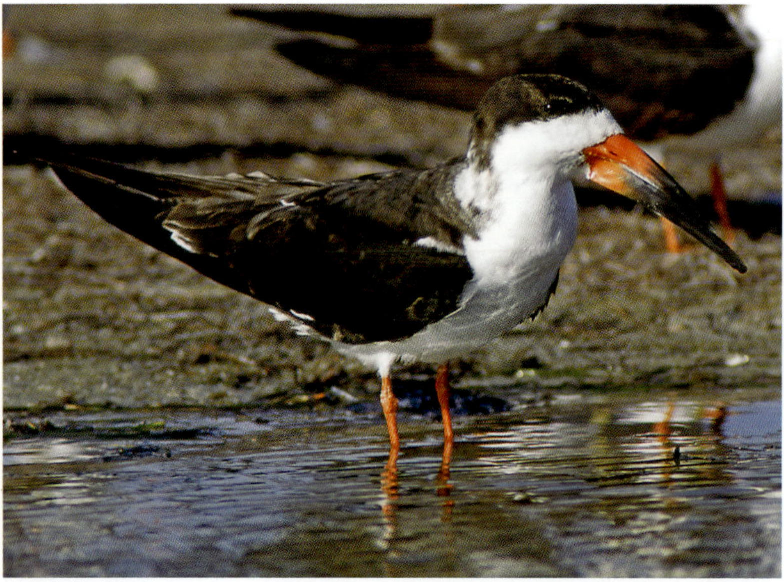

Adult. Florida, February.

Taxonomy Polytypic (3). Race in Jamaica is *R. n. niger*.
Description L 40–51cm (16–20in). Medium-sized tern, black above and white below, with characteristic massive orange bill with black tip, with upper mandible much shorter than the lower one.
Similar species None.
Habitat and behaviour Feeds by flying between waves, skimming the surface with its lower mandible to catch surface-dwelling small invertebrate prey.

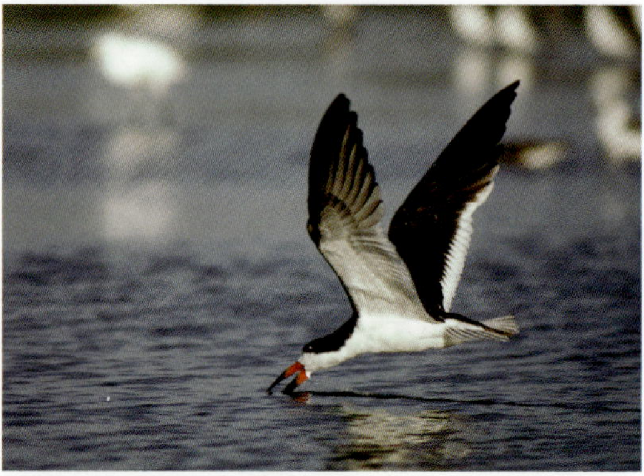

Adult in characteristic feeding flight. August.

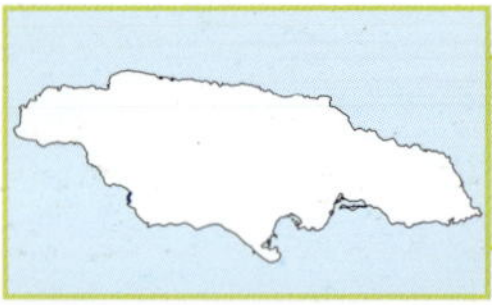

Range *R. n. niger* breeds on the coasts of North and South America.
Status in Jamaica Very rare passage migrant or vagrant.

Rock Pigeon
Columba livia

Taxonomy Polytypic. Birds on Jamaica are of domesticated stock (presumably *C. l. livia*).
Description L 33–36cm (13–14in). This is the domestic pigeon seen in flocks in most towns and cities. It is very variable in colour and may be white, blue, brown or parti-coloured.
Similar species White-crowned Pigeon lacks the white nostril-caps.
Voice Various coos.
Habitat and behaviour Usually found in open spaces in towns and cities. Flocks nest on window-ledges on high-rise buildings, especially near the coast, and feed on the ground in parks and gardens, generally on seeds and berries.

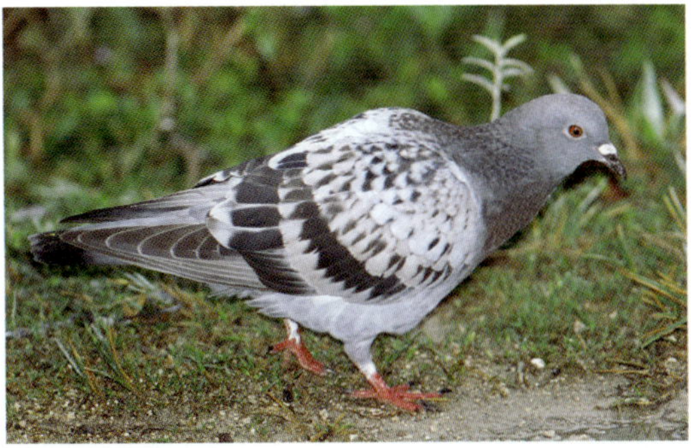

Adult. Cuba, May.

Range Originally native to the Old World. Introduced worldwide.
Status in Jamaica Common resident.

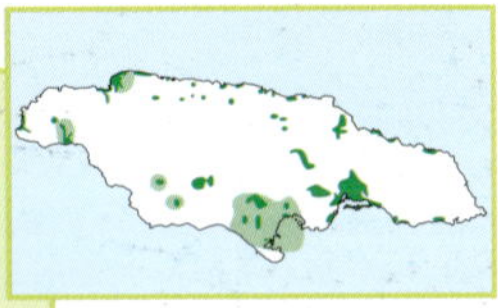

Scaly-naped Pigeon
Patagioenas squamosa

Taxonomy Monotypic.
Description L 36–40cm (14–16in). Large slaty-grey pigeon. Yellowish bare skin on orbital ring and reddish-purple wash on neck and breast visible at close range.
Similar species White-crowned Pigeon has white crown.

Range West Indies, islands off Venezuela.
Status in Jamaica Very rare. Breeding status unclear.

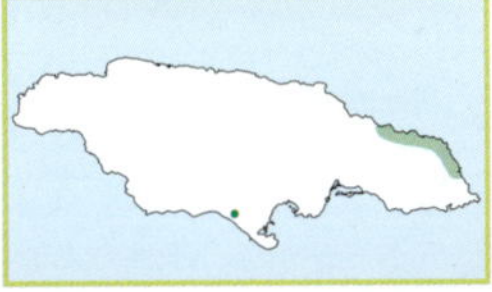

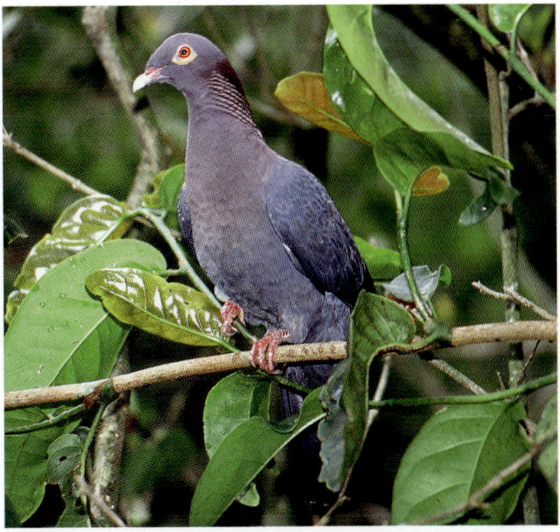

Adult. Note rufous on neck. Dominican Republic, April.

Ring-tailed Pigeon
Patagioenas caribaea

Local name Ringtail
Taxonomy Monotypic. Endemic to Jamaica.
Description L 41cm (16in). Large, sleek, light grey pigeon, with a dark band across the middle of the buffy-ash tail and an iridescent bluish-green patch on the hind neck. Iris and eye ring red, bill black. Chin white. Underparts are beige or lavender fading to buffy on the undertail-coverts. Legs red. Juvenile resembles adult but has yellow iris and white undertail-coverts.
Voice Fairly low-pitched *cru cru crooooo* with last note longer and lower.
Similar species None.
Habitat and behaviour Forested, moist mountain areas, mostly at 100–2000m. Descends to lower elevations in winter. Arboreal, rarely seen on ground. Often found in flocks of up to 20 on a single feeding tree, feeding silently on small fruits or berries. Nests (March–September) in mid-canopy of tall trees in old-growth forest. Lays two white eggs.

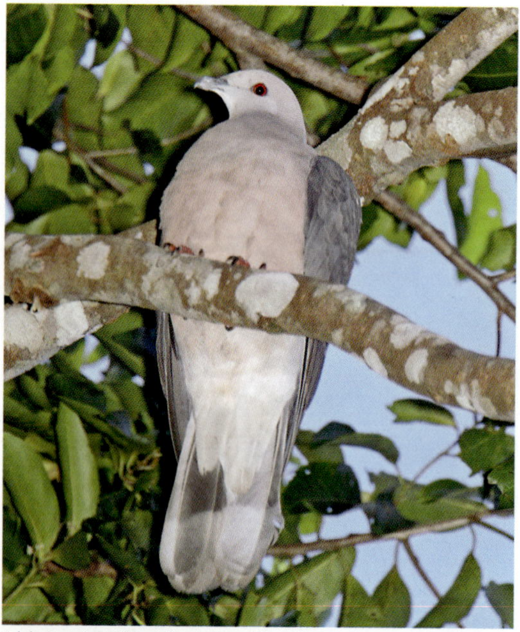

Adult. Port Royal Mountains, Jamaica, November.

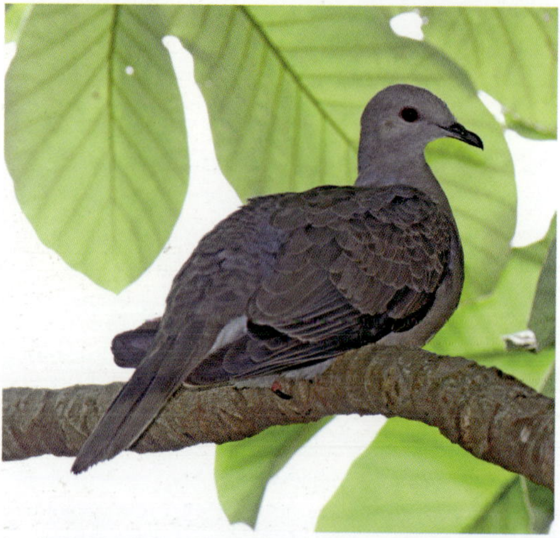

Juvenile. Port Royal Mountains, Jamaica, November.

Range Jamaica. Global conservation status: Vulnerable.
Status in Jamaica Generally rare but locally common and seasonal. Decreasing due to illegal hunting and forest destruction.

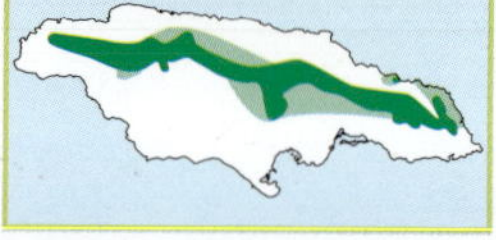

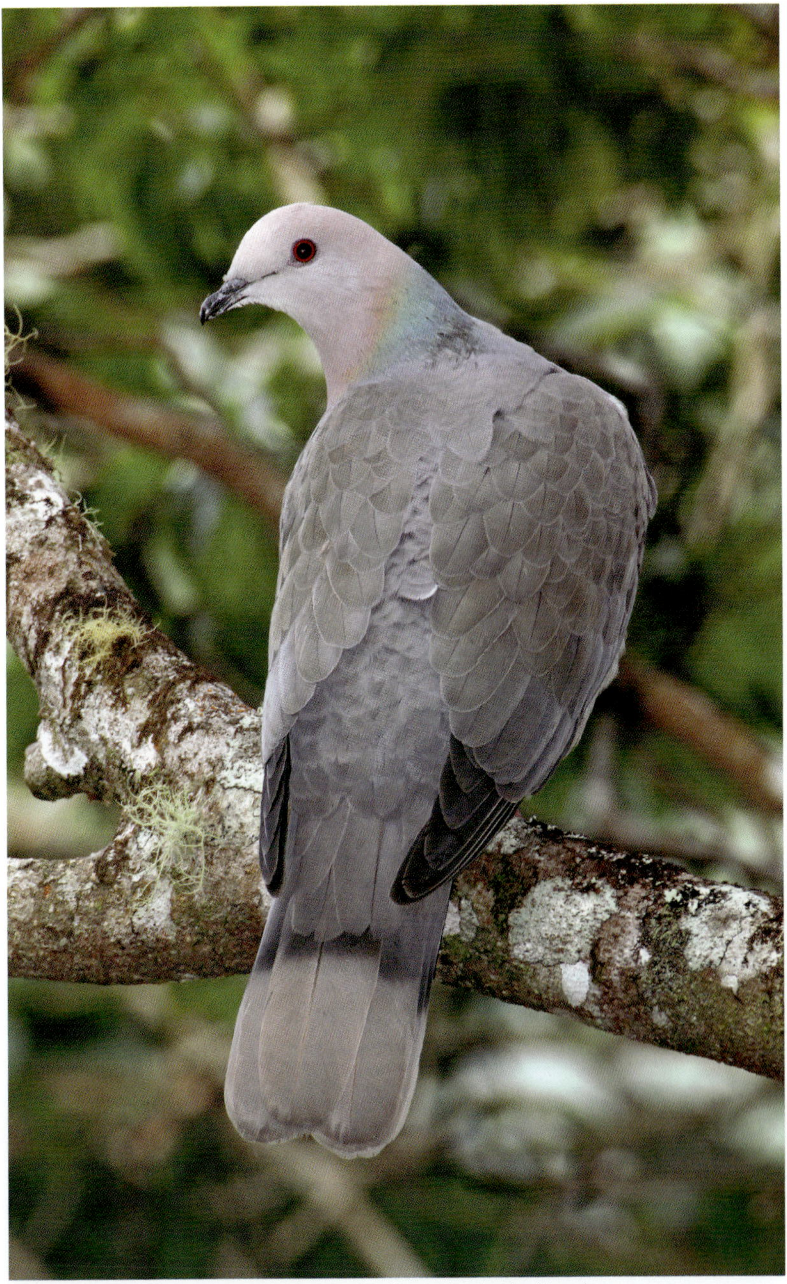

Adult. Note buffy undertail-coverts and red legs. Port Royal Mountains, Jamaica. November.

White-crowned Pigeon
Patagoienas leucocephala

Local names Baldpate, Ball Plate.
Taxonomy Monotypic.
Description Male L 31cm (12in). Female L 28cm (11in). Fairly large blue-black pigeon with a white crown, and white eyes. Legs red. Bill maroon, tipped pearl-grey. Male has a patch of iridescent blue-green on hind-neck and collar. Female smaller than male and has less iridescence on the hind-neck. Crown has greyish wash. Immature is dark brown with brown crown which becomes white in second year; pale fawn or whitish edging to the wing-coverts.
Similar species Plain Pigeon or Rock Dove are sometimes shot accidentally by hunters targeting this pigeon.

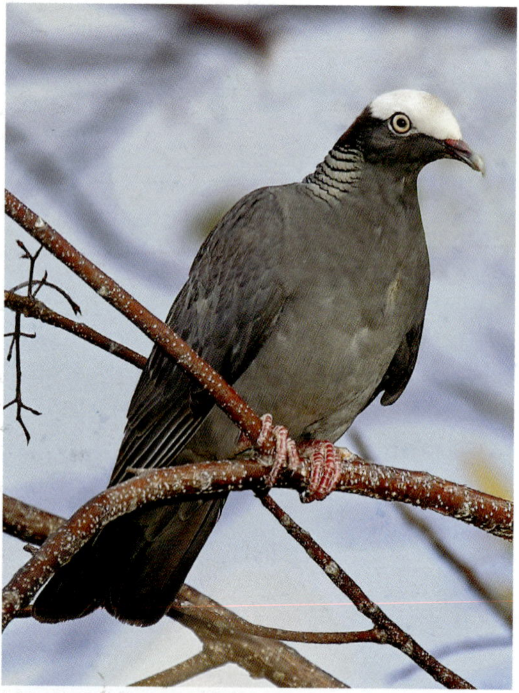
Adult male. Ecclesdown, Jamaica, November.

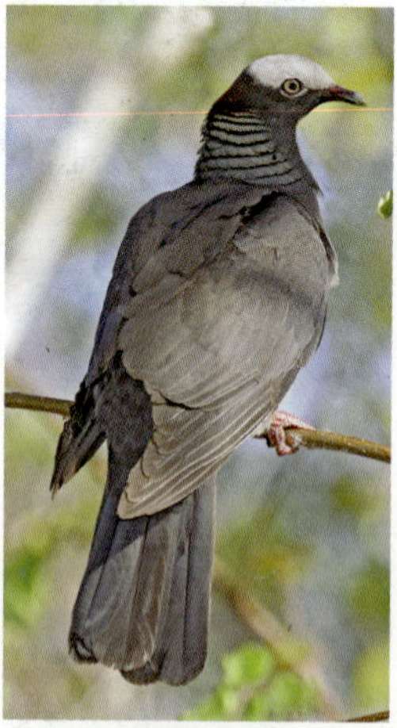
Adult male. Cayman Islands, April.

Voice A deep *whoooo? Who took two? Who took two? Who took two?* Also a purring *ccruuuuuuuuu*.
Habitat and behaviour Found in forests at all elevations. Coastal populations may fly to the hills to feed in the coastal forests in the morning and return to the mangrove forests late in the afternoon. Strong flier. Seldom seen on the ground. Feeds on fruit of Red Birch, Pimento, Burnwood, mangrove, Bird-Pepper, bullets and other trees and shrubs. Nests colonially elsewhere in range but singly in Jamaica (April–September) in well wooded areas, mangroves, logwood and other tall, spindly trees. Parents build open nest of twigs. Lays two white eggs.

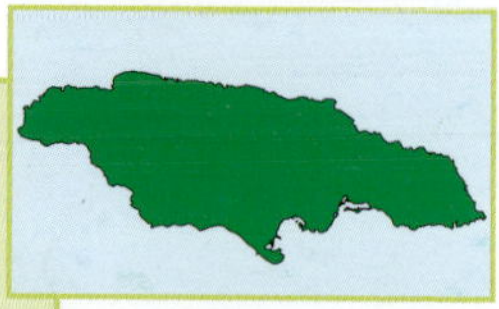

Range Breeds Florida Keys, Bahamas, Caribbean coast of Central America, throughout Greater Antilles and more rarely in southern Lesser Antilles. Global conservation status: Near Threatened.
Status in Jamaica Common resident.

Plain Pigeon
Patagioenas inornata

Local name Blue Pigeon.
Taxonomy Polytypic (3). *P. i. exigua* is endemic to Jamaica.
Description L 41cm (16in). A large, dark, robust pigeon. Head and breast beige. Bill light grey with a dark tip. Iris is white, eye-ring red. When perched, the folded wing has a rufous patch and white edges to the outer coverts. In flight, wings show indistinct white wing-bars. Upper back is drab, lower back and tail grey, with no white in tail. Legs crimson. Juvenile has a brownish head and breast. Abdomen and undertail-coverts are grey, the latter with darker tips.
Similar species None.
Voice Similar to White-crowned Pigeon but shorter and higher in pitch – *cruuuu cru cru*. Sounds like *who are you?*
Habitat and behaviour Coastal scrub, dry forests, mangroves and mountain areas. Feeds on fruits and berries (e.g. palm fruits). Moves inland to nest (April–July) in deeply wooded areas in the mountains. Winters (September–October) in coastal wetlands and dry forests, mainly in the Portland Bight area.

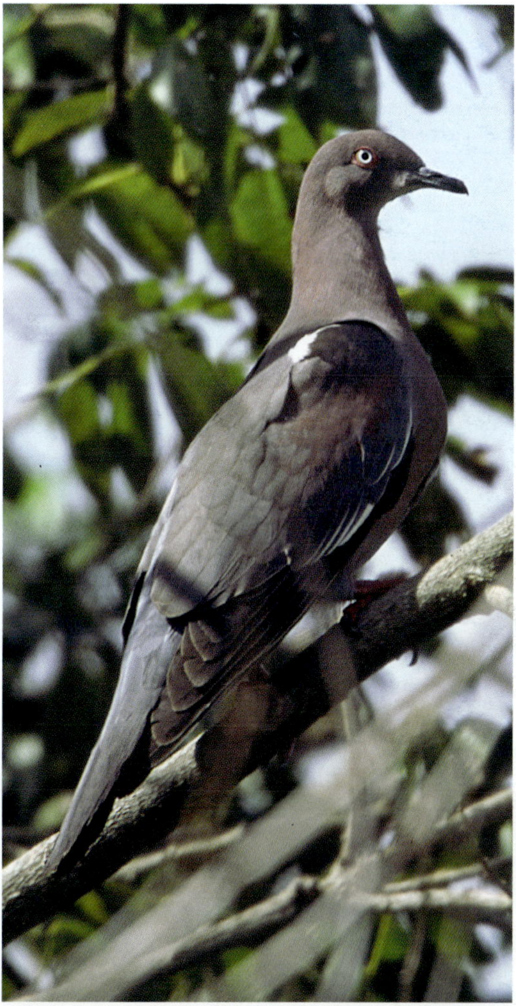

Adult *P. i. inornata*. Note rufous wing-patch. Dominican Republic, May.

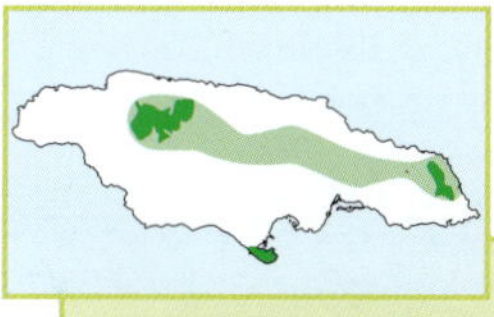

Range *P. i. exigua* is endemic to Jamaica. Other races in Cuba (and the Isle of Pines) and Hispaniola, and Puerto Rico.
Status in Jamaica Uncommon resident. Possibly declining, threatened by illegal hunting and hurricane damage to coastal habitats. Also rare elsewhere in its range.

White-winged Dove
Zenaida asiatica

Local name White-Wing, Lapwing.
Taxonomy Polytypic (3). Race in Jamaica is *Z. a. asiatica.*
Description L 25cm (10in). Brown on head, fading to grey on back. Iris red, surrounded by blue orbital skin. Tail dark grey, broadly tipped light grey, except on the two central tail feathers. Reddish-brown on throat becoming pale grey on lower underparts. In flight shows broad white wing-patch (appears as white line when bird is perched). Immature browner with narrow pale edges to wing-coverts and grey-tipped tail.
Similar species Zenaida Dove lacks large white edge to folded wing and is more likely to be seen on the ground.

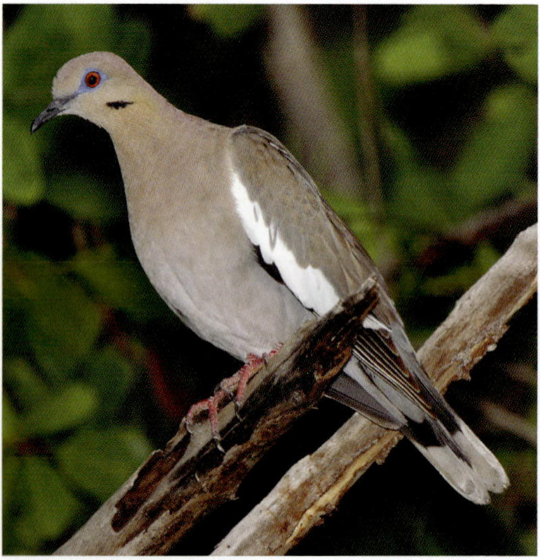

Adult. Jamaica, November.

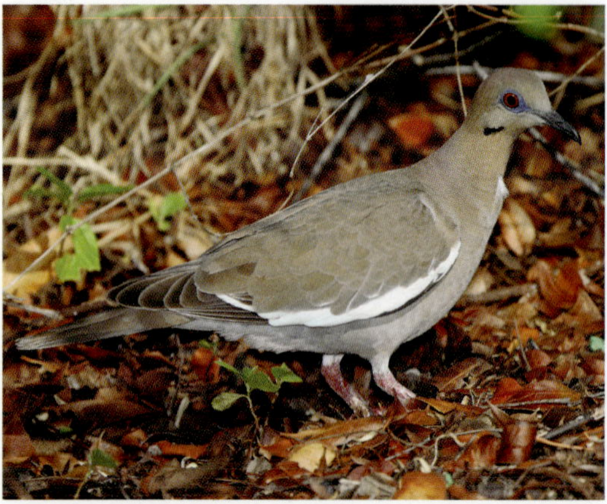

Adult. Cayman Islands, March.

Voice *Two bits for two*; and has a longer and more varied song, rendered as *since Mr. Gilpin dead, cow-head poil, cow-head poil, cow-head poil*.
Habitat and behaviour Flocks common in cultivatation and in mangroves. In flight wings appear to miss a beat, hence the common name 'lapwing.' Feeds in flocks in trees or on the ground on fruits, seeds and corn. Nest (March–June) is a cup of twigs in a shrub or tree. Lays two white eggs. Often raises two broods per year.

Range *Z. a. asiatica* on Greater Antilles, western Caribbean islands and the Bahamas. Other races North and South America.
Status in Jamaica Very common and widespread resident.

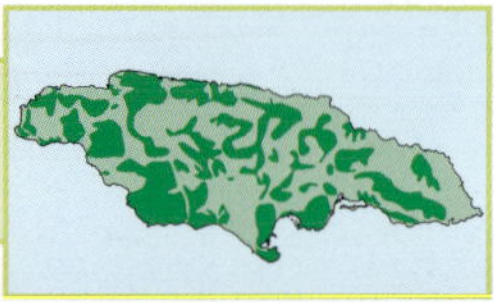

Mourning Dove
Zenaida macroura

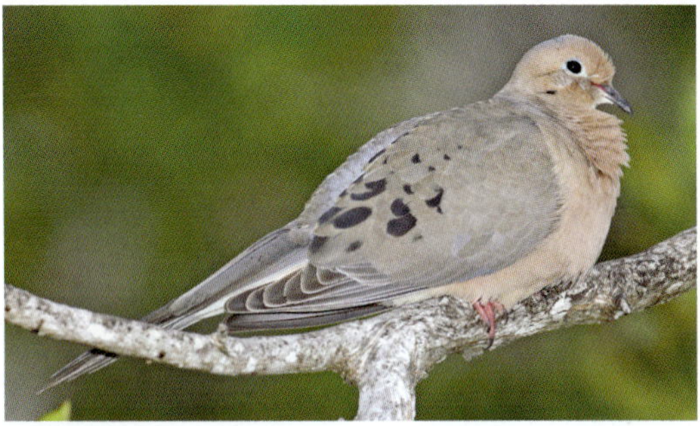

Adult. Note long, pointed tail. Hope Gardens, Jamaica.

Local names Paloma, Long-tail Pea Dove.
Taxonomy Polytypic (5). Race in Jamaica is *Z. m. macroura.*
Description L 27cm (11in). Medium-sized brown dove with a long tail. Head tawny washed with grey. Rosy wash on hind neck. Back and wings brown with dark spots on wing-coverts. Underparts brown on throat, pinkish buff on undertail-coverts. In flight long diamond-shaped tail tipped with white is diagnostic. Female paler than male; juvenile browner with paler edges to wing feathers, grey-tipped tail.
Similar species Zenaida Dove has short, rounded tail; white in wings; seen in pairs not flocks.
Voice Similar to Zenaida Dove but higher-pitched.
Habitat and behaviour Southern coastal lowlands in dry limestone scrub. Feeds on seeds and corn in flocks. Nest (April–June) is cup of twigs in bushes. Lays two white eggs.

Range *Z. m. macroura* Greater Antilles and the Bahamas. Other races in North and Central America.
Status in Jamaica Locally common resident.

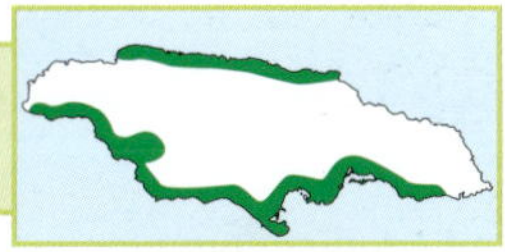

Zenaida Dove
Zenaida aurita

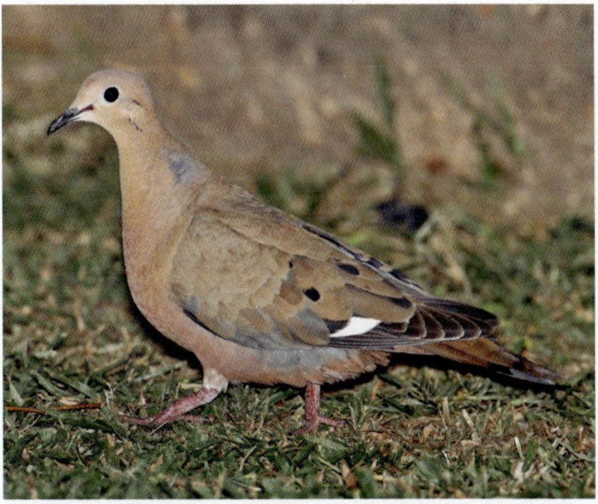

Adult. Hope Gardens, Jamaica, March.

Local name Pea Dove.
Taxonomy Polytypic (3). *Z. a. zenaida* is endemic to the Greater Antilles and the Bahamas.
Description L 28cm (11in). A ruddy-brown dove, with some purplish spots in the otherwise brown wing-coverts. Secondaries are tipped with white, giving a white wing-bar on the folded wing. Hind-neck is iridescent greenish. Short, thin, black streak below ear-coverts. Female is paler than the male, especially during the breeding season. Juvenile is paler than adults, with tail feathers tipped with white

or greyish-white.

Similar species White-winged Dove is rarely seen on the ground, and has striking white wing-patches. Mourning Dove has a distinctive long and pointed tail.

Voice *What AM I to do?* Often followed by *Oo-Ah-oo*. Traditional mnemonics include *Moses PREACH God's word* and *Sary sey coat-blue*.

Habitat and behaviour Wooded cultivations and gardens, from sea-level to the mountains. Usually seen walking in pairs or small family parties on paths, roadways or lawns. Wings make a distinctive creak on take-off. Feeds on fruits and seeds, often on the ground. Nests (May–June) at the base of coconut fronds or in bromeliads. When alarmed during the breeding season may put on a 'broken wing' display to distract predators and lead them away from the nest.

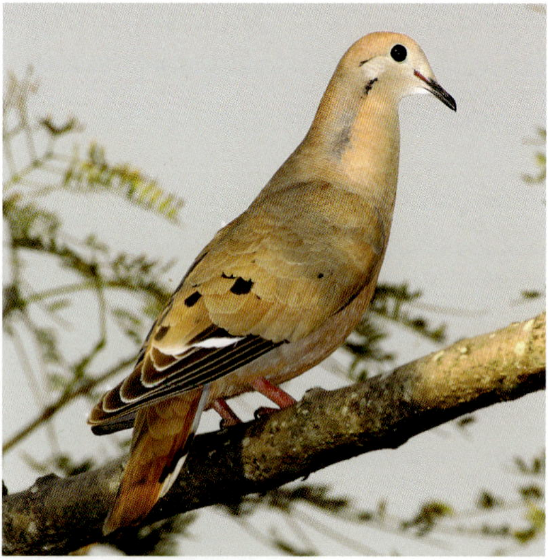

Adult. Hope Gardens, Jamaica, March.

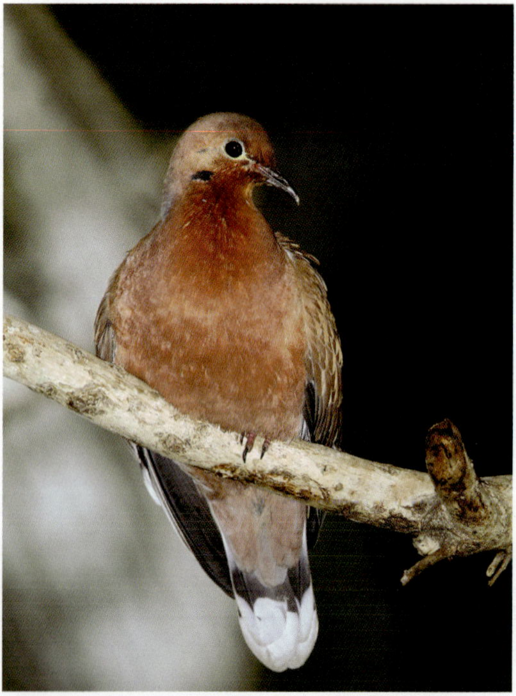

Adult rufous. Montego Bay, Jamaica.

Range *Z. a. zenaida* Greater Antilles and the Bahamas. Other races inhabit Yucatán Peninsula and offshore islands, and the Lesser Antilles. *Z. a. zenaida* is paler and greyer than the eastern Caribbean race.

Status in Jamaica Common and widespread resident that occurs from sea-level to the mountains.

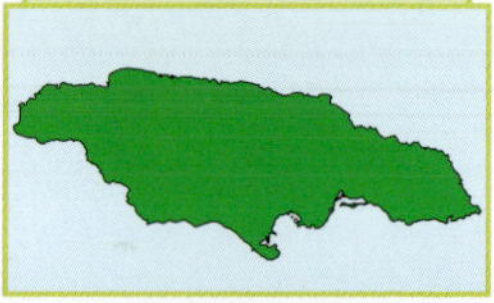

Common Ground Dove
Columbina passerina

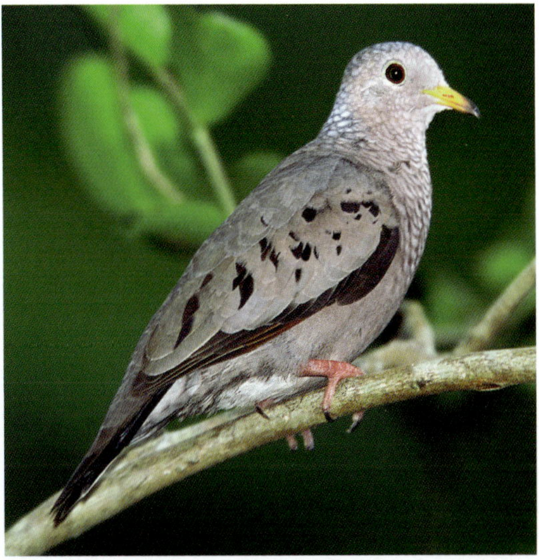

Adult male with pinkish breast. Rocklands, Jamaica, May.

Taxonomy Polytypic (18). *C. p. jamaicensis* is an endemic race to Jamaica.
Description L 15cm (6in). A small plump dove, greyish-brown above. Bill orange, tipped black. Wings dark rufous, edged with dark grey; wing-coverts have iridescent purplish-blue spots. Tail black with greyish brown centre. Feet pink. Male has pinkish-brown head and neck and dark fringes to breast feathers giving a scaly effect. Female has greyish head and breast. Juvenile similar to female but wing-covert spots rusty, becoming purplish-black within two months of hatching; bill pale yellowish.
Similar species Smaller than any other dove in the region.
Voice *Hoe-Ah, Hoe-Ah, Hoe-Ah woop, woop* and a growling call (like the twang of a plucked rubber band) made just before copulation.

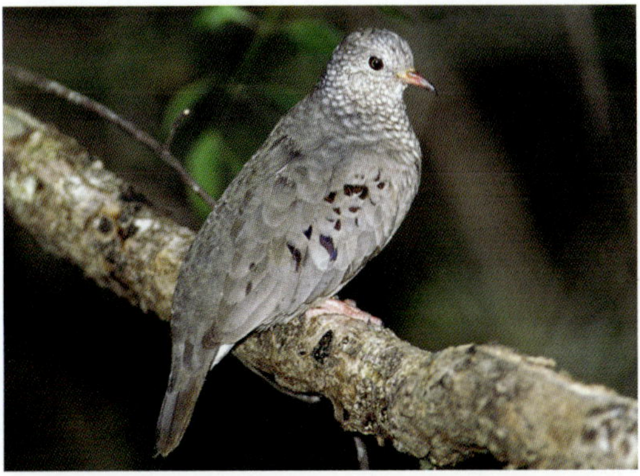

Adult female. Rocklands, Jamaica, May.

Habitat and behaviour Forest edges, clearings, gardens in the plains and foothills. Less common in the mountains. Pairs or small family parties (usually 2–4) are often seen walking on the ground, foraging on fallen seeds and fruits. When disturbed, they fly a short distance along the path. Pumps head back and forth when walking. Nests year-round, but chiefly March–June and November–December, in shrubbery near the ground and sometimes in small trees e.g. Pimento. Nest is cup-shaped and made of twigs. Lays two white eggs.

Range *C. p. jamaicensis* endemic to Jamaica. Other races in Bermuda and the Bahamas, throughout the West Indies and in North, Central and South America.
Status in Jamaica Very common resident.

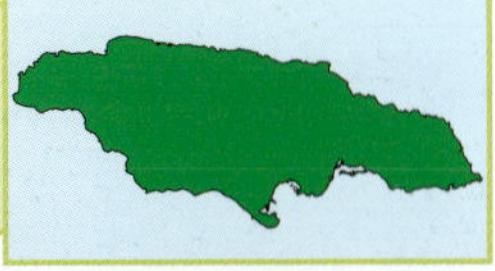

Caribbean Dove
Leptotila jamaicensis

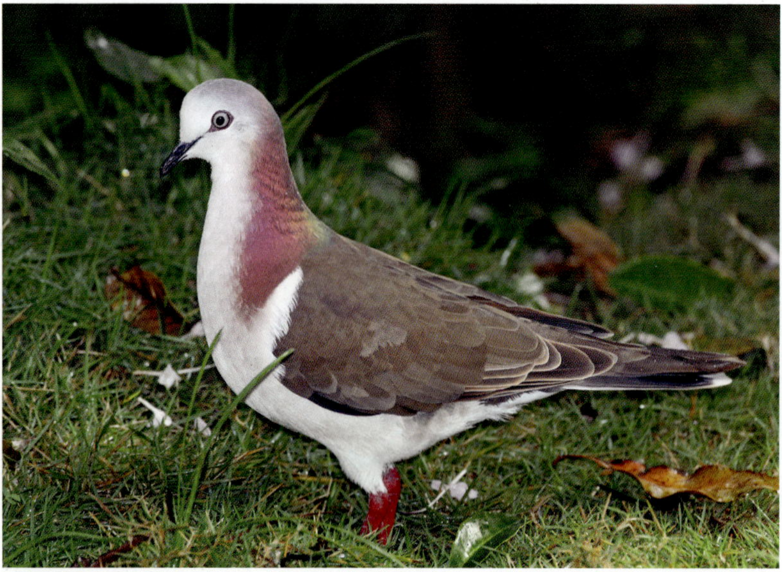

Adult male. Note extensive pink on neck. Rocklands, Jamaica, November.

Local name White-belly.
Taxonomy Polytypic (5). *L. j. jamaicensis* is an endemic race to Jamaica.
Description L 30cm (12in). A large, plump dove with a dark olive back and snowy white underparts. White forehead, pinkish-grey head and powder-grey face. Bare skin around eye pinkish. Bill black. Outer tail feathers tipped with white. Wing lining cinnamon. Crimson feet get brighter during the breeding season. Male has iridescent rosy-vinaceous patch on hind-neck and collar. Female and juvenile paler than the male with less iridescence on the hind-neck.
Similar species Crested Quail-dove has similar walk but has a distinct crest and lacks the white belly.
Voice A plaintive *who cooks for you-oo?* Pairs often duet together – the reply is interpreted as *what's that to you-oo?*
Habitat and behaviour Dry, mesic and wet forest, gardens, cultivated areas and orchards. Largely terrestrial. Usually seen singly. May be first detected by sound of rustling leaves, then the bird may emerge from underbrush into a clearing, roadway or path, pumping head forward and flicking tail up. Often perches on low branches. Feeds on seeds of fallen fruit such as Orange, Naseberry and Red Birch, as well as on small snails. Nests (March–June) builds a platform of sticks in low bushes e.g. Logwood. Lays two white eggs.

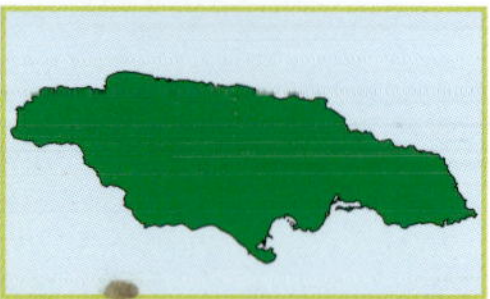

Range *L. j. jamaicensis* endemic to Jamaica but also introduced to New Providence in the Bahamas. Other races in Yucatán Peninsula, Grand Cayman and San Andres Island, Columbia.
Status in Jamaica Common resident.

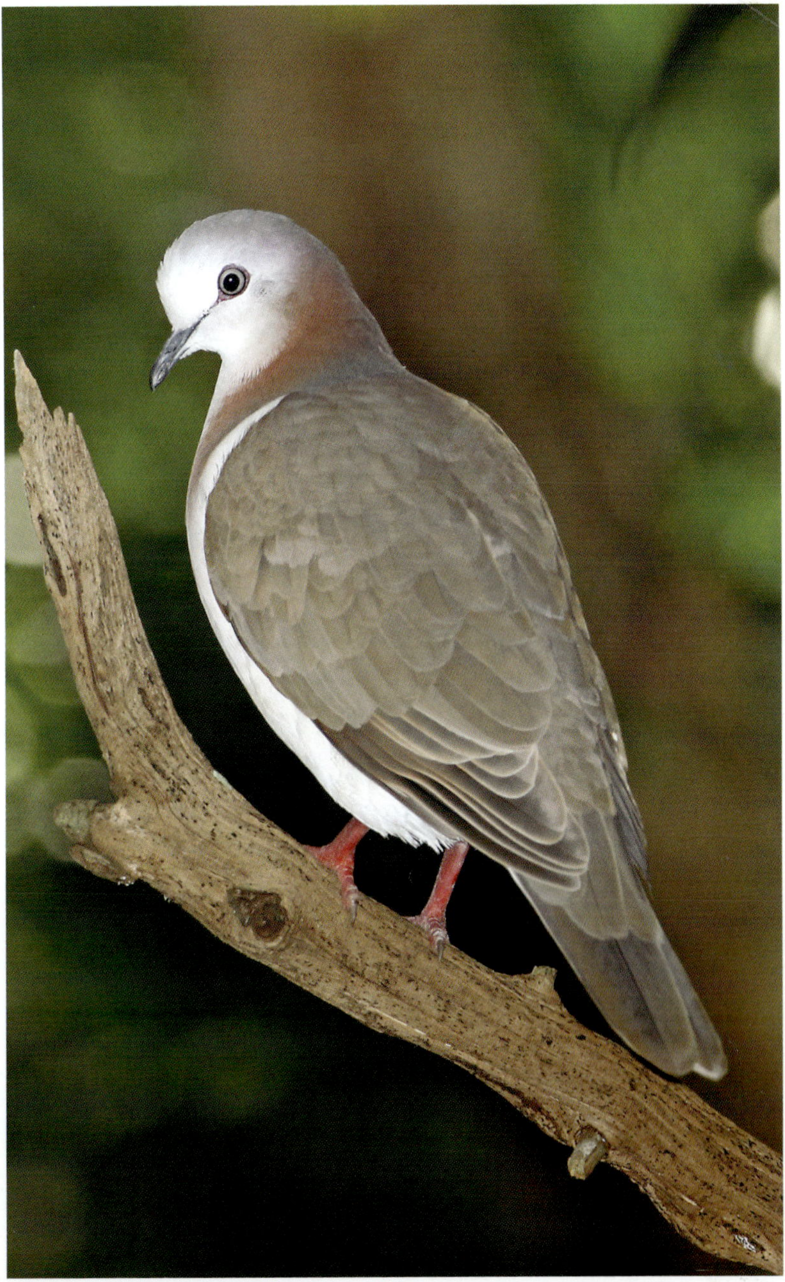

Adult female. Rocklands, Jamaica, November.

Crested Quail-dove

Geotrygon versicolor

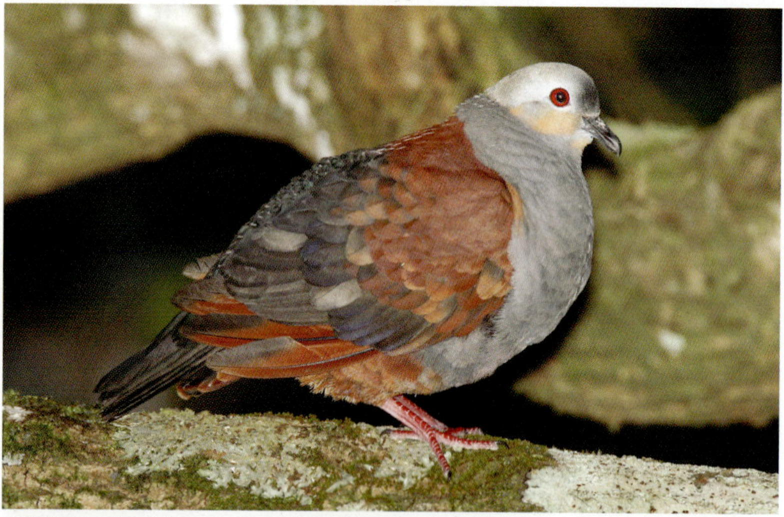

Adult. Note crest here is flattened. Port Royal Mountains, Jamaica, May.

Local name Mountain Witch, Blue Dove, Blue Partridge.
Taxonomy Monotypic. Endemic to Jamaica.
Description L 30cm (12in). A stocky grey and rufous dove with a small nape crest formed by a series of elongated grey feathers. Forehead black, iris and eye ring red, lower face and chin buffy. Bill reddish black. Wings, upper back and wing-coverts rufous and magenta. Lower back iridescent greenish black. Throat and breast grey, breast washed with magenta. Lower underparts and tail-coverts are bright rufous. Legs red. Female is duller than the male. Juvenile has browner underparts.
Similar species See Ruddy Quail-dove and Caribbean Dove.
Voice A two- or three-note *woof-wooo*, first note explosive, last note descending.
Habitat and behaviour Only occurs in the mountains in montane and wet limestone forest, in old-growth forest where the canopy is dense and the understorey is relatively open. Often first detected by rustling of leaves. May emerge from dense vegetation to feed on the edge of a road, a path or a clearing. When walking pumps head back and forth and flicks tail up and down. When flushed usually does not fly far. Perches in lower

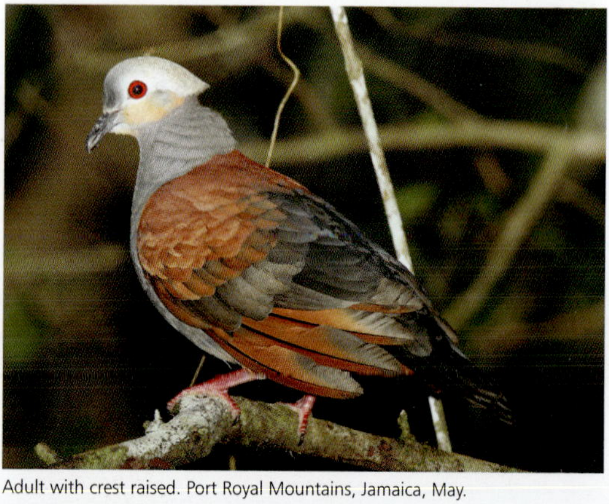

Adult with crest raised. Port Royal Mountains, Jamaica, May.

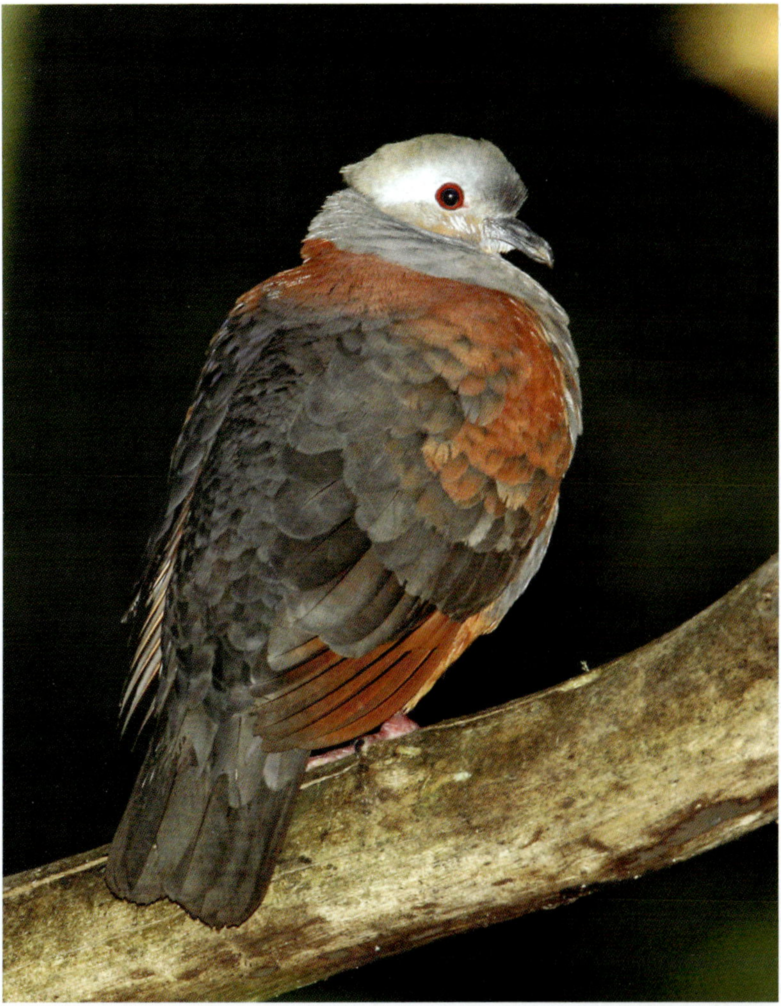

Adult. Port Royal Mountains, Jamaica, May.

branches of trees. In flight wings make a whooshing sound. Feeds on fallen fruits and seeds (e.g. of Wild Ackee). Nests (March–June) in the understorey, on top of bromeliads or in bushes such as coffee.

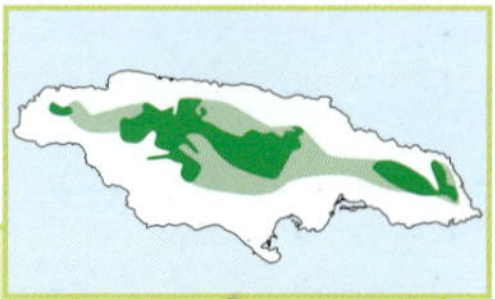

Range Endemic to Jamaica. Global conservation status: Near Threatened

Status in Jamaica Uncommon resident. Population is probably decreasing. Tends to decline after hurricanes due to the disturbance of old-growth forest.

Ruddy Quail-dove
Geotrygon montana

Local name Partridge.
Taxonomy Polytypic (2). Race in Jamaica is *G. m. montana*.
Description L 25cm (10in). A richly coloured ruddy-brown stocky dove. Bill maroon with a paler tip. Iris cinnamon. Wing-coverts and back are a mixture of brown and buffy. Underparts are mainly vinaceous with buffy undertail coverts. Legs red, brighter when breeding. Adult male rich ruddy brown with a distinctive buffy line on cheek. Adult female is duller with less distinct cheek stripe and pale throat. Juvenile olive brown with pale fringes to upperpart feathers; cheek stripe indistinct or absent.
Similar species In flight Crested Quail-dove looks similar, with rufous wings and underparts, but has grey head.
Voice A booming, low moaning *OOOah, OOOah* like a dying man, repeated for long periods.
Habitat and behaviour On the ground or in dense vegetation, in mountain and coastal forests, disturbed woodlands and citrus groves. Usually seen walking on paths, feeding on seeds. Pumps head when walking. Also flies very fast through the understorey, rarely above the canopy. Nest (April–June) is a loose, flattish cup made of sticks in dense shrubbery. Lays two buff-coloured eggs. At mid-levels populations appear to increase at the start of the breeding season, probably due to altitudinal movements.

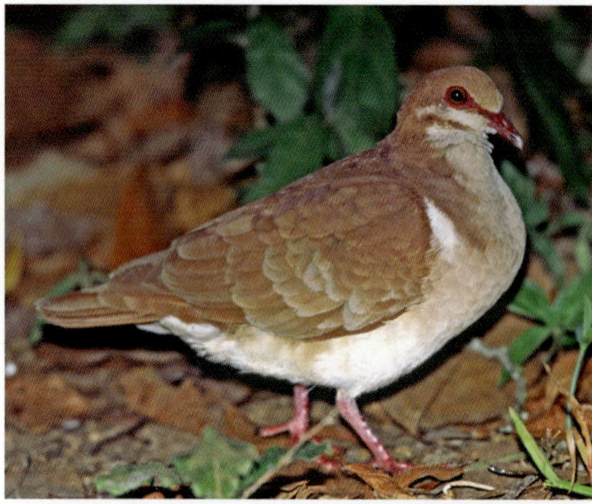

Adult male. Note stripe on face. Marshall's Pen, Jamaica, May.

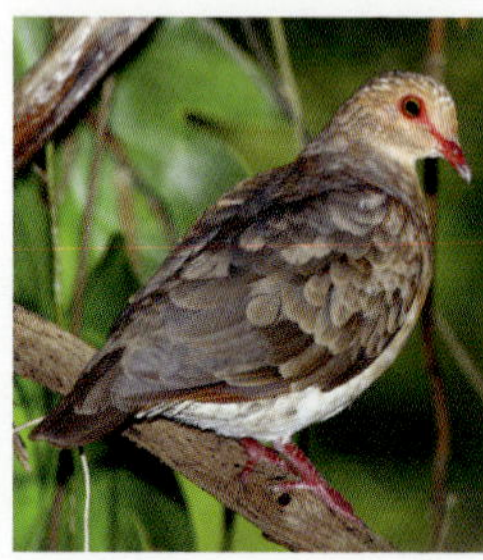

Adult female. Marshall's Pen, May.

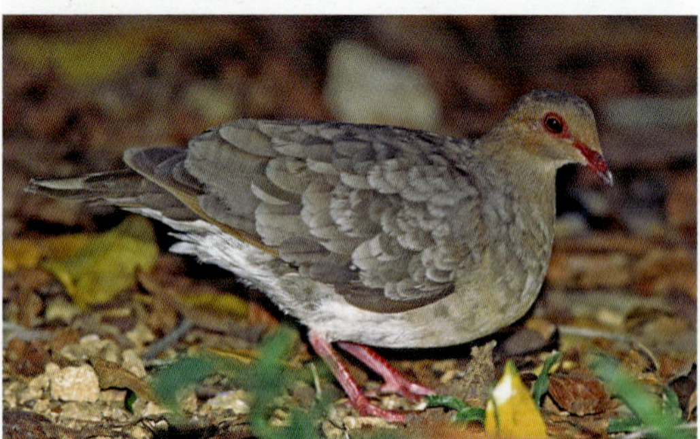

Juvenile. Rocklands, Jamaica, May.

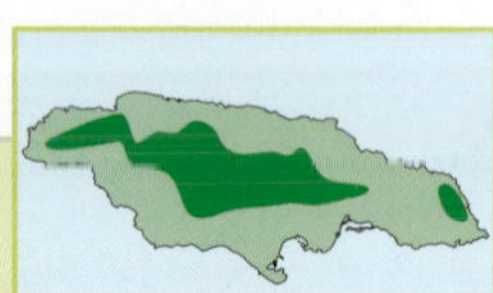

Range *G. m. montana* Greater Antilles, Grenada and Central and South America. Other race in Lesser Antilles.
Status Resident. Common at mid-levels during the breeding season. Otherwise uncommon.

Green-rumped Parrotlet

Forpus passerinus

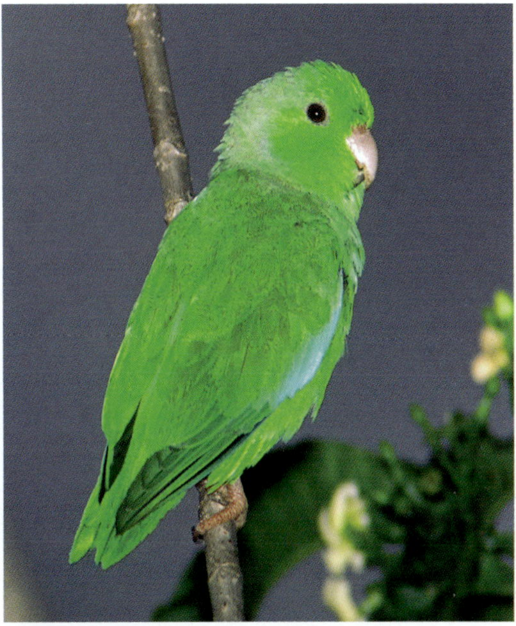

Adult male with blue in the wing. Montego Bay, Jamaica, May.

Local names Parrolet, Parakeet.
Taxonomy Polytypic (5). Introduced to Jamaica; race has not been established.
Description L 13cm (5in). Tiny green parakeet with a short pointed tail. Bill horn-coloured. Legs pinkish grey. Wings black with green coverts. Male has blue wing patches and blue-green rump. Adult female and juvenile lack blue wing patches and have yellower breasts.
Similar species Resembles a budgerigar with a short tail.
Voice A loud chattering when flock settles in a tree: *swee-swee-swee-sweetie*. In-flight song *phil-ip, phil-ip, phil-ip*.
Habitat and behaviour In forest habitats from the coast (including the edges of wetlands) to mid-level woodland. Usually seen flying in noisy flocks or feeding on fruits and seeds in trees and shrubs. Can sometimes be seen perching on power-lines. Feeds on Coconut blossom, Indian Corn, Orange, Logwood, figs, grass seeds, fruit and small berries. Nests in holes in trees, old woodpecker holes and under the eaves of houses.

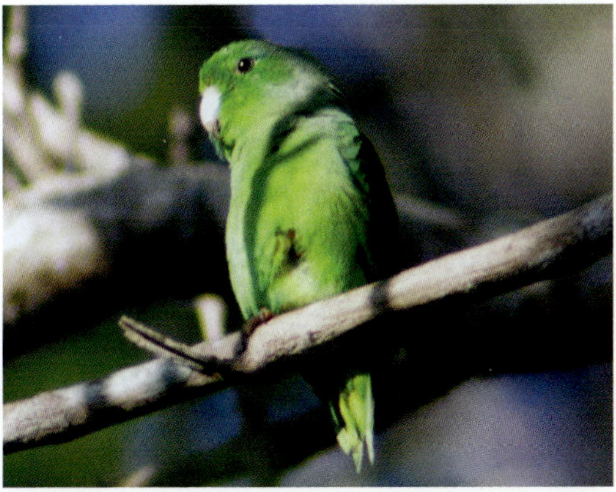

Adult female. Good Hope, Jamaica, May.

Range Native range is northern South America. Introduced to Jamaica in 1918.
Status in Jamaica Uncommon resident. Numbers decline steeply after hurricanes and recover relatively slowly afterwards.

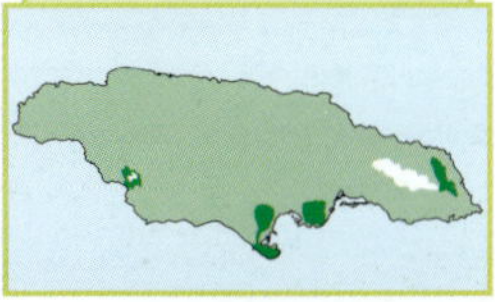

Olive-throated Parakeet
Aratinga nana

Local name Parakeet
Taxonomy Polytypic (3). *A. n. nana* is an endemic race to Jamaica.
Description L 31cm (12in). Small parakeet with a green head and back and long pointed tail. Outer webs of flight feathers blue. Dark olive-brown below. Iris orange, surrounded by cream-coloured bare skin. Bill pale horn.
Similar species Yellow-billed Parrot is much bulkier with short blunt tail; has shallower wing beats.
Voice Very noisy both in flight and when perched and feeding. Shrill *creek creek* in flight or a single *preeeit*, not unlike the call of a Jamaican Woodpecker.
Habitat and behaviour Widespread in wooded hills, mountain slopes, scrub, cultivations and gardens in humid or semi-arid areas from sea level to mid-level forests. Usually seen in noisy flocks of up to 20 birds, flying or feeding in the tops of trees. Flies fast, with deep wing beats. Feeds on buds and fruit of many trees, e.g. figs, Red Birch, Erythrina, Bauhinia, Spathodia; also on cultivated corn and gungo peas and other fruit, often becoming a pest. Nests (March–June) in holes in termite nests or trees.

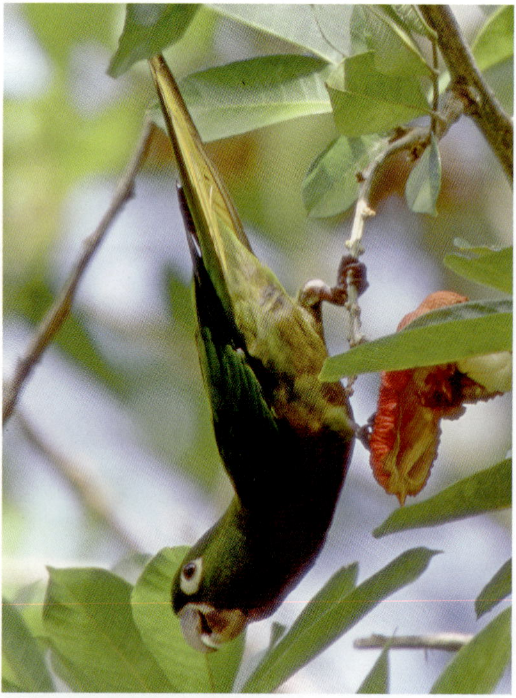

Adult. Montego Bay, Jamaica, April.

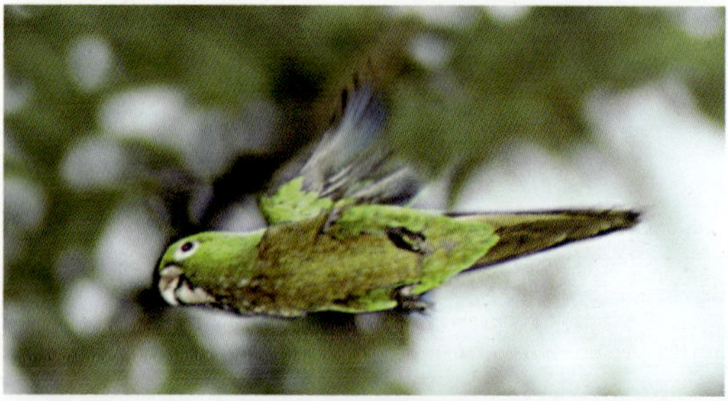

Adult in flight. Good Hope, Jamaica, April.

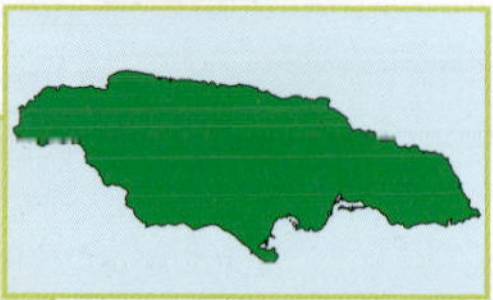

Range *A. n. nana* endemic to Jamaica. Other races in Central America.
Status in Jamaica Common and widespread.

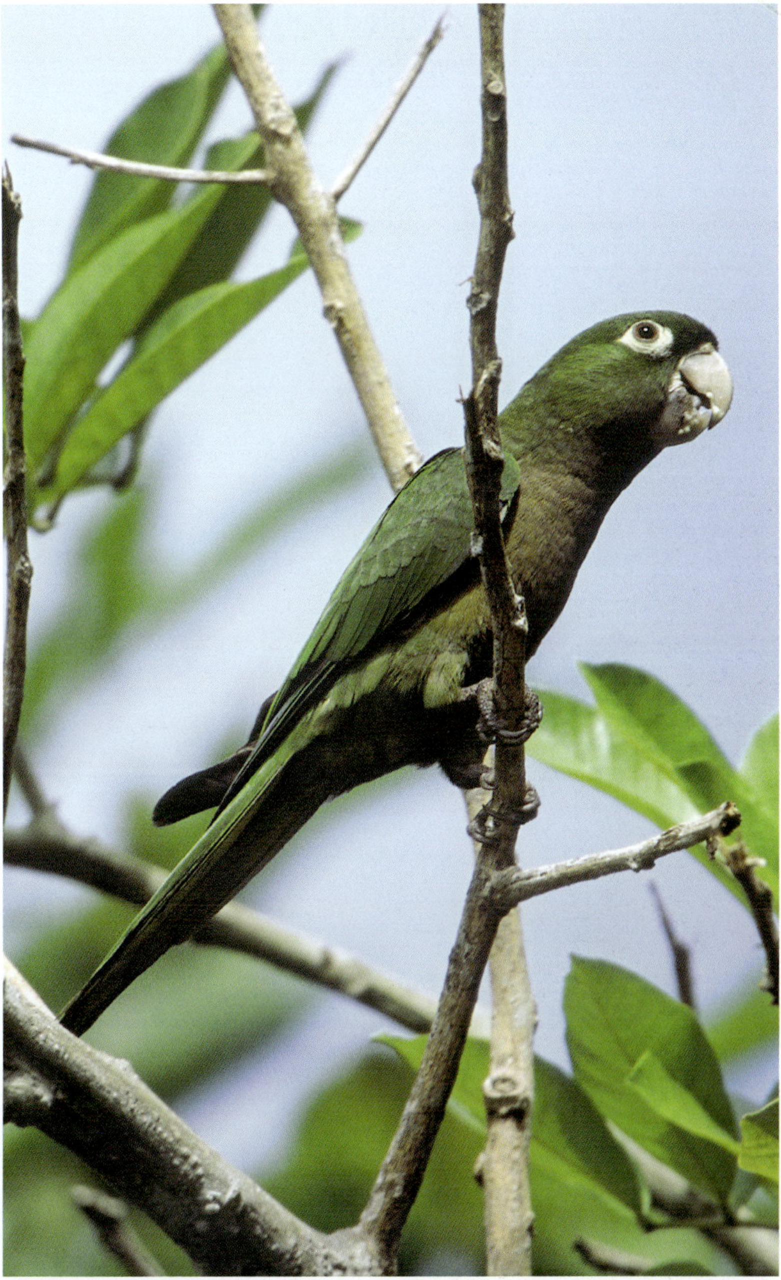

Adult. Montego Bay, Jamaica, May.

Yellow-billed Parrot
Amazona collaria

Local name Yellow-bill.
Taxonomy Monotypic. Endemic to Jamaica.
Description L 27cm (11 in). Green parrot whose yellow bill with a paler tip is conspicuous in flight. Head is green with a blue wash and there is a narrow white band above the bill. Bare skin around the eye is cream. Iris dark hazel. The throat and hind neck feathers are a deep rose-colour, edged dark grey-green to give a scaly effect. Wings are dark green with blue flight feathers. The short square tail is greenish-yellow with a rosy pink base. Legs and feet are yellow.
Similar species Olive-throated Parakeet is slender with a long and pointed tail. Black-billed Parrot has a black bill and a red patch in the wing; its wingbeats are shallower, stiffer and faster.

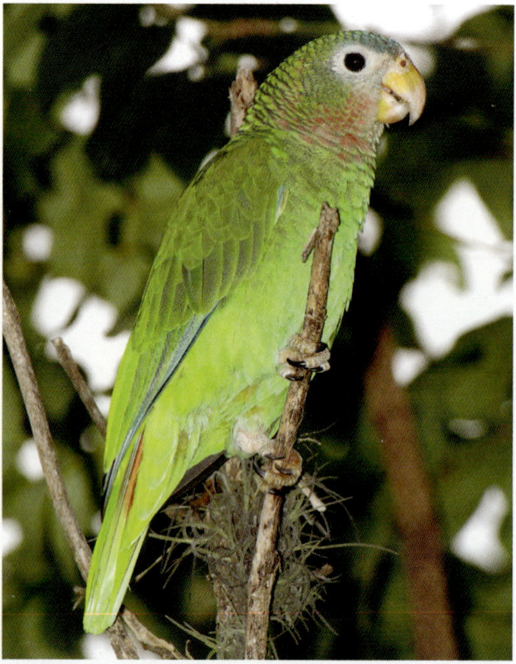

Adult. Hope Gardens, Jamaica, November.

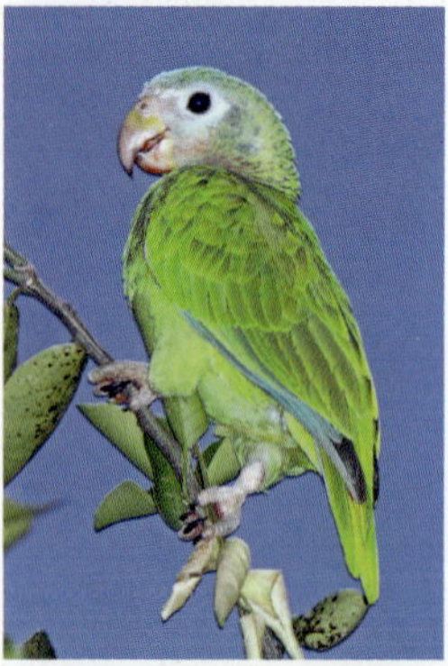

Adult. Hope Gardens, Jamaica, Nov.

Voice A high-pitched *ah-ah-eeeeek* that rises on the last note, and a *whip-whip-whip-waaaark* in flight; also many other squawking noises.
Habitat and behaviour Forested limestone hills and mountains. May be seen in pairs, in small groups or flocks of 20 or more birds. Flies with deep, slow wing beats. Flocks leave roosting areas in the forests in the morning, flying high above the forest on their way to feeding grounds, returning at dusk. Exploits cultivations and orchards where it may become a pest on oranges and corn. Feeds on fruit and seeds of many wild and cultivated plants, also flowers and leaf buds. Nests in holes in trees (e.g. abandoned woodpecker holes) or (rarely) in rock crevices.

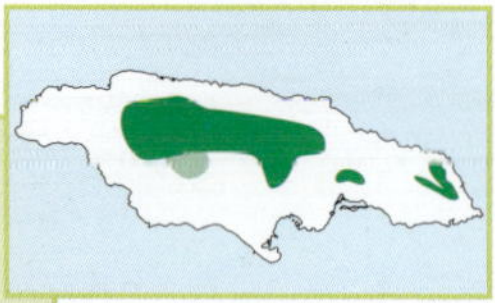

Range Endemic to Jamaica. Global conservation status: Vulnerable.
Status in Jamaica Fairly common but may be decreasing due to illegal hunting (for food and pest control), collecting for the pet trade and habitat destruction.

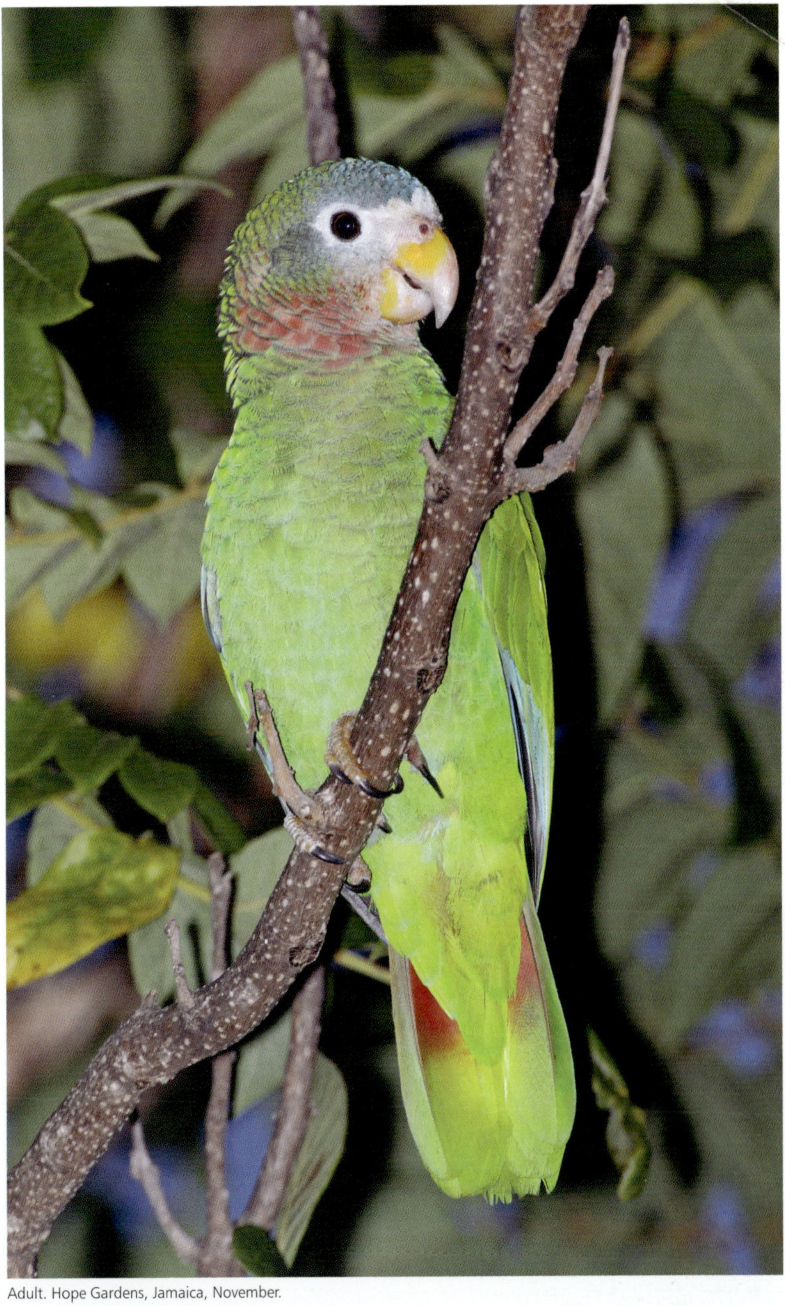

Adult. Hope Gardens, Jamaica, November.

Black-billed Parrot

Amazona agilis

Local name Black-bill.
Taxonomy Monotypic. Endemic to Jamaica.
Description L 26cm (10in). A plain green parrot. Bill is dark grey with a lighter tip. Bare skin around the eye is black. Iris is hazel. Ear-coverts are narrowly edged with black. Hind-neck feathers are edged dark grey, giving a scaly appearance. Primaries are black, with blue leading edges. The primary coverts are red, and are mainly visible in flight. Tail is short, square, and green; outer feathers are red basally and edged with blue. Legs are dark grey. Juvenile has green primary coverts.

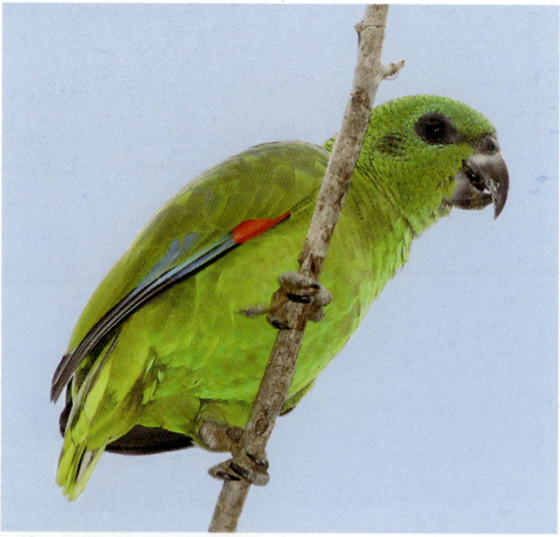

Adult. Cockpit Country, Jamaica, November.

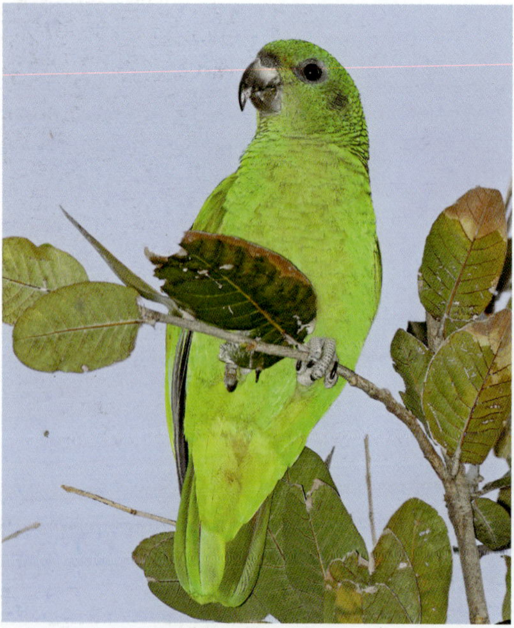

Adult. Hope Gardens, Jamaica, November.

Similar species Yellow-billed Parrot is a yellower green, and has a yellow bill, a pink throat and has no red in the wing; its wing beats are deeper and slower.
Voice Similar to Yellow-billed Parrot but tones are richer and more varied. In-flight calls are higher-pitched.
Habitat and behaviour Forests and woodlands on limestone, mainly at 100–1400m. Flocks leave roosting areas in the morning, flying high above the forest on their way to feeding grounds, returning at dusk. Flies with stiff, shallow, wingbeats. Nests (March–July) in holes in trees. Lays 2–3 eggs. Only the female incubates. She is fed by the male and does not leave the nest. Incubation takes 26 days. Fledging takes 55–60 days. The endemic Yellow Snake is a known predator of nestlings.

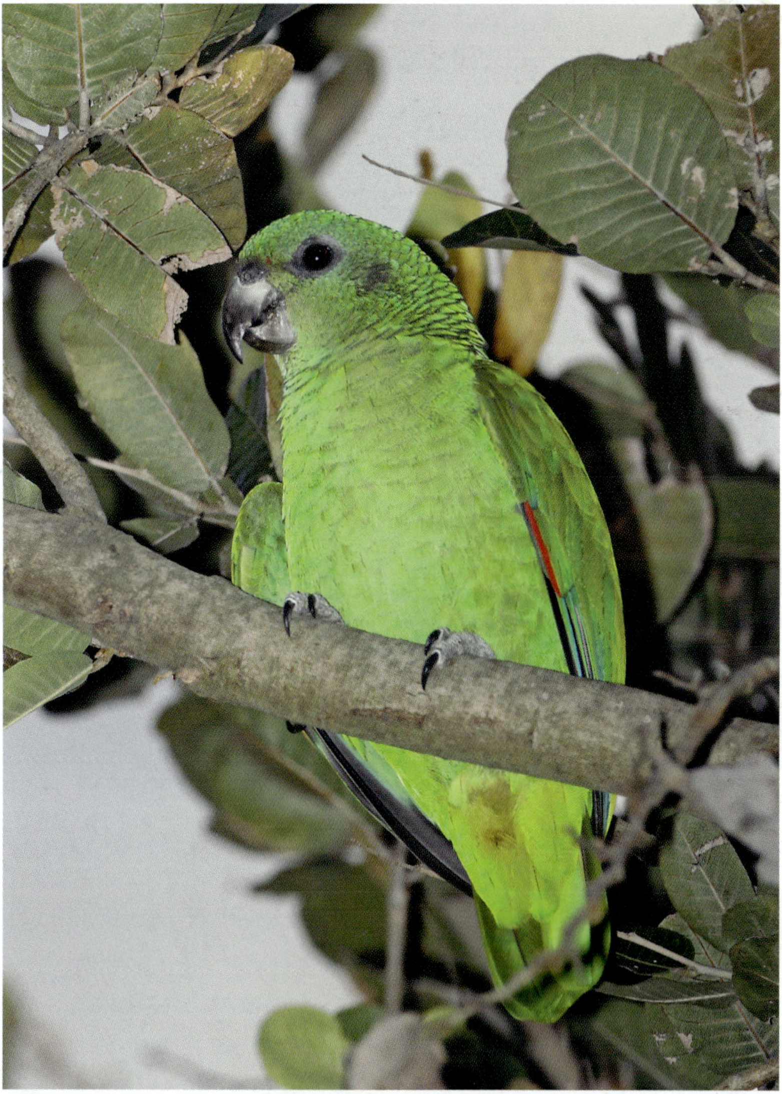

Adult. Note black bill and smudge on face and red primary coverts. Hope Gardens, Jamaica.

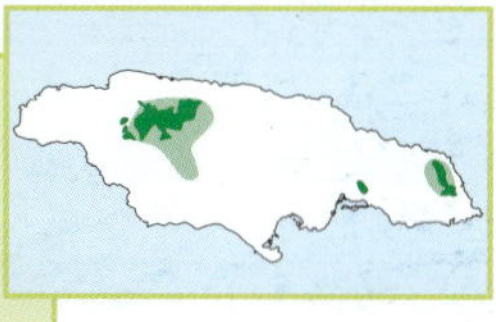

Range Endemic to Jamaica. Global conservation status: Vulnerable.
Status in Jamaica Fairly common but may be decreasing due to illegal hunting, collection for the pet trade and habitat destruction (including from bauxite mining).

Yellow-billed Cuckoo
Coccyzus americanus

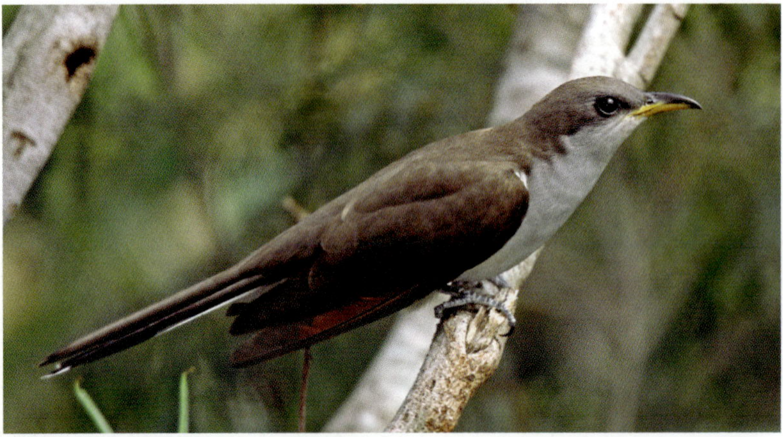

Adult. Note rufous flight feathers. Dominican Republic, April.

Taxonomy Monotypic.
Description L 28cm (11in). Slender. Bill slightly decurved; upper mandible yellow at base, with a dark tip; lower mandible yellow. Upperparts brown, underparts white. Flight feathers rufous. Tail long and graduated, all but two central feathers broadly tipped with white spots.
Similar species Mangrove Cuckoo lacks rufous in wings and has buffy underparts.
Voice Usually silent. Occasionally utters three or four guttural coos, quite unlike any other cuckoo in Jamaica.
Habitat and behaviour Mangroves, hedgerows and lowland dry scrub by the sea, chiefly on the south coast. Feeds on insects including caterpillars that it beats on a branch until only the skin remains which it swallows. Nests very rarely in Jamaica (May) in the lowlands in mangroves or cashaw trees.

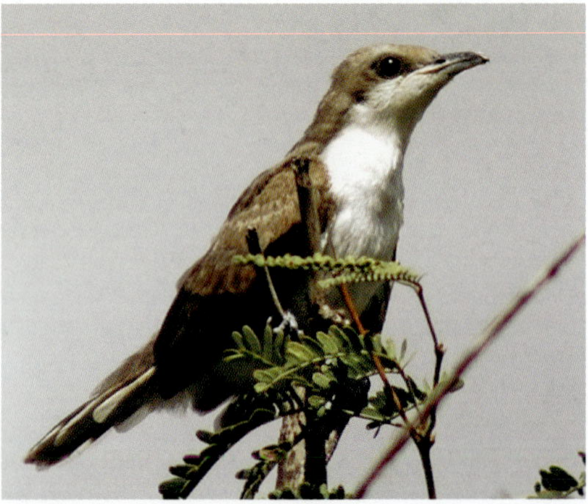

Juvenile. September.

Range Breeds in North and Central America, the Bahamas and Greater Antilles. Winters in South America.
Status in Jamaica Rare resident and passage migrant. More common in some years than in others.

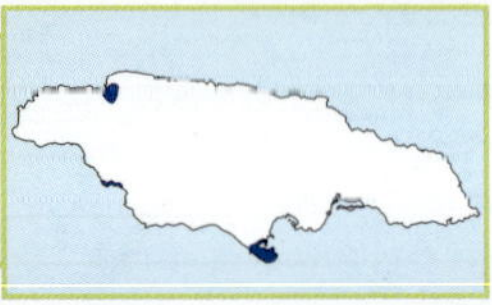

Mangrove Cuckoo
Coccyzus minor

Taxonomy Monotypic.
Description L 28cm (11in). Slender cuckoo with a broad, dark eye-line. Bill slightly decurved. Upper mandible is black. Lower mandible has a yellow base. Upperparts dark brownish grey, underparts light cinnamon. No rufous in wing. Tail is black, narrow, long and graduated; all but the two central feathers are broadly tipped with white spots. Juvenile has grey breast and cinnamon lower underparts.
Similar species Yellow-billed Cuckoo is very similar but has white underparts and rufous in the wings.
Voice Sometimes a short *kar-kar-kar*, but usually longer and ending *ca-ca-co-co-coa.*
Habitat and behaviour Usually seen in dense understorey or scrub, in low and mid-level dry limestone forest, coastal swamps, orchards and gardens from sea level to middle elevations. Rarely seen in mangroves, despite its name. Feeds on insects (including grasshoppers and mantises) found in hedges, scrub and woodlands from sea level to mid-levels. Nest (March–June) is a loose platform of sticks in shrubs or small trees.

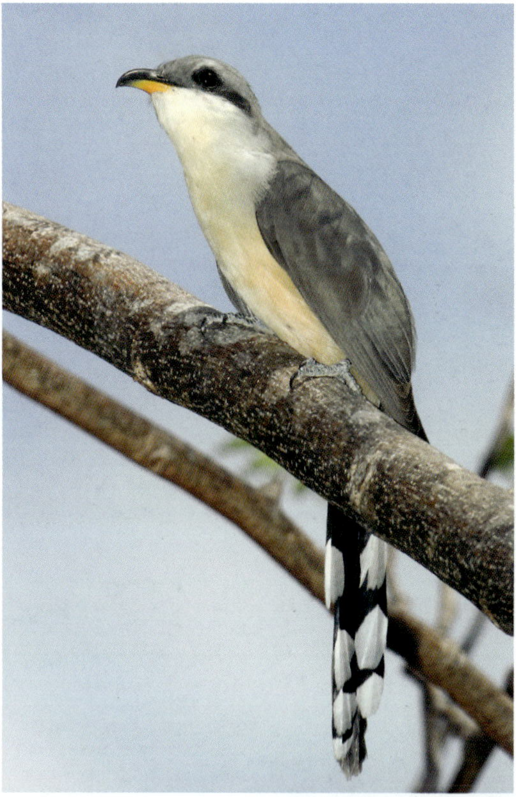

Adult. Cayman Islands, March.

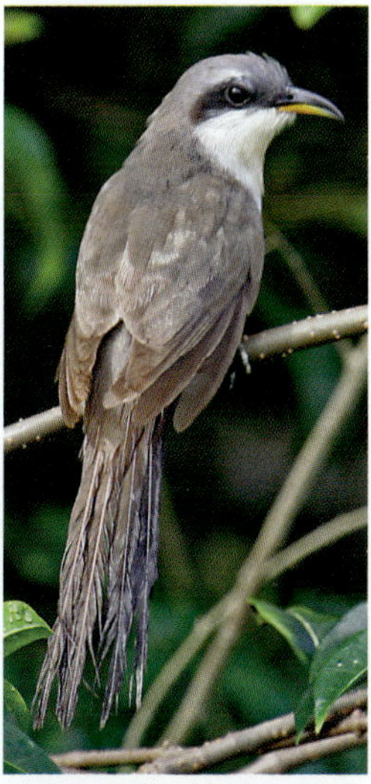

Adult. Puerto Rico, March.

Range Northern Florida and the West Indies to coast of northern Brazil.
Status in Jamaica Fairly common resident and migrant.

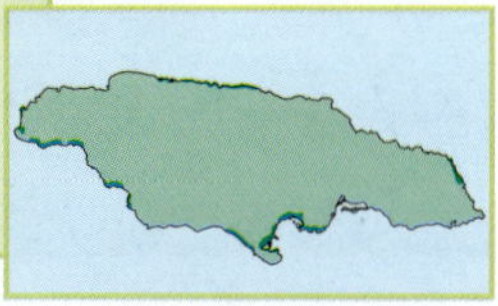

Jamaican Lizard Cuckoo

Coccyzus vetula

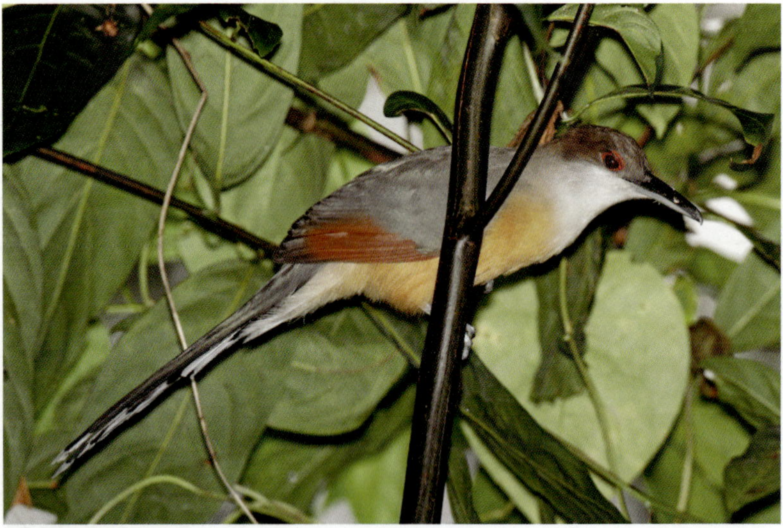

Adult. Ecclesdown, Jamaica, November.

Local name Old Woman Bird.
Taxonomy Monotypic. Endemic to Jamaica.
Description L 83cm (15in). Medium-sized, slender cuckoo. Long straight bill; upper mandible dark grey, lower mandible light grey with dark tip. Iris hazel. Eye surrounded by red orbital skin. Head and back reddish brown. Chin is white, throat is greyish. Breast to vent yellow-ochre fading gradually to cream on undertail-coverts. Wings short, grey with rufous, grey-tipped primaries. Back and central tail feathers light grey. Other tail feathers dark grey, graduated; all broadly tipped with white. Legs light grey.
Similar species Chestnut-bellied Cuckoo lacks red orbital skin and is larger, with a decurved bill. Jamaican Lizard Cuckoo more likely to be found in lower canopy or shrub layer than Chestnut-bellied Cuckoo.

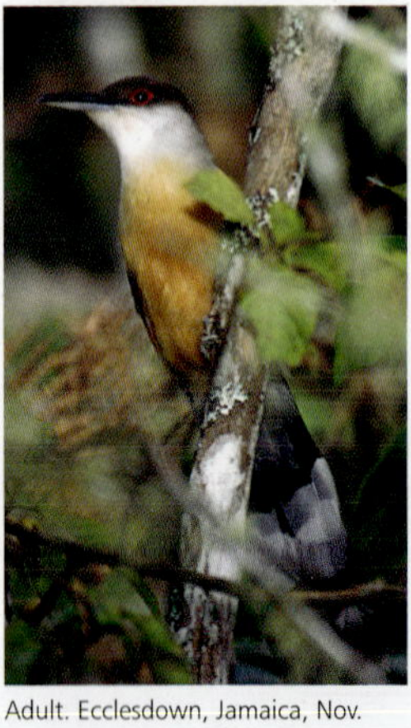

Adult. Ecclesdown, Jamaica, Nov.

Voice Rapid, low, trailing *cak-cak-cak-ka-ka-ka-k-k*, like a machine gun, faster than call of Chestnut-bellied Cuckoo.
Habitat and behaviour Widespread in wooded areas. Often perches erect or horizontal (with head lower than feet) or runs along branches like a rat. Feeds on lizards, caterpillars, large grasshoppers and nestlings of other birds. Nests (March–June) in wet mountain areas. Nest is a shallow platform of criss-crossed twigs lined with leaves, placed in a tangle of branches.

Range Endemic to Jamaica.
Status in Jamaica Fairly common resident.

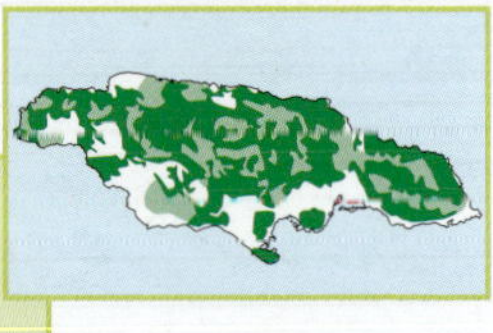

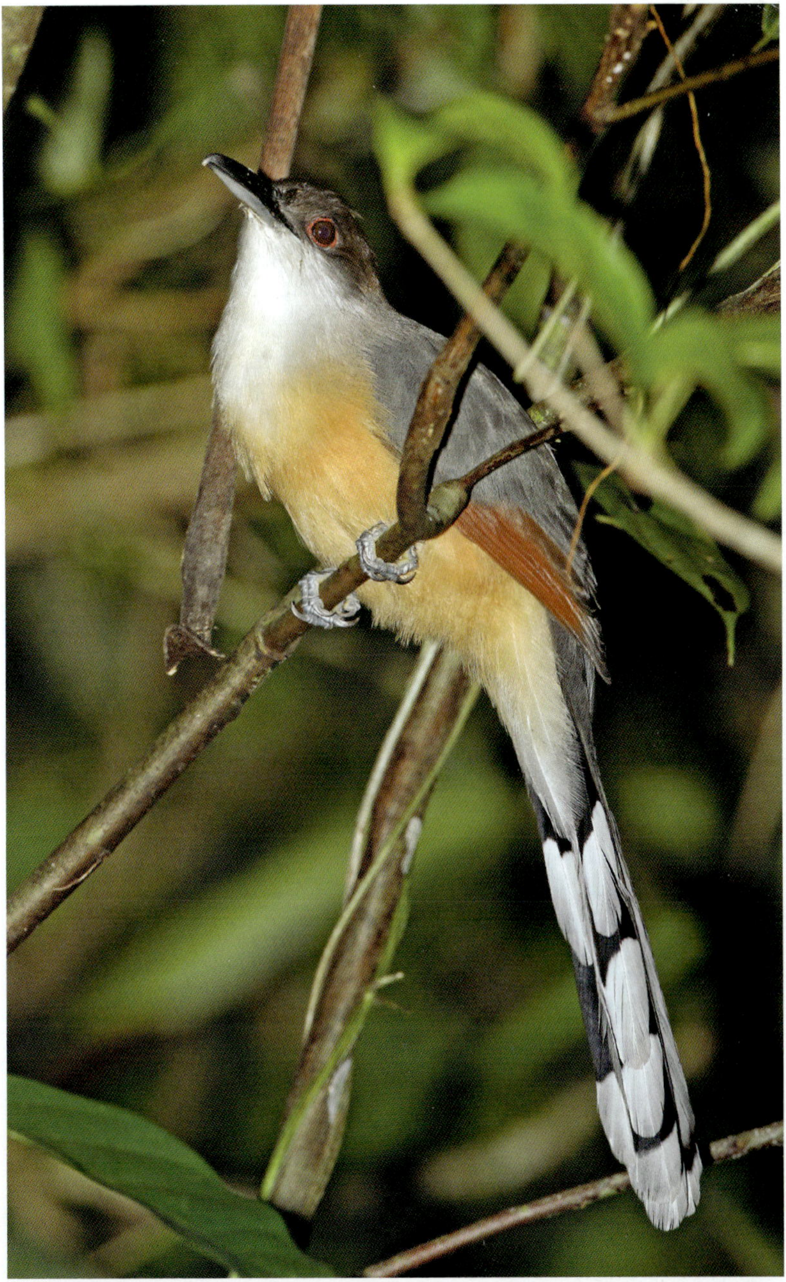

Adult. Note straight bill, yellow-ochre belly and red orbital skin. Ecclesdown, Jamaica, November.

Chestnut-bellied Cuckoo
Coccyzus pluvialis

Local name Old Man Bird.
Taxonomy Monotypic. Endemic to Jamaica.
Description L 48cm (19in). Large cuckoo with a thick, dark grey decurved bill. Throat creamy white fading into pale grey on breast, contrasting with chestnut belly and underparts. Dark grey tail is broad and long; tail feathers are tipped with large white spots. Legs are grey.
Similar species Jamaican Lizard Cuckoo has a red eye ring, straight bill and rufous wing-patches.
Voice Most frequently heard April–June. Hoarse *quak-quak-quak-ak-ak-ak-ak-ak,* slow at first and accelerating towards the end, is often the first indication of the presence of this bird. Also a variety of growls and grumbles, like a hungry stomach.

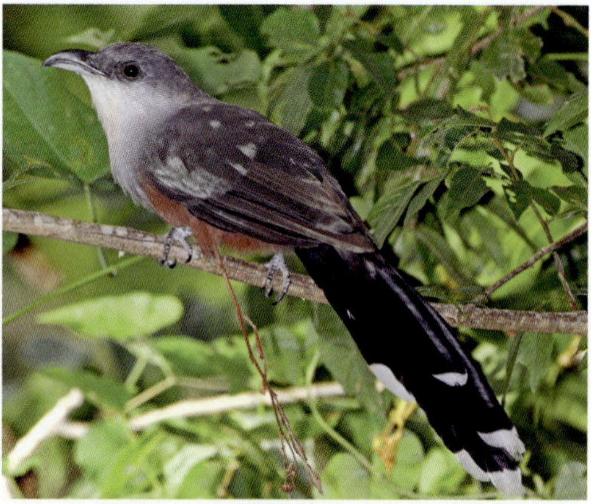
Adult. Note dark, decurved bill and chestnut belly. Port Royal Mountains, Nov.

Habitat and behaviour Moist woodland, wooded cultivation or open thickets in the hills and mountains. Descends to lower elevations in winter but only found near the coast in the wettest places. Runs along branches like a large rat or sails silently on extended wings from one tree to another. Feeds on lizards, mice, insects, caterpillars, nestlings and eggs. Other birds often mob it. Nest (March–June) is an untidy platform made of sticks in the middle or upper canopy of tall mature trees.

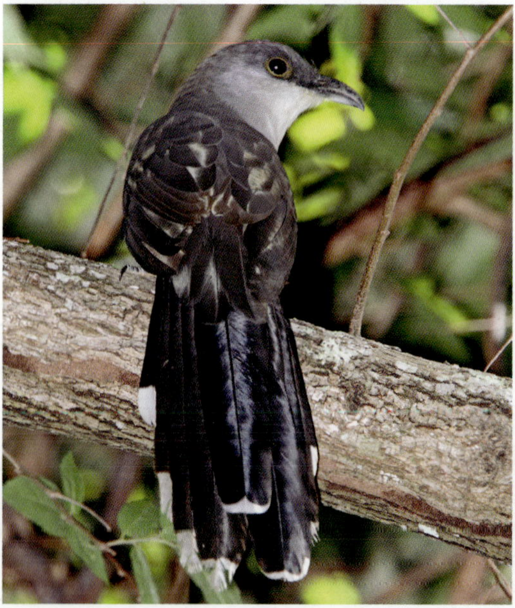
Adult. Port Royal Mountains, Jamaica, November.

Range Endemic to Jamaica.
Status in Jamaica Fairly common and widespread.

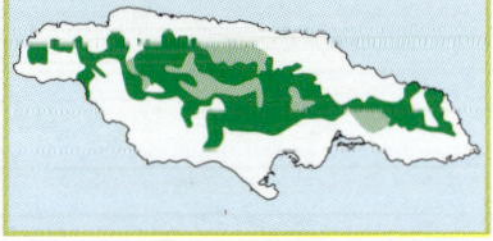

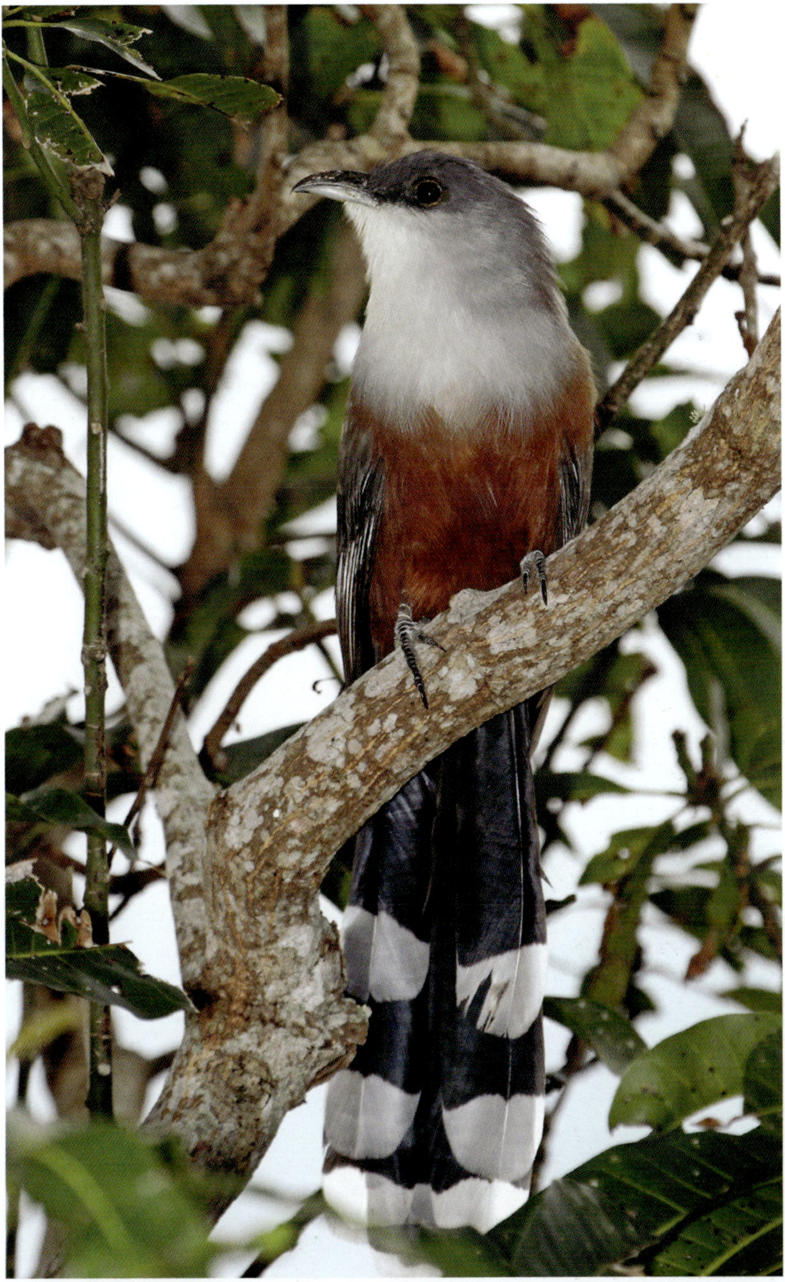

Adult. Port Royal Mountains, Jamaica, November.

Smooth-billed Ani

Crotophaga ani

Local name Tick Bird, Savanna Blackbird.
Taxonomy Monotypic.
Description L 35cm (14in). Medium-sized, glossy and iridescent black bird with an enormous thick bill that rises above the top of the head in a spade-like ridge. Long, rounded tail is graduated and flattened. Juvenile browner and more scaly than adult, with bill level with top of head.
Similar species Greater Antillean Grackle has a pointed bill, yellow iris and keel-shaped tail.
Voice A high-pitched *anee-eeek* rising, repeated two or three times, sounding like its name. Also growls and clucks.
Habitat and behaviour Found in cultivations, pastures, gardens, roadsides and wet meadows. One of the most common birds in Jamaica, seen flying across country roads or in scrubby vegetation. Often occurs in flocks, one bird acting as a sentinel. Walks rather than hops when on the ground. Often appears to overbalance on landing. Groups bathe noisily in bird-baths. Feeds on insects (especially termites), lizards, frogs, nestlings of other birds and fruit (including Fiddlewood, Pudding Withe and oranges). Before the arrival of Cattle Egret on Jamaica it was the most likely bird to be seen in association with cattle. Nests year-round (mainly April–June) in a large communal nest, in bamboo, mangroves or overgrown creepers such as Bougainvillea, from sea level to the mountains. Eggs are chalky blue. Clutches are separated by layers of leaves but usually only the top layer hatches.

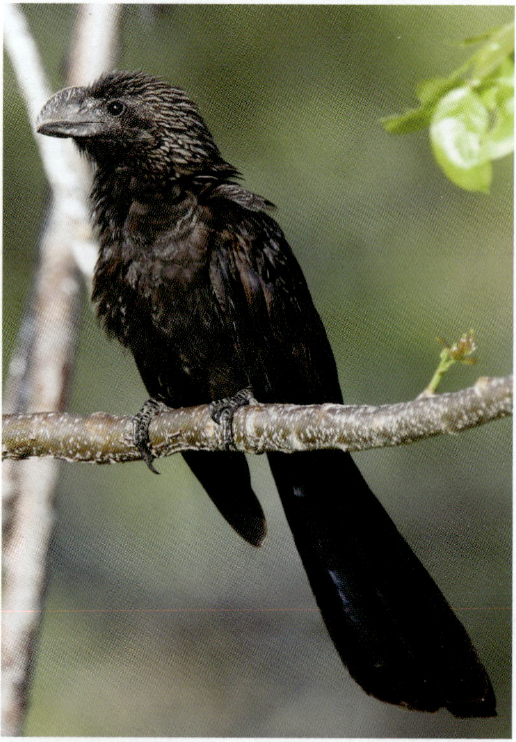
Adult. Cayman Islands, April.

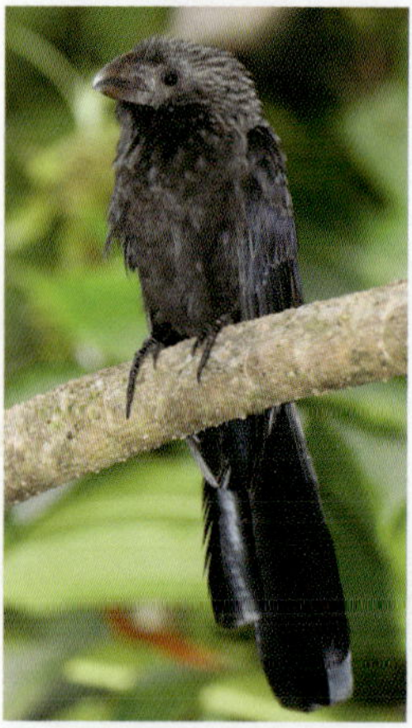
Adult. Ecclesdown, Jamaica.

Range Most Caribbean islands, Bahamas, mainland America.
Status in Jamaica Common resident.

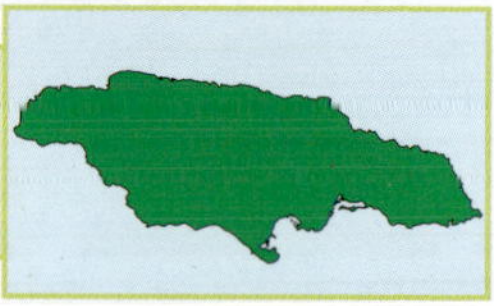

Barn Owl
Tyto alba

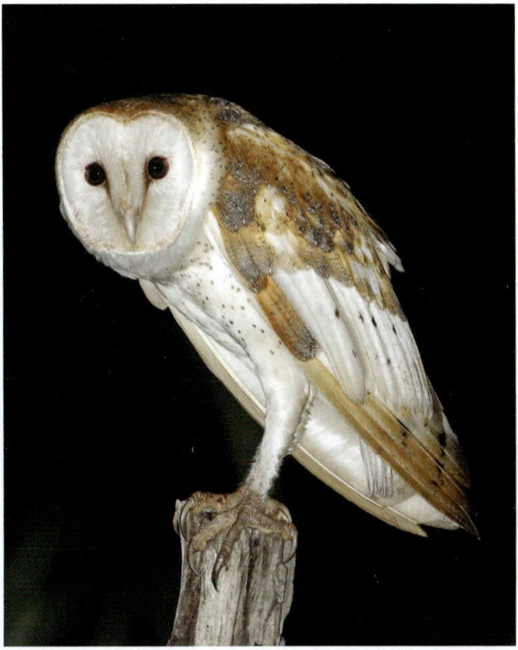
Adult. Black River, Jamaica, November.

Local name Patoo, Screech Owl, Scritch Owl, White Owl
Taxonomy Polytypic (28). *T. a. furcata* a West Indian endemic.
Description L male (41cm, 16in); female (43cm, 17in). Large pale owl with heart-shaped white facial disc, outlined with brown-tipped feathers. Iris dark brown. Upper mandible is hooked and horn-coloured at the tip. Head and back cinnamon, mottled brownish. Wing-coverts same colour as the back, but leading edge of outer primaries washed rufous, with 1–3 dark spots in centre of each (number is age-dependent). Underparts white with grey spots on breast. Tail cinnamon above, white underneath, with darker brown speckled bands. Adult male smaller and paler than female. Juvenile browner than female and facial disk is outlined with buffy-tipped feathers; head, back to tail-coverts darker than in adults; underparts white, washed with cinnamon on breast and heavily spotted dark grey; wings white, with dark spots in centre of feathers. Nestling white; there are two downy stages before fledging.
Similar species Jamaican Owl is brown and smaller.
Voice *Creek, creek, creek* and a harsh shriek that is said to sound like the tearing of cloth.
Habitat and behaviour Found from sea level to the mountains, in open or partly open situations, often near human habitations, including large cities. This is the large white owl commonly seen flying or perched on fence posts at dusk, or heard shrieking at night. Feeds chiefly on rats, mice, sometimes birds. Nest (year-round but chiefly July–December) is a shallow hollow without nest material, placed on a ledge in a cave, sink-hole, tree cavity, large barn or church. Lays three eggs.

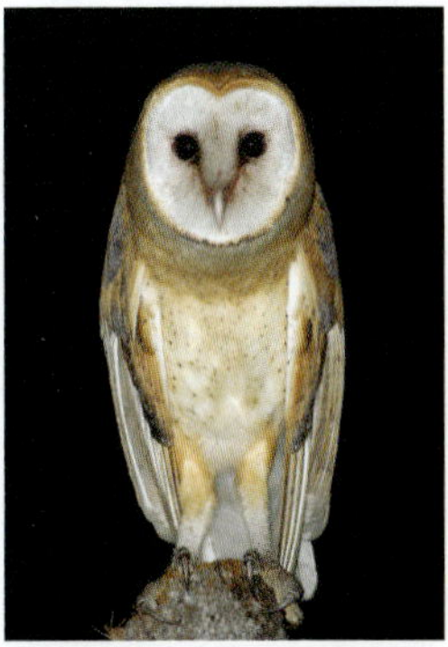
Juvenile. Cayman Islands, April.

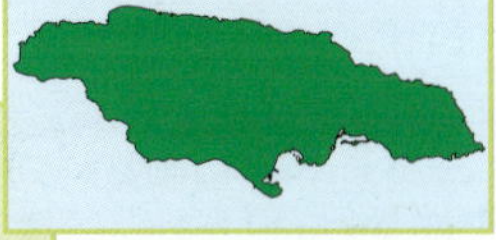

Range *T. a. furcata* occurs on Cuba, Isle of Pines, Jamaica and (rarely) the Cayman Islands. Other races worldwide.
Status in Jamaica Common resident, despite some persecution.

Jamaican Owl
Pseudoscops grammicus

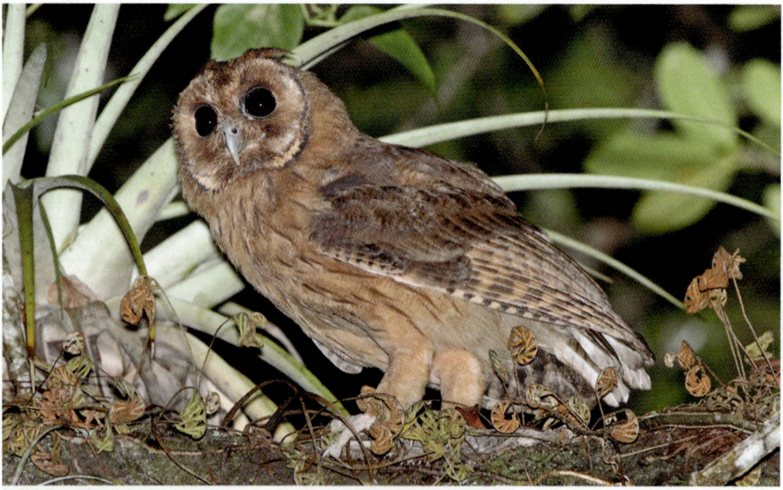

Adult. Note the ear-tufts are lowered. Port Antonio, Jamaica, November.

Local name Patoo, Brown Owl.
Taxonomy Monotypic. Endemic to Jamaica.
Description L 29cm (11in). Small brown owl with dark brown and black flecks in centre of each feather. Feathers on fore-crown are elongated into erectile ear-like tufts. Back and wings are darker than underparts. Wings and tail barred and mottled with black. Bill light bluish grey. Large, protruding eyes have hazel irises that reflect red in a spotlight at night. A blue nictitating membrane may be drawn across the eye in daytime. Legs and feet tawny. Juvenile like adult but lighter brown. Newly-hatched nestling is downy white, moulting to downy buffy brown before fledging. Newly fledged birds retain some downy feathers.
Similar species Barn Owl is much larger and appears white.
Voice A hoarse, throaty *whow*, sometimes *to-whoo*, usually heard at dusk and just before dawn. First calls before it leaves the roost. Voice of female lower-pitched than male. Juvenile begging call a piercing *wheee-eee*. When chick first hatches it calls softly but as it ages the call gets much louder and more frequent. Young continue to use this loud call up to around 12 months to attract parents.
Habitat and behaviour Widespread at all elevations wherever there are large trees, e.g. at forest edges in gardens, or on golf courses. Usually associated with large, mature, trees e.g. West Indian Cedar. Often found near houses and in open spaces with isolated trees. Daytime roosts are always well shaded but may be in dense tangled vines or in any part of the canopy low down or high up, close to the trunk or towards the end of branches. Often uses the same roost for months at a time. Feeds on mice, lizards, tree frogs, beetles, grasshoppers and other insects, and occasionally bats and small birds. Nests (December–March) in holes in trees, under bromeliads, or in overgrown vines on trees. Usually one chick, occasionally two. Parents feed young for 10–12 months. When ready to nest again, the parents drive the young out of their territory.

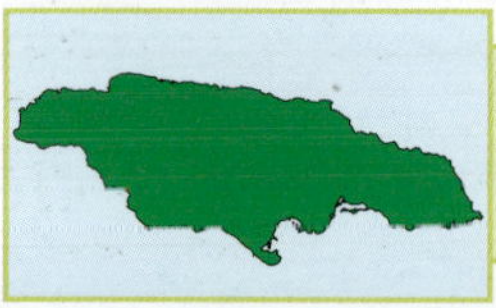

Range Endemic to Jamaica.
Status in Jamaica Common and widespread resident, but hard to find.

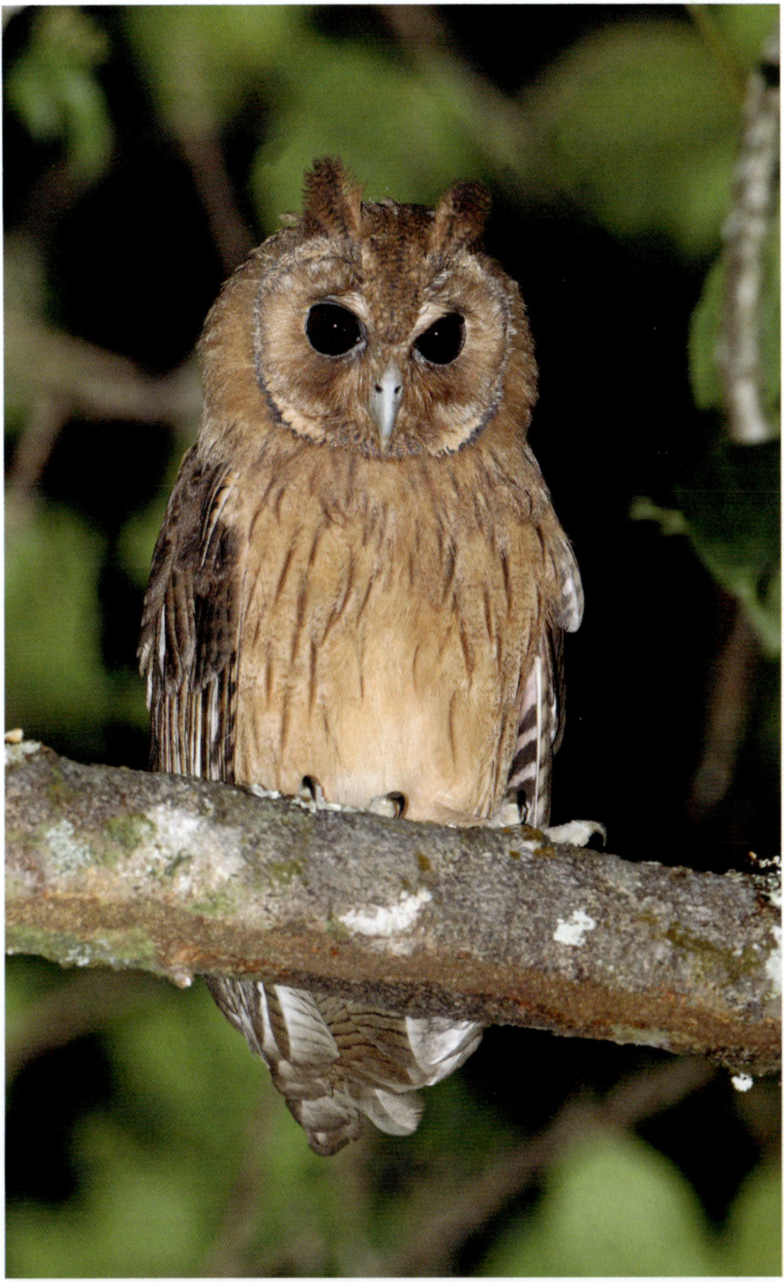

Adult. Note the raised ear-tufts. Port Antonio, Jamaica, November.

Northern Potoo
Nyctibius jamaicensis

Taxonomy Polytypic (5). *N .j. jamaicensis* is an endemic race.
Description L 41cm (16in). A superbly cryptic bird that sits erect, resembling a piece of dead wood. Overall dark brown and cinnamon, streaked and mottled with cream. Wings and tail very long, irregularly barred greyish cream and dark brown. Black hooked bill is small but the gape is enormous. Iris yellow, reflecting red in a spotlight at night. Nestling white with dark brown feather-shafts; Juvenile brown, much paler than adult
Similar species None. The name 'patoo' is used in Jamaica for owls.
Voice *Qwaa-a-a-a-a qwa-qwa-qwa-qwa*, also *wow*.
Habitat and behaviour Roosts during the day on lightly shaded snags at the edges of clearings in forests

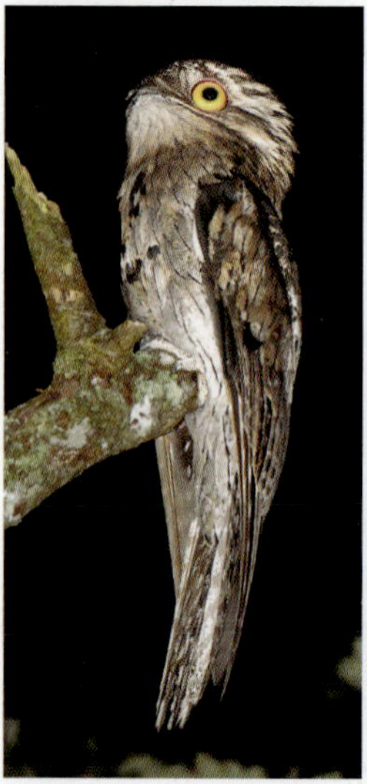

Adult. Marshall's Pen, Jamaica, Nov.

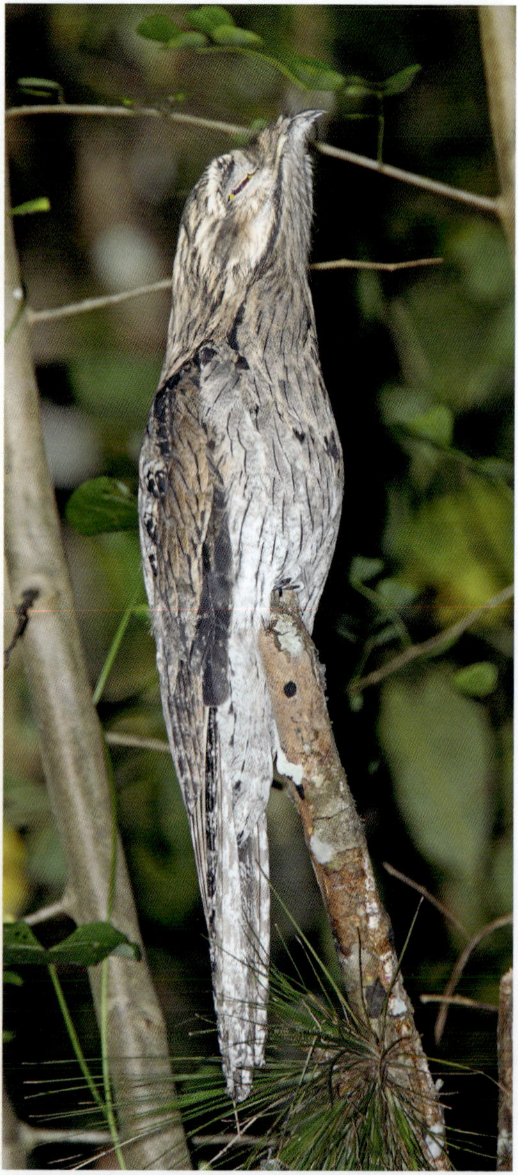

Adult. Note the cryptic 'alarm' pose. Cockpit Country, Jamaica, November.

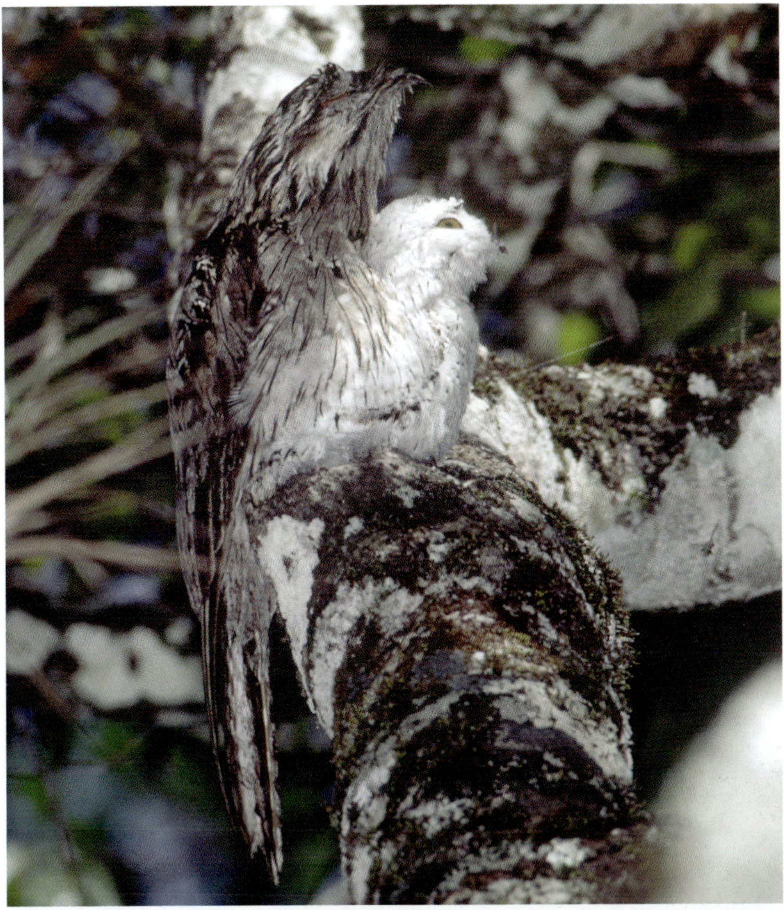

Adult with downy chick. Cockpit Country, Jamaica, May.

and woodlands. Often found beside pastures with wooded hills, on wooded golf courses, in mangroves or in trees on open lots in towns. Generally perches with head horizontal but if alarmed may snap abruptly into the cryptic pose, with bill elevated skyward. Otherwise head movements are imperceptibly slow. Uses the same daytime roost regularly for months or years. Feeds mainly on scarab beetles (or tumblebugs), which it catches by hawking from a favourite feeding perch. Nest (February–November) is a shallow hollow on a bare broken limb or branch. Lays one egg. Both parents incubate the egg and feed the chick, which sits beside or under the adult as long as possible. When alarmed the chick gradually melds into the parent's feathers, until totally concealed.

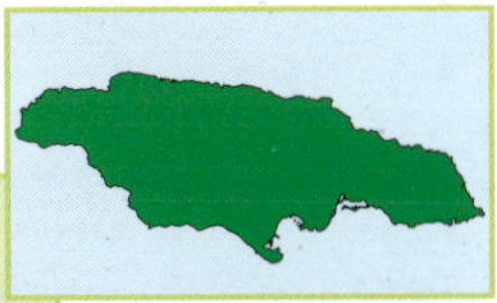

Range *N .j. jamaicensis* Jamaica. Other races on Hispaniola, Mexico, Central and South America, Trinidad and Tobago.
Status in Jamaica Common resident.

Antillean Nighthawk
Chordeiles gundlachii

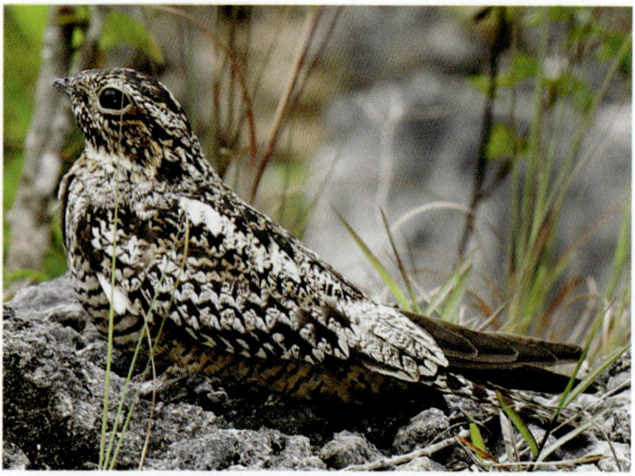
Adult. Mandeville, Jamaica, June.

Local name Gimme-Me-Bit
Taxonomy Polytypic (2). Race on Jamaica is *C. g. gundlachii.*
Description L 23cm (9in). Usually seen in flight when it appears as a slender, brown-mottled, hawk-shaped bird with a tiny dark bill and a long and slightly forked tail. Both sexes have a white band across the primaries in the dark, pointed wing. When perched, the white bar on the throat and white patch on the wings are prominent. The wing-tips reach the tail-tip. Male has a white band across the tail and a white throat. Female has a smaller, duller bar on throat. Juvenile resembles the female but lacks the white wing and throat marks.
Similar species Swallows and martins are smaller and lack the white bands across wings. Common Nighthawk best distinguished by call.
Voice Repeated *gimme-me-bit* from which its local name is derived.

Habitat and behaviour Urban, suburban, rural open or wooded areas island-wide. Usually seen and heard (in groups or singly) at or shortly before dusk. In well-lit urban areas they often continue to call throughout the night. The mating display (June) includes a booming dive from a great height. These birds nest (April–September) on open ground in a wide variety of habitats – on, for example, sandy beaches, grassy commons, on exposed earth in bauxite mines, on pine needles, or in cities and towns on flat, shingly rooftops. Lays two white eggs. Does not build a nest.

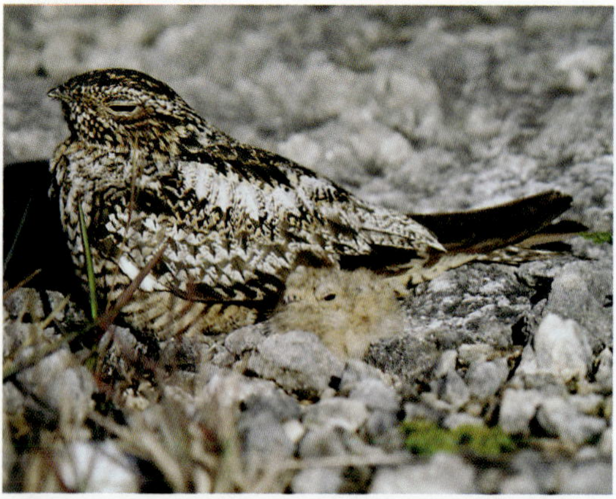
Adult with a chick. Mandeville, Jamaica, June.

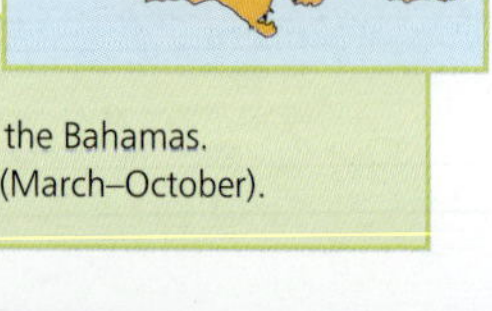

Range *C. g. gundlachii* also breeds on Cuba, Hispaniola and Puerto Rico; another race breeds in southeastern United States and the Bahamas.
Status in Jamaica Very common and widespread summer resident (March–October).

Chuck Will's Widow

Caprimulgus carolinensis

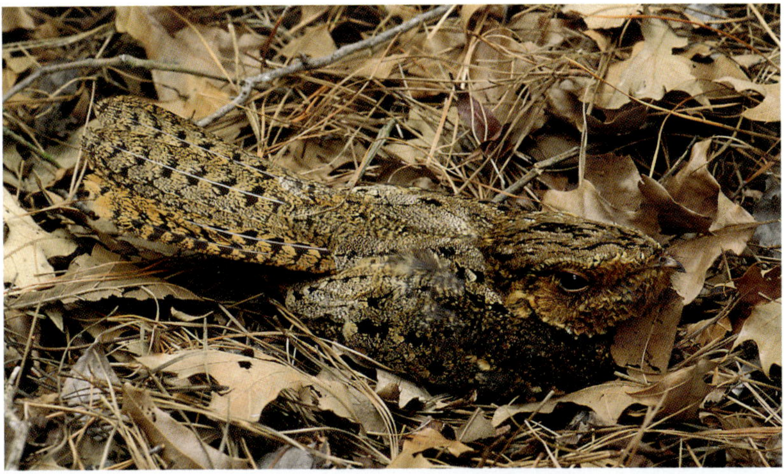

Adult roosting. September.

Taxonomy Monotypic.
Description L 28cm (11in). Cryptically coloured nightjar. Male buff-brown with a whitish throat and outer three tail feathers white on inner vane. Female resembles male but lacks white in tail.
Similar species Antillean Nighthawk is smaller, has white on wing and is unlikely to be in Jamaica in the winter.
Voice Named after the call it makes, but in Jamaica it is usually silent. Prior to departure in March it sometimes calls – *will's widow*.
Habitat and behaviour Low and mid-level woodland. Crepuscular. Roosts in daytime, usually sitting lengthwise on a limb or on the ground.

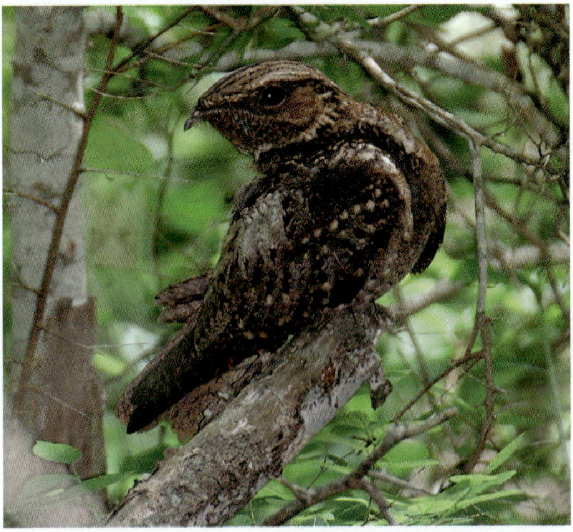

Adult perching. April.

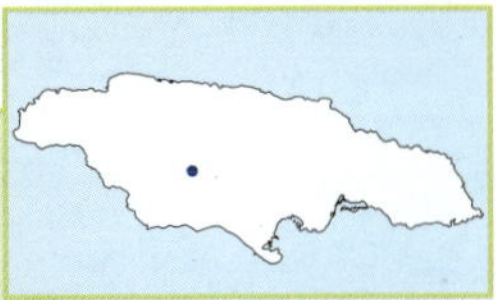

Range Breeds in North America. Winters from Texas to northern South America via Central America and the West Indies.
Status in Jamaica Uncommon winter visitor and passage migrant (October–March).

Black Swift
Cypseloides niger

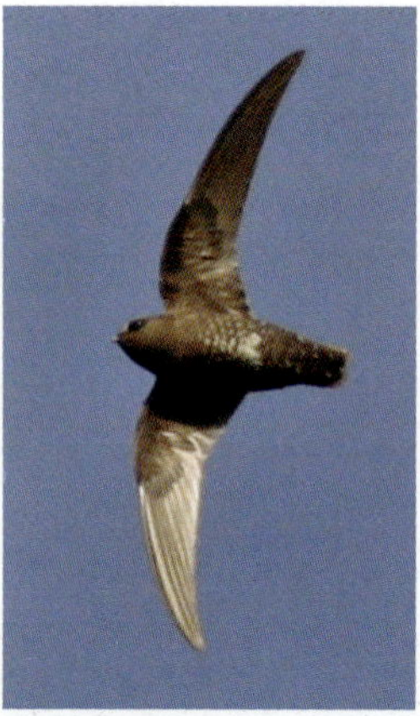

Adult *C. n. borealis*. California, July.

Local name Black Swallow.
Taxonomy Polytypic (3). Race in Jamaica is *C. n. niger.*
Description L 18cm (7in). Black with a slightly notched tail and long, pointed wings, held slightly below horizontal in flight. White on forecrown visible at close range.
Similar species Collared Swifts are larger with white collars and a more gliding flight.
Voice Clear *t-sip, t-sip*. Often detected by calls.
Habitat and behaviour Flocks range widely, often appearing in advance of rain clouds; feeds on aerial insects. Nests in clefts in rock faces near waterfalls.

Range *C. n. niger* West Indies, Trinidad, Guyana. Other races in North and Central America.
Status in Jamaica Uncommon resident.

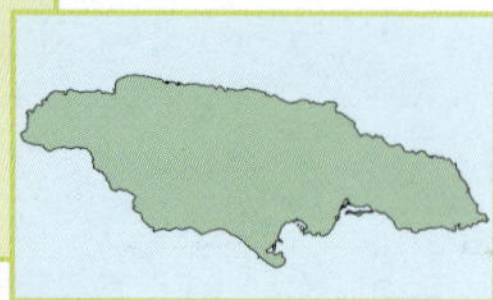

White-collared Swift
Streptoprocne zonaris

Local name Ringed Gowrie
Taxonomy Polytypic (9). *S. z. pallidifrons* is a West Indian endemic.
Description L 22cm (9in). Large black swift with a white collar and slightly forked tail. In flight looks scimitar-shaped.
Similar species Black Swift smaller, lacks the white collar. Antillean Nighthawk brown with white wing-patches.
Voice A loud shrill *wee-eet, wee-eet*, but often silent.
Habitat and behaviour Montane forest and interior valleys in the mountains; descends to the lowlands on cloudy days. Flocks of up to 50 occur. Flight alternates between flaps and glides. Holds wings stiffly below body when gliding. Nests (May–June) in clefts in steep rock faces.

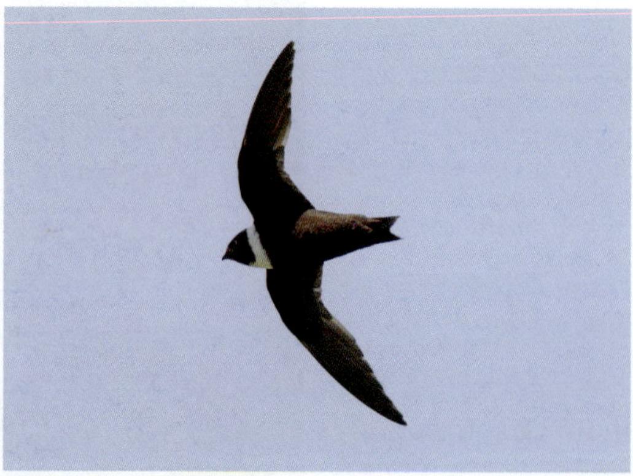

Adult. Dominican Republic, May.

Range *S. z. pallidifrons* Jamaica, Cuba and Hispaniola and the Lesser Antilles. Other races Central and South America.
Status in Jamaica A fairly common resident.

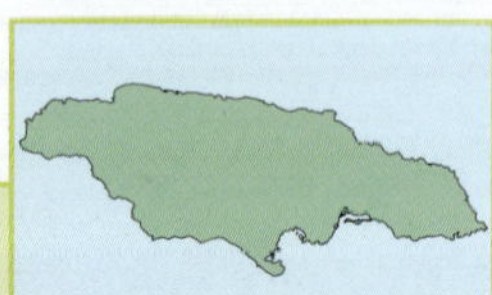

Antillean Palm Swift
Tachornis phoenicobia

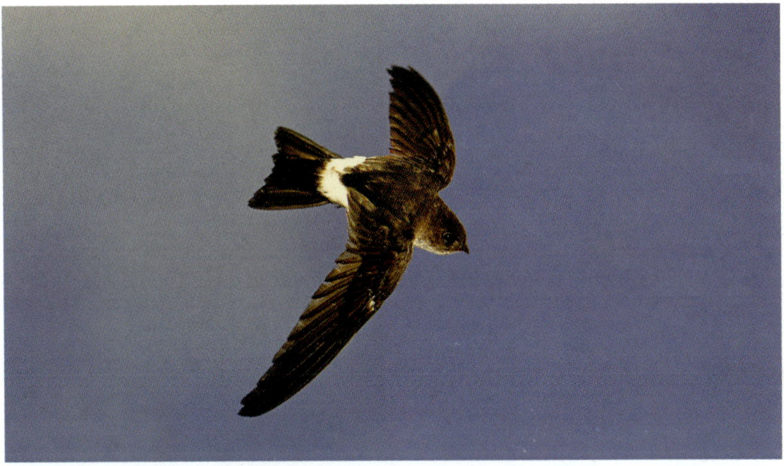

Adult. Dominican Republic, May.

Taxonomy Polytypic (2). *T. p. phoenicobia* is endemic to the West Indies.
Description L 10cm. (4in). Tiny. Adult is black with white throat and centre of belly and a conspicuous white rump. Bill is black and small with a wide gape. Tail is forked. Juvenile is buffy below.
Similar species None.
Voice A twittering.
Habitat and behaviour Lowlands; common near human habitations, over golf courses, dry swamps and canefields wherever there are mature native thatch palms in the vicinity. Found in the mountains in hot weather. Flight is very active and batlike, alternating between rapid wing beats and gliding with wings held stiffly below body. Wheels, dives and twists from side to side. Nests (year-round) colonially in mature thatch palms. Nests are made of leaves and feathers glued together with saliva under the dead fronds.

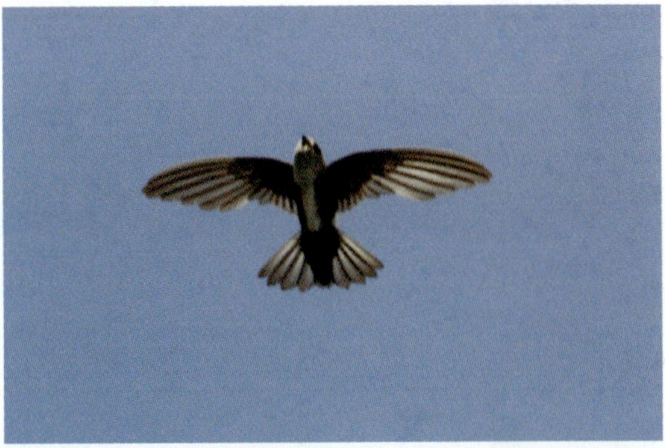

Adult. Dominican Republic, November.

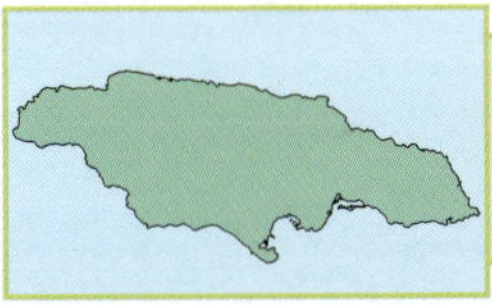

Range *T. p. phoenicobia* occurs on Jamaica and Hispaniola. Other race in Cuba.
Status in Jamaica Very common resident but has declined since Hurricanes Ivan and Dean in 2003 and 2007 respectively.

Jamaican Mango
Anthracothorax mango

Local name Mango Hummingbird.
Taxonomy Monotypic. Endemic to Jamaica.
Description L 14cm (5in). The largest and darkest of Jamaica's hummingbirds, with a distinctive long, black, slightly decurved bill. Adult male has greeny-bronze head and back, washed with magenta on the upper back and wing-coverts. Iris is dark brown. Sides of the head and neck flash with iridescent magenta in sunlight. Underparts are black. The two central tail feathers are black, the rest of the tail feathers are an iridescent, rusty-purple-blue, appearing red in flight, and tipped with black scallops. Undertail-coverts bluish-green. There are sometimes two white puffs on either side of the vent. Female is paler than the male. Juvenile has a blue gorget and brownish underparts.

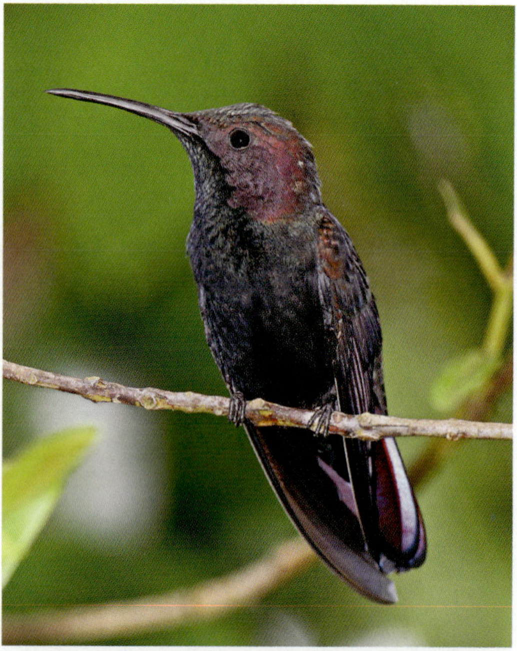

Adult male. Rocklands, Jamaica, November.

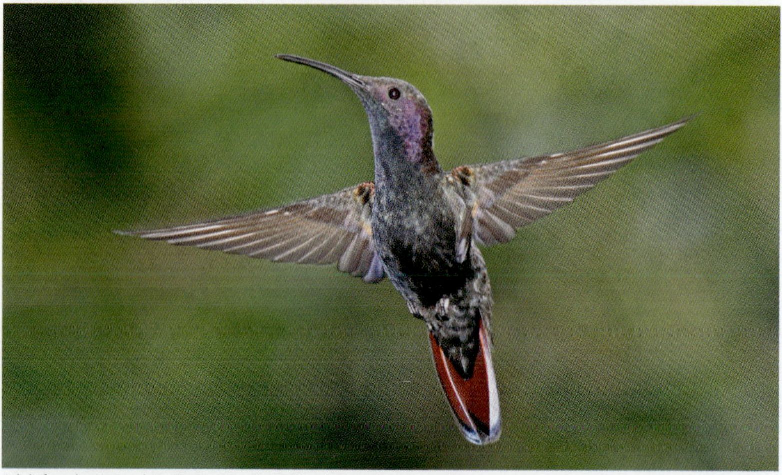

Adult female. Note white tips to the tail. Rocklands, Jamaica, November.

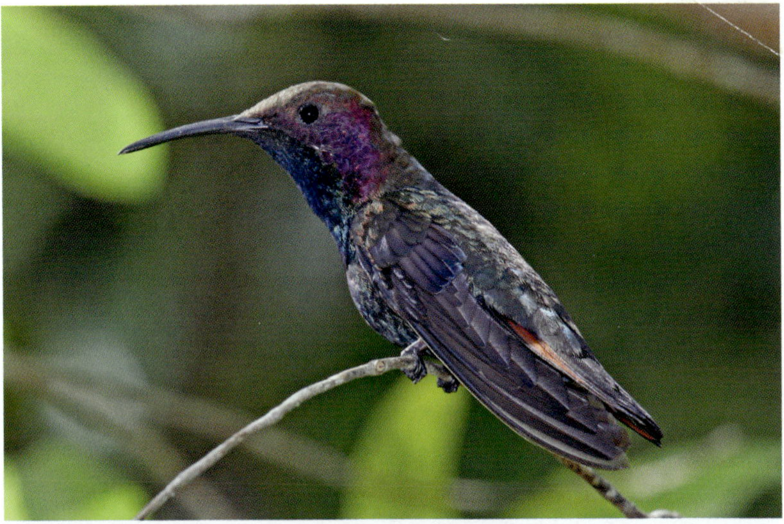

Adult male. Rocklands, Jamaica, May.

Similar species Black-billed Streamertail (when tail feathers have been moulted) is smaller, green underneath and has a short black tail.

Voice A raspy, tuneless *chi-chi-chi-chi-chi* (like *chip* note of Black-throated Blue Warbler but louder) or a single loud *chip*.

Habitat and behaviour Abundant in open, arid areas in the lowlands and near the coast. Less common in upland areas. Feeds on nectar of flowers of many trees and plants such as banana, coffee, cactus and Bauhinia; will also readily visit hummingbird feeding stations. Also eats small insects such as fruit flies. Nest (year-round) is a deep cup made of down from seeds (e.g. Tillandsia), held together with cobwebs and decorated with lichen. Nest placed above eye-level, saddled on branches of trees such as Casuarina, mangroves or Cashaw.

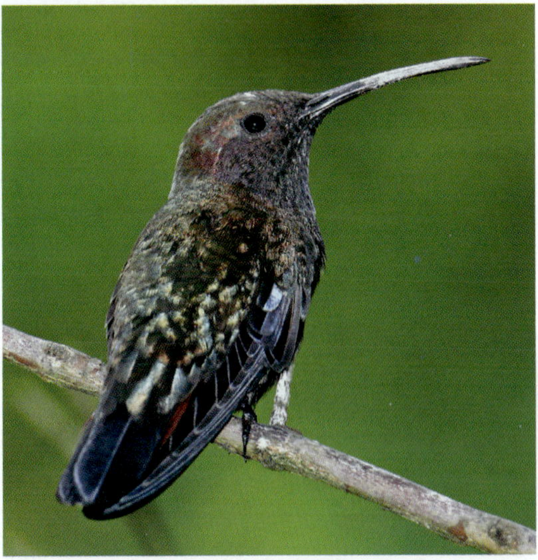

Adult male. Note the pollen on the bill. Ecclesdown, Jamaica, November.

Range Endemic to Jamaica.
Status in Jamaica Common resident.

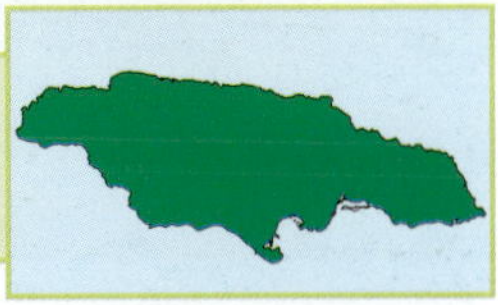

Red-billed Streamertail
Trochilus polytmus

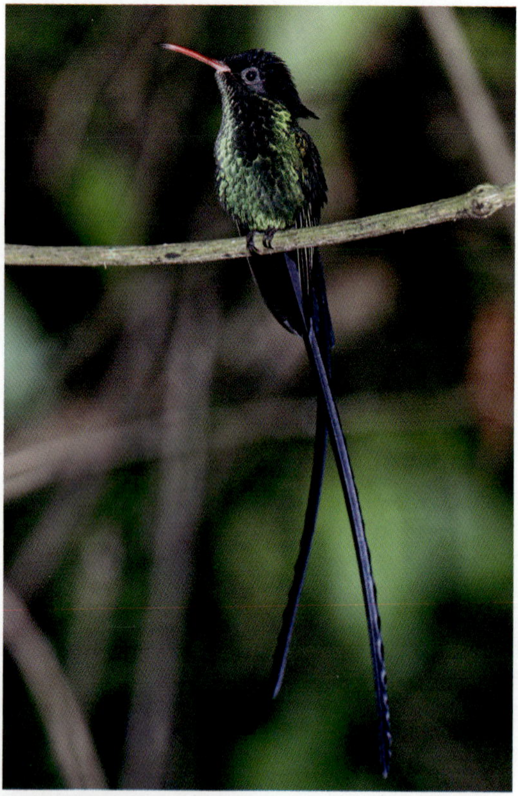

Adult male. Rocklands, Jamaica, November.

Local names Doctorbird, Scissors Tail, Longtail Hummingbird, God Bird.
Taxonomy Endemic genus and species.
Description Male L (excluding streamer) 13cm (5in); including streamer 31cm (12in). Female: L 8cm (3in). Male head black, with lateral crown feathers and ear-coverts elongated beyond nape, forming a crest that is extended to the sides during mating display. Bill red with a black tip and sides, the black decreasing with age. Body bright iridescent emerald green, yellower on back when faded. Wings black edged with green; the 10 tail feathers are black, shot with green. The next to outermost tail feathers are elongated to form 'streamers', which are scalloped and fluted on the inside to create a whining hum in flight. The streamers are often crossed at the base when the bird is perched. Adult male in moult lacks the streamers. Female head green washed grey. Upper mandible is reddish black, the amount of red on the lower mandible increasing with age. Back yellowish-green; underparts white, lightly spotted green at breast-sides. Lacks streamers, but outer three tail feathers are broadly tipped with white. Immature male has black upper mandibles, which become red from the centre as the bird ages; underparts green, and streamers do not develop until second year. Immature female similar to adult female but has brownish crown and

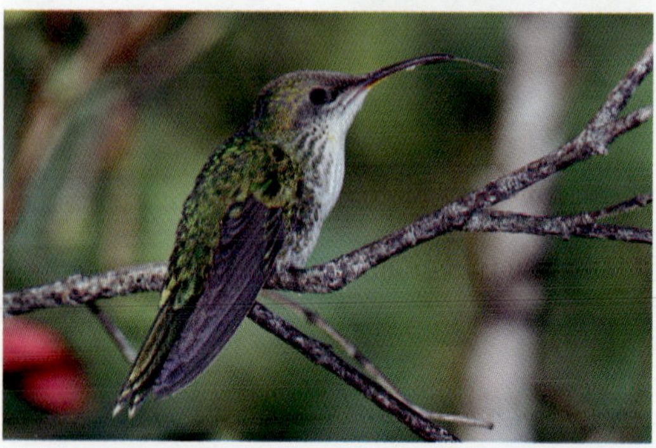

Adult female. Note white underparts. Port Royal Mountains, Jamaica, Nov.

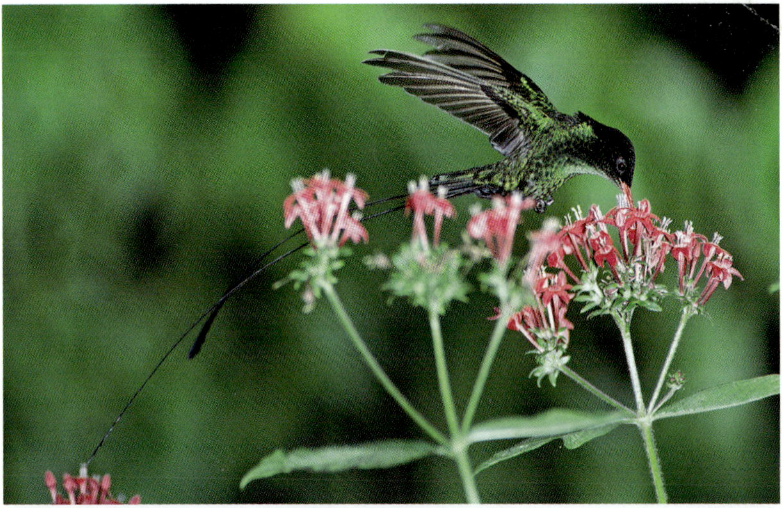

Adult male feeding. Marshall's Pen, Jamaica, May.

only the two outer tail feathers are tipped with white; upper mandible brownish black with a small area of red at the base of the lower mandible.
Similar species Black-billed Streamertail has completely black bill at all ages and is restricted to east of the island.
Voice Loud repetitive *chink, chink* or *tsee, tsee* or *teet, teet*.
Habitat and behaviour Found in gardens, at roadside flowers and trees in bloom. Abundant and widespread wherever there are flowers, regardless of altitude. Feeds on nectar and small insects. Nest (year-round, chiefly April–June) is a tiny cup of plant fibres bound with cobwebs (sometimes collected from inside houses) and decorated with lichen, attached to or straddling a horizontal limb. Only the female builds the nest, incubates and feeds the young. Lays two white eggs, which apparently always produce one male and one female chick.

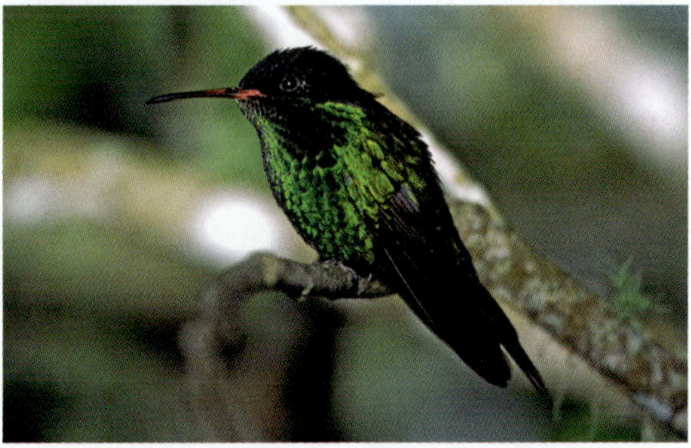

Juvenile male with mostly dark bill. Port Royal Mountains, Jamaica, November.

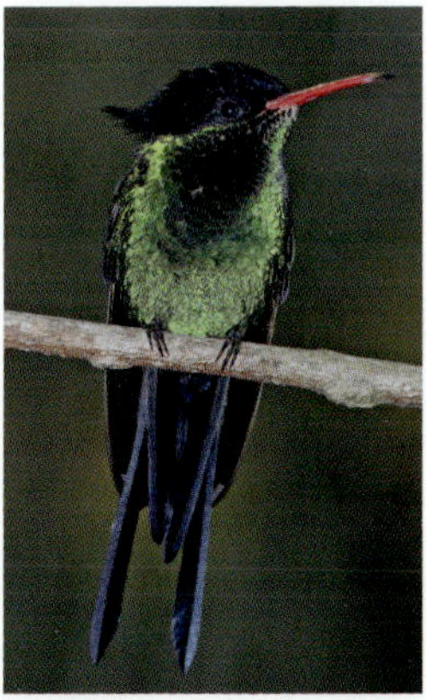

Adult male with streamers partly grown.

Range Western and central Jamaica. Rare, but possibly increasing, in eastern Jamaica.
Status in Jamaica Common and widespread.

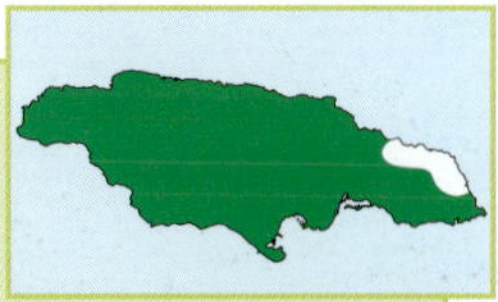

Black-billed Streamertail

Trochilus scitulus

Local name Doctorbird, God Bird.
Taxonomy Endemic genus and species.
Description L 23–26cm (9–10in). Resembles Red-billed Streamertail but slightly smaller, more blue-green in colour and has completely black bill at all ages. Bill is narrower at base.
Similar species Juvenile Red-billed Streamertail has black upper mandibles for a few months after fledging.
Voice Higher-pitched but similar to that of Red-billed Streamertail.
Habitat and behaviour Forests and wooded gardens. Feeding and nesting habits similar to Red-billed Streamertail.

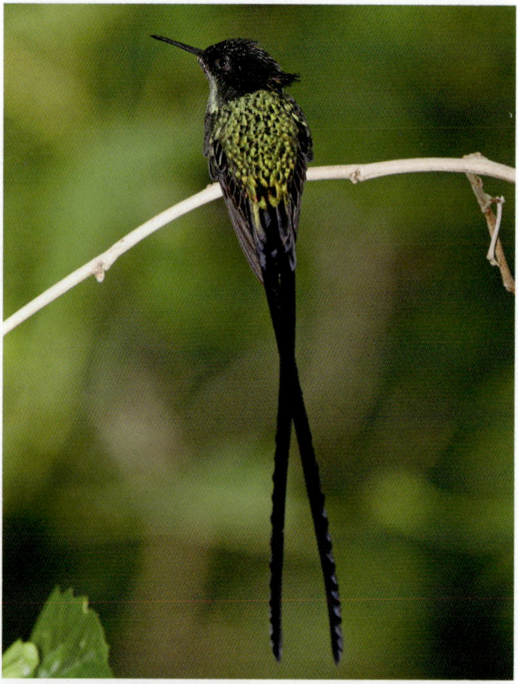

Adult male. Ecclesdown, Jamaica, November.

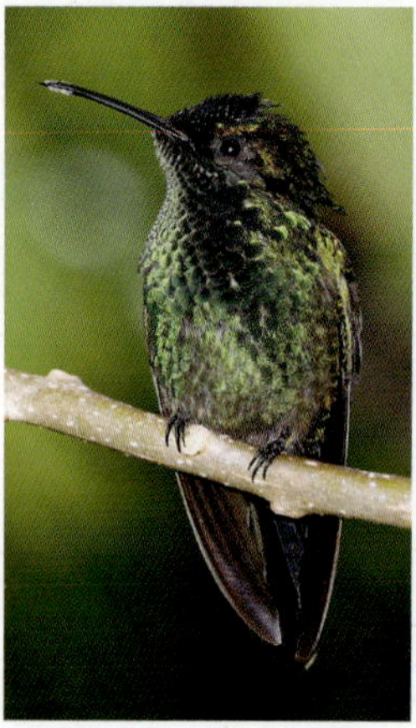

Immature male. Ecclesdown, Nov.

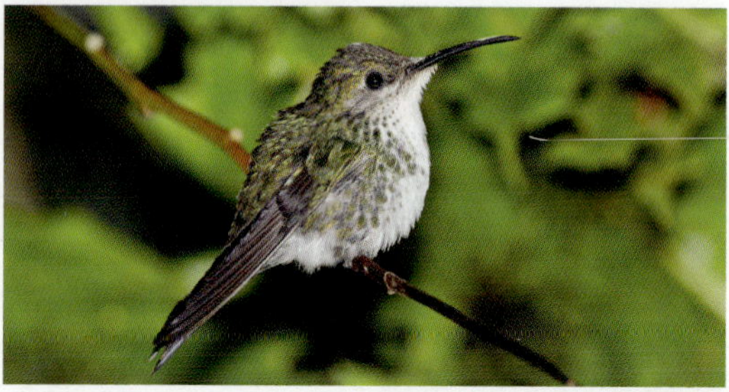

Adult female. Ecclesdown, Jamaica, November.

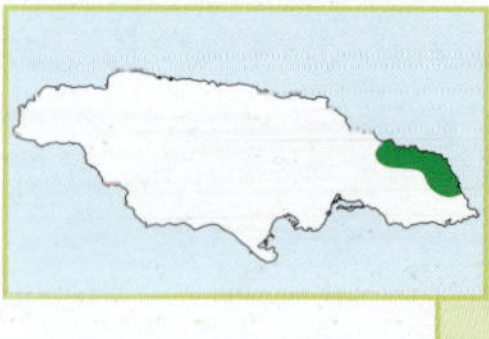

Range Eastern Jamaica. Only found in humid eastern parts of Jamaica, from the San San area in Portland east to Bowden on the southeast coast and in the John Crow Mountains, Cornpuss Gap and Bath.
Status in Jamaica Locally fairly common.

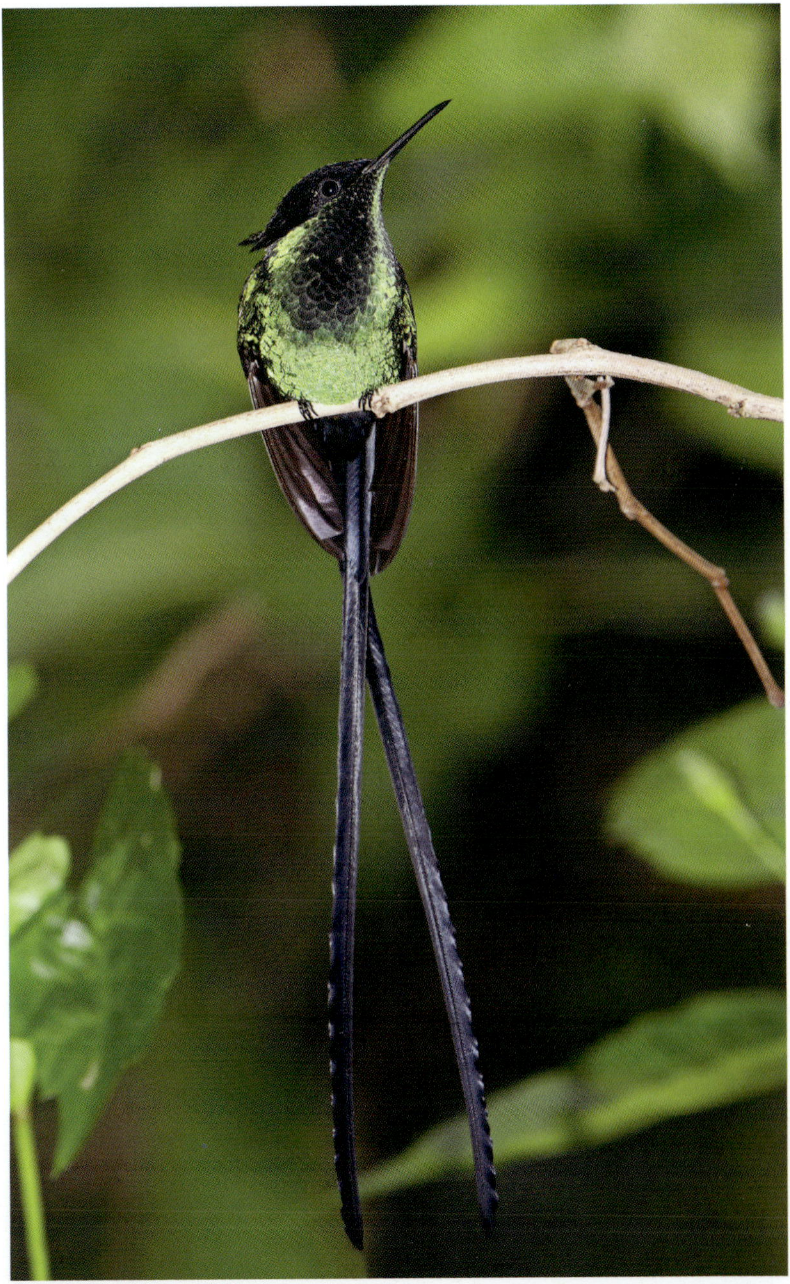

Adult male. Ecclesdown, Jamaica, November.

Vervain Hummingbird
Mellisuga minima

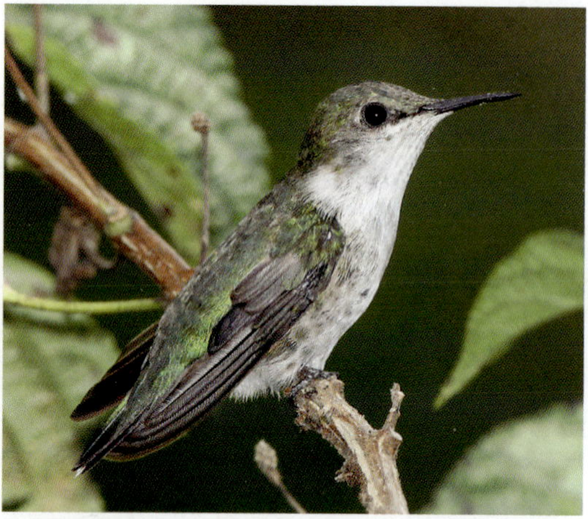

Adult male. Marshall's Pen, Jamaica, November.

Local name Little Doctorbird
Taxonomy Polytypic (2). *M. m. minima* is an endemic race.
Description L 5cm (2in). Minute, brownish-green hummingbird with a relatively short, straight, black bill. Male slightly smaller than female and has a forked tail. Head and back green, underparts white, sides of breast lightly spotted green. Female has rounded tail broadly tipped white on outer three tail feathers. Juvenile resembles adult female but white tail tips are smaller.
Similar species Female streamertails are larger, have longer bills (which may have some red at base of lower mandible) and longer tails.
Voice Song is a distinctive, prolonged twittering, sung by the male from exposed perch.
Habitat and behaviour Common and widespread in gardens, forest edges and roadsides. Wings sound like a buzzing bee. Feeds on smaller flowers than the other two hummingbirds, e.g. Vervain, Tamarind, Pentas and Chinese Hat. Rarely seen at hummingbird feeders due to competition with larger species. Nests (year-round, mainly December–February) in a tiny cup made of cobwebs and decorated with lichen, usually attached to a twig in a low shrub. Sometimes nests inside houses. Usually lays two eggs. Courtship display sees male and female rise together face to face to a great height, twittering and sometimes clutching each other by the feet then tumbling to the ground, or parting at the top of flight and falling away in opposing arcs. Territorial defence is similar to courtship behaviour but is performed by two birds of the same sex.

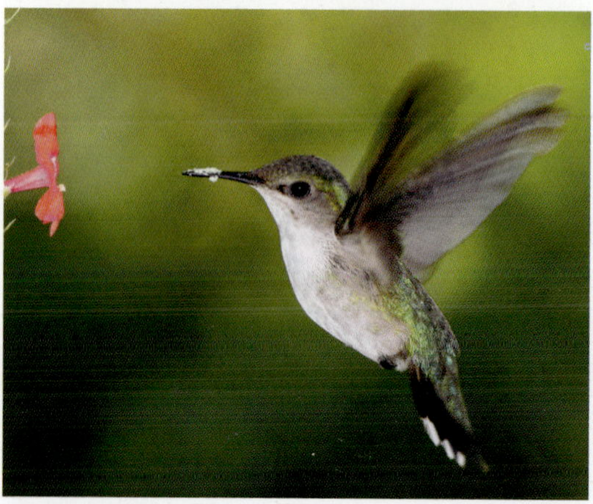

Adult female hovering with pollen on bill. Marshall's Pen, Jamaica, November.

Range *M. m. minima* endemic to Jamaica. Other race on Hispaniola.
Status in Jamaica Very common resident.

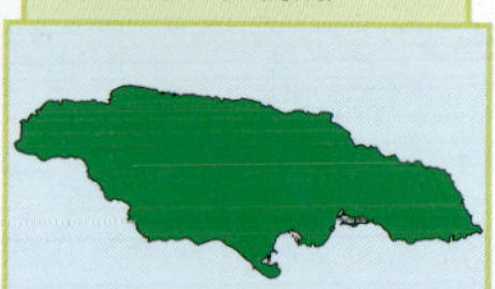

Belted Kingfisher
Ceryle alcyon

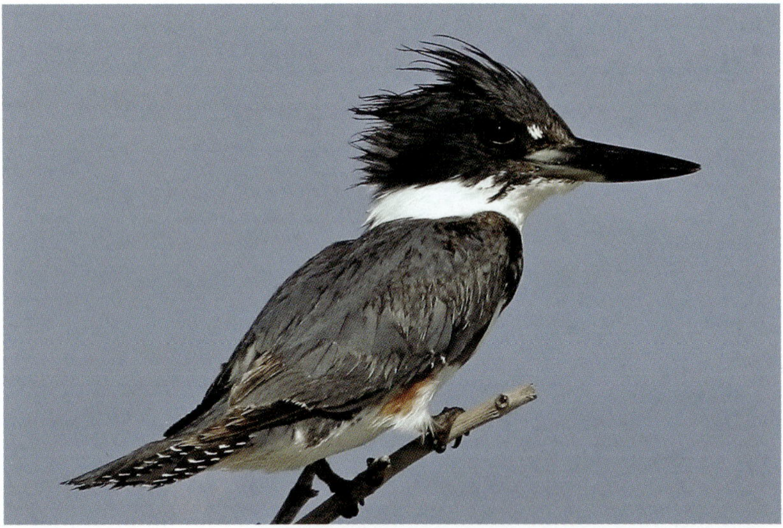

Adult female. Cayman Islands, April.

Taxonomy Monotypic.
Description L 31cm (12in). Male has a large grey-blue head with a shaggy double crest, grey back and wings, white underparts and a short dark tail. A single grey band crosses the white breast. Bill large and grey.

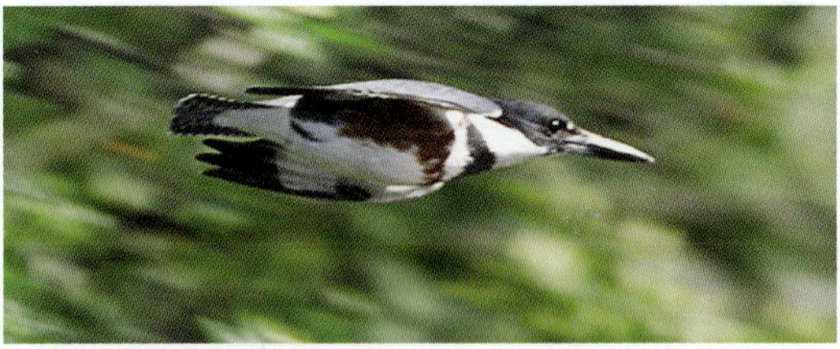

Adult female in flight. Florida, May.

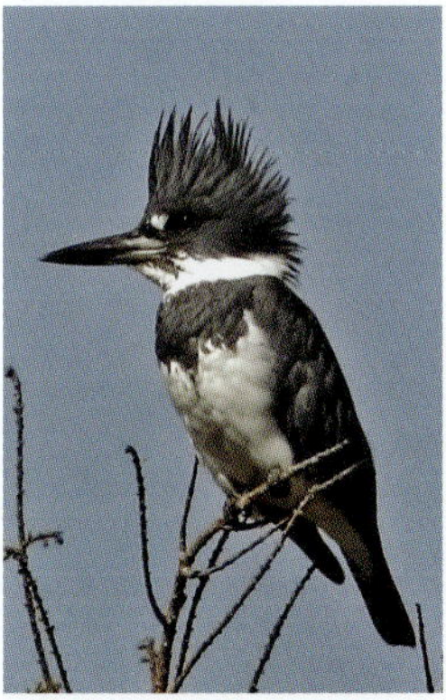

Adult male. Florida, February.

Female is more colourful than male, with two breastbands, one grey and one rufous.
Similar species None.
Voice A loud rattle similar to call of Loggerhead Kingbird, but harsher.

Habitat and behaviour Found near fresh or salt water, along the coastline, near ponds, streams and reservoirs. Aggressively territorial. Feeds mainly on fish. Fishes from a perch, or hovers and plunge-dives.

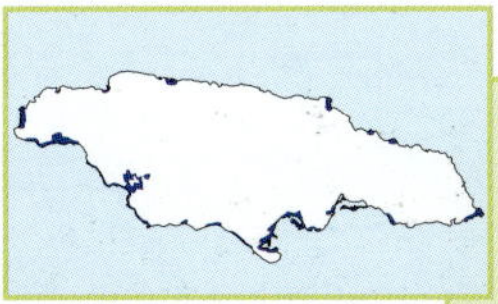

Range Breeds North America, wintering south to South America via West Indies.
Status in Jamaica Common winter visitor (August–April).

Jamaican Tody
Todus todus

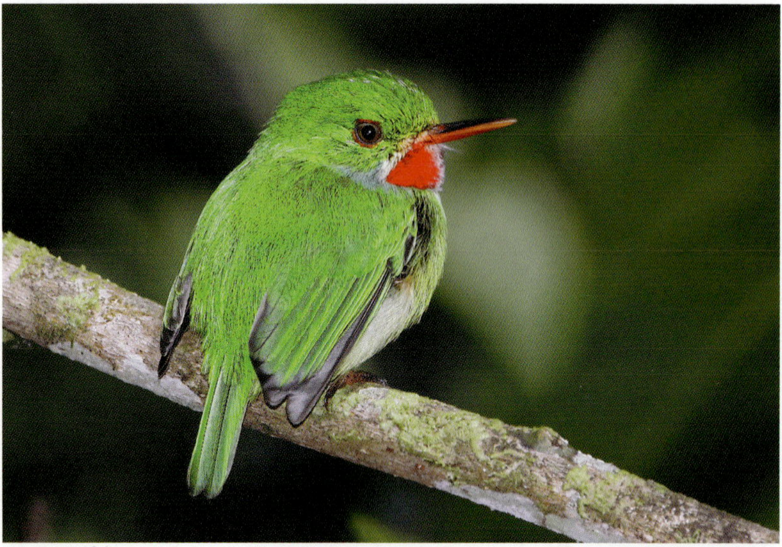

Adult. Port Royal Mountains, Jamaica, November.

Local names Robin Redbreast, Rasta Bird.
Taxonomy Monotypic. Endemic to Jamaica.
Description L 9cm (3in). Unmistakeable small, chunky, leaf-green bird with a red throat and long flattened bill. Upper mandible brown, lower orange-red. Iris blue-grey or brownish-white or rarely half and half. Breast white, washed green on the sides. There are variable amounts of pink on the flanks and bright yellow or green on the lower belly. Wings and tail black, edged green. Legs brown. Immature has less colourful underparts and the greenish wash extends over the whole breast.
Similar species None.
Voice Alarm is a loud, hissing *cheep*. Also a rattly *frrrup*.
Habitat and behaviour Found

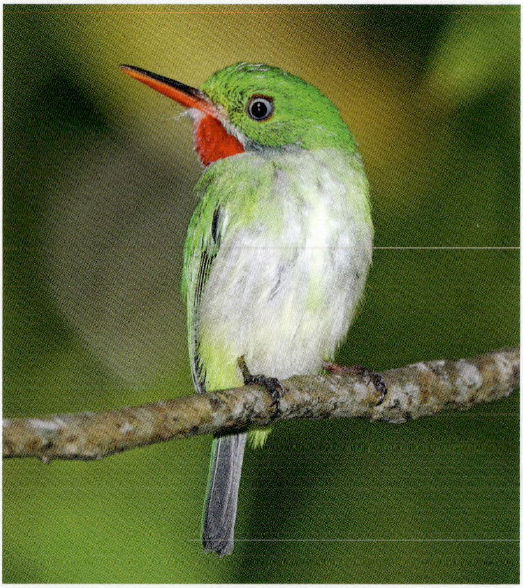

Adult. Port Royal Mountains, Jamaica, November.

in all types of forest from mangroves to the mountains but most common at mid-levels. In flight the wings make a diagnostic buzzy noise. Feeds on moths and caterpillars (which it kills by battering), small flies and small fruits. A foliage gleaner; perches on a thin twig under foliage, with long, broad bill tilted upwards. Insects are gleaned from the underside of leaves. Nest (December–July) is in a chamber at the end of a tunnel dug in a bank, drystone wall or horizontal rotten tree-trunk. Both parents excavate nest.

Range Jamaican endemic.
Status in Jamaica Common resident.

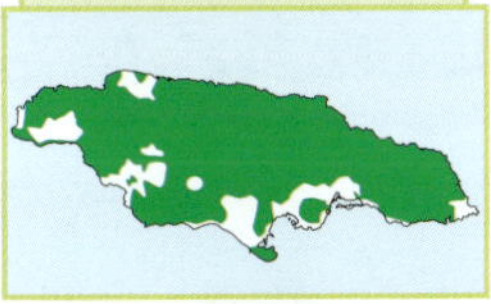

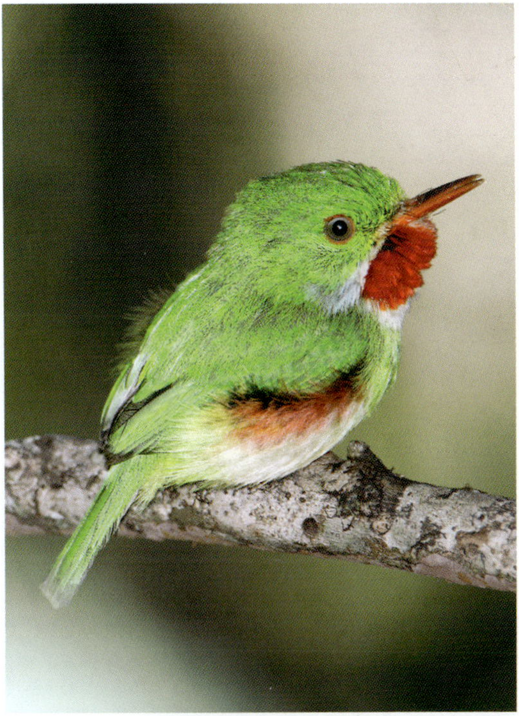

Adult. Port Royal Mountains, Jamaica, May.

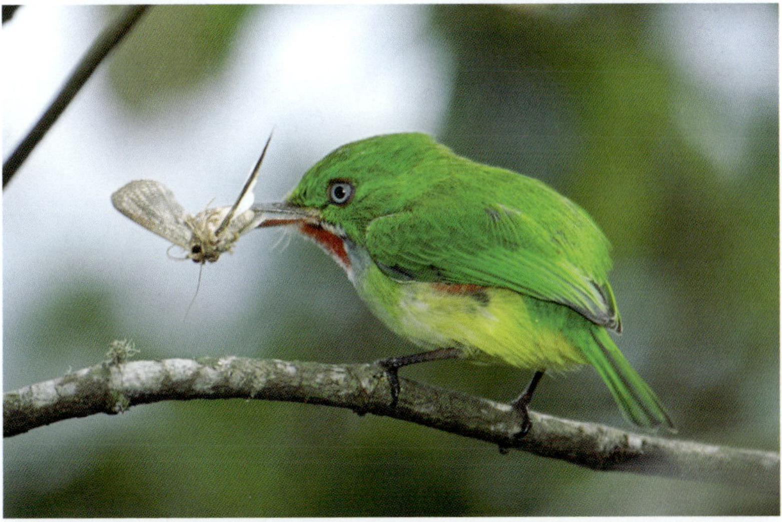

Adult. Port Royal Mountains, Jamaica, November.

Jamaican Woodpecker
Melanerpes radiolatus

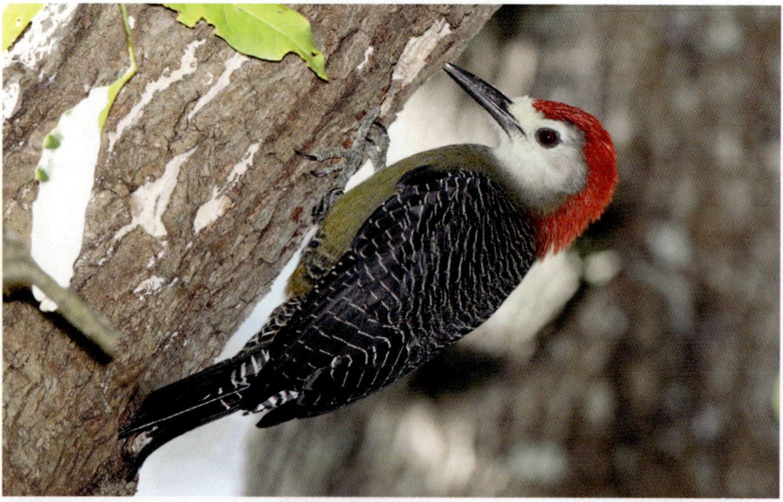

Adult male. Note the red crown; female (below) crown is olive. Port Royal Mountains, Jamaica

Taxonomy Monotypic. Endemic to Jamaica.
Description L male 24cm (10in); female: 23cm (9in). Upper back and wings are black with narrow white bars, wider on rump. Breast is brownish-olive, centre of abdomen is bright orange bordered with orange-yellow and buff. Flanks and undertail-coverts are barred black-and-white, washed orange-yellow. Tail black. Legs and bill black. Iris crimson. Male has cream forehead, face and throat. Nape and crown are bright scarlet. Female has cream face with brownish olive crown and scarlet nape. The sex of the juveniles can be determined by the head colour, even before they leave the nest.
Similar species Yellow-bellied Sapsucker is an uncommon winter visitor, much smaller and with a different facial pattern.

Voice Loud, rolling *chee-ee-erp*, not unlike a parakeet. Sometimes gives a fairly rapid *churp-churp-churp*. Also marks territory with a loud drumming on hollow trees.
Habitat and behaviour Forests, woodlands, cultivations and gardens from sea level to the mountains. Flight is undulating. Feeds mainly on insects excavated from under the bark of trees or from rotten wood, but will also eat fruit and nectar. Nests (December–August) in holes excavated in trees or rotten telegraph poles. Two broods per year are raised in the same nest, which may be used year after year (unless taken over by Eurasian Starlings).

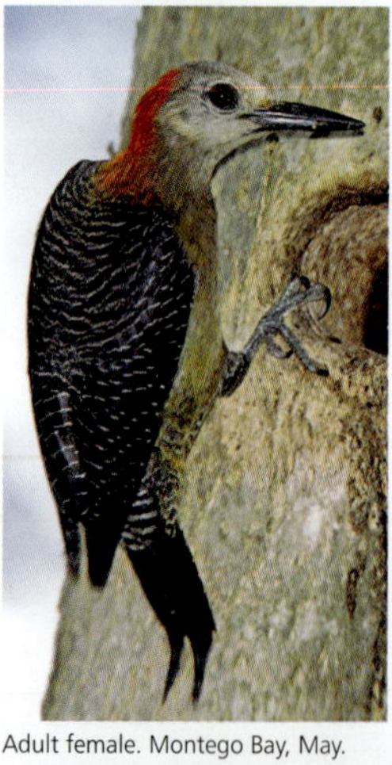

Adult female. Montego Bay, May.

Range Endemic to Jamaica.
Status in Jamaica Common and widespread resident.

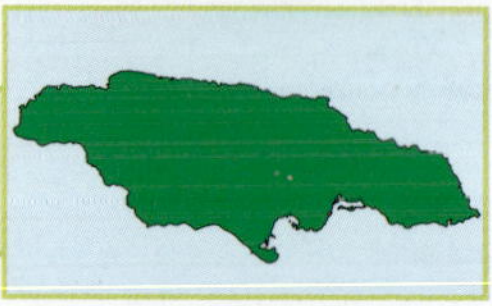

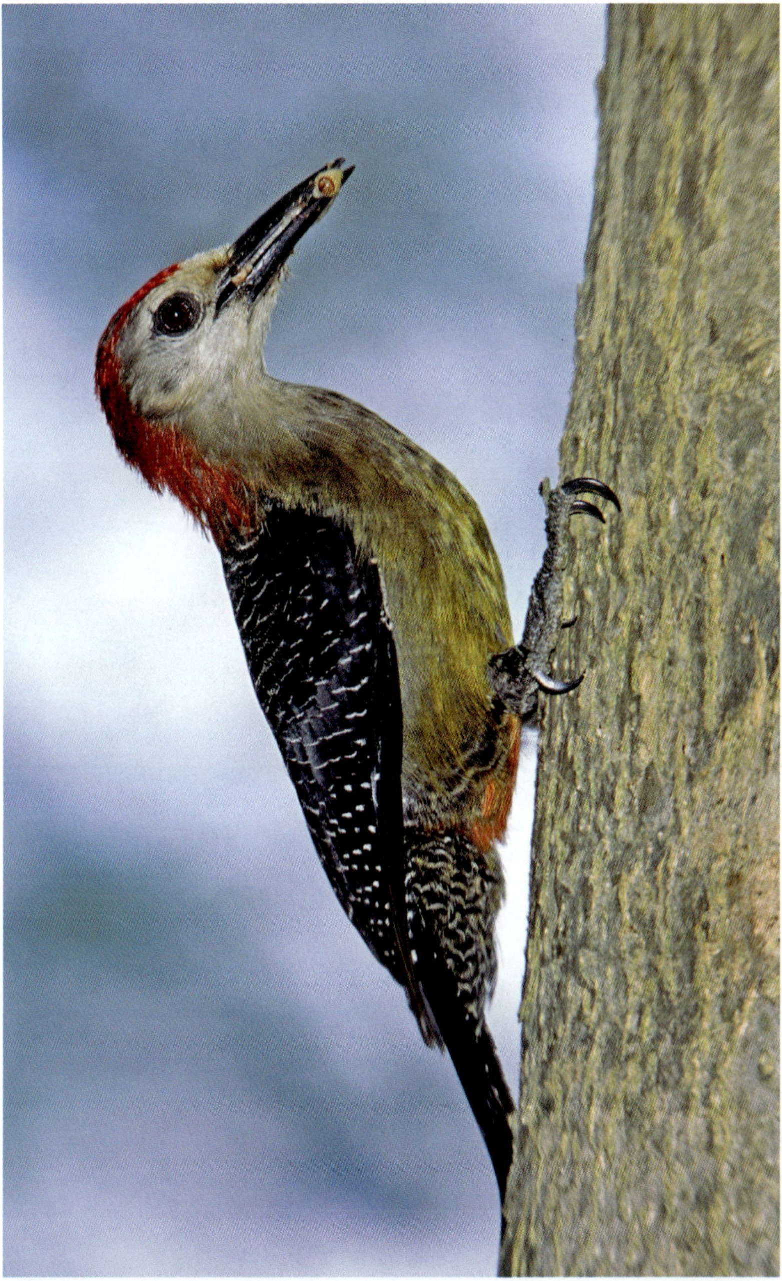

Adult male. Port Royal Mountains, Jamaica, May.

Yellow-bellied Sapsucker
Sphyrapicus varius

Local names Sapsucker, Spanish Woodpecker.
Taxonomy Monotypic.
Description L 20cm (8in). Small black, white and yellow bird with red forehead and crown, and black-and-white facial stripes. Underparts yellow with a black bib on breast. Wings black with prominent white bar due to white tips to the wing-coverts. Male has red throat. Female throat white. Juvenile browner and spotted, generally less distinctly marked.
Similar species Jamaican Woodpecker is larger, has buffy underparts, lacks stripes on face. Back is black, finely barred white.
Voice A loud *mew*. Also gentle tapping.
Habitat and behaviour Seen mainly at forest edges and gardens in the hills. Drills holes around the trunks of trees. Feeds on sap, which it obtains by boring holes into the bark of suitable trees, leaving characteristic lines of holes to which it returns regularly (and annually).

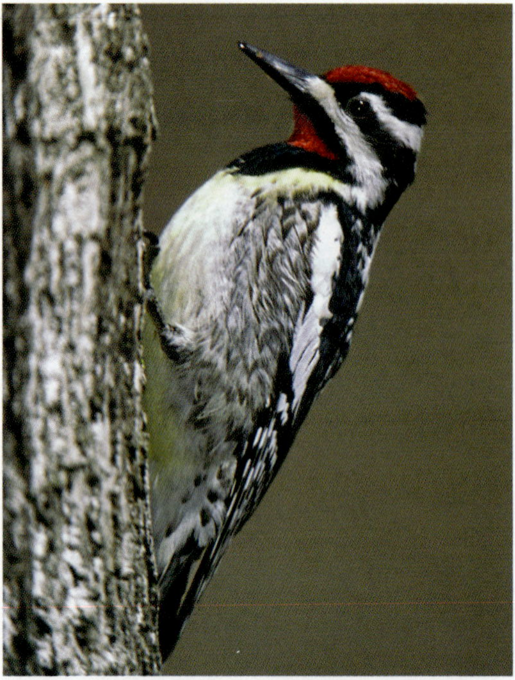

Adult male. Note the red throat. May.

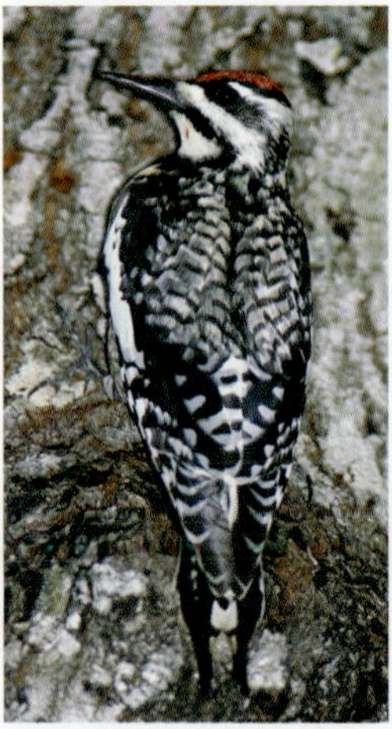

Adult female. Florida, March.

Range Breeds North America, wintering in Greater Antilles, Bahamas, South and Central America.
Status in Jamaica Uncommon winter visitor (Oct–March).

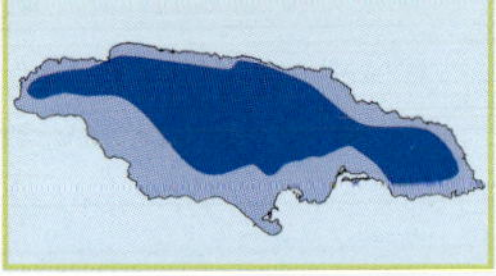

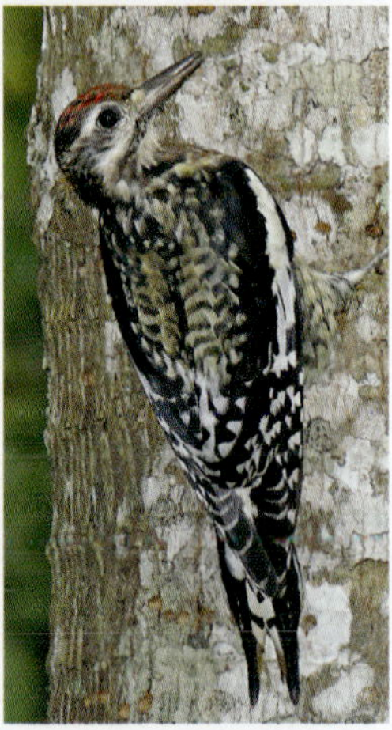

Juvenile female. Jamaica, November.

Greater Antillean Elaenia
Elaenia fallax

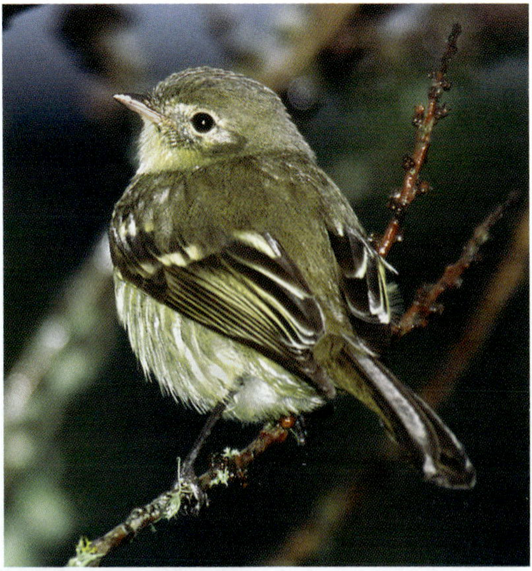

Adult. Note indistinct facial markings. Port Royal Mountains, Jamaica, June.

Local name Sarah Bird.
Taxonomy Polytypic (2). *E. f. fallax* is endemic to Jamaica.
Description L 15cm (6in). Small, plumpish, short-billed tyrant flycatcher with dark, olive-brown upperparts and two distinct wing-bars. Head dark with only faint suggestions of a supercilium and dark eye-line. Bill mostly dark grey but base of the lower mandible pale pinkish. White crown-patch usually concealed. Neck and breast pale yellow, indistinctly streaked grey; rest of underparts pale yellow. Wings grey, edged white forming 'V's on back. Wing-coverts edged greenish-white. Tail long, grey, slightly forked. Outer feathers are almost translucent. Juvenile lacks crown-patch.
Similar species Jamaican Elaenia lacks wing-bars and has more distinct facial markings.
Voice *Pwee-chi-chi-chiup* or *see-ere, chewit-chewit.*
Habitat and behaviour Forest edges. In summer (April–October) abundant when breeding in the Port Royal Mountains, Blue Mountains and higher hills of St. Andrew. In winter (November–March) scattered thinly throughout the island and is inconspicuous, silent and rarely seen. Feeds like a warbler on insects but also eats berries. Nests (May–June) high in trees. Builds a cup-shaped nest of moss.

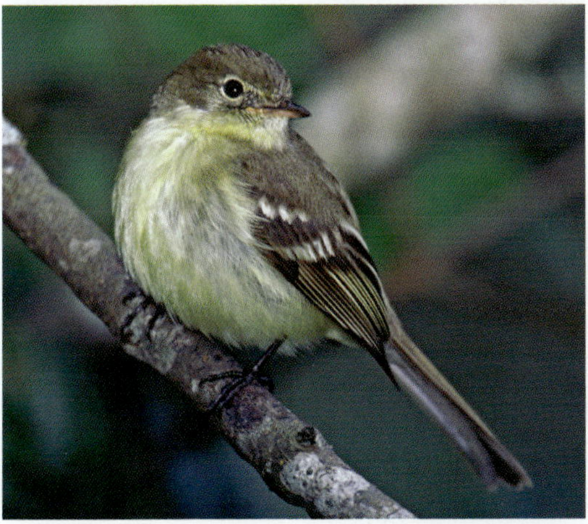

Adult. Note the two distinct wing-bars. Port Royal Mountains, June.

Range *E. f. fallax* endemic to Jamaica. Other race in Hispaniola.
Status in Jamaica Common resident; very hard to see in winter.

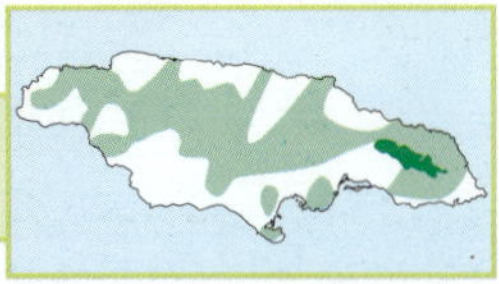

Jamaican Elaenia
Myiopagis cotta

Local name Sarah Bird.
Taxonomy Monotypic. Endemic to Jamaica.
Description L 13cm (5in). Small tyrant flycatcher with brownish olive upperparts, darker crown, grey throat and very pale yellow underparts. Very short black bill. Face has dark grey eye-line and pale yellowish-white supercilium. Iris brown with thin broken eye ring. Orange-yellow crown-patch often concealed. Wings brownish-olive, edged greenish-olive, more noticeable in centre of wing. No wing-bars. Tail long, square, brownish-olive, edged olive-yellow. Legs dark brown. Juvenile has greyish white underparts, yellow in ventral area. Lacks crown-patch.
Similar species Greater Antillean Elaenia has two whitish wing-bars.
Voice A fast *ti-si-si-sip* or *si-si-si-sip* or *si-sip*.
Habitat and behaviour Forest and bushy areas, coffee fields, from lowlands to the mountains. Inconspicuous and difficult to locate. Usually first detected by the call. Feeds by picking insects off leaves and twigs or catching them in mid-air. Also eats berries. Nest (March–June) is cup-shaped, well-hidden in bunches of seed-pods or in Spanish Moss.

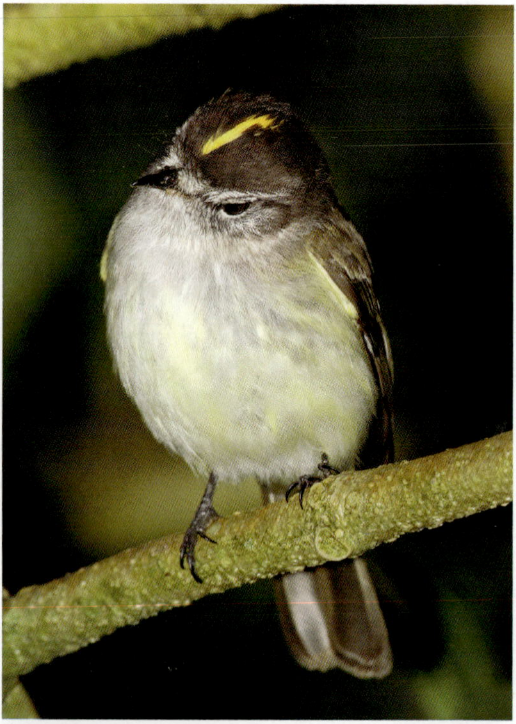
Adult displaying crown-patch. Port Royal Mountains, Jamaica, November.

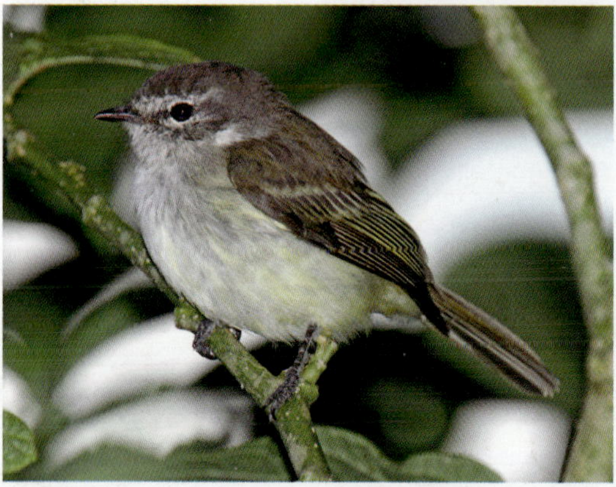
Adult. Note absence of wing-bars. Port Royal Mountains, November.

Range Jamaica.
Status Locally common and widespread.

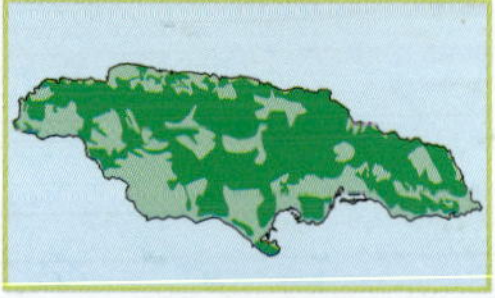

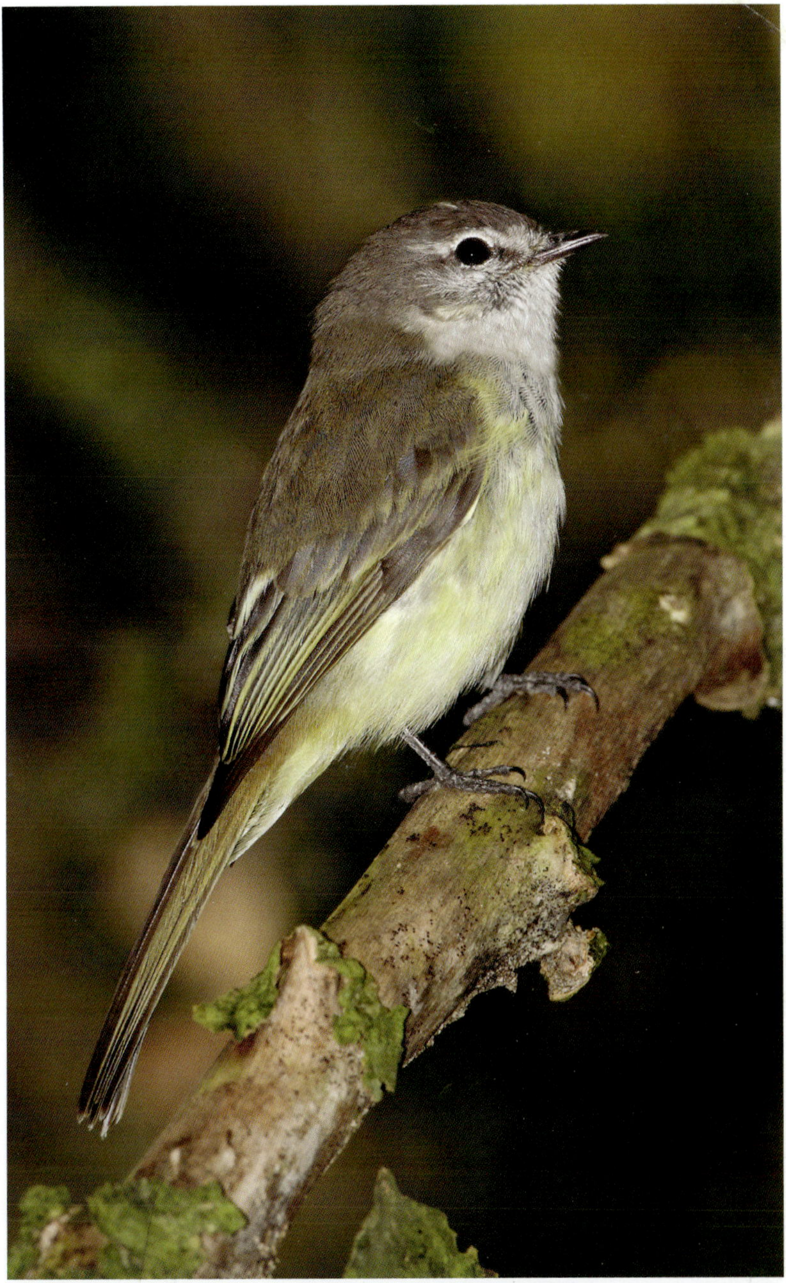

Adult. Note light supercillium and dark line through eye. Port Royal Mountains, November.

Jamaican Pewee
Contontopus pallidus

Local names Willie Pee, Little Tom Fool.
Taxonomy Monotypic. Endemic to Jamaica.
Description L 15cm (6in). Very plain, erect, small brown tyrant flycatcher, with a long, thin, notched tail. Adult upperparts dark brown, head darker with a crest that can be raised. Upper mandible brown, lower paler and pinkish-brown. Breast and flanks light brown. Centre of belly lighter. Wings brown with two indistinct buffy wing-bars. Tail as long as body, usually looks slightly wider at the end, with a distinct central notch. Legs brown. Juvenile throat to vent grey, paler centrally; lower mandible pale horn.
Similar species Greater Antillean Elaenia has distinct wing-bars. Sad Flycatcher is chunkier, has a shorter tail and a lemon-yellow belly.
Voice A plaintive *pewuee* at varying tonal levels. Rarely *pee-wee* is heard. Also *weep, ee-oo, chick*.
Habitat and behaviour Chiefly mid-level and moist forest edges. Perches in a very erect pose. Characteristically flicks tail on landing. Feeds by sallying on flying insects, which it catches with an audible snap of the bill. Tends to return to the same or a nearby perch. Nests April–June, possibly twice a year; builds a cup of grass, straw and *Tillandsia* roots in the fork of a tree.

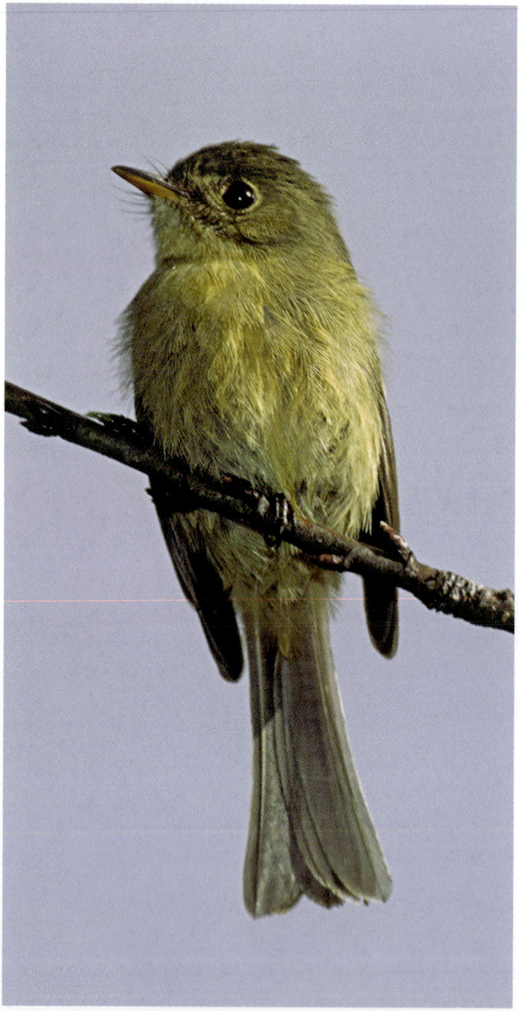

Adult. Tail is as long as body, broader at the tip, and slightly notched. Port Royal Mountain, Jamaica, November.

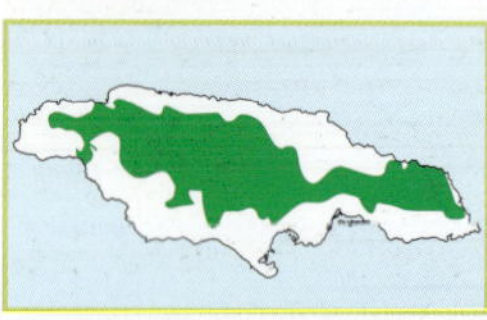

Range Endemic to Jamaica.
Status in Jamaica Locally fairly common resident.

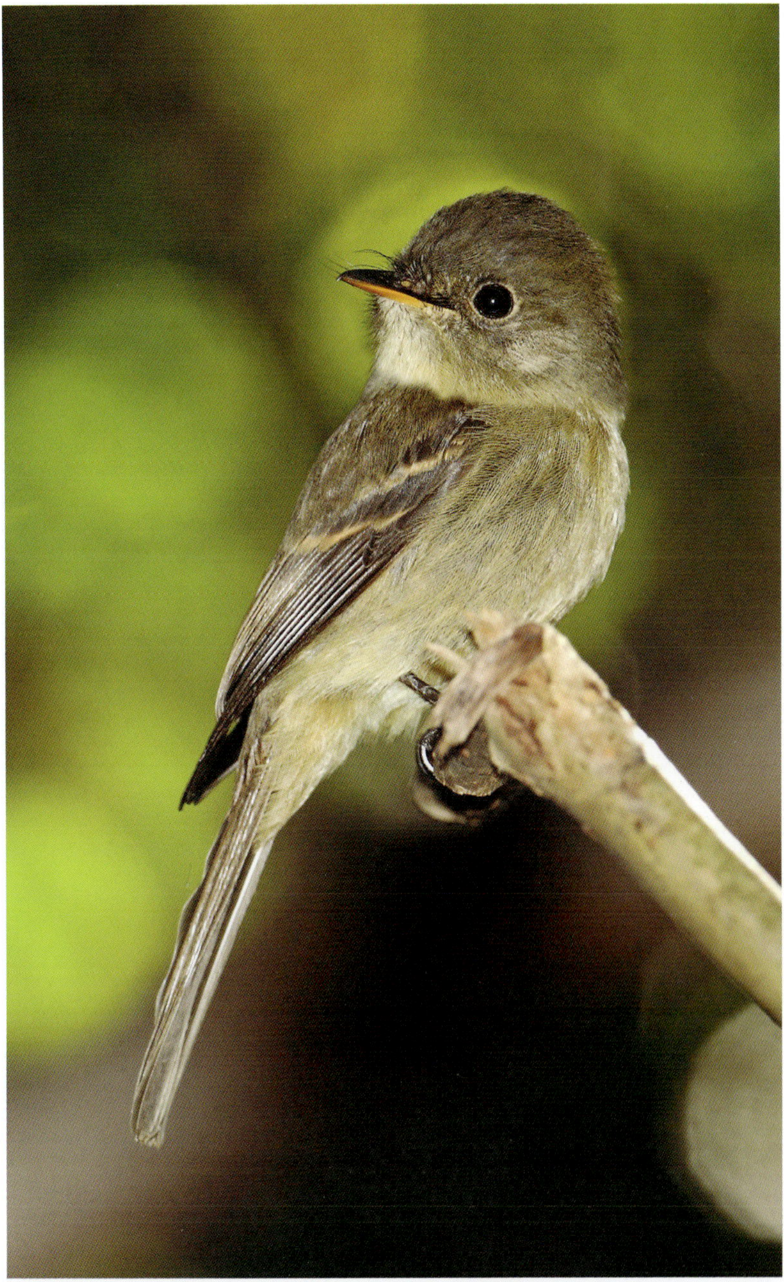

Adult. Note the two indistinct wing-bars and dark upper mandible. Port Royal Mountain, Jamaica, May.

Loggerhead Kingbird
Tyrannus caudifasciatus

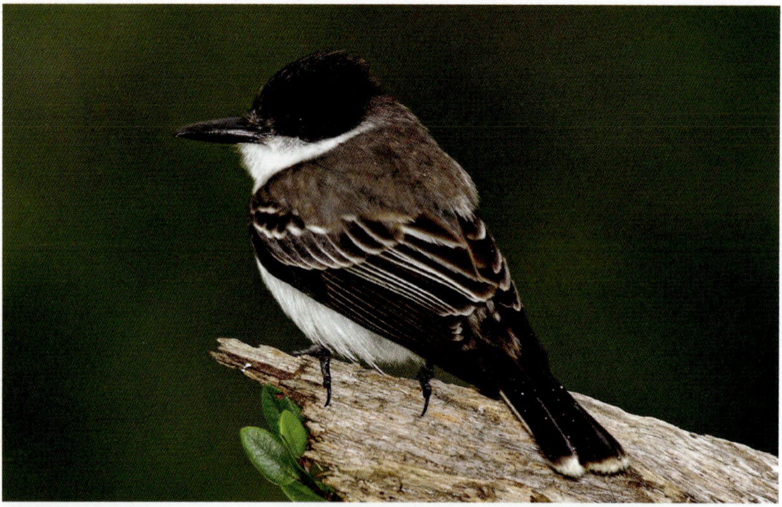

Adult. Note the black cap. Hope Gardens, Jamaica, May.

Taxonomy Polytypic (7). *T. c. jamaicensis* is endemic to Jamaica.

Description L 21cm (8 in). Stocky tyrant flycatcher with heavy bill. Black cap extends to below the eye; contrasts with grey back and snowy white throat and breast. Orange-yellow crown-patch only exposed during displays. Iris dark brown. Wings black. Secondaries and secondary coverts have narrow white edges. Tail square, black with white base, and broadly tipped off-white. Undertail-coverts and bases of tail feathers are tinged pale yellow. Juvenile lacks crown-patch; when this first develops it is white, becoming yellow by maturity. Tail is tipped buff. Wings are mostly black; secondaries and secondary coverts are narrowly edged with yellow or cream.

Similar species Grey Kingbird

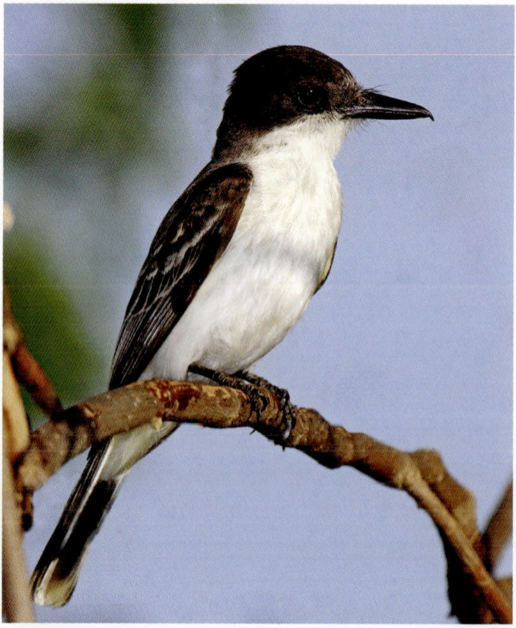

Juvenile. Cap is brownish. Hope Gardens, Jamaica, May.

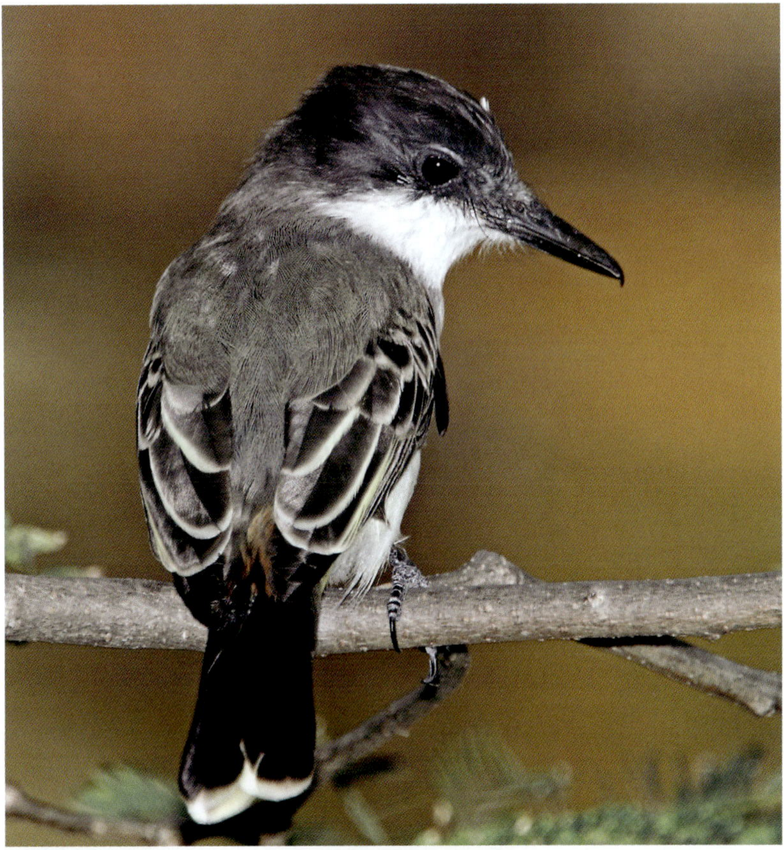

Adult. Note the 'squared-off' shape of the cap. Hope Gardens, Jamaica, November.

has a forked tail without white tip; cap less contrasting. Eastern Kingbird has similar white tips to tail feathers but is only a very rare vagrant in Jamaica (see Appendix 2).
Voice Call is a loud rolling chatter. Song (mainly in breeding season) *P-P-Q* or *P-P-U*. The display song is a continuous chittering associated with much wing- and tail-flicking and changing of position from side-to-side.
Habitat and behaviour Usually seen on exposed perches on forest edges, orange groves, gardens and roadsides. Often perches on power lines. Very aggressive but has to cede highest perches to Grey Kingbirds when they arrive in spring. Feeds on lizards and large insects, which it batters on a tree branch before eating or feeding to nestlings. May swoop low over water to take insects from the surface. Nest (November–July) is cup-shaped, woven from seed-pods, stems, grass or string, often in a fork of a small tree or shrub. Usually lays three eggs.

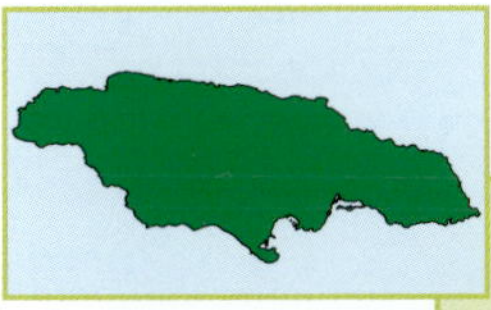

Range *T. c. jamaicensis* endemic to Jamaica. Other races occur in other Greater Antillean islands and the Bahamas.
Status in Jamaica Common and widespread resident.

Grey Kingbird
Tyrannus dominicensis

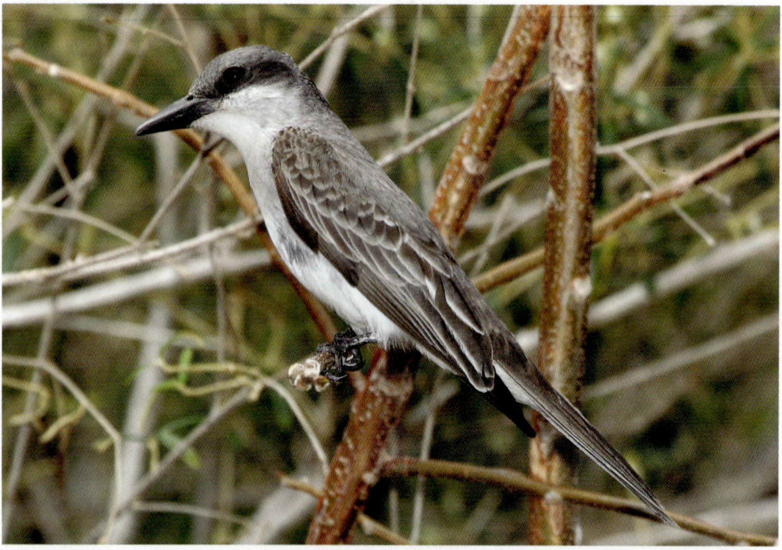

Adult. Note the broad, dark line through the eye. Cayman Islands, May.

Local name Petchary.
Taxonomy Polytypic (2). Race in Jamaica is *T. d. dominicensis.*
Description L 23cm (9in). Large and very visible flycatcher, grey above and white below with a dark head. Bill large and black. Crown grey, contrasting with broad darker mask extending under eye and over ear-coverts. Yellow crown-patch exposed in aggressive displays. Chin and underparts white with a greyish wash on upper breast. Wing feathers dark with light edges. Tail slightly forked. Juvenile resembles adult but has buffy edges to the wing feathers.
Similar species Loggerhead Kingbird has rounder head with a distinct black cap, and tail square not notched.
Voice Loud, emphatic and very often repeated *pit-cherrie,* from which its local name is derived,

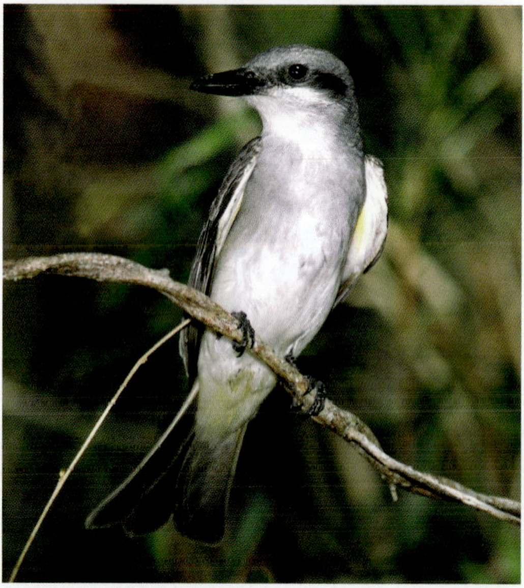

Adult. Note the slightly notched tail. Cayman Islands, May.

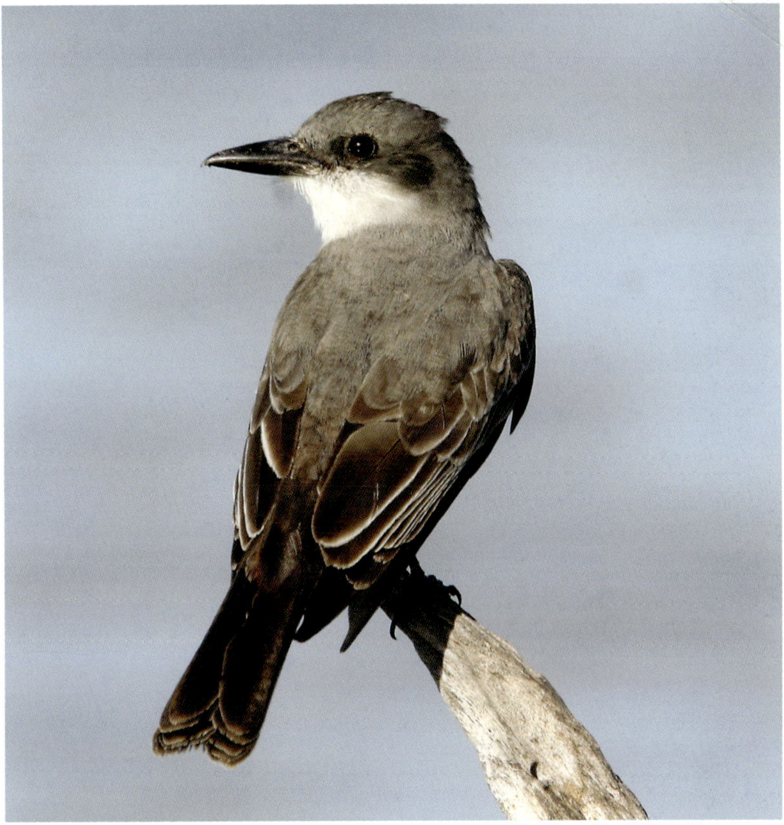

Adult. Cayman Islands, May.

plus a wide variety of other notes.

Habitat and behaviour Conspicuous in open and wooded areas e.g. gardens, golf courses, pastures and disturbed habitats, where it often perches on wires and posts (taking over the highest points from Loggerhead Kingbirds in the summer). Noisy and aggressive. Arrives singly in March, leaves in early October, when large groups assemble on power lines near the coast. A few individuals may stay until November or December. Feeds on insects and small fruit. Sallies from exposed perches to feed on insects, returning to its perch to batter its prey before swallowing it. Nest (April–June) is a loosely constructed cup, often at the base of fronds of coconut or other palm. Aggressively chases Turkey Vultures and Red-tailed Hawks away from the nest area, or while they are riding thermals, sometimes even landing on their backs. May rear two broods in a season.

Range *T. d. dominicensis* breeds in the West Indies, Bahamas and mainland coasts around the Caribbean. Winters from Hispaniola and Puerto Rico south to Venezuela and northern Brazil. Another race breeds in the Lesser Antilles.

Status in Jamaica Abundant summer visitor and breeding species.

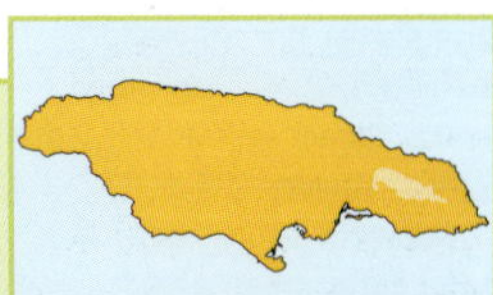

Sad Flycatcher
Myiarchus barbirostris

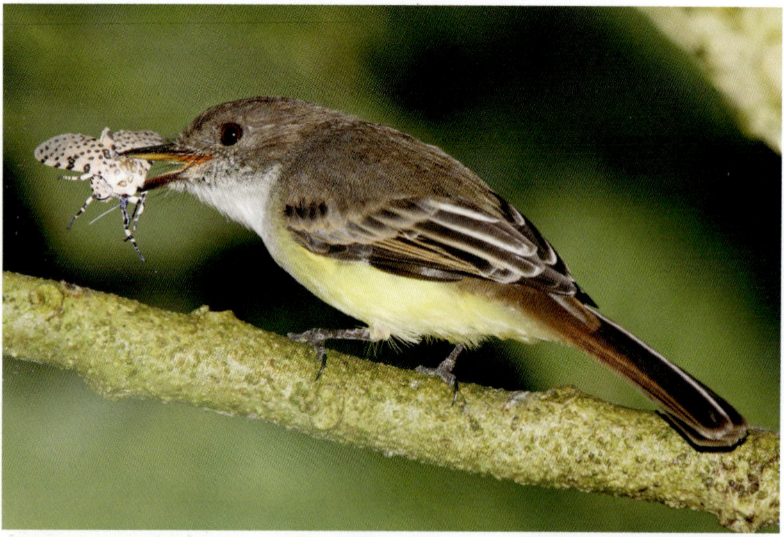

Adult. Port Royal Mountains, Jamaica, November.

Local name Little Tom Fool.
Taxonomy Monotypic. Endemic to Jamaica.
Description L 17cm (7in). The smallest of the three Jamaican *Myiarchus* flycatchers. Upperparts are buffy-brown. Head darker than body, rounded, and looks peaked when crest is erected. Bill dark and broad, lower mandible slightly paler at base. Throat and upper breast greyish-white, other underparts lemon-yellow. Wings and wing-coverts edged buff (more indistinctly than in Stolid Flycatcher). Tail finely edged with rufous. Legs black. Juvenile has greyish-white breast and throat; only the ventral area is yellow.
Similar species Stolid Flycatcher is larger and whiter, has a flatter head, broader rufous edges to tail, more white in wings and a completely different call.
Voice Its frequent calls *pip* or *pip-pip* or a rising *pip-pip-pireee* are diagnostic.
Habitat and behaviour Forests, woodlands and pasture edges at all elevations. Feeds by sallying out from a perch to feed on small insects. Also eats small fruits. Nests (April–June) in holes in trees, rotten fence posts, sometimes in eaves of houses.

Range Jamaica.
Status in Jamaica Locally common resident.

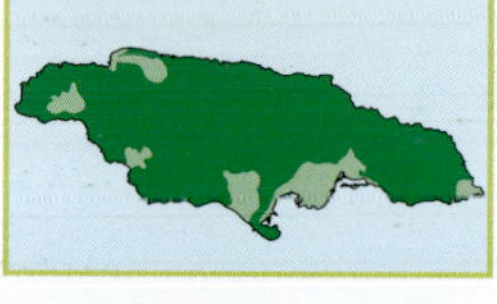

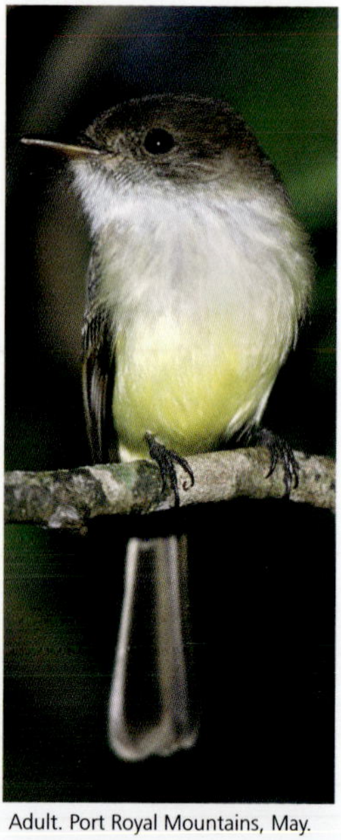

Adult. Port Royal Mountains, May.

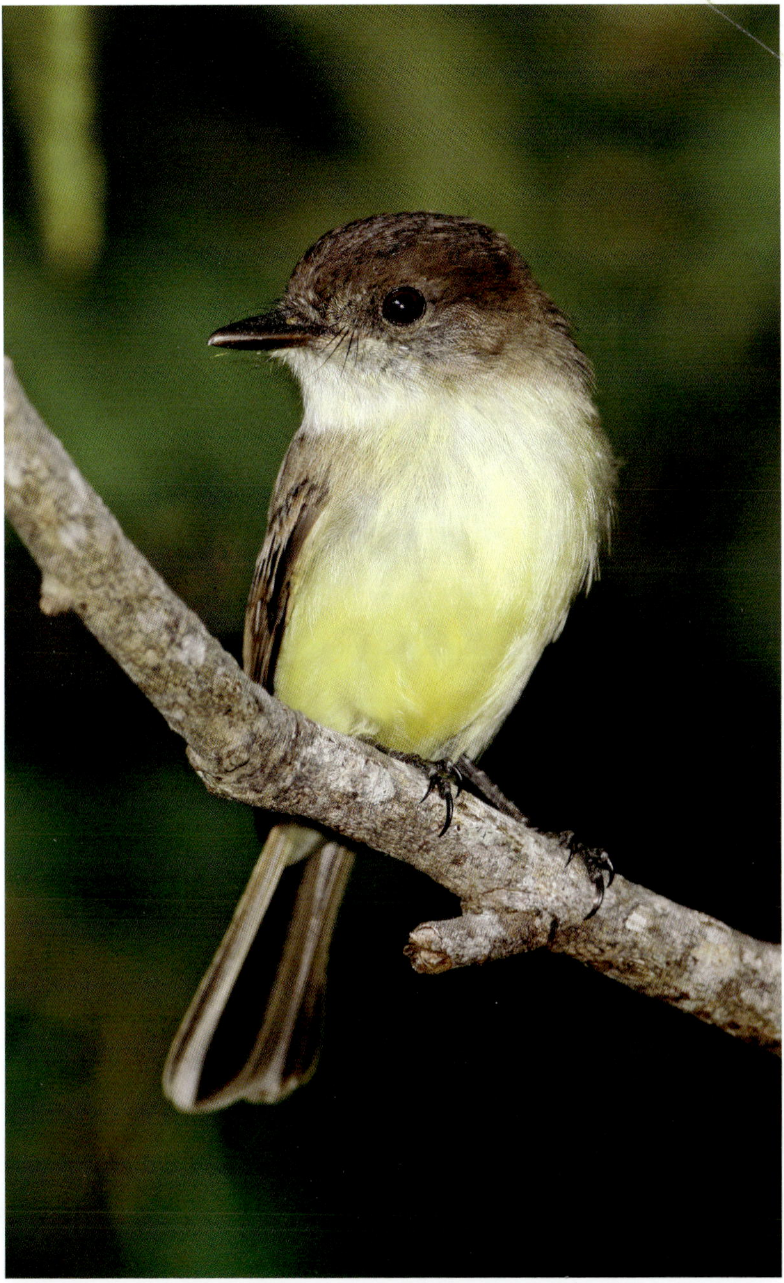

Adult. Note the lemon-yellow lower belly and rounded head. Marshall's Pen, Jamaica, November.

Rufous-tailed Flycatcher
Myiarchus validus

Local name Big Tom Fool.
Taxonomy Monotypic. Endemic to Jamaica.
Description L 24cm (10in). The largest *Myiarchus* flycatcher in Jamaica, with distinctive rufous wings and tail. Brown head rounded or peaked (when crest erected for display). Bill large, wide and flattened; dark brown upper mandible contrasts with orange base of lower mandible. Back brown. Wings and tail are olive-brown, broadly edged with orange-rufous. Tail browner below. Throat and upper breast pale grey, yellowish on the belly. Undertail-coverts light brown. Legs greyish-brown. Juvenile greyer with lighter fringes to secondary coverts and more orange on lower mandible.
Similar species Sad and Stolid Flycatchers are noticeably smaller and lack rufous on wings and tail.
Voice A fast, rolling *pree-ee-ee-ee-ee* reminiscent of a horse neighing, or *chi-chi-chiup*.
Habitat and behaviour Wooded hills and mountains. Feeds on insects and berries (e.g. Red Birch, Sweetwood and Bitterwood). Nests (April–July) in vertical cavities in trees and rotten fence posts

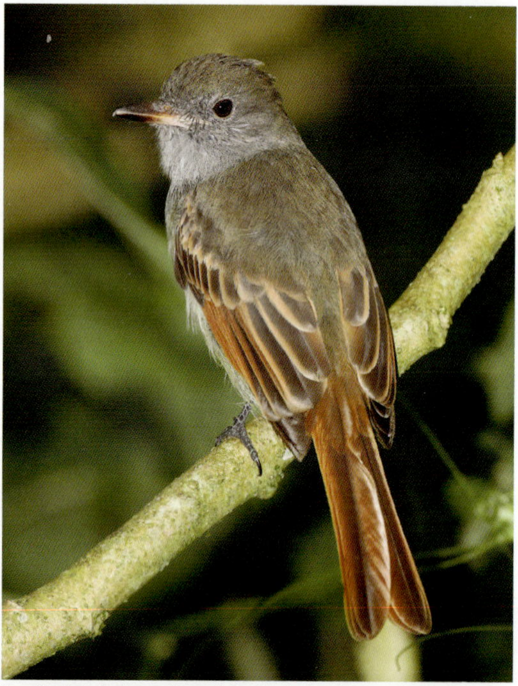
Juvenile. Note the long rufous tail. Rocklands, Jamaica, November.

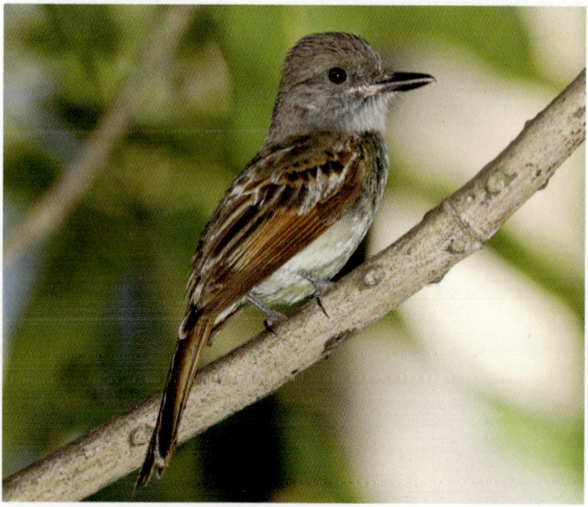
Adult. Rocklands, Jamaica, November.

Range Jamaica.
Status in Jamaica Locally fairly common resident.

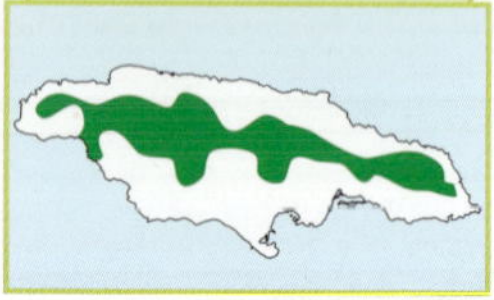

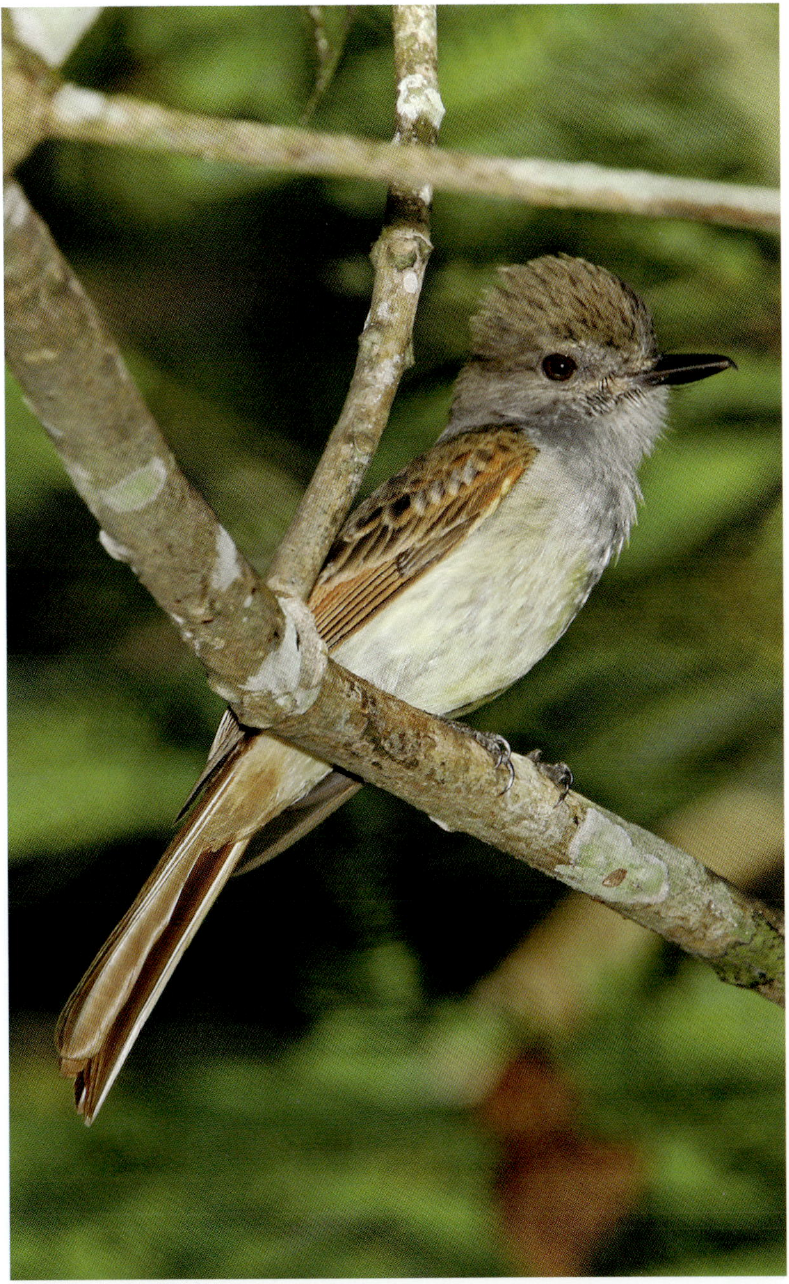

Adult with raised crest. Rocklands, Jamaica, November.

Stolid Flycatcher
Myiarchus stolidus

Local name Tom Fool.
Taxonomy Polytypic (2). *M. s. stolidus* is endemic to Jamaica.
Description L 20cm (8in). A brown tyrant flycatcher with two greyish-white wing-bars. Head large compared to body, olive-brown with darker crown. Crest often raised. Bill black, long and very broad. Upperparts olive-grey. Tail is long, brown, broadly edged with rufous. Wings brown with white edges to primaries (more distinctly marked than Sad Flycatcher). Throat and breast greyish-white, rest of underparts pale yellow, darker on centre of belly. Juvenile resembles adult but inner webs of tail feathers clay-coloured.
Similar species Sad Flycatcher is more likely to be found in the hills and mountains. It is smaller, has more yellow on underparts, indistinct wing-bars and a very different call.
Voice Calls frequently. A prolonged, rolling whistle, *whee-ee-ee*, *swee-ip*, or *bzzzrt*.
Habitat and behaviour Woodland, mangrove forest, coastal scrub, dry limestone coastal forests, wet limestone forest (e.g. Cockpit Country) and montane forest (Port Royal and Blue Mountains). Feeds on butterflies and other insects, often plucking them from leaves; also fruit. Nest similar to that of Sad Flycatcher.

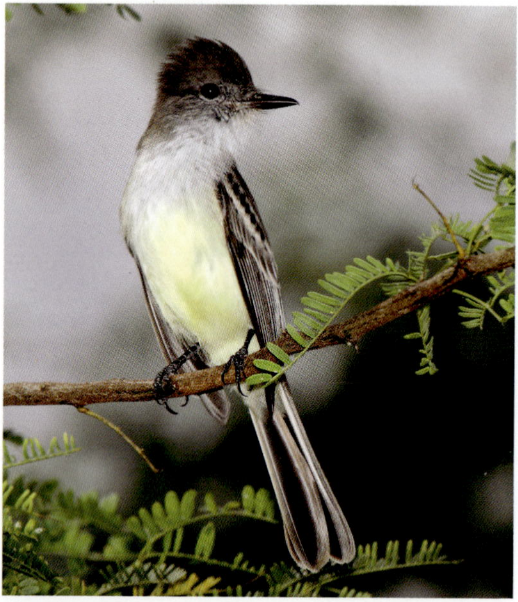

Adult with crest raised. Portland Ridge, Jamaica, November.

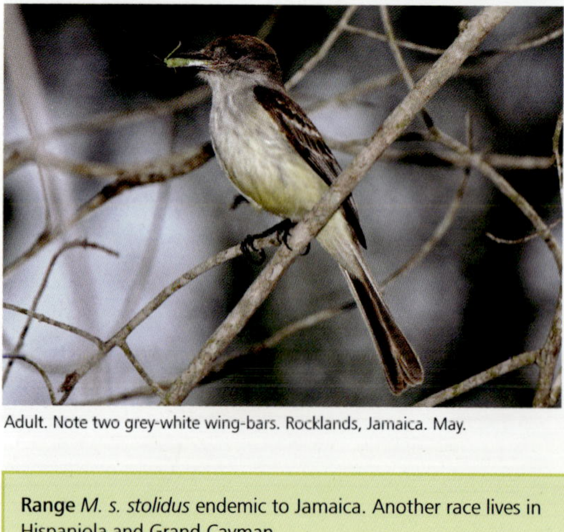

Adult. Note two grey-white wing-bars. Rocklands, Jamaica. May.

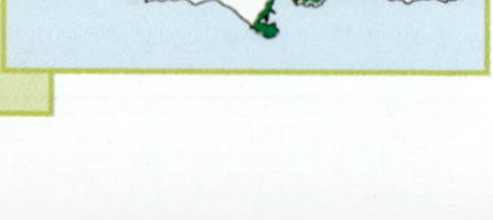

Range *M. s. stolidus* endemic to Jamaica. Another race lives in Hispaniola and Grand Cayman.
Status in Jamaica Fairly common resident.

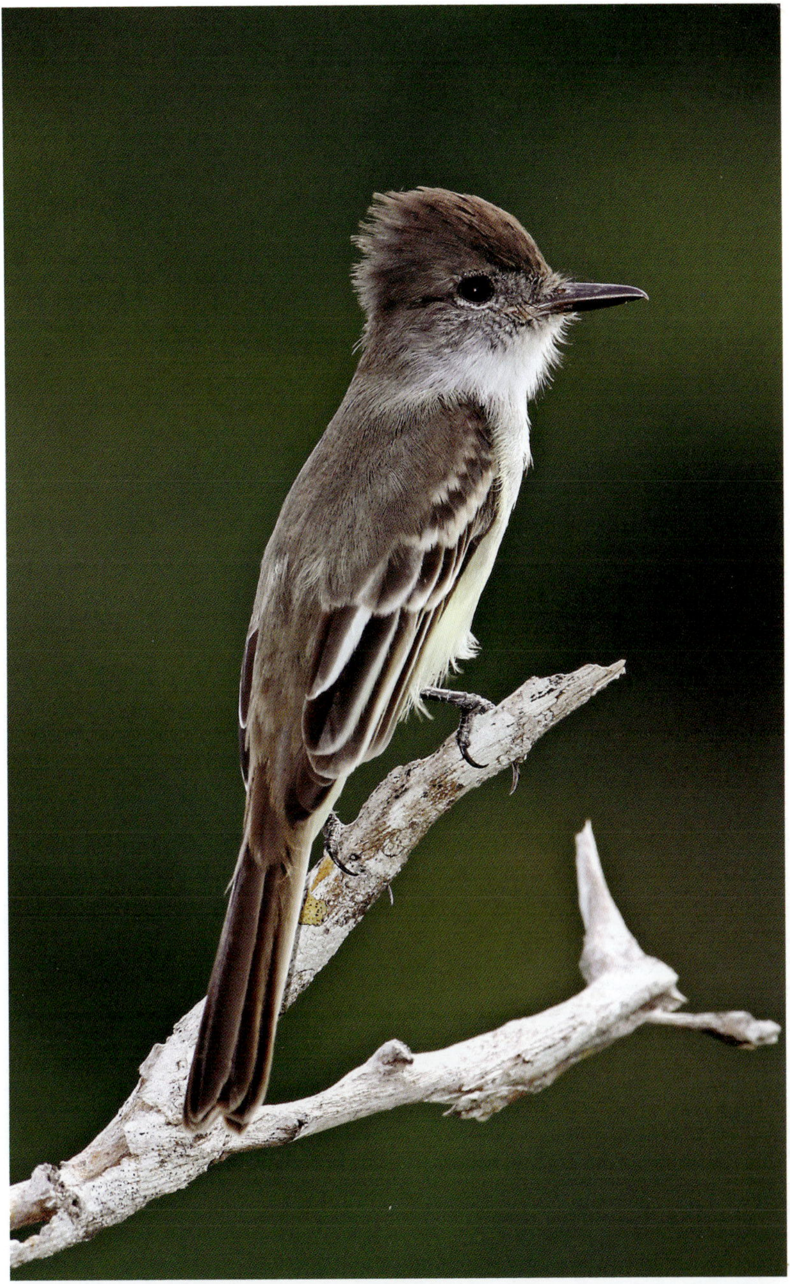

Adult. Note brown crown and all-black bill, Jamaica, November.

Jamaican Becard
Pachyramphus niger

Local names Judy (male), Mountain Dick (female), Rickatee, London City
Taxonomy Monotypic. Endemic to Jamaica.
Description L 18cm (7in). Resembles a flycatcher with a large, square head and heavy black bill. Eye large, iris brown. Tail is slightly forked. Male glossy black above, duller below, with a concealed white spot in scapulars (usually visible only in flight). Female and juvenile have bright rufous heads, throats and wings. Backs and tails grey. Underparts pale pinkish buff shading to grey, with a yellowish wash on the undertail-coverts.

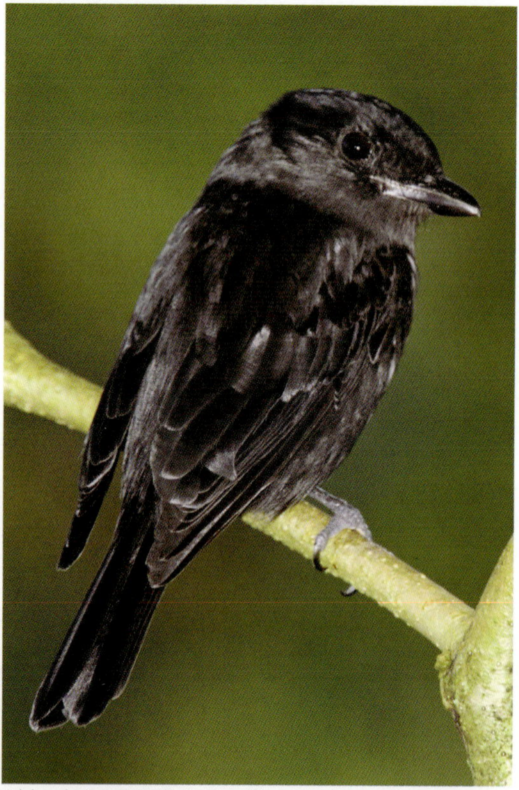

Adult male. Note thick bill. Port Royal Mountains, Jamaica, November.

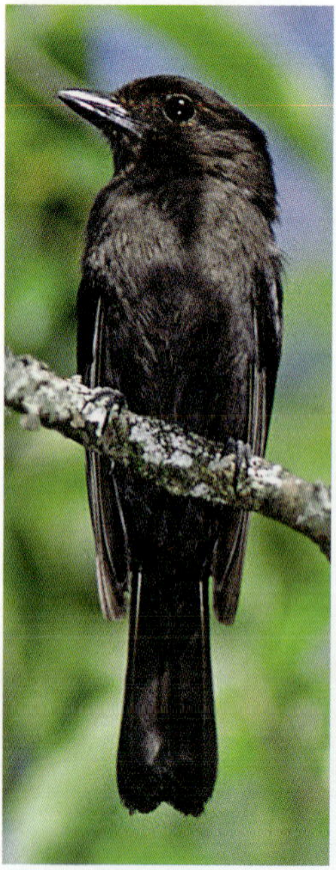

Adult male. Port Royal Mts, May.

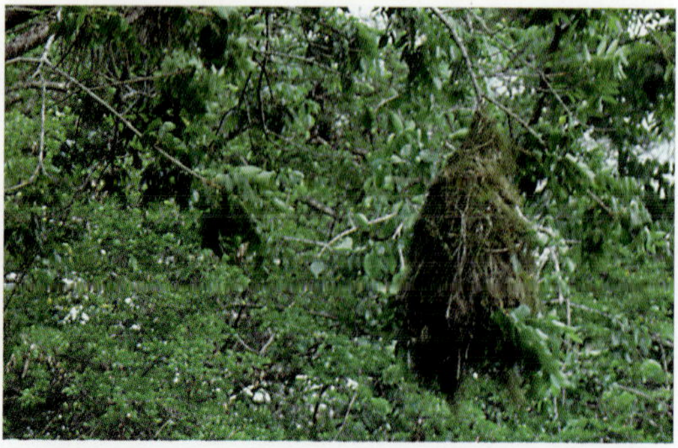

Nest. Port Royal Mountains, May.

Similar species Male recalls Jamaican Blackbird (whose head is less square and bill is thinner) or male Shiny Cowbird (which has purple iridescence and a smaller head and eye). Female might be mistaken for Rufous-tailed Flycatcher but has heavier black bill and head is rufous, not olive-brown.

Voice A melodious trailing *ricka-ticky-ti-tee*. Usually first located by call.

Habitat and behaviour Edges or open parts of wooded hills, or pastures with large trees. Most common at mid-levels but also found in the mountains. Not usually found in dry coastal forests. Feeds on insects and berries in upper and middle canopies of tall trees such as Sweetwood and Prickly Yellow. Nests (March–June) in a very large nest, up to 1.2m (4 feet) in circumference, made of sticks, skeletonised leaves, grasses, ferns and vines, and sometimes completely of Old Man's Beard, draped and hung at the end of a slender branch. The main entrance is usually at the bottom but there are often several false entrances to baffle predators. Many other species steal the nest material to make their own nests.

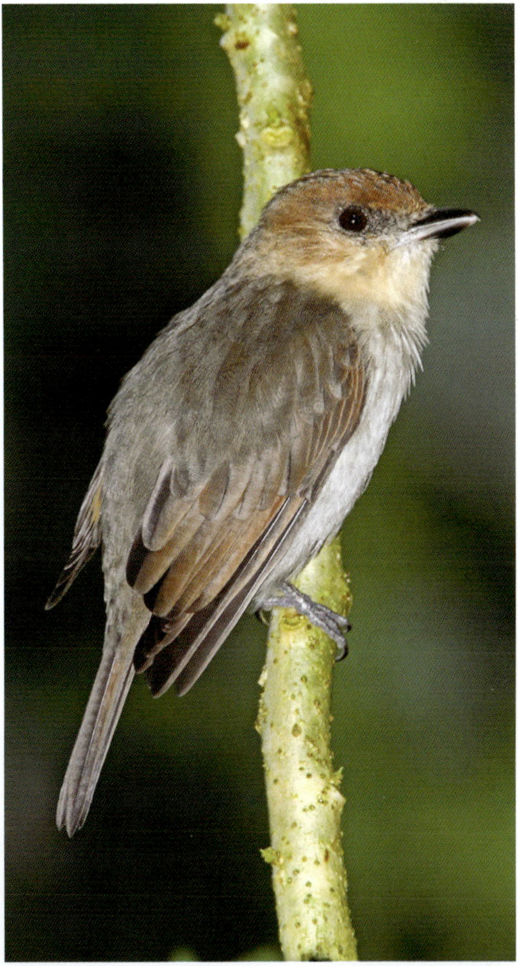

Adult female. Note rufous head, dark eye. Port Royal Mountains, Jamaica, May.

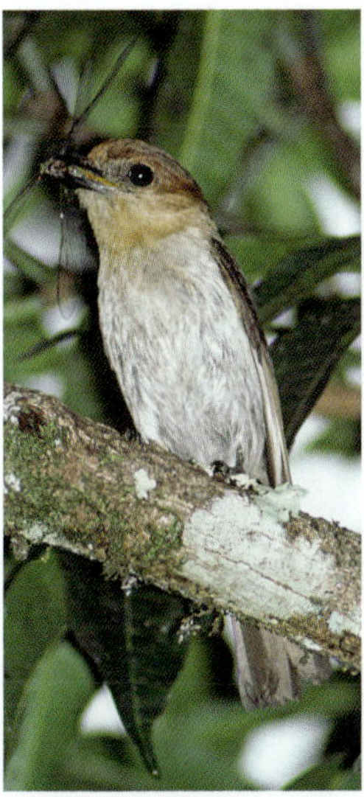

Adult female. Port Royal Mts, Nov.

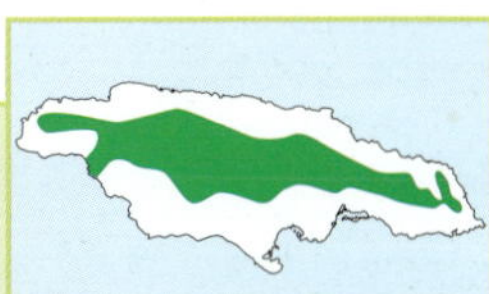

Range Endemic to Jamaica.
Status in Jamaica Fairly common and widespread resident.

Sand Martin
Riparia riparia

Taxonomy Polytypic (5). *R. r. riparia.*
Alternative name Bank Swallow.
Description L 13cm (5in). Upperparts dark brown. Underparts white with a distinct brown band across the breast. Tail forked. Wing beats shallow and quick.
Similar species Northern Rough-winged Swallow has brown throat and no breast-band.
Habitat and behaviour Often in mixed flocks, flying or perched on power lines.

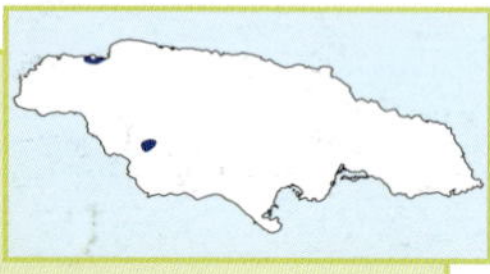

Range *R. r. riparia* breeds in North America, and Europe. North American migrants winter in South America via the West Indies.
Status in Jamaica Rare passage migrant (September, January, May).

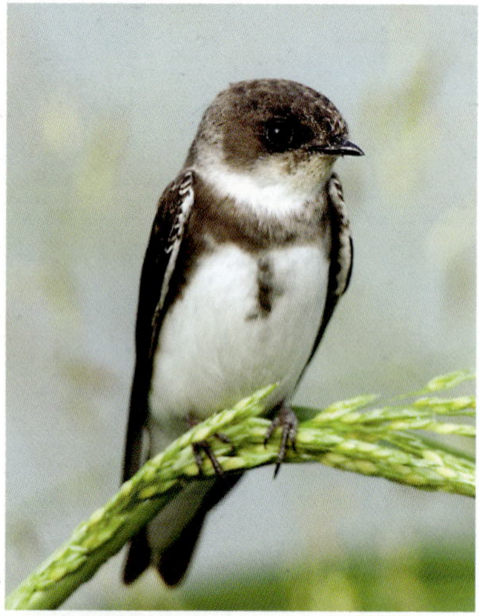

Juvenile. Note dark breast-band. September.

Northern Rough-winged Swallow
Stelgidopteryx serripennis

Taxonomy Polytypic (6).
Description L 13cm (5in). Upperparts brown, slightly paler on rump. Throat and breast light brown, the rest of the underparts are dirty white. Tail square.
Similar species Sand Martin has white throat and contrasting brown breast-band.
Voice Generally silent in Jamaica.
Habitat and behaviour Usually seen over wet fields or freshwater marshes or on electric wires near the coast. Often in mixed flocks with Cave Swallows.

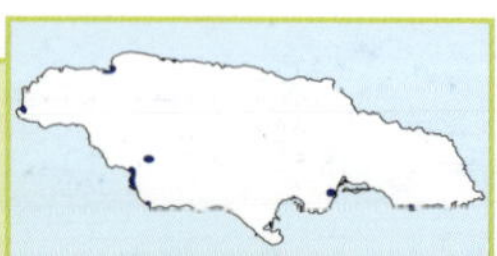

Range Breeds North America, wintering in Central America and casually in the Greater Antilles.
Status in Jamaica Rare passage migrant (Aug–Nov, Jan–April) in small numbers.

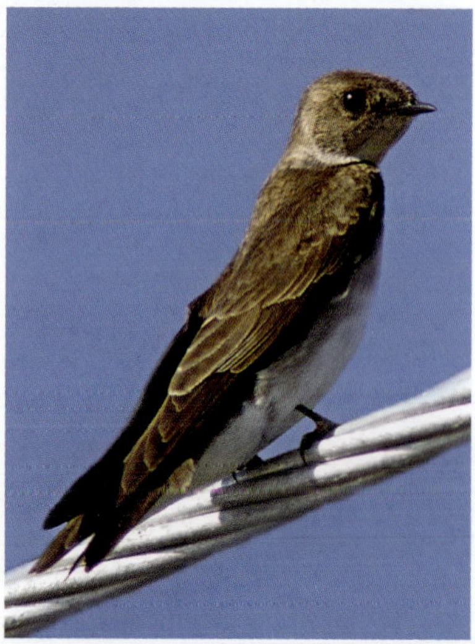

Adult. Note drab throat. May.

Tree Swallow
Tachycineta bicolor

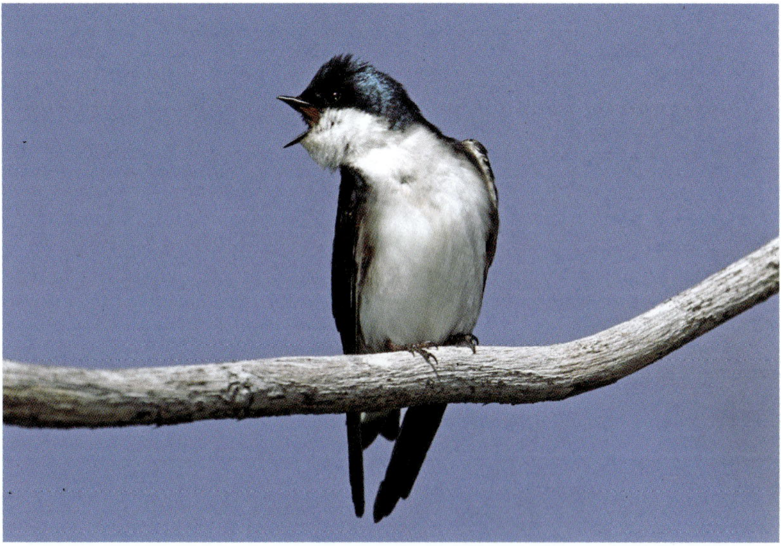

Adult male. Note iridescent head. May.

Taxonomy Monotypic.
Description L 13cm (5in). A smart bluish-green swallow. Back is bluish-green, with underparts contrasting bright white. Tail slightly forked. Immature has browner upperparts, white underparts.
Similar species Golden Swallow has golden iridescence on the back, a more deeply forked tail and is not likely to be found close to the coast.
Voice Usually silent.
Habitat and behaviour Usually seen catching insects in flight over coastal ponds.

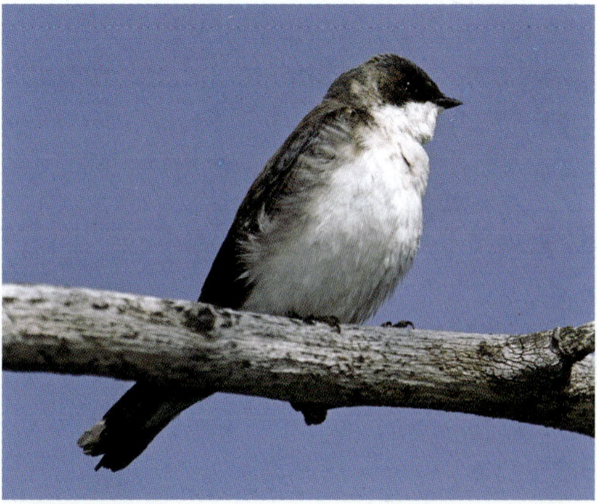

Adult female. May.

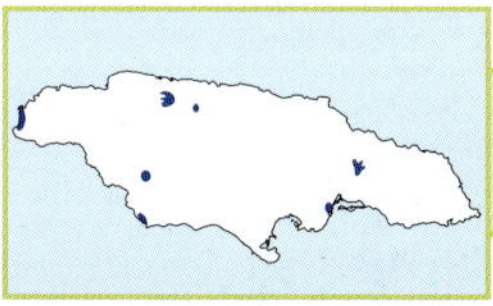

Range Breeds in North America, wintering to northern South America via Central America, Bahamas and Greater Antilles.
Status in Jamaica Common passage migrant (January–April).

Golden Swallow
Tachycineta euchrysea

Taxonomy Polytypic (2). *T. e. euchrysea* is endemic to Jamaica.
Description L 13cm (5in). Small, iridescent-green swallow with snowy white underparts (including chin, ear-coverts and malar). Golden glints on lower back. Female underparts mottled greyish brown. Juvenile resembles female but less glossy, with grey on sides of head.
Similar species Antillean Palm Swift has white rump and fine dark breast-band. Tree Swallow has a less deeply forked tail, shorter wings and less graceful flight.
Voice A soft twittering (April–June) but usually silent.
Habitat and behaviour In the past was often seen swooping low over hills and canefields on the northern edge of the Cockpit Country, mainly in Nov–Jan and July–Aug. Feeds on insects. Nests in caves.

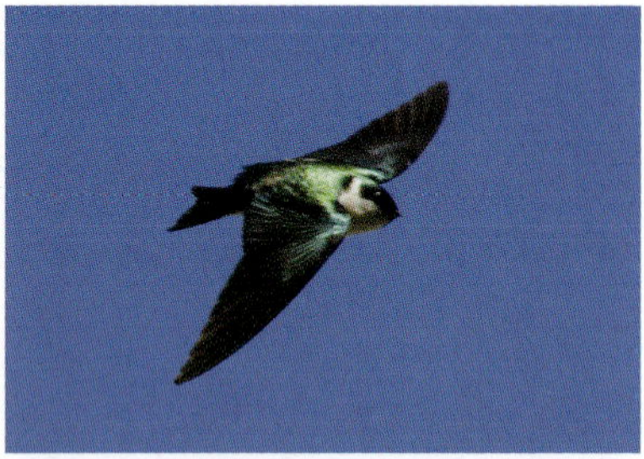

Adult *T. e. sclateri*. Note iridescence. Hispaniola, February.

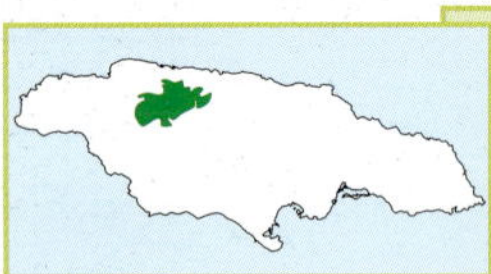

Range *T. e. euchrysea* on Jamaica. Other race in Hispaniola, where declining. Global conservation status: Vulnerable.
Status in Jamaica Resident race is very rare, possibly extinct. It was reportedly plentiful up to the end of 19th century but there have been no confirmed sightings since the 1980s.

Barn Swallow
Hirundo rustica

Taxonomy Polytypic (7). *H. r. erythrogaster*.
Description L 15cm (6in). Dark-backed swallow, with long wings and deeply forked tail. Back glossy blue-black, white underparts. Forehead and throat rich chestnut. Tail has elongated outer feathers and band of small white spots. Adult male has deep orange underparts, longer tail streamers. Adult female paler below. Juvenile like female but has short outer tail feathers.
Similar species Very uncommon Purple Martin has dark head and square tail.
Voice Repeated high *che-chet*.
Habitat and behaviour Usually seen over grassy coastal areas. Flies low to feed on small insects.

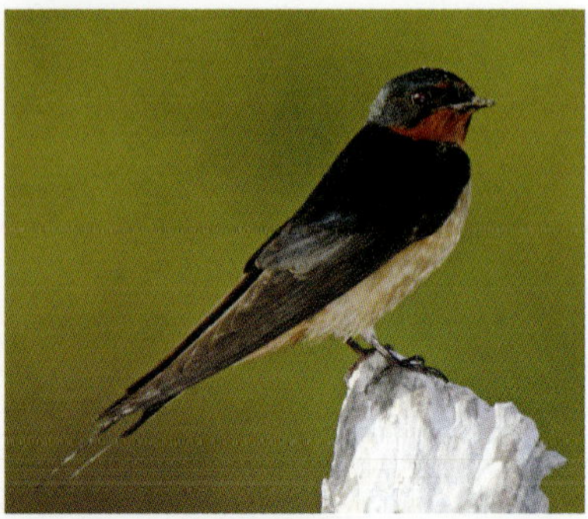

Adult. Cayman Islands, May.

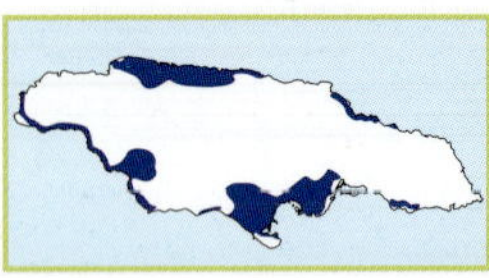

Range *H. r. erythrogaster* in Americas; winter South America via West Indies and Central America. Other races worldwide.
Status in Jamaica Common passage migrant (mainly August–November and January–April).

Caribbean Martin
Progne dominicensis

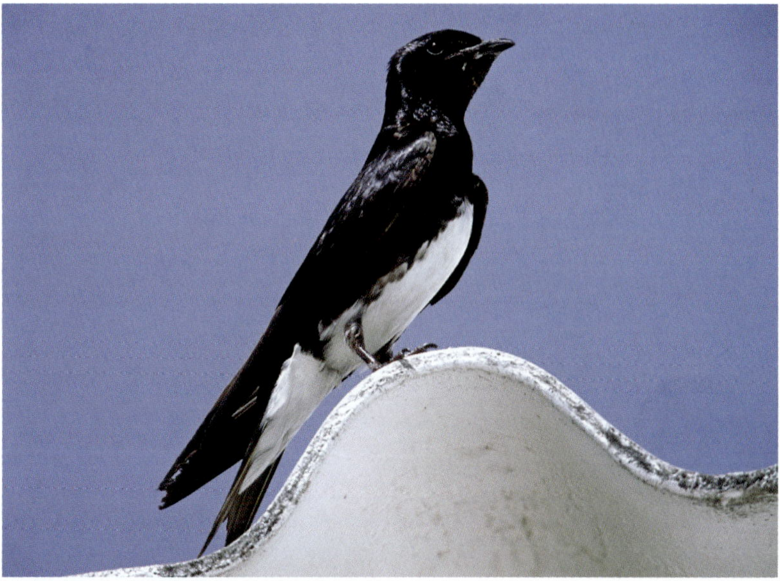

Adult male. Note glossy back and throat. Dominican Republic, May.

Taxonomy Monotypic.
Description L 20cm (8 in). A large swallow with a slightly forked tail and white lower breast and belly. Male is blue-black above and on the upper breast and flanks. Undertail-coverts are lightly tipped blue-black. Female and juvenile are brown above with some blue in the wing-coverts.
Similar species Purple Martin (see Appendix 2) is bluish-black all over.

Voice A liquid *chileet, chur-chur, chi-chi-chiwee*.
Habitat and behaviour Often seen near the coast, in flocks over wetlands or perched on power lines. Never seen in the hills. Feeds on flying insects. Nests (April–June) in old woodpecker holes in palms or telegraph poles, or at the top of dead coconut trees. More than one pair may use a hole.

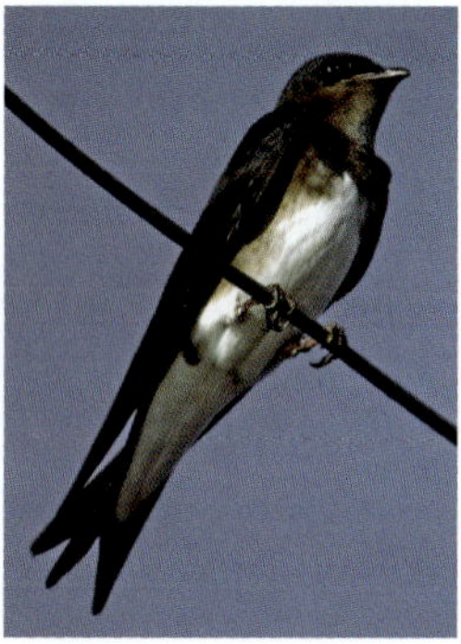

Adult female. Dom. Republic, May.

Range West Indies (except Cuba and the Isle of Pines), Tobago and Mexico.
Status in Jamaica Locally common summer visitor and breeding species (February–October).

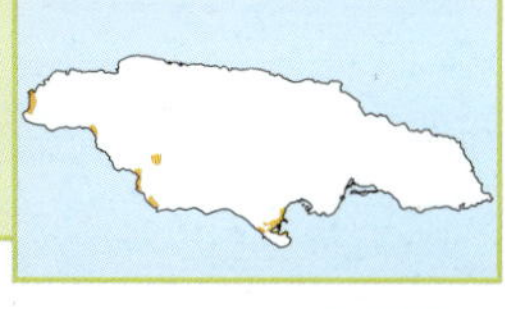

Cave Swallow
Petrochelidon fulva

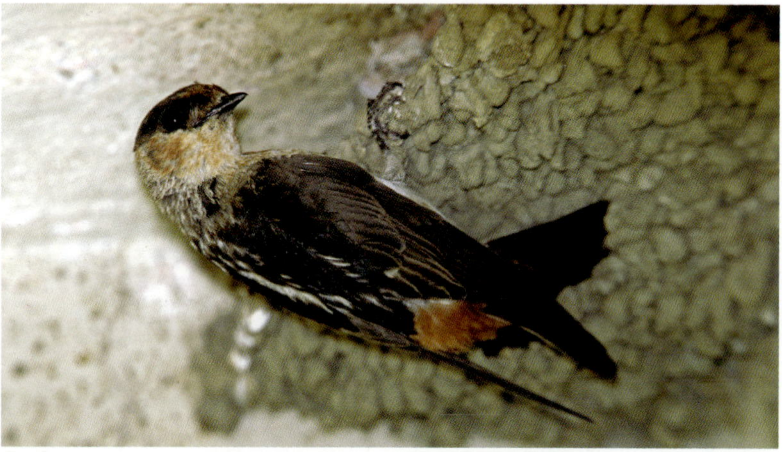

Adult *P. f. fulva* at the nest. Dominican Republic, May.

Taxonomy Polytypic (6). *P. f. poeciloma* is an endemic race.
Description L 13cm (5in). In flight has characteristic short-necked silhouette. Head has dark crown, rufous forehead and rich buffy throat. Belly light creamy white. Wings dark. Rump rufous-brown. Tail dark and very shallowly forked or square.
Similar species Vagrant Cliff Swallow (Appendix 2) has pale forehead; may be overlooked because of its similarity to Cave Swallow.
Voice Sweet-sounding twittering *twit-swee* or *wit-wit*.
Habitat and behaviour Seen over wetlands, forests and wooded areas islandwide. Feeds in dense flocks on mosquitoes and other small insects. Nests colonially (October–May), nest is made of mud and saliva, often horizontally striped brown and white. It is shaped like a flattened cup, attached to a wall or on horizontal beam close to a roof, under an overhanging cliff, in a cave, under a bridge, on a veranda or in a barn. Nesting colonies are used as roosts year round.

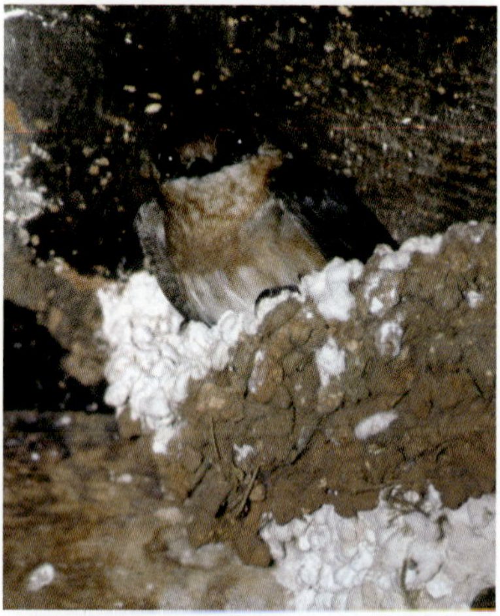

Adult *P. f. poeciloma*. Jamaica, May.

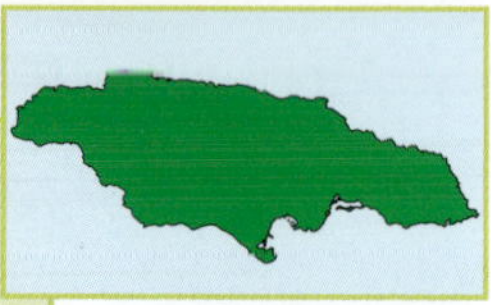

Range *H. f. poeciloma* endemic to Jamaica. Other races in Cuba, Hispaniola, Puerto Rico and North and South America.
Status in Jamaica Common and widespread resident. Local populations are apparently increased by migrants in winter.

Cedar Waxwing
Bombycilla cedrorum

Taxonomy Monotypic.
Description L 15cm (6in). Adult is a small grey and brown bird with a crest behind the crown. Black mask and throat. Secondaries have red wax-like tips. Tail tipped yellow. Juvenile duller, grey head, streaked underparts.
Similar species None.
Voice Usually silent in Jamaica.
Habitat and behaviour Hedges, scrub and disturbed woodland. Usually in flocks of up to 150, flying in close formation or perching on tall trees. Feeds on small figs, privet and other berries and flowers.

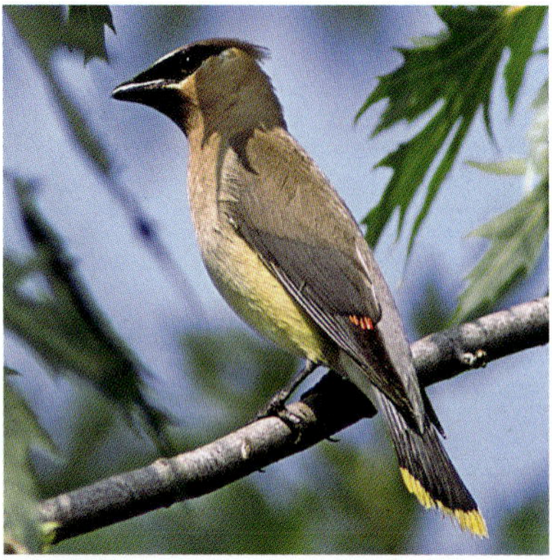

Adult. May.

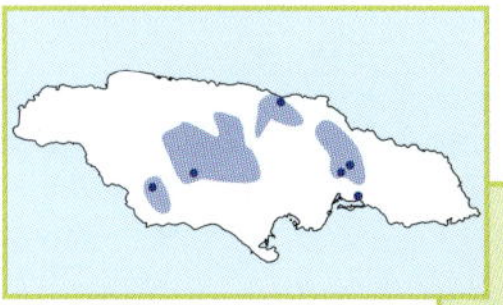

Range Breeds North America. Winters south to northern South America.
Status in Jamaica Rare northbound passage migrant (December–April). Irruptive – common in some years, absent in others.

Grey Catbird
Dumetella carolinensis

Taxonomy Monotypic.
Description L 20cm (8in). A plain grey bird with black crown and rusty undertail-coverts. Tail black, sometimes cocked upwards.
Voice *Me-eeew* – like a cat. More often heard than seen. Does not sing in winter until just before departure in April.
Similar species None.
Habitat and behaviour Lowlands, in thick undergrowth such as bushy open lots or disturbed scrubby margins of herbaceous wetlands. Shy and secretive. Feeds on insects, seeds and berries in the undergrowth.

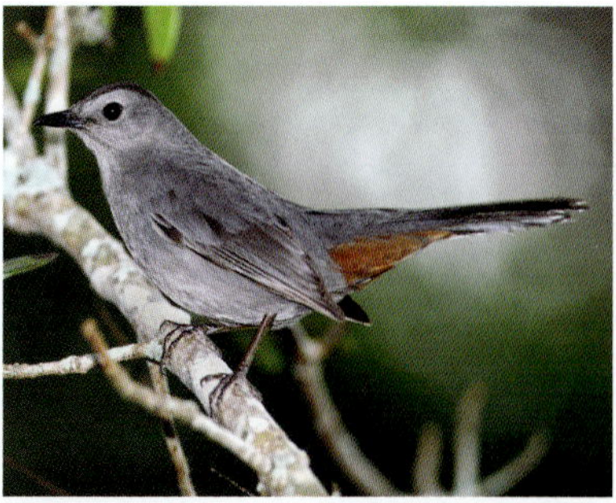

Adult. Florida, February.

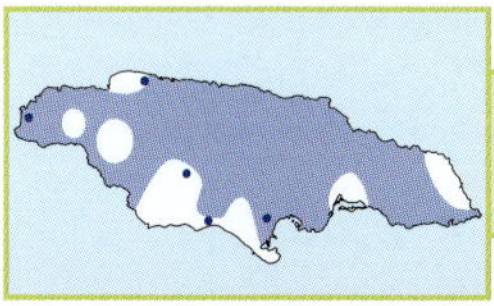

Range Breeds in North America and Bermuda. Winters south to South America via the Greater Antilles and Central America.
Status in Jamaica Uncommon winter visitor (September–April).

Bahama Mockingbird
Mimus gundlachii

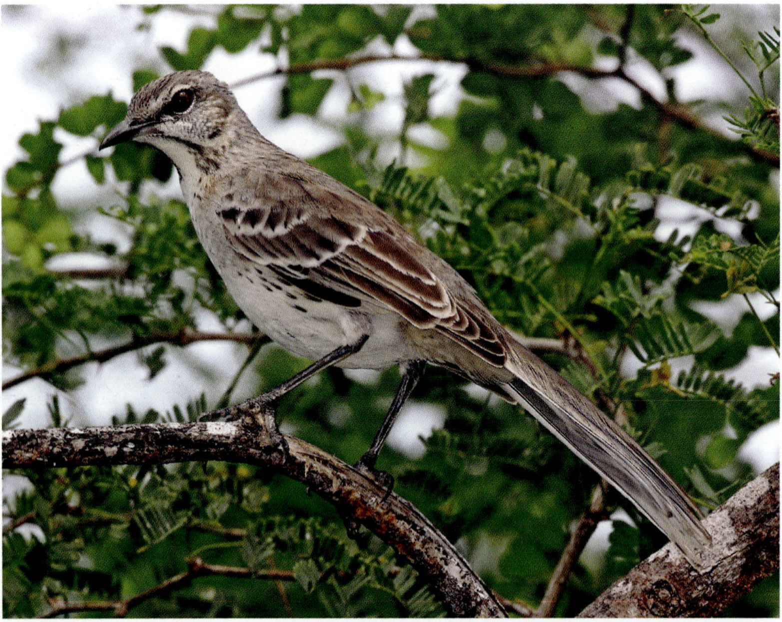

Adult. Note lack of white flash in wing. Portland Ridge, Jamaica, November.

Local names Hill's Mockingbird, Salt Island Nightingale, Spanish Nightingale.
Taxonomy Polytypic (2). *M. g. hillii* is endemic to Jamaica.
Description L 28cm (11in) Brownish grey above with fine dark streaks. Underparts whitish, finely streaked brownish on flanks. Undertail-coverts mottled. Face mottled, with blackish malar stripes. Wing-coverts edged white. Tail white-tipped except for the two central feathers. Legs are dark brown. Juveniles are similar.
Similar species Northern Mockingbird is smaller and greyer, has a white flash in the wing and lacks mottling on face; it also has white outer-tail feathers.

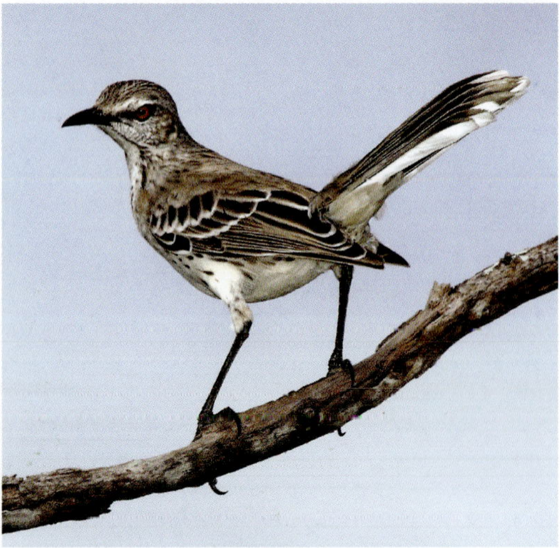

Adult. Note white-tipped tail. Portland Ridge, Jamaica, November.

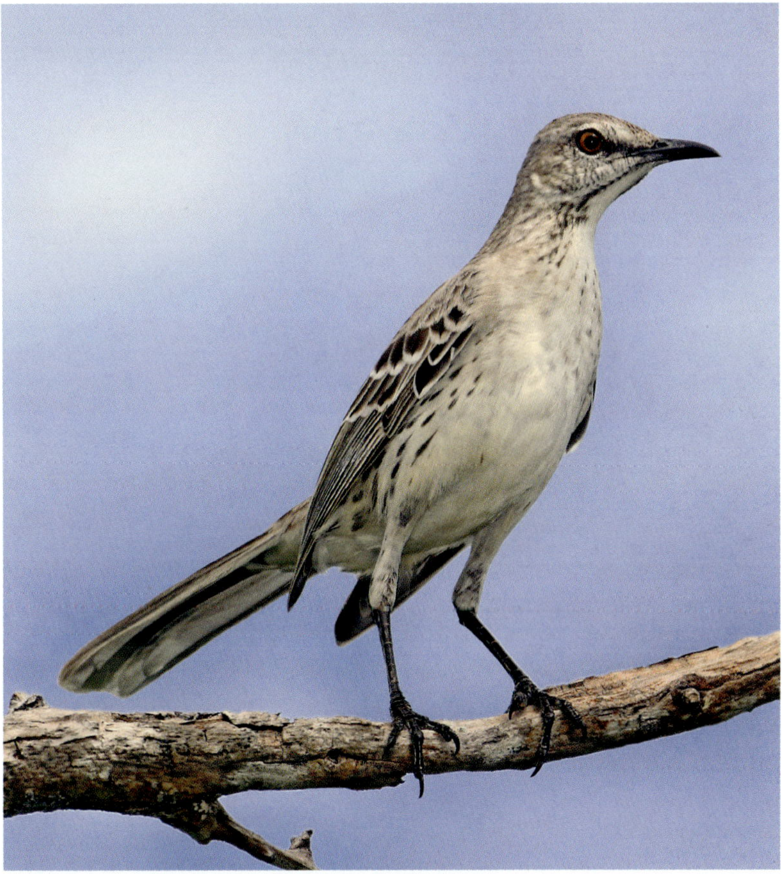

Adult. Note streaks on breast and flanks. Portland Ridge, Jamaica, November.

Voice Similar to Northern Mockingbird but richer. Does not mimic other birds. Call is *torkey-torkey-torkey-tork, chup chup chup*. During breeding season, sings from an exposed perch, throwing itself vertically into the air at the crescendo of the song.

Habitat and behaviour Restricted to coastal dry limestone forest in the Portland Bight Protected Area – Hellshire Hills, Portland Ridge and Salt Island Lagoon; there is overlap with Northern Mockingbird on the edges of the range. Behaviour of Bahama Mockingbird similar to Northern but it rarely feeds on the ground. Nest (mainly February–June) is an open cup, built in a scrubby tree such as cashew or cactus.

Range *M. g. hillii* endemic to Jamaica. Other race in Bahamas, including Great Inagua, the cays off the northern coast of Cuba, and the Caicos.

Status in Jamaica Very locally common, range-restricted resident. May be declining due to serious hurricane damage to its main habitats in 2003 and 2007.

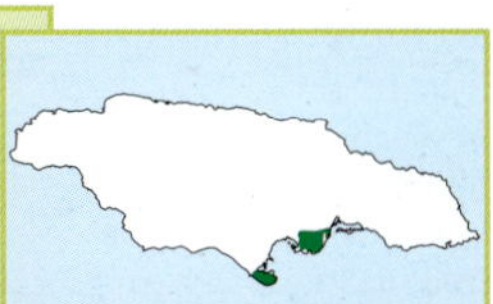

Northern Mockingbird
Mimus polyglottos

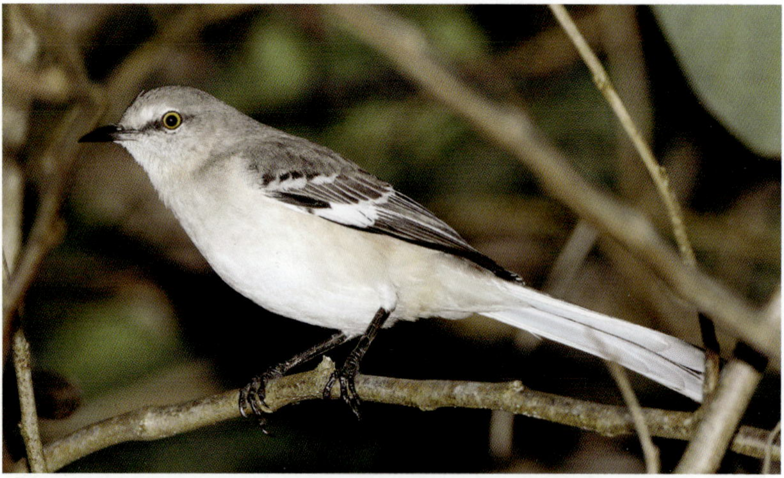

Adult. Note unstreaked underparts and white flash in wing. Ecclesdown, Jamaica, November.

Local name Nightingale
Taxonomy Polytypic (3). Race in Jamaica is *M. p. orpheus.*
Description L 21cm (8in). Grey above, contrasting with white underparts. Wings dark with conspicuous white patches. Tail dark with conspicuous white on outer three feathers. Immature browner than adult with less white on wing; throat streaked.
Similar species Bahama Mockingbird similar but larger, and has a white tipped tail.
Voice A beautiful songster and expert mimic of other birds and ambient sounds, including car alarms in towns and shorebirds in swamps. May repeat *Peter, Peter* frequently in its song. After nesting does not sing again until Oct–Nov but has a loud *chet* call. Begging call of young is a penetrating and oft-repeated *pseeent*.
Habitat and behaviour Disturbed areas such as gardens, forest edges, also wetlands. Year-round in the lowlands, also in higher areas in summer. Tail usually cocked upward. Males sing from the highest perch in the area. Feeds on insects and small fruits. When foraging on the ground jerk its wings frequently to display the white wing-patches. Nest (Nov–June) is a roughly built, thorny, open cup constructed in trees or hedges.

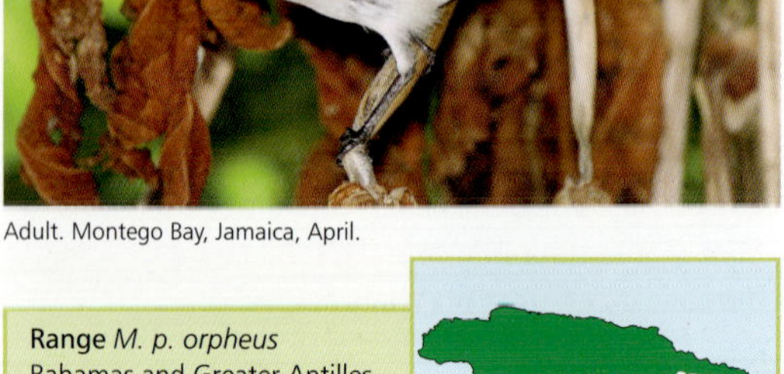

Adult. Montego Bay, Jamaica, April.

Range *M. p. orpheus* Bahamas and Greater Antilles. Other races in North America.
Status in Jamaica Common and widespread resident.

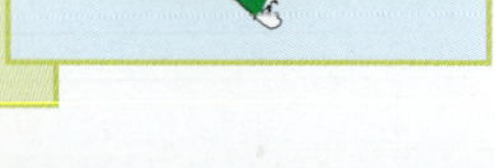

Rufous-throated Solitaire
Myadestes genibarbis

Local name Solitaire, Mountain Whistler, Fiddler.
Taxonomy Polytypic (6). *M. g. solitarius* is endemic to Jamaica.
Description L 19cm (8in). A bluish-grey thrush with a reddish-brown throat and bright white crescent below eye. Head and back dark bluish-grey with a tiny white spot under the chin and faint greyish-white marks on either side of the bill. Iris brown. Throat and undertail-coverts chestnut. Belly pale grey. Wings dark blackish grey edged white. Tail blackish-grey with three outer feathers mostly white. Legs yellow. Juvenile has rufous spots on back and underparts, cinnamon-tipped wing-coverts and salmon crescent below eye.
Similar species None.
Voice Song in summer variable slow flute-like whistles and trills. Winter call is a single long toot like a distant car horn, as well as other short calls. Ventriloquial and therefore difficult to locate by call
Habitat and behaviour In summer (Apr–Nov) breeds in mountain forests above 1000m. Distinct seasonal movement, descending to spend the winter (Dec–Mar) in mid-level woodlands but never found on the coastal plains or in dry forests. Feeds on fruits and insects. Nest (Apr–Aug) is an open cup in a cavity in a tree fern, creeper or bromeliad.

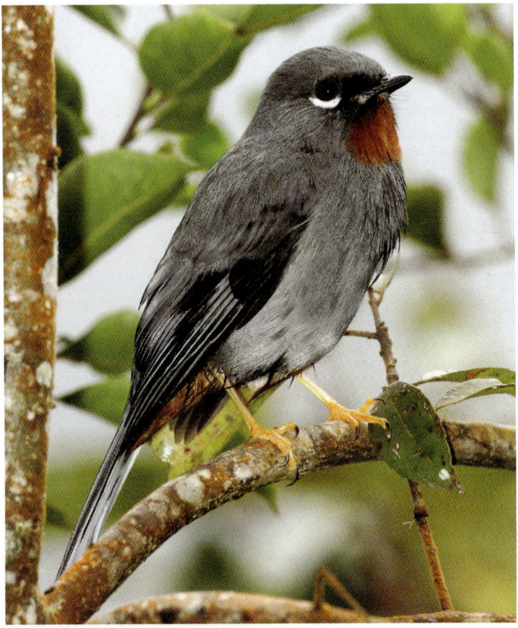

Adult. Port Royal Mountains, Jamaica, February.

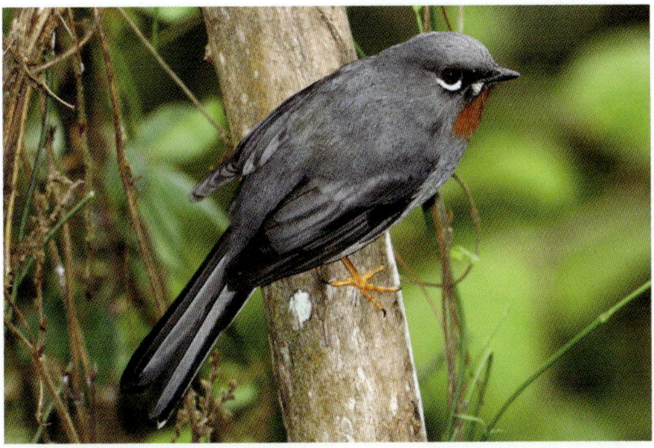

Adult. Port Royal Mountains, Jamaica, February.

Range *M. g. solitarius* endemic to Jamaica. Other races in Hispaniola, Dominica, Martinique, St Lucia and St Vincent.
Status in Jamaica Fairly common resident.

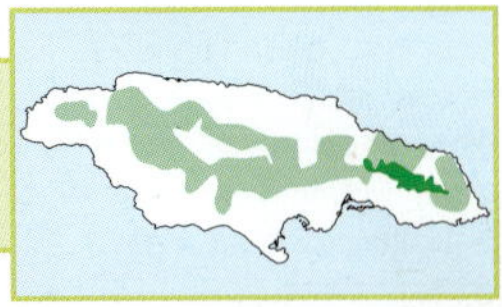

Veery
Catharus fuscescens

Taxonomy Polytypic (4).
Description L 15cm (6in). Medium-sized thrush. Upperparts uniformly reddish brown. Throat and upper breast buffy with faint spots and streaks.
Similar species None.
Voice Rarely calls in Jamaica.
Habitat and behaviour Frequents undergrowth and shrubs in forests, woodlands and gardens.

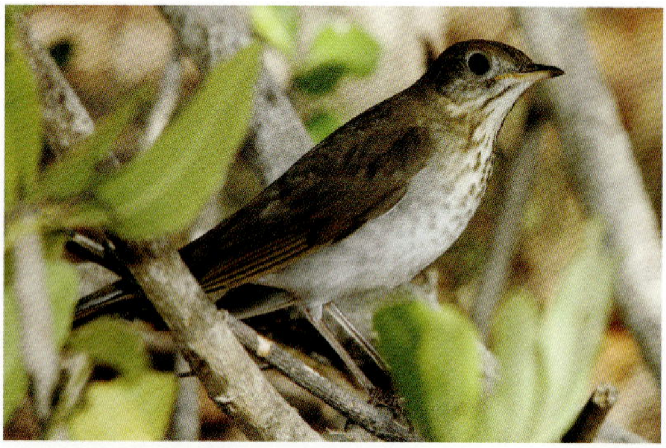

Adult. Note faint spots on upper breast. May.

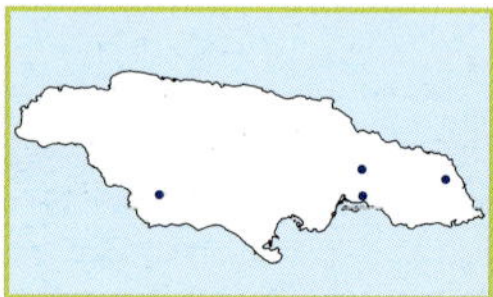

Range Breeds North America. Winters South America usually via Mexico but rarely through Bahamas and Greater Antilles.
Status in Jamaica Rare passage migrant (January–April).

Grey-cheeked Thrush
Catharus minimus

Taxonomy Polytypic (2). Race in Jamaica is *C. m. aliciae.*
Description L 16–19cm (6–7in). Back and tail dull olive-brown. Boldly spotted breast and light underparts. Cheek grey. Eye-ring pale. Upper mandible dark, lower mandible has a pinkish base.
Similar species Bicknell's Thrush has a yellowish-orange base to lower mandible; best distinguished by response to playback.
Voice Rarely calls or sings in Jamaica but responds to playback.
Habitat and behaviour Seen in coastal and midlevel forests, where it skulks in dense undergrowth and is generally found only with playback or when caught during ringing.

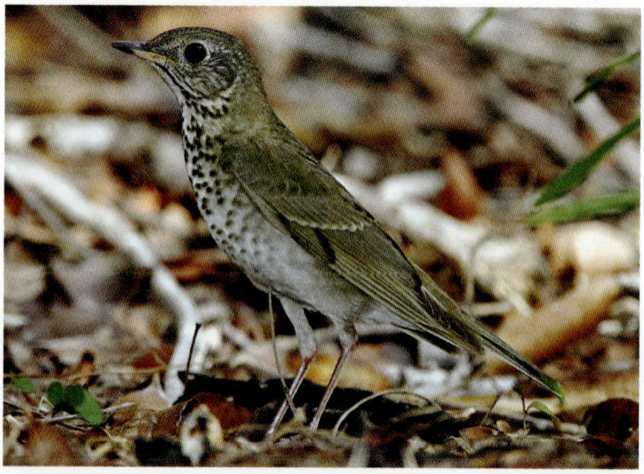

Adult. Note pinkish base to lower mandible and heavily spotted breast. Cayman Islands, April.

Range *C. m. aliciae* breeds southeast Canada and winters south to the West Indies.
Status in Jamaica Regular but scarce passage migrant (Oct–Dec).

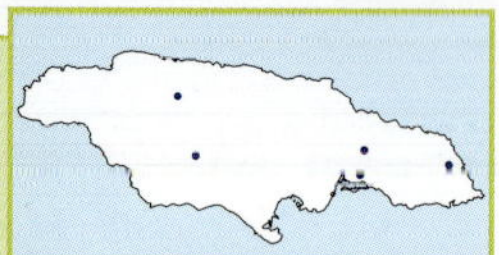

Bicknell's Thrush
Catharus bicknelli

Taxonomy Monotypic.
Description L 16–17cm (6in). Slender thrush with olive-brown back contrasting with rich reddish-brown tail. Cheeks grey. Eye-ring indistinct. Upper mandible dark, lower has a yellowish-orange base. Breast creamy-buff, boldly spotted with black on sides of throat. Underparts otherwise white.
Similar species Very similar Grey-cheeked Thrush best distinguished by response to playback. Swainson's has distinctive eye ring. Veery has finer spots and redder back.

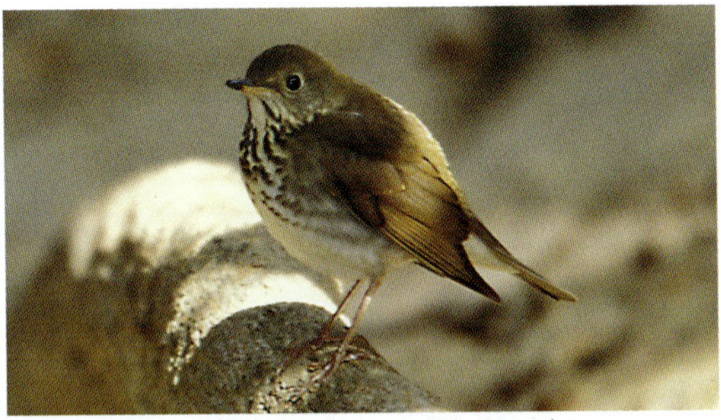

Adult. Note yellow-orange base to lower mandible. September.

Voice Rarely calls in Jamaica except in response to playback.
Habitat and behaviour Restricted to the open understorey of old-growth forests above 1000m. Very secretive, may be overlooked.

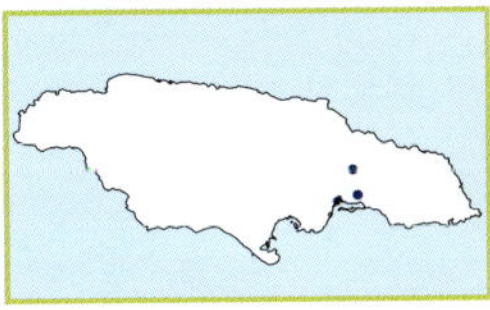

Range Breeds in eastern N America. Winters in the West Indies, mainly in Hispaniola. Global conservation status: Vulnerable.
Status in Jamaica Extremely rare and hard to see.

Swainson's Thrush
Catharus ustulatus

Taxonomy Polytypic (4). Race in Jamaica is *C. u. swainsoni.*
Description 15cm (6in). Olive-brown above, white below, but throat and breast are buffy, streaked darker. Iris dark. Buffy eye ring.
Similar species Ovenbird is smaller and has a crown-stripe.
Voice Silent in Jamaica.
Habitat and behaviour Usually seen in dry forests near mangroves.

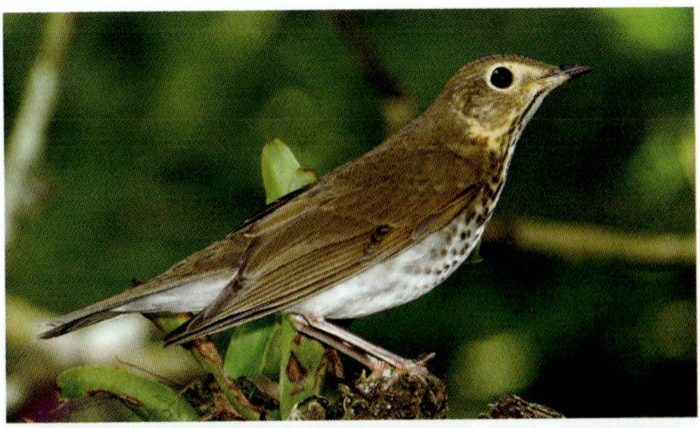

Adult. Note buffy 'spectacles'. Florida, October.

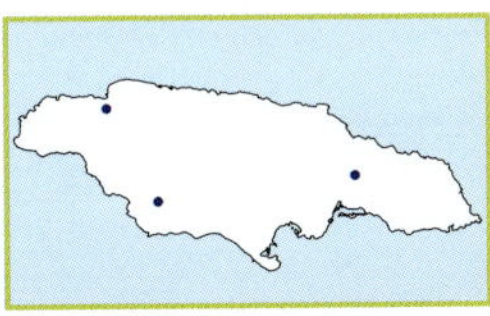

Range *C. u. swainsoni* breeds in eastern North America. Winters south to the West Indies and to northern Argentina.
Status in Jamaica Uncommon winter passage migrant (September–October, April)

White-eyed Thrush
Turdus jamaicensis

Local names Glass-Eye, Shine Eye.
Taxonomy Monotypic. Endemic to Jamaica.
Description L 23cm (9in). Entire head bright rufous-brown with conspicuous greyish-white iris, upperparts otherwise dark grey. Bill black. Throat white, streaked rufous-brown. White upper breast-band conspicuous. Lower breast and sides grey, fading to whitish on mid-abdomen. Undertail-coverts white, spotted grey. Legs and tail medium brown. Juvenile lacks streaking on throat but heavily streaked on breast; eye is grey.
Similar species None.
Voice Song is varied and musical, each phrase repeated two or three times. Has a two-note location call. Alarm notes are harsh and shrill.
Habitat and behaviour Forested gullies, hills and mountains. Apparently more abundant at mid-levels in February–July but altitudinal movement has not been documented. Feeds on fruit and insects from tops of trees to ground level. Nest (April–June) is cup-shaped.

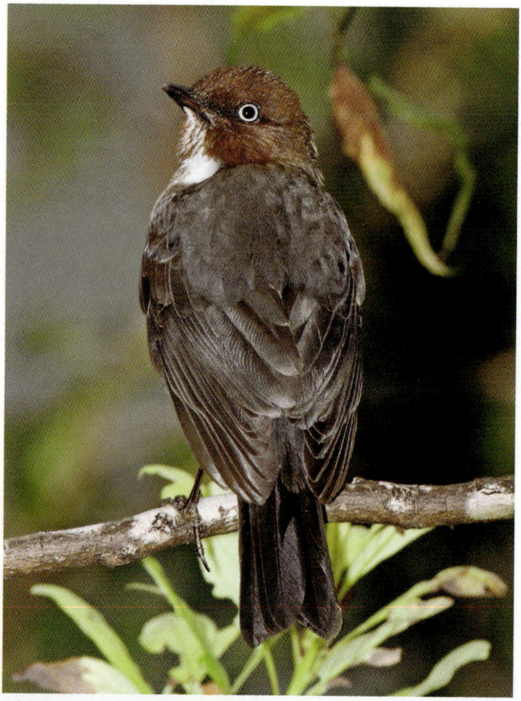

Adult. Port Royal Mountains, Jamaica, November.

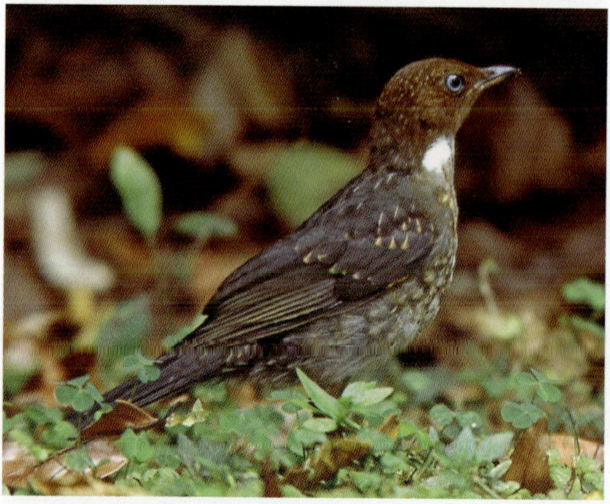

Subadult. Marshall's Pen, Jamaica, May.

Range Endemic to Jamaica.
Status in Jamaica Locally fairly common resident.

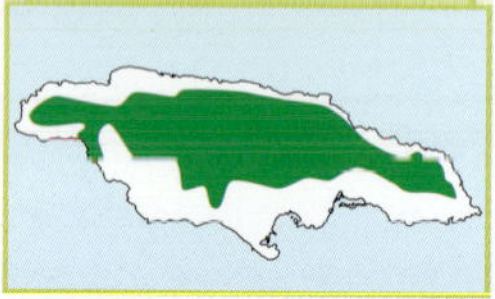

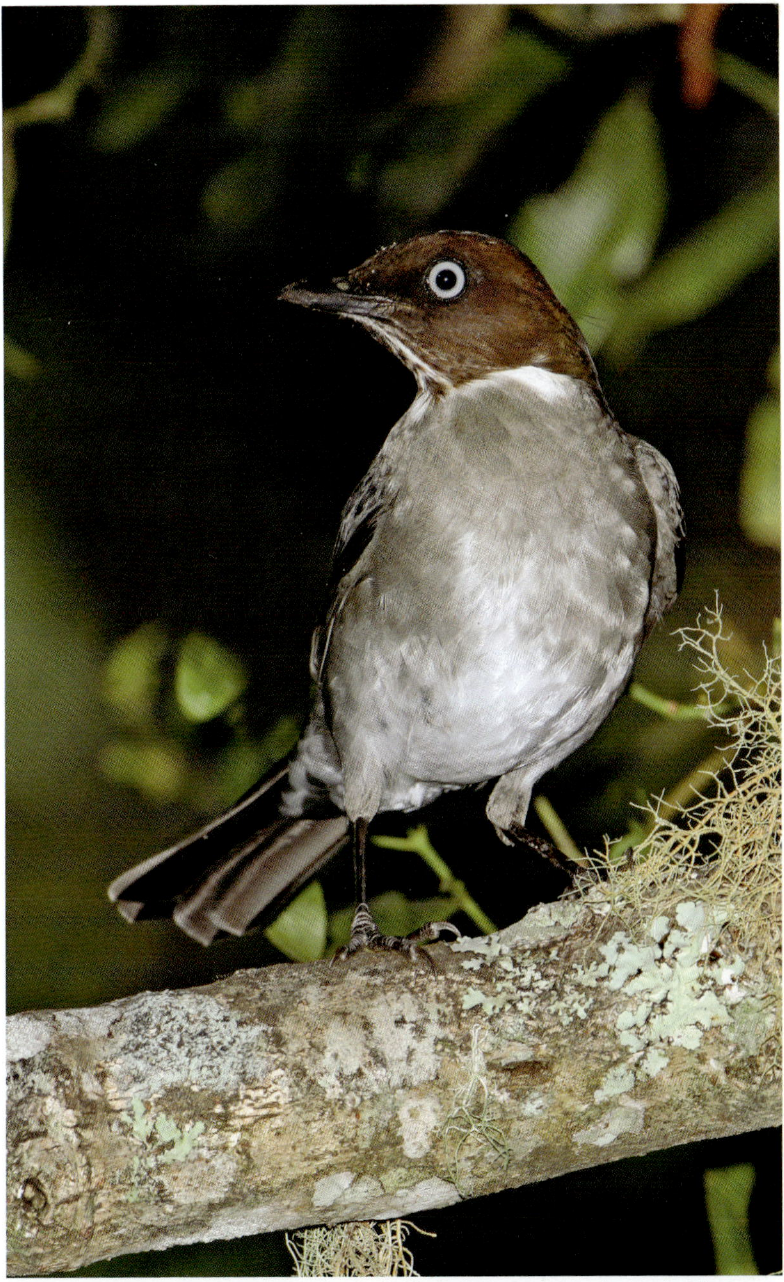

Adult in typical alert pose. Note white iris and spotted undertail-coverts. Port Royal Mountains, Jamaica, February.

White-chinned Thrush
Turdus aurantius

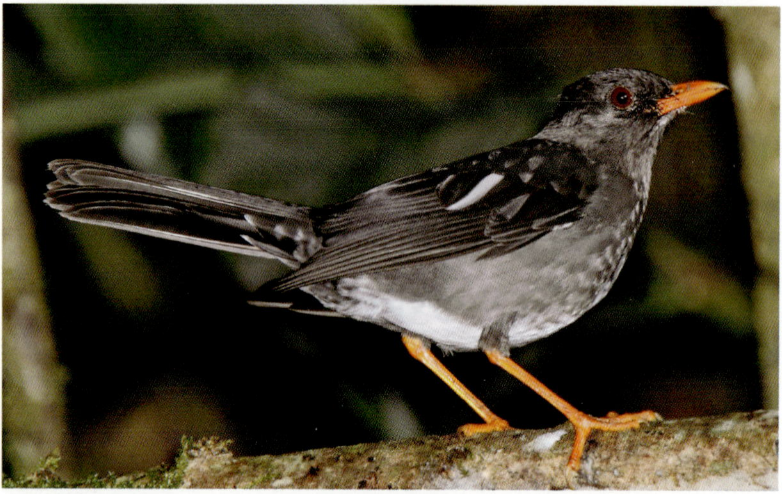

Adult. Note bright orange bill, legs and feet. Port Royal Mountains, Jamaica, November.

Local name Hopping Dick, Chapman-Chick.
Taxonomy Monotypic. Endemic to Jamaica.
Description L 24cm (9.5in). A stocky, long-tailed, dark-backed thrush. One or two of inner greater coverts variably edged white, forming a white patch. The 'white-chin' is small and easily overlooked. Bill bright orange with a black tip. Legs orange, brighter in the breeding season. Iris chestnut. Upperparts and wings dark grey. Underparts paler, mostly grey with centre of belly white, undertail-coverts grey with white tips. Tail plain black and often held erect. Juvenile darker browny-grey on breast; with streaked belly, grey lower underparts and a few white flecks in undertail-coverts.
Similar species None.
Voice Varied, from a melodious lullaby in the breeding season to shrill whistles *p'lice, p'lice,* and chicken-like clucking, repeated for long periods.
Habitat and behaviour Wooded hills and mountains, but occurs at sea level on the north and southwestern coasts. Most often seen hopping along the verges of country roads. Feeds on lizards, insects, slugs and berries. Nests (May-July) at the base of coconut fronds and in shrubs.

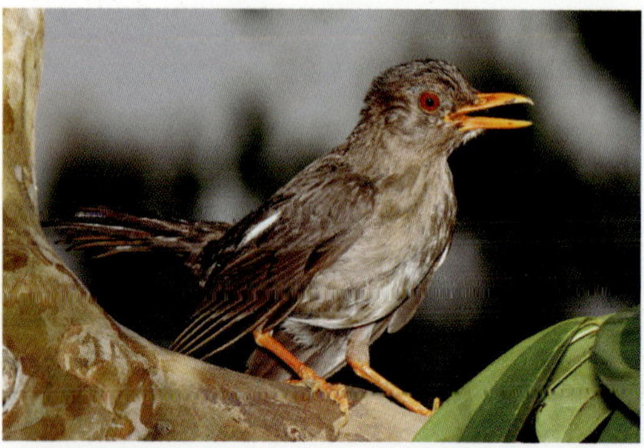

Juvenile brownish with dull orange bill and legs. Rocklands, Jamaica, May.

Range Jamaica.
Status Common resident.

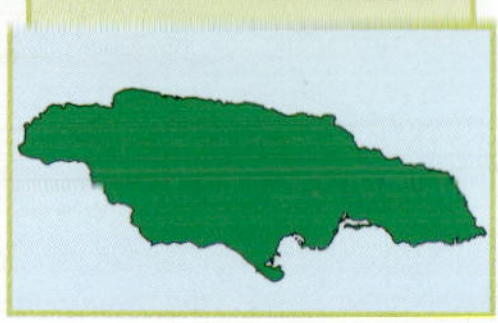

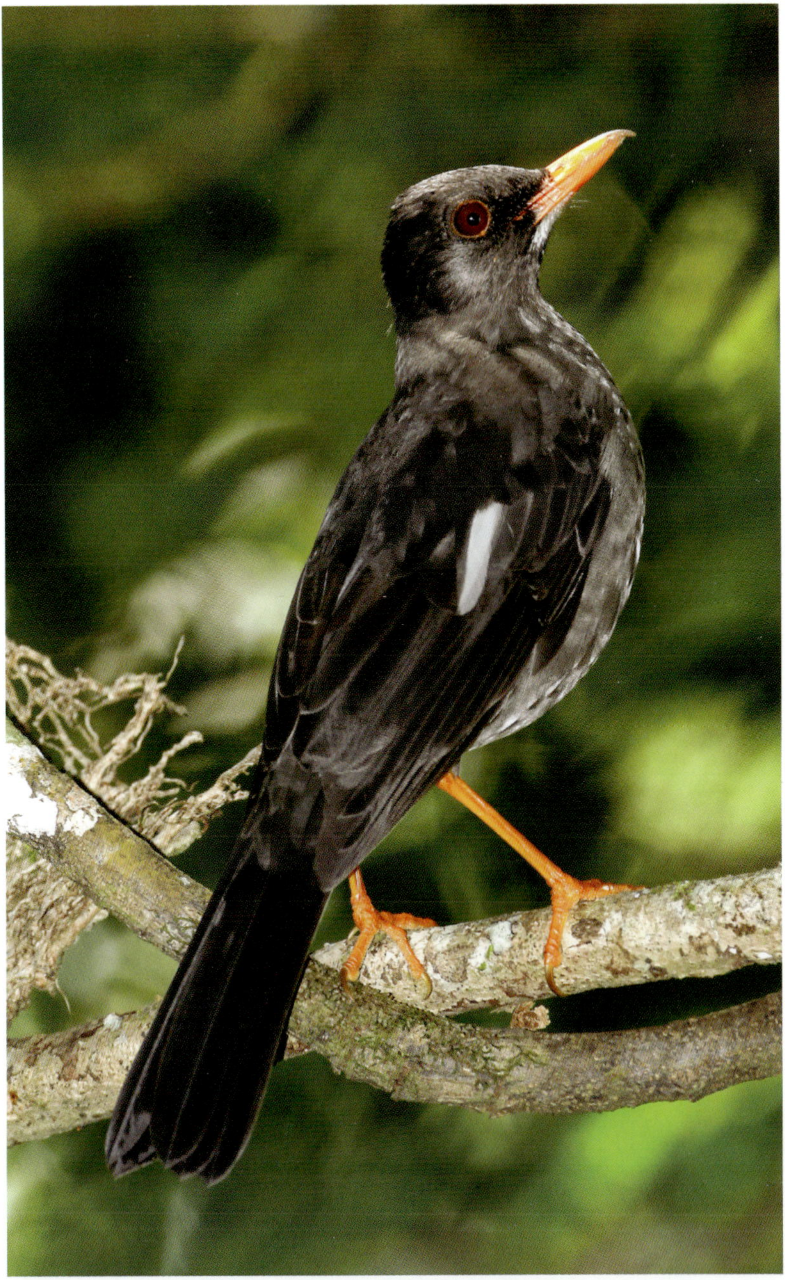

Adult. Note white patch in wing and inconspicuous white chin. Port Royal Mountains, Jamaica, November.

Jamaican Crow

Corvus jamaicensis

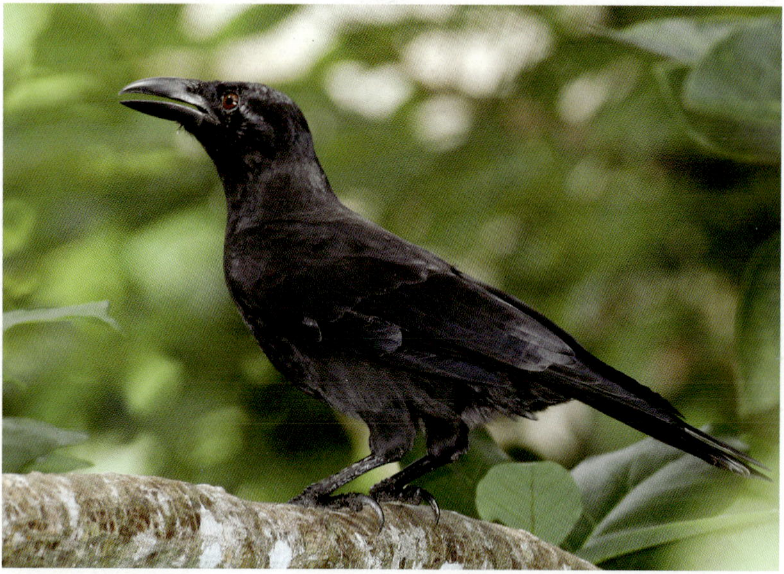

Adult. Rocklands, Jamaica, November.

Local names Jabbering Crow, Jamming Crow, Jamicrow.
Taxonomy Monotypic. Endemic to Jamaica.
Description L 39cm (15in). Unmistakeable large, sooty-black crow with a heavy black bill. Immature browner than adult.
Similar species None.
Voice Harsh *caw-caw* calls and loud jabbering and gobbling – like a turkey being strangled. Usually first located by calls.
Habitat and behaviour Wet mid-level limestone forest. Feeds omnivorously on fruit, eggs, small birds and lizards. Also forages by probing in bromeliads and under bark for amphibians, crabs and insect larvae. Hides seeds in bromeliads for later use. May fly out of the forest, ranging over wide areas to feed in gardens, cultivations and commercial plantations e.g. on pawpaw where it may be a pest. Nest (April–June) is a roughly built platform high in a tall tree.

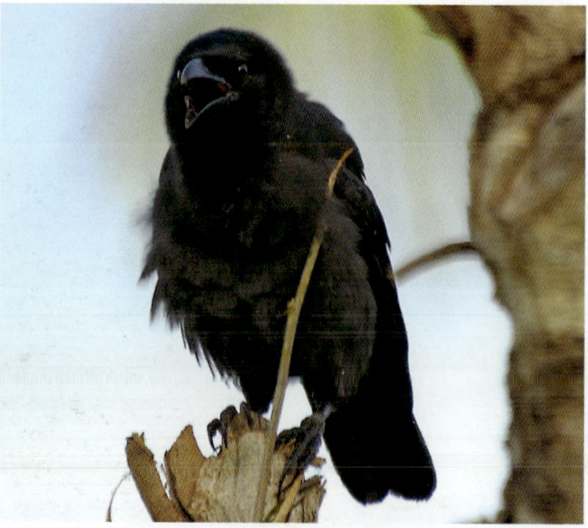

Adult. Hanover, Jamaica, September.

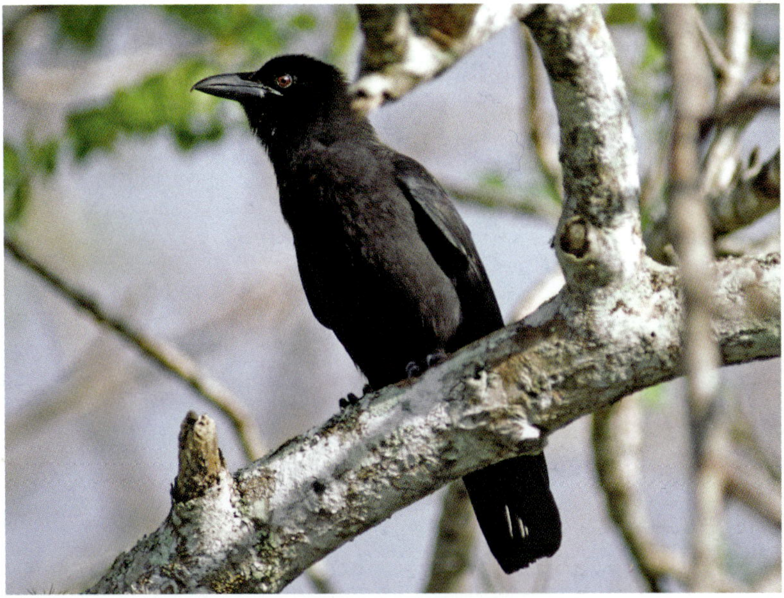

Adult. Good Hope, Jamaica, May.

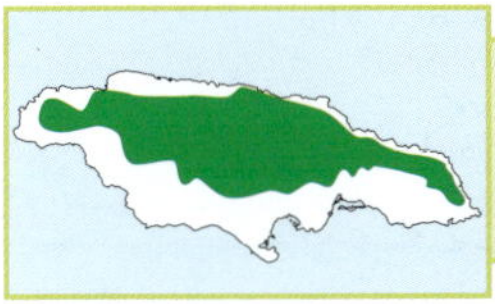

Range Jamaica.
Status in Jamaica Locally common resident. May be extending its range into agricultural lands island-wide.

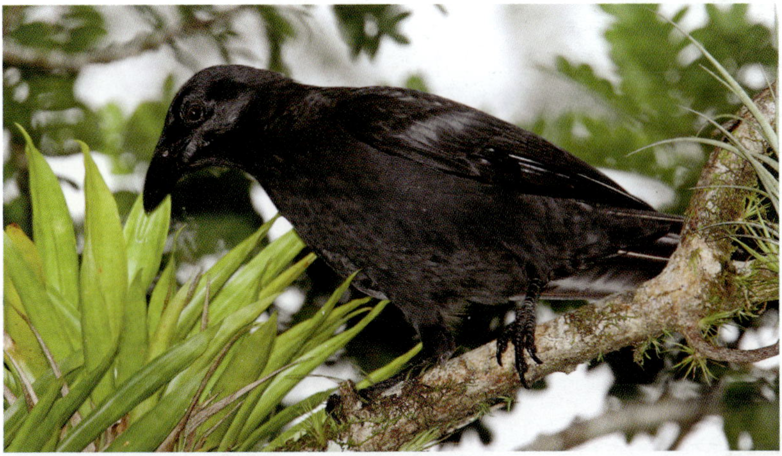

Adult. Ecclesdown, Jamaica, November.

European Starling
Sturnus vulgaris

Local name Starling
Taxonomy Polytypic (12).
Description L 15cm (6in). Adult in breeding plumage is black and glossy with small white spots on the back and glossy purplish iridescence on the head and neck. Bill is bright yellow. Legs pink. Non-breeding adult heavily spotted white overall, bill dark. Juvenile is grey-brown with a dark bill.
Similar species None.
Voice A soft descending whistle. Also clicks or mimics sounds such as creaking gates.
Habitat and behaviour Found chiefly in the lowlands and mid-levels, in parks, gardens and pastures. Moves around the island, and at a particular place can sometimes be abundant and at other times absent. Usually seen in flocks, which can be quite large. Wings appear pointed in flight. Roosts colonially e.g. in thousands at electricity sub-stations in Kingston. Nests (April–June) in natural and artificial cavities and woodpecker holes (sometimes persistently and aggressively evicting the woodpeckers).

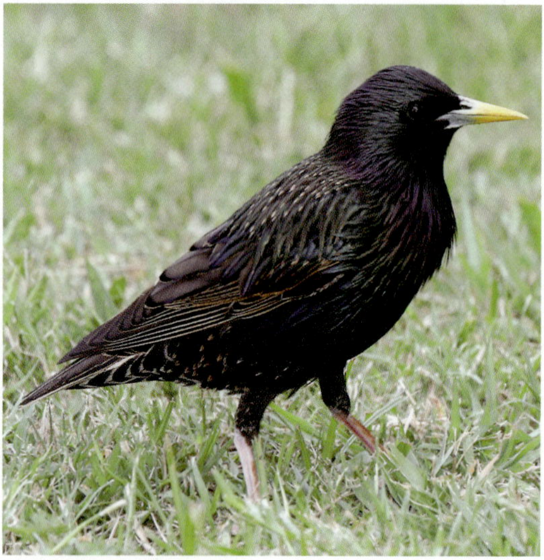

Adult breeding. Montego Bay, Jamaica, May.

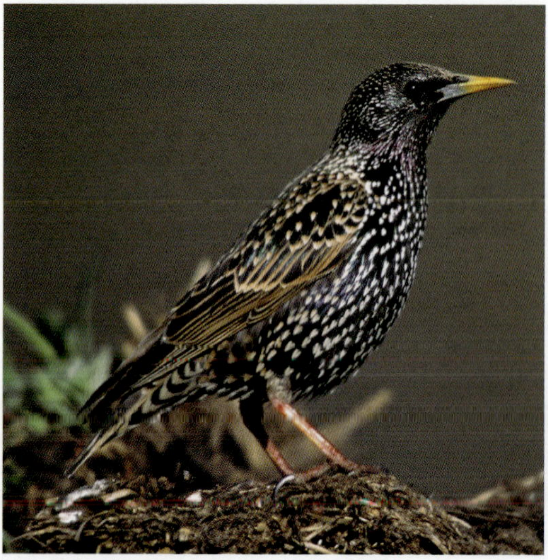

Adult non-breeding. Montego Bay, Jamaica, November

Range Old World. Introduced to Jamaica in about 1903. After a slow start its numbers increased rapidly in the 1960s and 1970s but now seem to have stabilised.
Status in Jamaica Common resident.

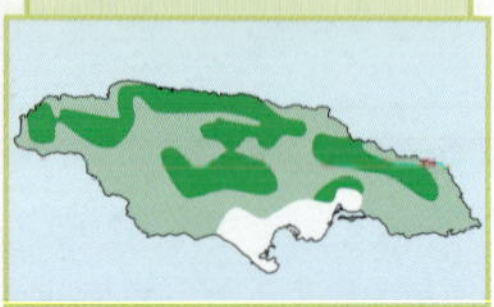

House Sparrow
Passer domesticus

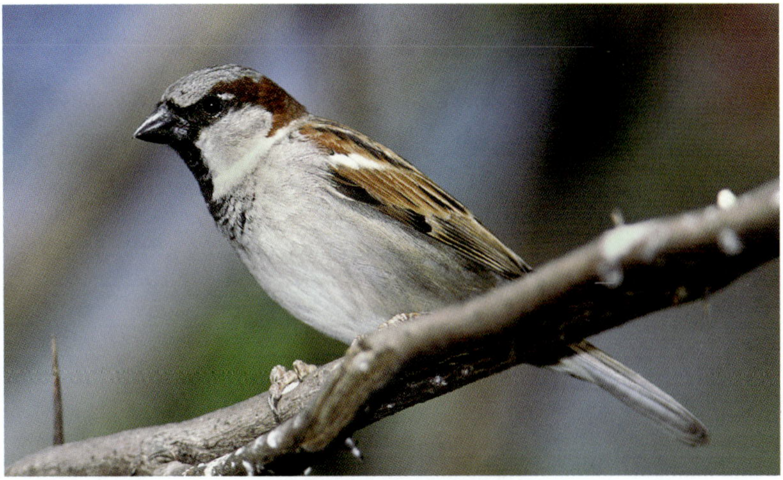

Adult male. Florida, May.

Taxonomy Polytypic (12).
Description L 16cm (6in). Small finch-like brown bird with a short, thick bill. Breeding male has a distinctive black bib. Crown grey, cheek whitish. Back rich brown with darker streaks. Wings are brown with a light wing-bar. Underparts greyish. Non-breeding male paler and black on breast may be obscured. Adult female and juvenile lack black bib, and are generally dusky brown with a buffy stripe above the eye.
Similar species None.
Habitat and behaviour Coastal scrub and garden. Feeds on seeds on the ground. Nests (of dried grasses lined with feathers) in coconut palms.

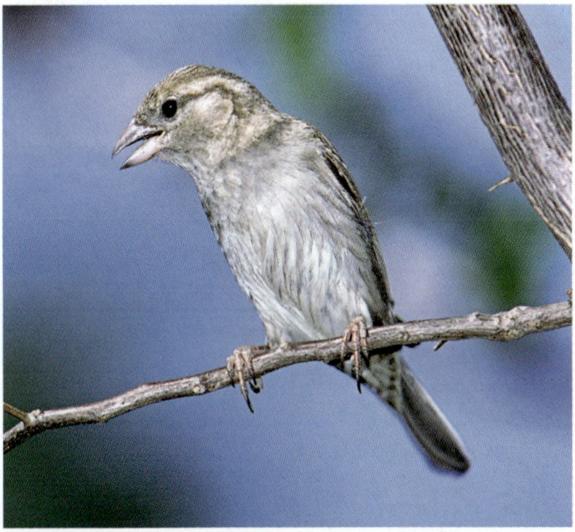

Adult female. Bahamas, May.

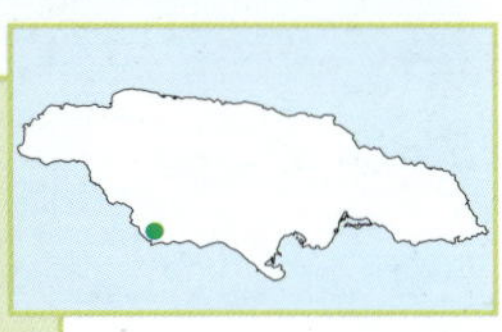

Range Eurasia and Africa. Introduced to Jamaica in 1903 but did not become established until after Hurricane Gilbert, since when it has been seen regularly in Treasure Beach, St Elizabeth (but not elsewhere). Also established Hispaniola and several other islands.
Status in Jamaica Very rare and local.

Yellow-crowned Bishop
Euplectes afer

Taxonomy Polytypic (3).
Description L 14cm (6in). Breeding male is bright yellow on the crown, nape and lower back. The lower face, neck, upper breast and centre of the belly are jet black. The wings are brown, short and rounded. Tail is dark brown, short and square. Legs are pinkish. Adult female, non-breeding male and juvenile are brown and sparrow-like.
Similar species Females and non-breeding males are similar to the Grasshopper Sparrow but the latter has a prominent crown stripe. Juvenile and immature birds resemble grassquits.
Voice A high-pitched and squeaky song, reminiscent of that of Vervain Hummingbird.
Habitat and behaviour Trees and reeds growing in or near water courses. Feeds on seeds, setting off on short, fluttery flights but regularly returning to the same perch.

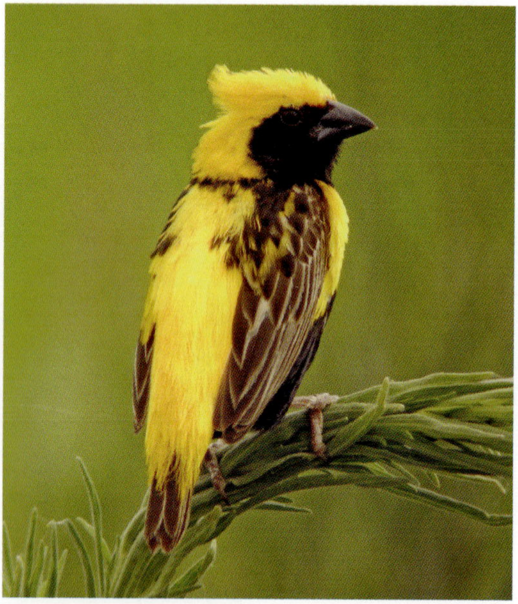
Adult male breeding. July.

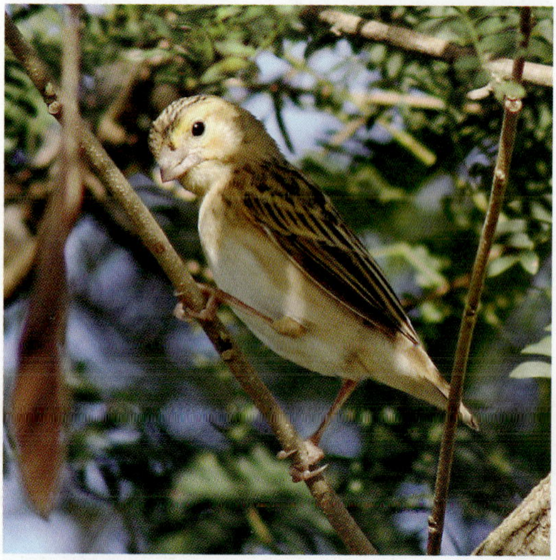
Juvenile. August.

Range Central Africa. Accidentally introduced to Jamaica in 1988. No recent information on status. In the West Indies also established in Puerto Rico.
Status Locally common resident.

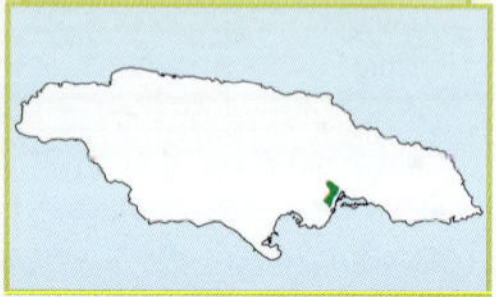

Orange Bishop
Euplectes franciscanus

Taxonomy Monotypic.
Description L 13cm (5in). Breeding male has orange collar and back. Crown, upper face and underparts black. Adult females and non-breeding males are sparrowlike.
Similar species Females and non-breeding males are similar to the Grasshopper Sparrow which has superciliary stripes instead of a crown stripe. Immatures resemble grassquits.
Voice Not described.
Habitat and behaviour Flocks feed on sorghum.

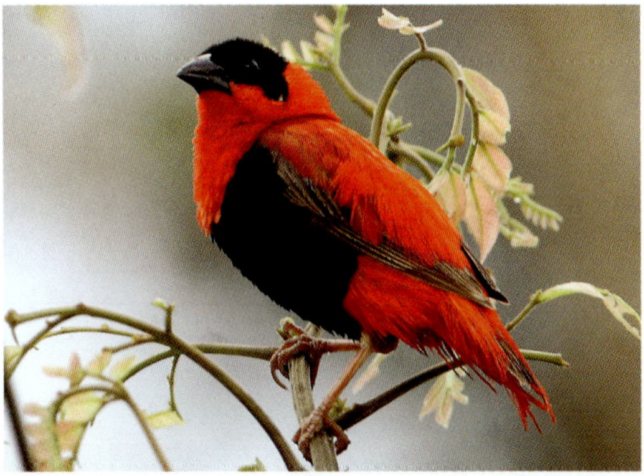
Adult male breeding. October.

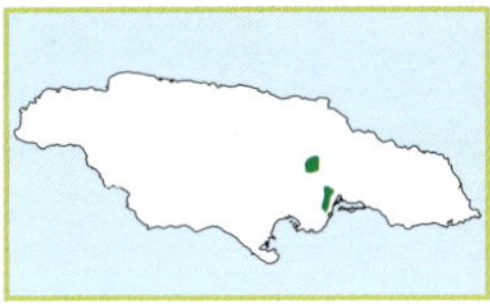

Range Africa. Accidentally introduced to Jamaica in 1988. In West Indies also established in Puerto Rico.
Status Rare and local resident.

Bronze Mannikin
Spermestes cucullatus

Taxonomy Polytypic (2).
Description L 10cm (4in). A small finch with a purple-black hood extending to centre of the white breast. Rich brown upperparts with iridescent green patch on shoulder. White below with dark brown barring on flanks. Tail dark. Heavy grey bill with darker upper mandible. Juvenile brownish.
Similar species Black-headed Munia is larger.
Voice A frequent *treep* given in flight.
Habitat and behaviour Open fields. Feeds in flocks on sorghum and grass seeds. Nest is a bulky dome.

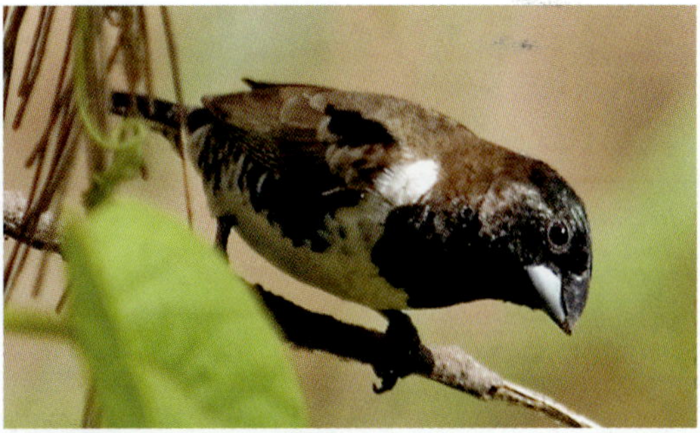
Adult. May.

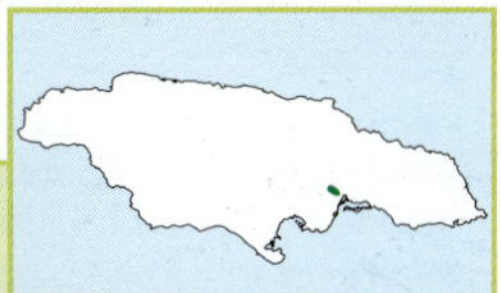

Range Africa. In the W Indies in Jamaica and Puerto Rico.
Status Rare and local resident. Accidentally introduced to Jamaica in 1988. No recent information on status.

Scaly-breasted Munia
Lonchura punctulata

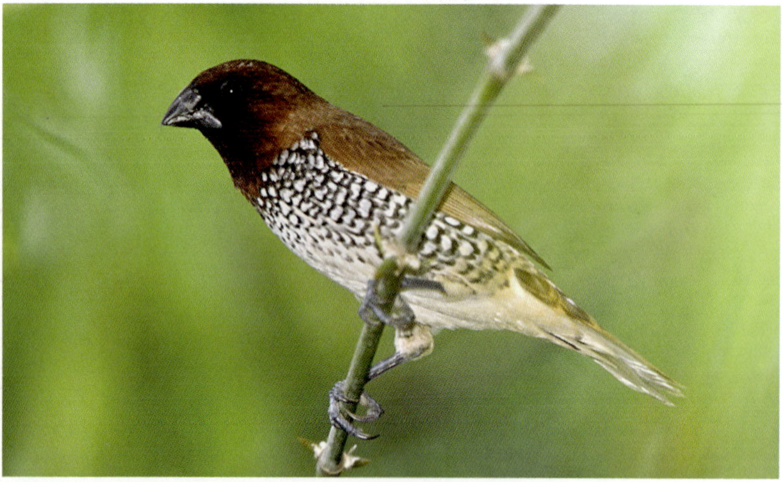

Adult. Dominican Republic, May.

Taxonomy Polytypic (12).
Alternative names Nutmeg Mannikin, Spice Finch.
Description L 12cm (5in). Brownish above, with whitish underparts. Head has darker brown hood and heavy, conical blackish bill. White below, breast and flanks finely scalloped blackish. Juvenile very plain buffish (not scaly), paler below.
Similar species Juvenile Black-headed Munia paler below than juvenile Scaly-breasted.
Voice Soft whistling.
Habitat and behaviour Open grassy fields and road margins, edges of canefields, and disturbed moist lowland forest. Usually seen in small flocks. Feeds mainly on the ground on grass seeds. Nest is a bulky dome with a side entrance.

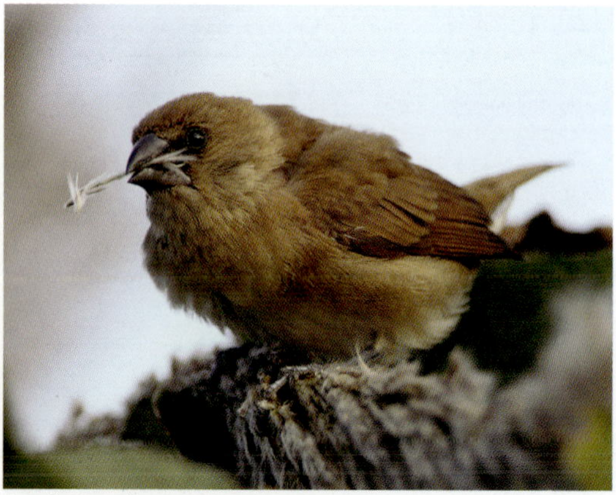

Juvenile. December.

Range Asia. Introduced to Jamaica as an escaped cage bird. In West Indies also introduced to Hispaniola, Cuba, Puerto Rico, Guadeloupe. Vagrant in Virgin Islands and Martinique.
Status in Jamaica Locally common resident.

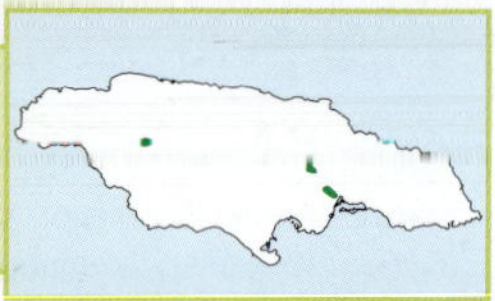

Black-headed Munia
Lonchura malacca

Taxonomy Polytypic (2).
Description L 12cm (5in). Black hood, chestnut upperparts. Underparts white with a black belly and undertail-coverts. Heavy pale grey bill. Juvenile brown above and buffy below.
Similar species Scaly-breasted smaller, darker, lacking rufous.
Voice Has a chirping flight-note.
Habitat and behaviour Seen in flocks on the south coastal plains in woods and on grassy margins of fields and freshwater marshes. Feeds on seeds on the ground or on seed heads; potentially a pest of rice and sorghum. Nest bulky and domed with a side entrance.

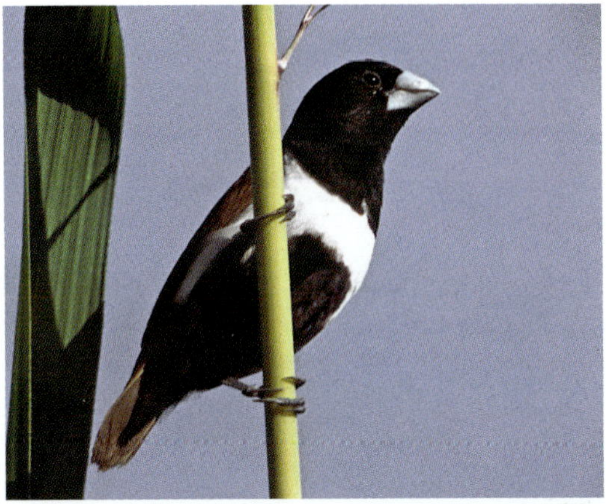

Adult. Dominican Republic, May.

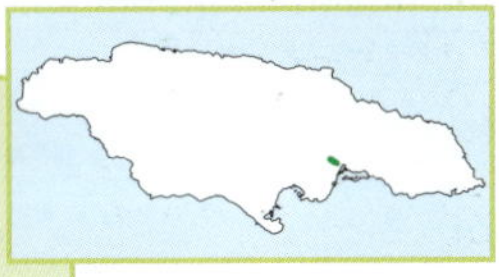

Range India, Nepal and Sri Lanka. In the West Indies in Jamaica, Hispaniola, Cuba, Puerto Rico and Martinique.
Status in Jamaica Locally common resident. Accidentally introduced in 1988 and became widely established in the 1990s.

White-eyed Vireo
Vireo griseus

Taxonomy Polytypic (7).
Description L 15cm (6in). Stocky, greenish with a grey head. Iris whitish. Yellow lores and eye ring resemble spectacles. Flanks are yellow. Wings dark grey with two distinct white wing-bars.
Similar species Jamaican Vireo is plainer with a greenish head.
Habitat and behaviour Usually seen in coastal woodlands. Feeds in shrubbery, close to the ground on insects and occasionally fruit e.g. of Red Birch.

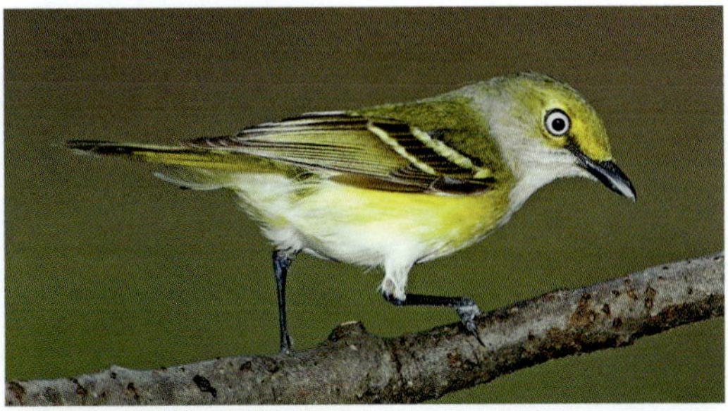

Adult. Note two white wing-bars and yellow 'spectacles'. March.

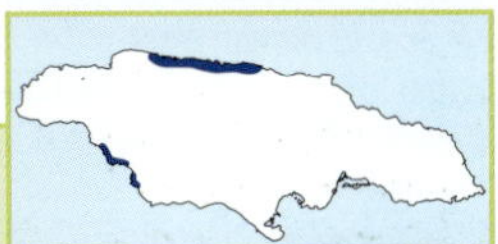

Range Various races North America, Bermuda and Mexico; winter south to Bahamas, Greater Antilles and Central America.
Status in Jamaica Rare passage migrant (January–March).

Jamaican Vireo

Vireo modestus

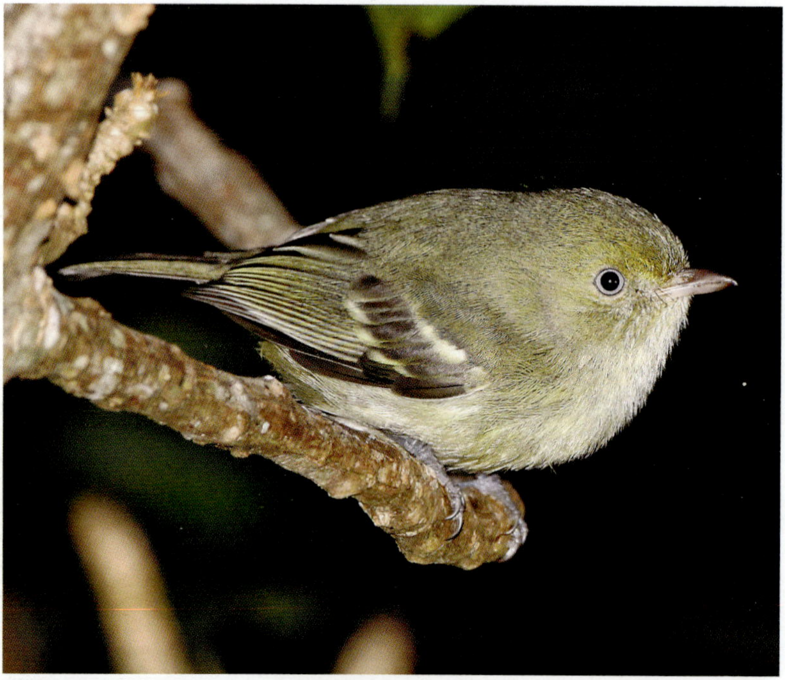

Adult. Note olive back, greenish wings and grey underparts. Port Royal Mountains, Jamaica, November.

Local name Sewi-Sewi.
Taxonomy Monotypic. Endemic to Jamaica.
Description L 11cm (5in). A small, compact, vireo. Wings dark greyish-olive edged with green, with two distinct greenish-white wing-bars. Head plain greenish-olive. Iris very pale grey (appearing white). Back olive. Underparts entirely lemon yellow. Tail square with greenish edges. Bill pinkish brown, paler below. Legs blue-grey. Juvenile greyish with yellow belly, iris is greyish-brown.
Similar species Juvenile Arrowhead Warbler has a dark

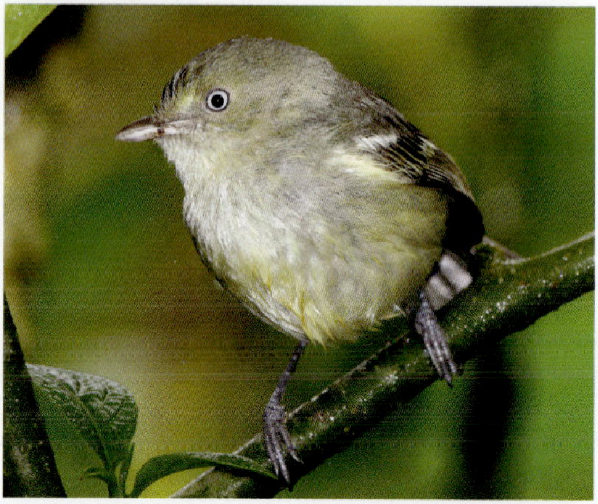

Adult. Note pinkish-brown bill. Port Royal Mountains, Jamaica, February.

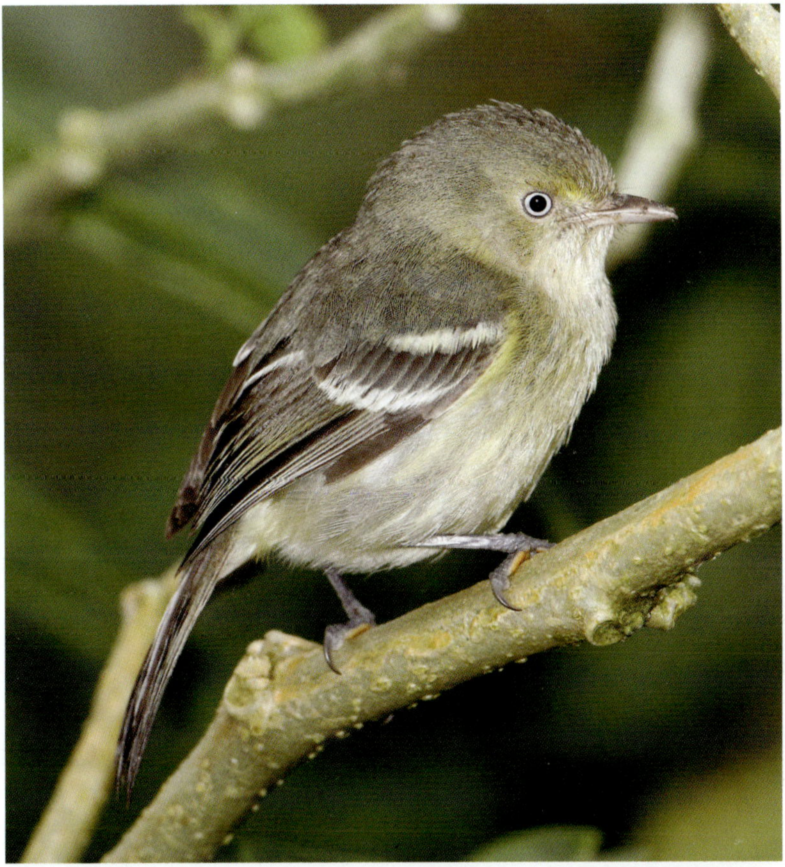

Adult. Note white iris and two distinct white wing-bars. Port Royal Mountains, Jamaica, February.

eye line and dark bill. Female Black-throated Blue Warbler lacks wing-bars.

Voice A wide variety of characteristic songs e.g. *sewi-sewi*, *wichy-wichy-woo* that it changes frequently but repeats for minutes at a time. Songs and calls are often echoed in long duets by a mate foraging nearby. Also has a loud alarm note, a fast *chi-chi-chi-chi-chi*.

Habitat and behaviour Found in bushy areas, forest edges and roadsides at all elevations. Does not occur in urban gardens. Very active and therefore difficult to see as it flits rapidly about in the understorey or overgrown bushes. Feeds on insects and small fruits. Nest (April–June) is a densely woven cup often suspended from a forked branch, constructed of fern or *Tillandsia* stalks and decorated with lichen.

Range Endemic to Jamaica.
Status in Jamaica Common and widespread resident.

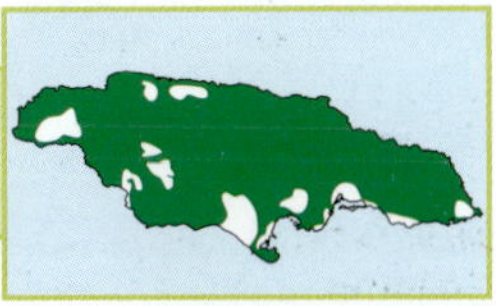

Blue Mountain Vireo

Vireo osburni

Taxonomy Monotypic. Endemic to Jamaica.
Description L 13cm (5in). A rather plain, chunky vireo. The blue-grey head contrasts with olive back, wings and tail. Underparts pale yellow. Lacks wing-bars. Bill heavy, black. Iris reddish brown. Juvenile browner above and has light grey underparts grading to pale yellow towards the tail.
Similar species Jamaican Vireo is smaller and has white iris and wing-bars. Black-whiskered Vireo is larger and has stripes on face. Female Orangequit is greyer.
Voice A rolling whistle. Alarm call is a harsh trill.
Habitat and behaviour Mountain forests in moist areas and among ginger lilies at roadsides (500–2000m). Feeds among leaves and is more often heard than seen. Nest resembles that of Jamaican Vireo but larger.

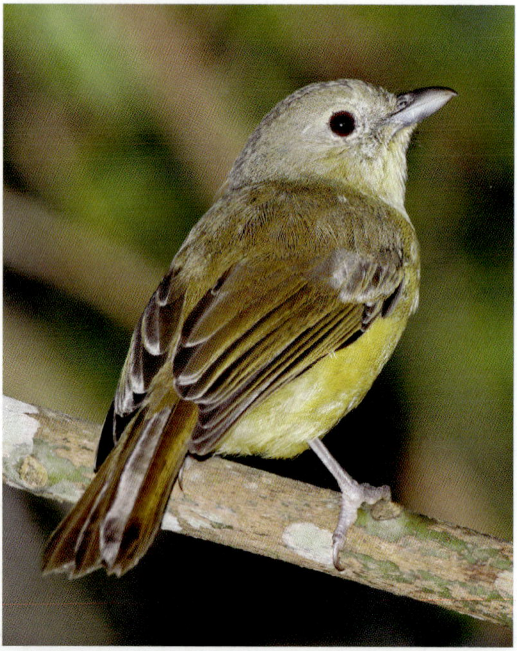

Adult. Port Royal Mountains, Jamaica, November.

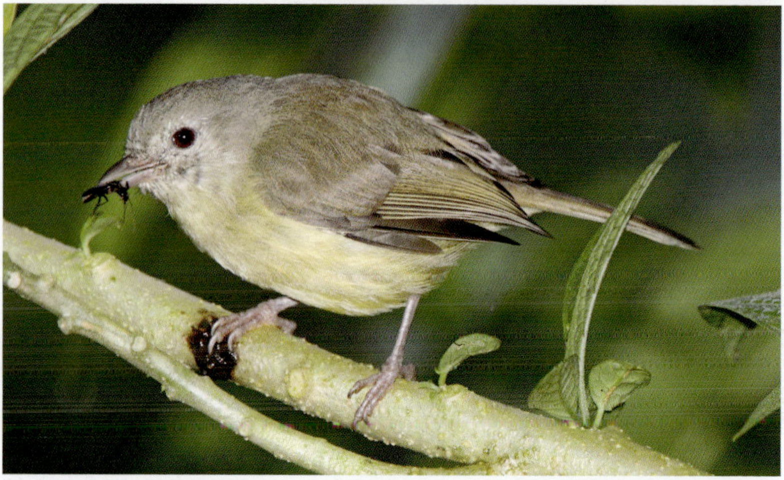

Juvenile. Note greyish cast to underparts and back. Port Royal Mountains, Jamaica. November.

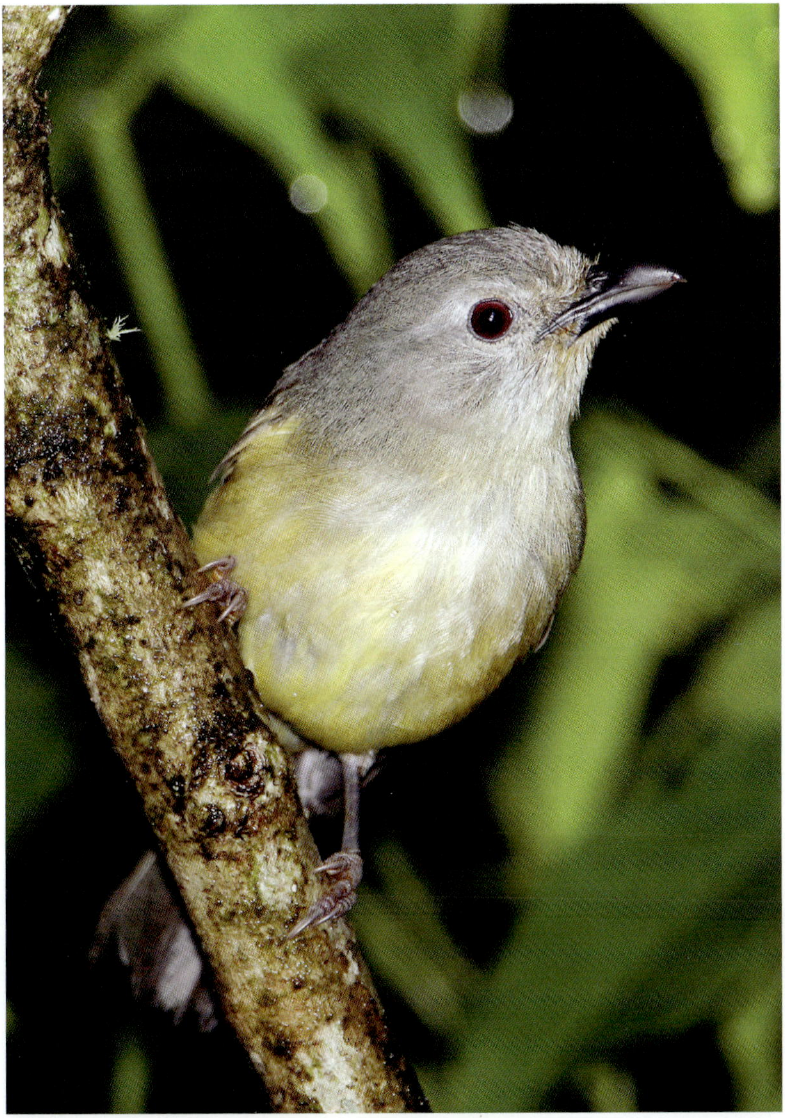

Adult. Note heavy, pointed black bill. Port Royal Mountains, Jamaica, May.

Range Jamaica. Global conservation status: Near Threatened.
Status in Jamaica Uncommon and local resident. May be decreasing as a result of habitat loss and disturbance to montane forests.

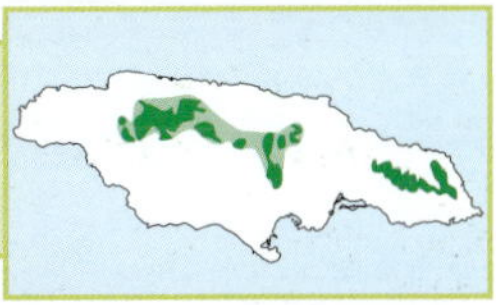

Yellow-throated Vireo

Vireo flavifrons

Taxonomy Monotypic.
Description L 13–15cm (5–6in). Yellow lores connected to yellow eye ring (resembling spectacles). Breast and throat bright yellow. Head is pale grey on the sides with a darker crown. Back olive. Wings are dark with two bold, white wing-bars.
Similar species None.
Voice Alarm call is a chatter.
Habitat and behaviour Mid-level woodlands. Tends to forage in deep cover. Skittish and hard to see.

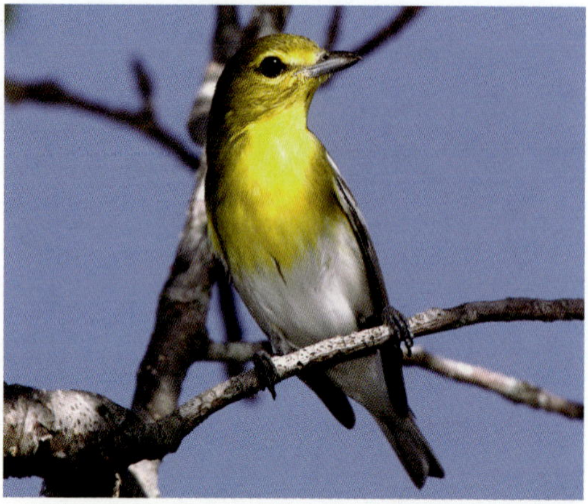

Adult. Note yellow 'spectacles' and bright yellow throat. Bahamas, April.

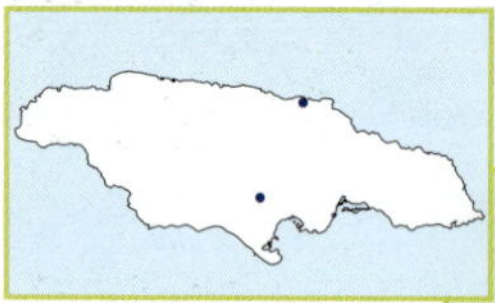

Range Breeds in eastern North America. Winters south to the West Indies and Venezuela.
Status in Jamaica Rare winter resident.

Red-eyed Vireo

Vireo olivaceus

Taxonomy Polytypic (12).
Description L 13cm (5in). Olive-green vireo with contrasting grey crown and white underparts. White supercillium edged with dark lines above and through eye. Iris red.
Similar species Black-whiskered has a distinctive call, shorter bill and black whiskers.
Habitat and behaviour Sometimes seen in coastal scrub and mangroves.

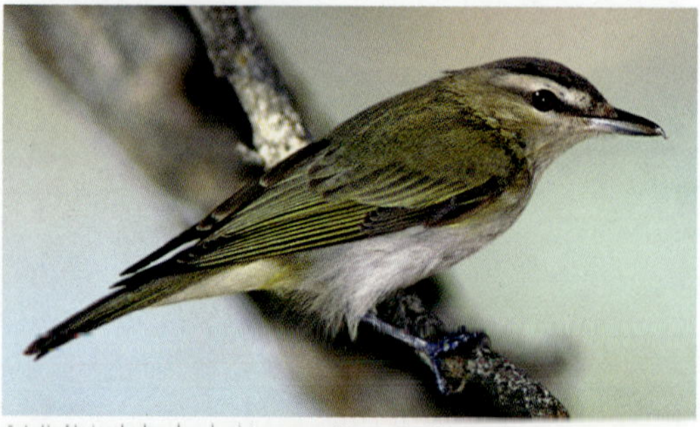

Adult. Note dark-edged white supercillium. Florida, April.

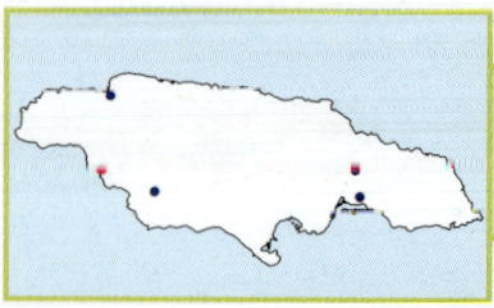

Range Breeds in North America. Winters in South America via Central America, the Bahamas and Greater Antilles.
Status in Jamaica Rare but regular passage migrant (Sept–Nov).

Black-whiskered Vireo

Vireo altiloquus

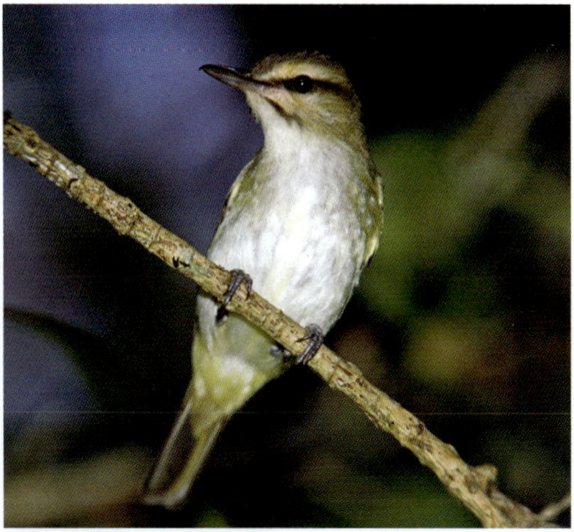

Adult. Note black whisker on throat. Montego Bay, Jamaica, May.

Local names John-To-Whit, John Chewit.
Taxonomy Polytypic (6). Race in Jamaica is *V. a. altiloquus*.
Description L 13cm (5in). An olive-greenish grey vireo with a dull white supercilium and a dark eye-stripe. Throat edged by fine dark 'whiskers'. Head is darker than yellowish-green back. Iris reddish. Wings brown, lacking wing-bars. Underparts white. Bill is large and thick, grey. Legs grey. Juvenile duller and its 'whiskers' may be very indistinct.
Similar Species: Red-eyed Vireo lacks the 'whiskers'.
Voice *John-chewit* or *sweet-John*, or *chewit-John* and sometimes the full *sweet-John-chewit* is repeated incessantly, dominating the 'soundscape' in April and May. From hatching in July until the birds leave in early October the typical song is often replaced by a *skwee*.
Habitat and behaviour Widespread throughout the island in open woodland and gardens at all elevations. Although common and vocal it is not always easy to see. Feeds in the canopy on insects and berries. May hover to pluck fruit from trees. Nest (Apr–Aug) is a pendant cup made of fine roots and moss, lined with down or cottony fibres and suspended from a forked branch. May raise two broods in a season.

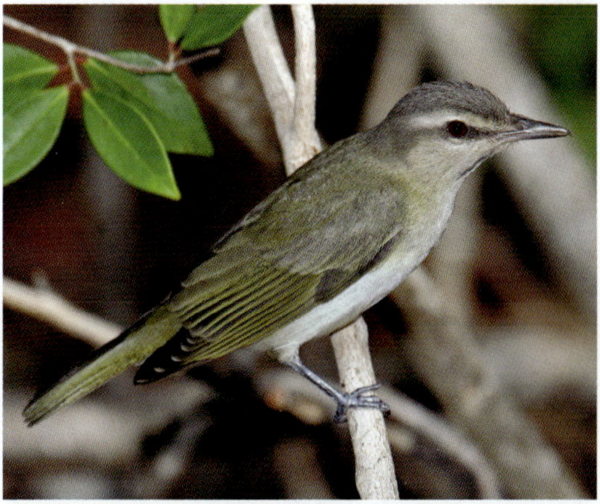

Adult. Note contrast between head and back. Cayman Islands, April.

Range *V. a. altiloquus* breeds in Greater Antilles (except Cuba and the Isle of Youth, Cayman, the Swan Islands and St. Croix) and winters in northern Brazil. A few are permanently resident in Hispaniola. Other races occur on other W Indian islands, islands in the western Caribbean and Netherlands Antilles.
Status in Jamaica Very common and widespread summer visitor and breeding species.

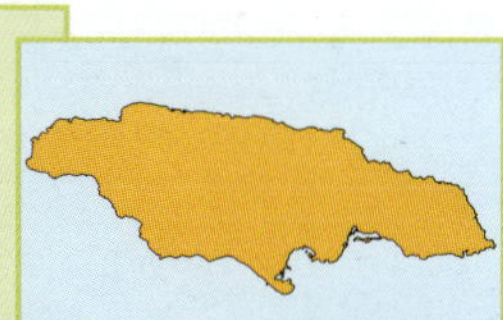

Jamaican Euphonia
Euphonia jamaica

Local names Cho-Cho Quit, Short-Mouth Bluequit.
Taxonomy Monotypic. Endemic to Jamaica.
Description L 11cm (4.5in) A small, chunky, blue-grey bird with a short, stubby, grey bill that is darker on top and at the tip. Iris reddish brown. Male has bright yellow mid-belly, beige undertail-coverts and greenish-yellow flanks. Wings and tail black edged blue, wing linings pale yellow. Legs grey. Female and juvenile olive-green above. Mostly pale grey below, shading to cream, but rump and flanks greenish yellow and undertail-coverts buff. Wings and tail dark grey edged green.
Similar species None.
Voice A rapidly repeated *chur-chur-chur-chur-chur* sounding rather like a car that cannot start, but sometimes

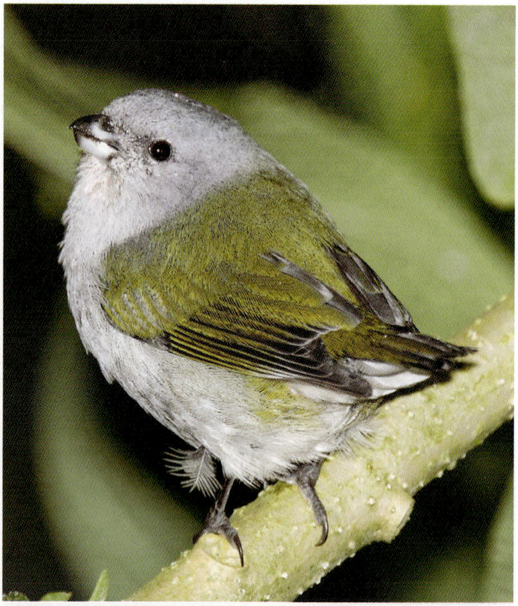

Adult female. Port Royal Mountains, Jamaica, November.

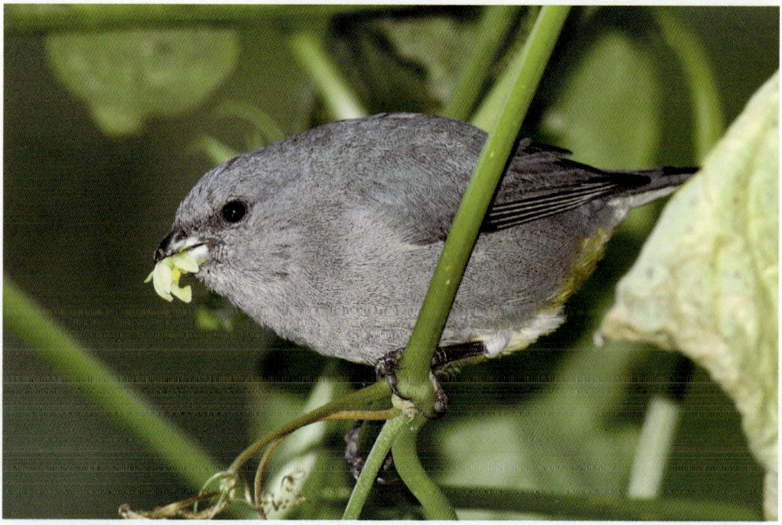

Adult male. Port Royal Mountains, Jamaica, November.

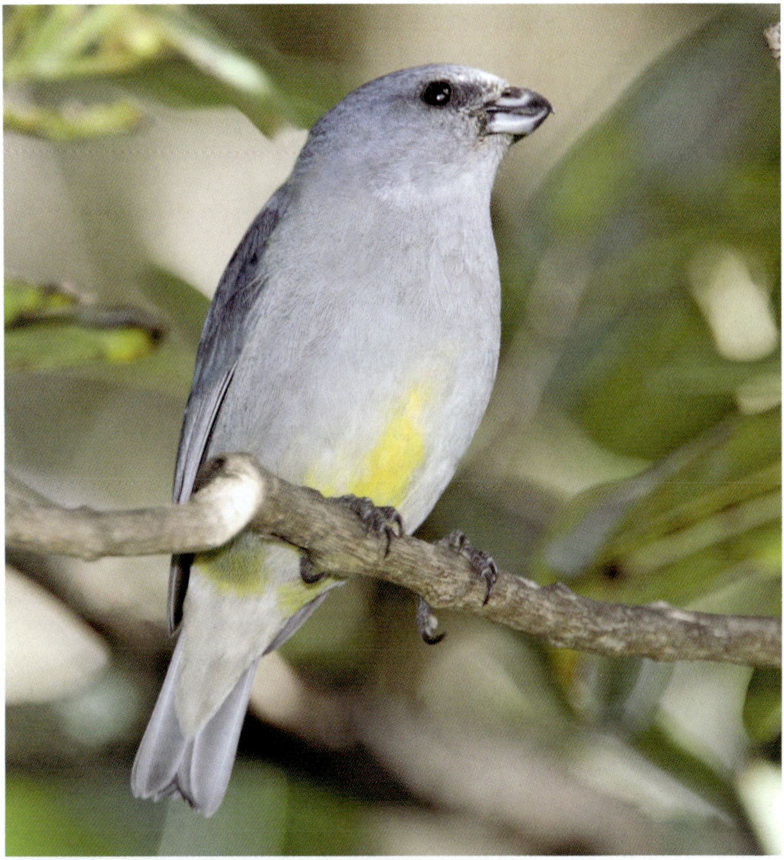

Adult male. Note yellow on belly and beige undertail-coverts. Port Royal Mountains, Jamaica, November.

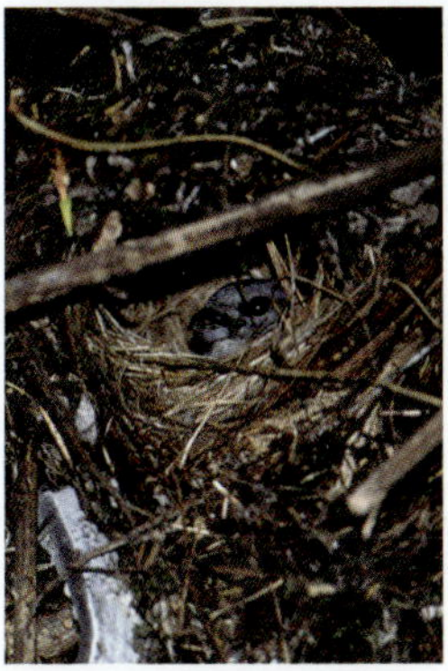

Male on nest. Montego Bay, May.

ending in a rising *chip*. Song is an attractive and very varied squeaky warble.

Habitat and behaviour
Woodlands, gardens and open areas with large trees, from sea level to the mountains. Feeds on soft berries especially wild figs, Jamaican Mistletoe, soft fruits such as soursop, guava, and peach, tender young shoots of cho-cho, and many different buds and flowers. Nest (March–May) is usually globular, made of *Tillandsia* down with a side entrance, hidden in a bunch of *Tillandsia*, Spanish Moss, or under other bromeliads.

Range Jamaican endemic.
Status in Jamaica Common and widespread resident.

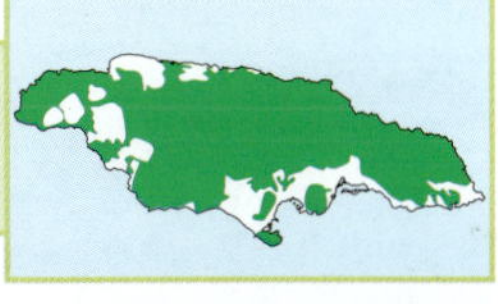

Northern Parula
Parula americana

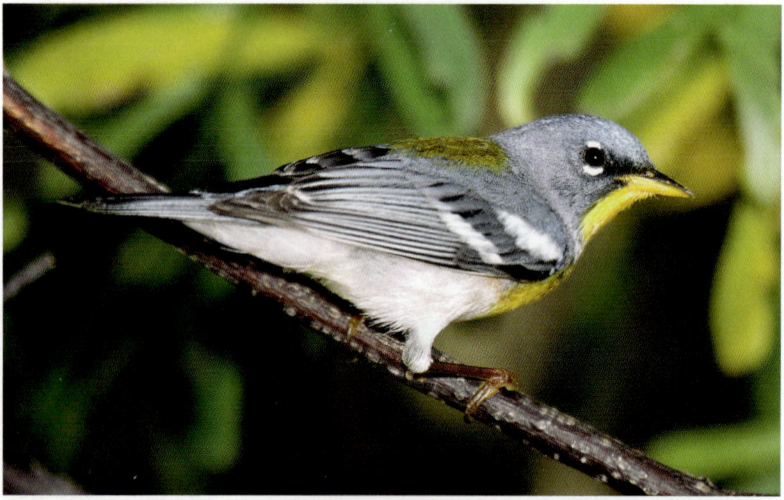

Adult male. Cayman Islands, April.

Taxonomy Monotypic.
Description L 10cm (4in). A brightly coloured warbler with a broken white eye-ring, yellow lower mandible, blue-grey wings with two white wing-bars, blue or greenish rump, and a bright yellow throat and breast contrasting with white underparts. Seen in a variety of plumages related to age and sex. Adult male in breeding plumage (March–May) has black lores, and chestnut on upper breast.
Similar species None.
Voice Various chips.
Habitat and behaviour Gardens and woodlands island-wide. Forages in shrubbery.

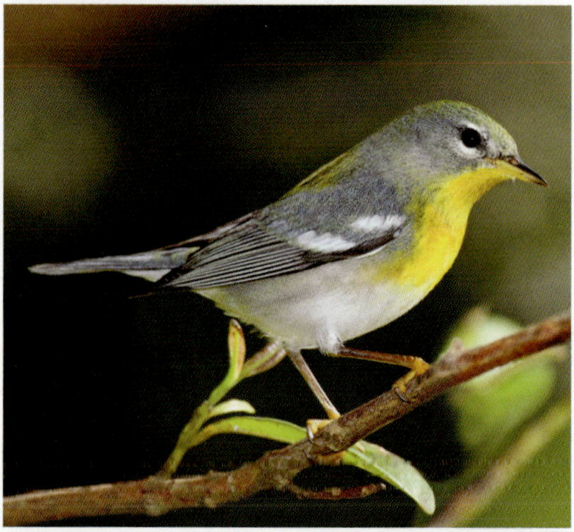

Adult female. Note two white wing-bars and broken eye-ring. Jamaica, November.

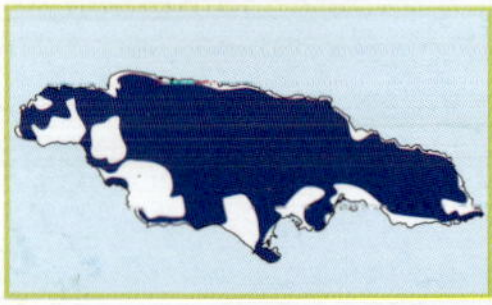

Range Breeds in North America. Winters in Central America and the W Indies.
Status in Jamaica Fairly common winter visitor (August–May).

Yellow Warbler
Dendroica petechia

Local name Mangrove Canary.
Taxonomy Polytypic (34). Race in Jamaica is *D. p. eoa*.
Description L 10cm (4in). Bright yellow, including the tail. Large black eye stands out against the yellow face. Adult male in breeding plumage has a reddish-brown cap and reddish streaks on the breast. Adult female lacks the cap and has fainter streaks. First-year is pale yellow and may be blotched with grey. Juvenile has greyish yellow body, lemon-yellow upperparts and wings.
Similar species First-year female resembles first-year female Hooded Warbler which has white tail-spots.
Voice Very vocal and often first detected by song, which can carry for 100m or more. Sings year-round. Mnemonic for song is *sister sue is sweet-eer*. Also exhibits a range of loud chips.
Habitat and behaviour Mangroves, other wooded wetlands and cays. Feeds on small insects in the canopy. Nest (April–June) is a large open cup made of grasses and fine twigs in the forks of trunks or large branches, close to the ground or in low canopy in mangroves and scrub forest.

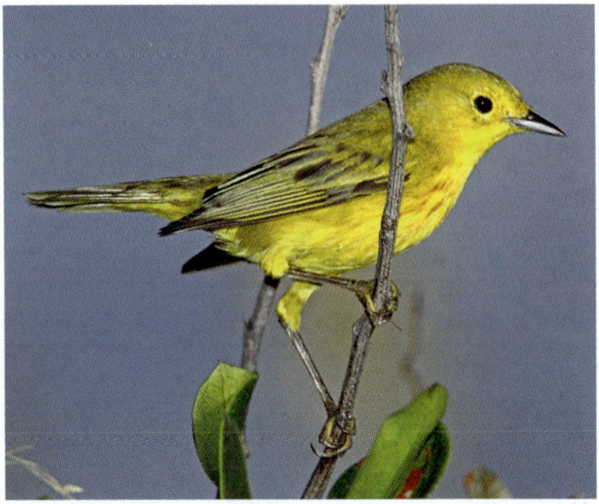

Adult male. Cayman Islands, April.

Range *D. p. eoa* on Jamaica and the Cayman Islands. Migrants are probably *D. p. amnicola* or *D. p. aestiva* from North and Central America. Other races in the Bahamas and throughout the Caribbean.
Status in Jamaica Common resident and uncommon passage migrant

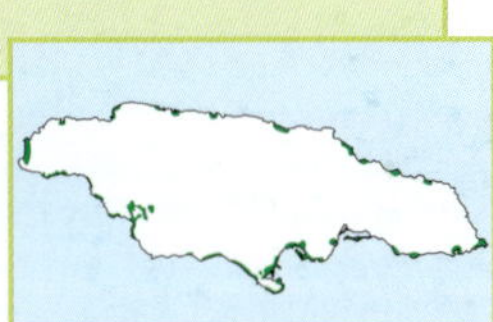

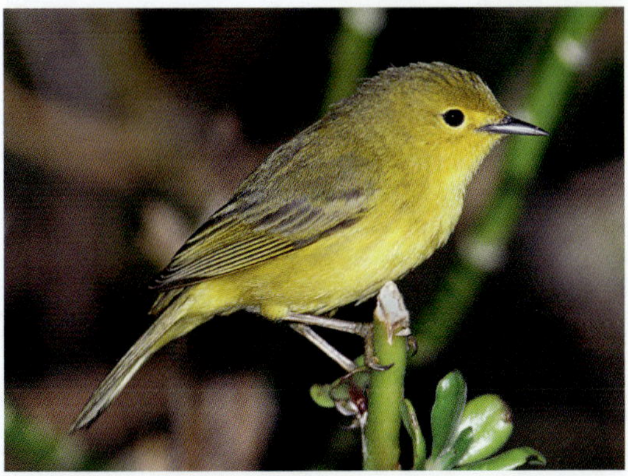

Adult female. Cayman Islands, April.

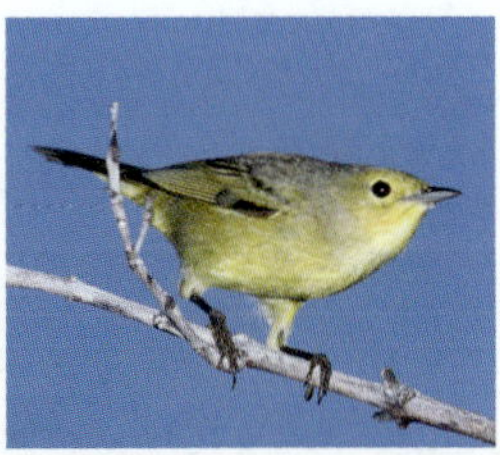

Juvenile. Cayman Islands, September.

Chestnut-sided Warbler
Dendroica pensylvanica

Taxonomy Monotypic.
Description L 11cm (4.25in). Male in breeding plumage has characteristic chestnut flanks and sides. Crown yellow. Broad black eye stripe extends into 'moustache'. Two yellowish wing-bars. Back dark and streaked, underparts white. Female breeding like male but less strongly marked. Non-breeding adult and first-winter have lime-green back, white eye-ring, grey face. Winter male has less chestnut than summer; female lacks chestnut on sides.
Similar species Bay-breasted Warbler has brown crown and yellow throat.
Voice Various chips.
Habitat and behaviour Mid-level forests, often beside roads. Feeds high up in the canopy.

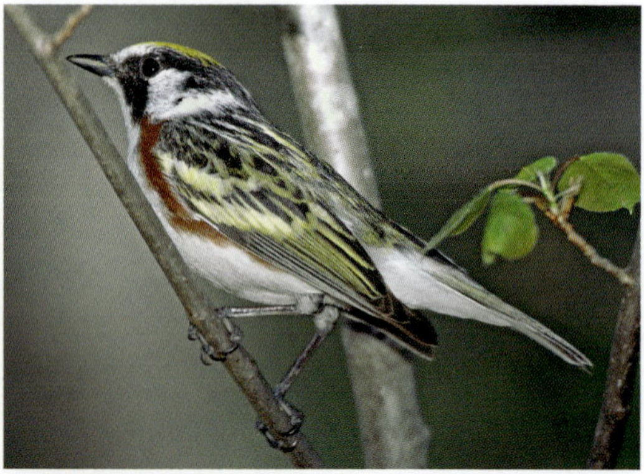

Adult male. May.

Range Breeds North America. Winters Central America.
Status in Jamaica Rare autumn passage migrant (Sept–Oct).

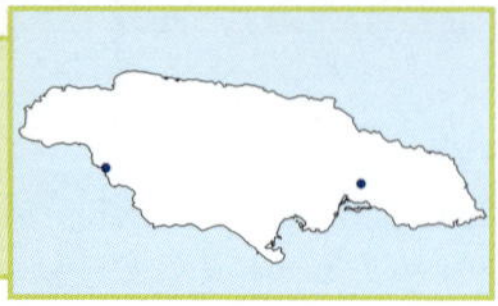

Blackburnian Warbler
Dendroica fusca

Taxonomy Monotypic.
Description L 11cm (4.25in). Adult male unmistakeable with striking black and bright orange head and throat. Breast and flanks yellowish streaked black. White wing-patch. Duller in winter. Adult female and first-winter yellow where the male is orange; two white wing-bars.
Similar species First-winter Black-throated Green Warbler resembles First-winter Blackburnian but has greenish crown and cheeks, unstreaked back.

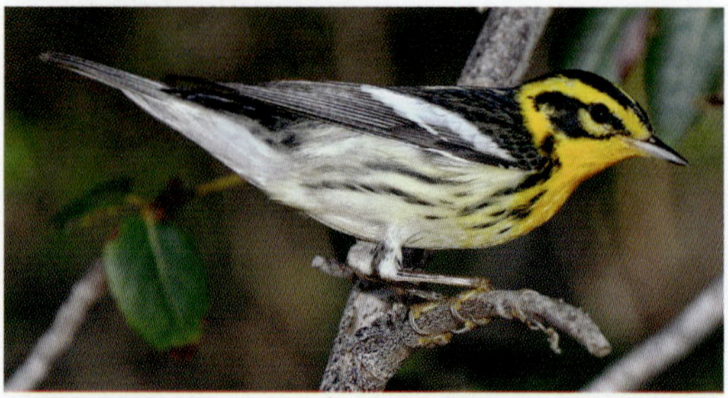

Adult male. April.

Voice Chips.
Habitat and behaviour Usually seen in coastal scrub.

Range Breeds in eastern North America. Winters south through Central America, Bahamas, Greater Antilles to South America.
Status in Jamaica Rare winter passage migrant (October–November and April–May).

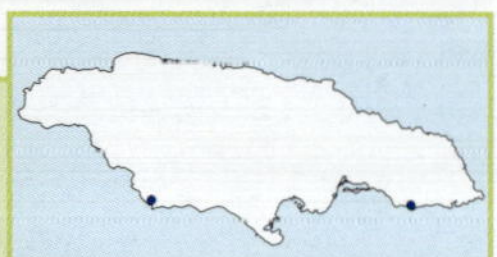

Magnolia Warbler
Dendroica magnolia

Taxonomy Monotypic.
Description L 11cm (4.25in). Seen in Jamaica in a variety of plumages and transitional forms, all of which have yellow rumps and dark tails with white flashes, which are conspicuous when seen from above. From below the tail looks white halfway down. White on wings ranges from narrow wing-bars (in first-winter) to a white patch (in adult male). Undertail-coverts white. Dark band across breast. Adult male in spring has yellow underparts, broadly streaked with black. Back black. Crown grey, supercilium white, eye-patch black. Adult female in spring greyer on head and back than male. First-winter very plain with greyish back, lacks black and white head markings, streaking on flanks indistinct.

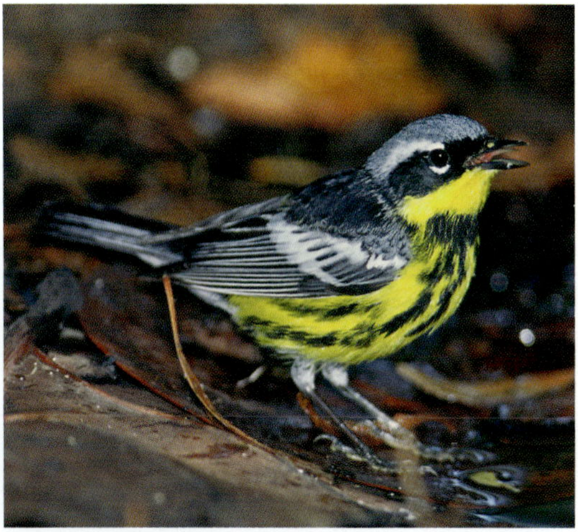
Adult male. Florida, April.

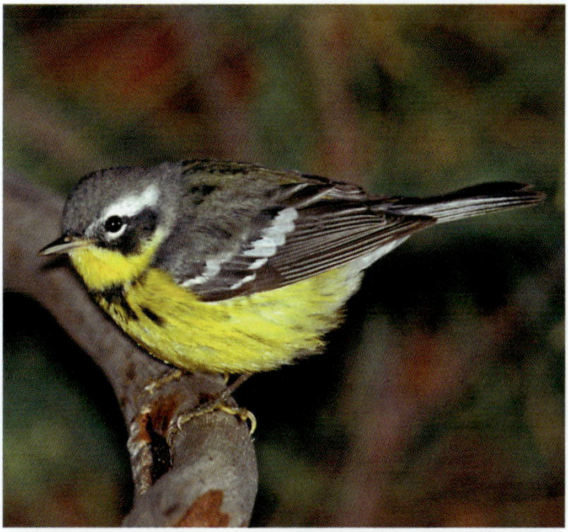
Adult female. Florida, April.

Similar species Several species, including Prairie, Black-throated Green and Blackburnian Warblers, have yellow underparts and resemble non-breeding adult or first-winter Magnolia Warbler, but they all lack the grey breast-band and have white tail-sides. Blackburnian has a yellow supercilium. Black-throated Green has a yellow face. Palm Warbler regularly pumps tail.
Voice Chips and buzzes.
Habitat and behaviour Gardens and orchards. Feeds on insects.

Range North America, wintering in the Bahamas, Greater Antilles, Central and South America.
Status in Jamaica Uncommon winter visitor and passage migrant (November–May). Irruptive.

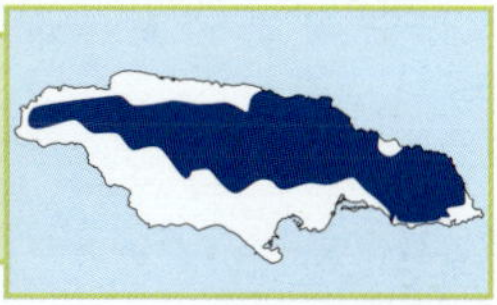

Cape May Warbler
Dendroica tigrina

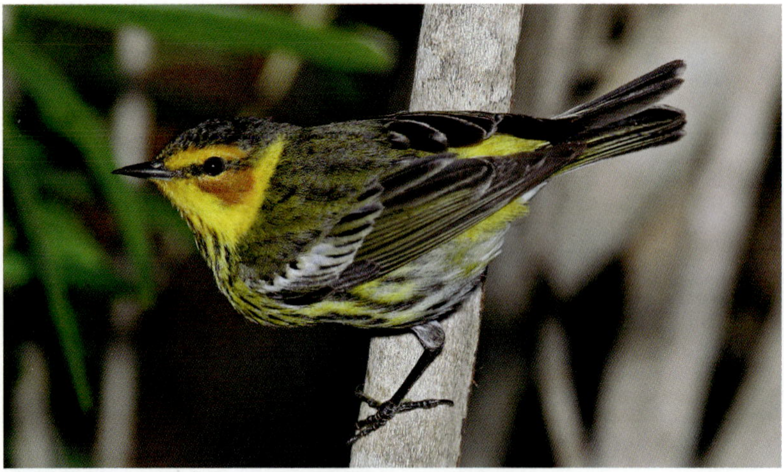

Adult male. Note rufous cheeks. Cayman Islands, April.

Taxonomy Monotypic.
Description L 11cm (4.25in). Small warber with a short tail. Thin, slightly decurved dark bill; narrow, dark eye-stripe, yellow on sides of neck, dull yellowish rump and white spots in tail. Occurs in a range of plumages and transitional forms on Jamaica. Adult male has bright rufous cheeks. Neck and throat bright yellow. Breast and sides are bright yellow, boldly streaked with black. Adult female duller and lacks rufous cheeks. First-winter olive above and greyish yellow to pale grey below, with faint dark streaks.
Similar species First-winter resembles that of Palm Warbler, which pumps its tail and has yellow undertail-coverts and a yellow supercilium.
Voice High-pitched chip.
Habitat and behaviour Open wooded areas at all elevations. Feeds on nectar and small fruit and honeydew from aphids. Sometimes comes to hummingbird feeders (which it defends aggressively) and may enter houses in search of spiders and their prey. Semi-tubular tongue is adapted for sipping nectar.

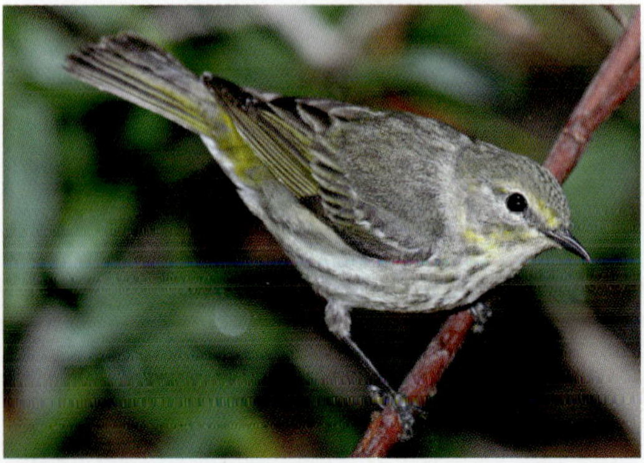

First-year female. Cayman Islands, April.

Range North America, wintering in the Bahamas and Greater Antilles and rarely in Central America.
Status in Jamaica Very common winter visitor (August–May).

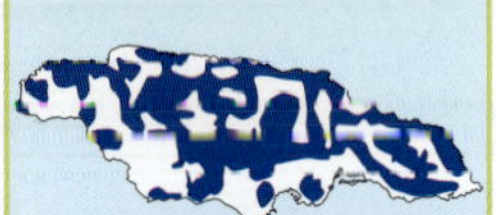

Black-throated Blue Warbler
Dendroica caerulescens

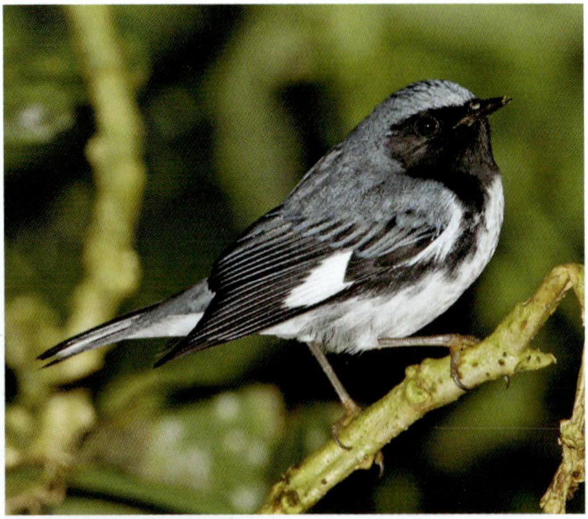

Adult male. Port Royal Mountains, Jamaica, February.

Taxonomy Polytypic (2). Both *D. c. caerulescens* and *D. c. cairnsi* occur in Jamaica.
Description L 11–13cm (4–5in). Adult male is a dapper warbler with a deep blue head and back, black throat and flanks, and white underparts. Small white patch at base of primaries diagnostic. White tail-spots. *D. c. caerulescens* has blue mantle, *D. c. cairnsi* has blackish or black-streaked mantle. Adult female and first-winter dull brownish with small white patches at base of primaries (sometimes very indistinct or absent in first-winter females). Pale supercilium contrasts with dark cheek patch. White lower eye-ring. Upperparts brownish-olive, underparts are buffy. Whitish spots in tail.
Similar species Female resembles juvenile Bananaquit, which is smaller, more slender and yellower, has a red spot at the base of the curved bill and lacks white lower eye-ring.
Voice Loud chips.
Habitat and behaviour Gardens and woodlands at all elevations. Males tend to be found in mid-level forests. Females tend to occur in scrubbier habitats and at higher elevations. Feeds on insects and small berries in open woods, mostly in the understorey.

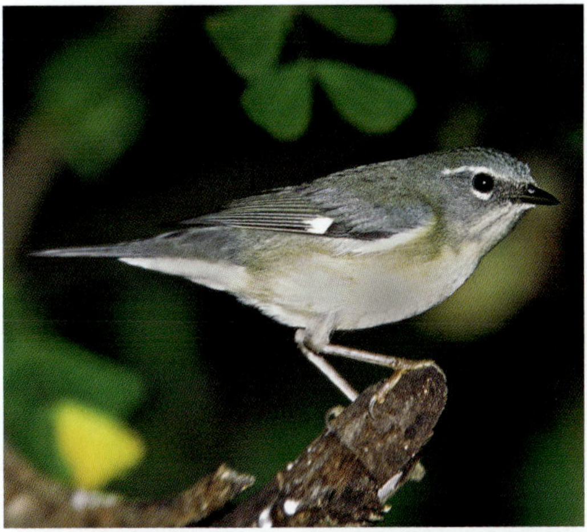

Adult female. Note white wing-patch. Cayman Islands, April.

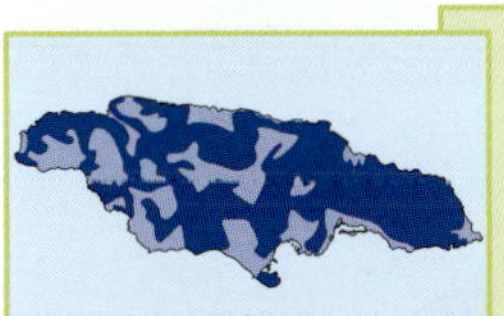

Range *D. c. caerulescens* breeds in southeastern Canada to northeastern United States and winters in the Bahamas and West Indies. *D. c. cairnsi* breeds in east and central United States and winters in Bahamas, West Indies and Central America.
Status in Jamaica Very common winter visitor (September–May). One of the most abundant migrant warblers in Jamaica.

Yellow-rumped Warbler
Dendroica coronata

Taxonomy Polytypic (4). Race in Jamaica is *D. c. coronata*.
Description L 12cm (4.75in). A large, dark-looking warbler with bright yellow rump and white spots in tail. Broken eye ring. Indistinct black streaks on back. Two faint white wing-bars. In winter usually seen in brownish non-breeding and first-winter plumages but yellow rump always distinctive. Adult non-breeding males have yellow on crown and breast sides and bold black streaking above and below. Females and first-winter birds are browner.
Similar species Adult male is distinctive. Females and first-winter birds have bright yellow rumps unlike those of Magnolia, Palm and Cape May Warblers.
Voice Various chips.
Habitat and behaviour Scrub in dry limestone forest; sometimes in gardens on acacia and cashew trees, where their yellow rumps are eye-catching

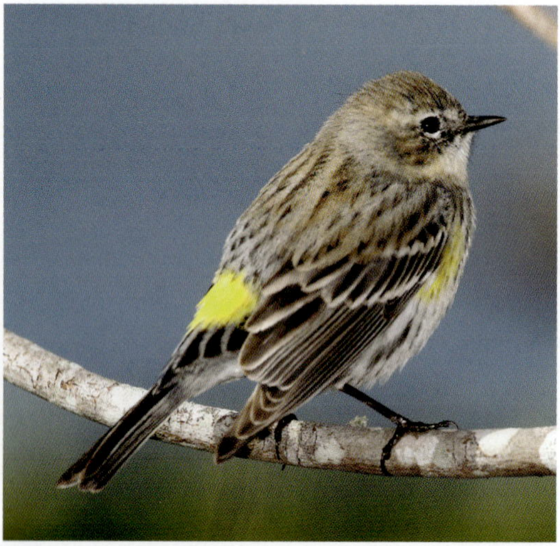

Adult winter. Note the yellow rump. Florida, February.

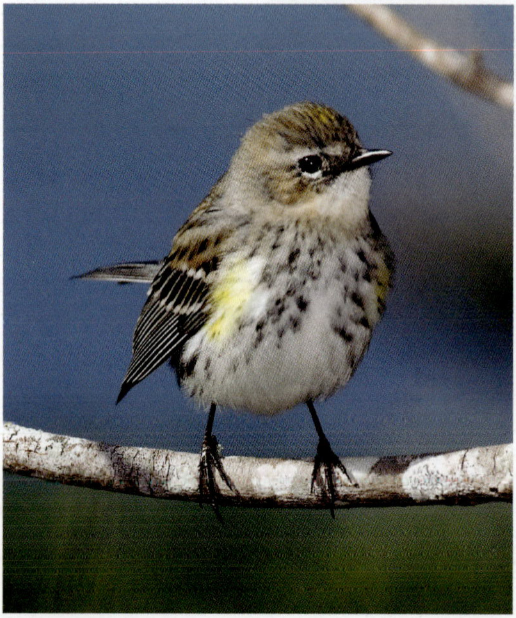

Adult winter. Florida, February.

Range *D. c. coronata* breeds in northern North America, wintering in West Indies and Central America south to Panama.
Status in Jamaica Winter visitor (October–May). Irruptive.

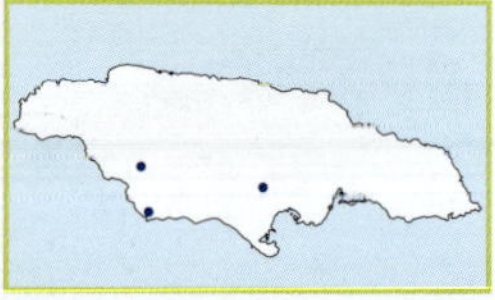

Black-throated Green Warbler

Dendroica virens

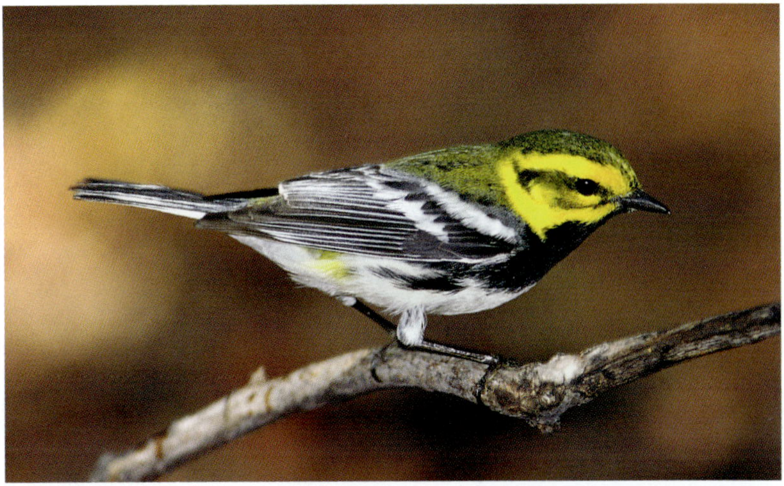

Adult male. Jamaica, April.

Taxonomy Monotypic.
Description L 11–12cm (4–5in). Adult male brightly coloured with bright yellow face and greenish cheek patches. Wing dark with two white wing-bars. Throat black, flanks streaked black. Rest of the underparts white or whitish. Back olive-green. Adult female duller, throat mostly yellow. First-winter has lightly streaked breast and flanks.
Similar species None.
Voice Various chips.
Habitat and behaviour Upland forests and scrub. Arboreal. Feeds on insects mainly by leaf-gleaning.

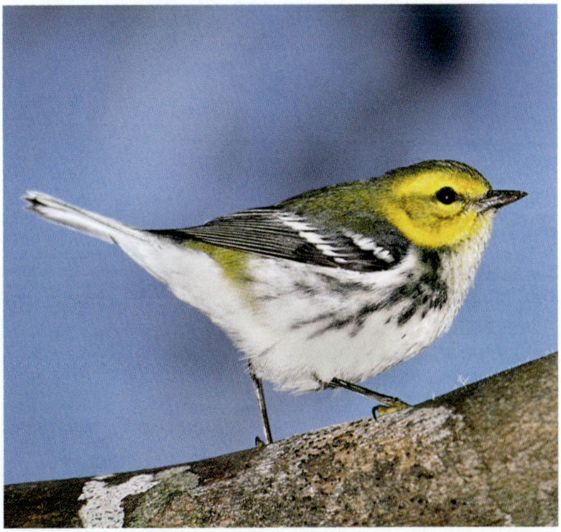

First-winter female. Note black limited to neck-sides. Jamaica, November.

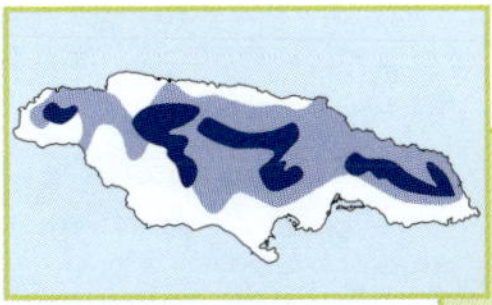

Range Breeds in eastern North America. Winters south to Central America, also Bahamas and the Greater Antilles.
Status in Jamaica Uncommon winter visitor (September–March) and passage migrant (March–April).

Yellow-throated Warbler
Dendroica dominica

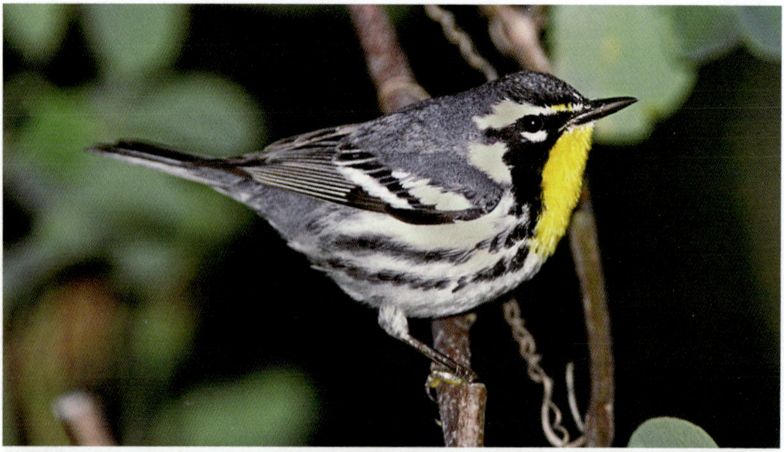
Adult female. Note yellow throat, white crescent under eye. Cayman Islands, April.

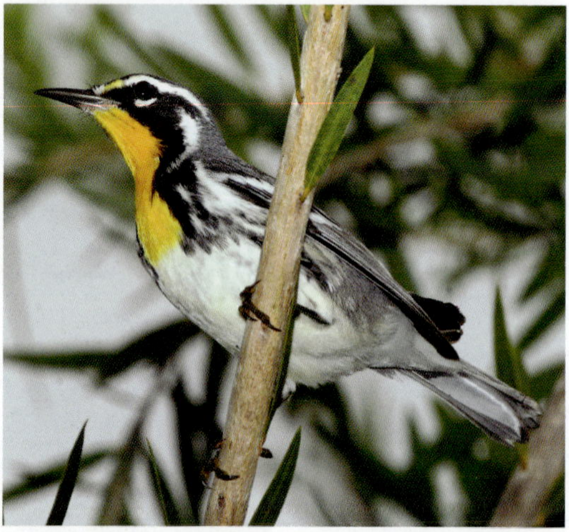
Adult. Florida, November.

Taxonomy Polytypic (4). Race in Jamaica is *D. d. albilora*.
Description L 13–14cm (5–6in). Dark grey upperparts. Throat and breast bright yellow. Large, roughly triangular black cheek-patch and black margins to throat. White crescent under eye. White supercilium. White patch on neck. Underparts white, streaked black on flanks. Two white wing-bars. Non-breeding and first-winter plumages duller.
Similar species No other warbler has a similar pattern on the face.
Voice Various chips.
Habitat and behaviour Forests, chiefly in the lowlands. Usually seen in the canopy. Feeds on insects.

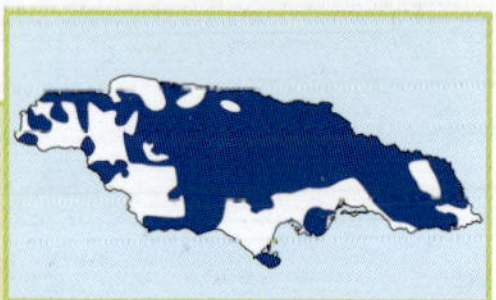

Range *D. d. albilora* breeds in east central and southeast United States. Winters Costa Rica, Cuba and Jamaica. Other races winter in the Bahamas, Greater Antilles and Central America.
Status in Jamaica Locally common winter visitor (Sept–May).

Prairie Warbler
Dendroica discolor

Taxonomy Polytypic (2). Race in Jamaica is *D. d. discolor.*
Description L 10cm (4 in). A small warbler, olive above and all-yellow below, distinctly streaked on sides and with a distinctive dark face pattern. Seen in Jamaica in a variety of transitional plumages. Adult male has a distinctive yellow and black face, with a yellow line above the eye and a wide yellow crescent below it, fringed by a broader black moustachial stripe. Flanks streaked black. Back and wings are brownish olive, rufous streaks on back. Two indistinct yellow wing-bars. Female duller. First-winter greyish above, yellow streaked olive below, with dark moustachial stripe.
Similar species None.
Voice Various chips.
Habitat and behaviour In mangroves, woodlands, gardens and orchards at all elevations. Pumps tail regularly but less frequently than Palm Warbler.

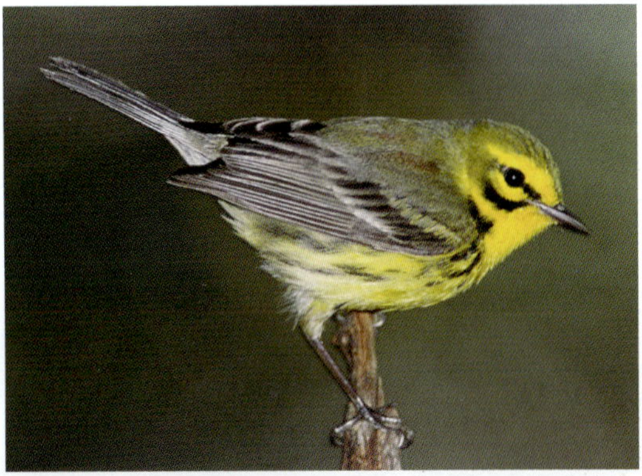
Adult male. Note bright yellow and black pattern on face. Cayman Islands, April.

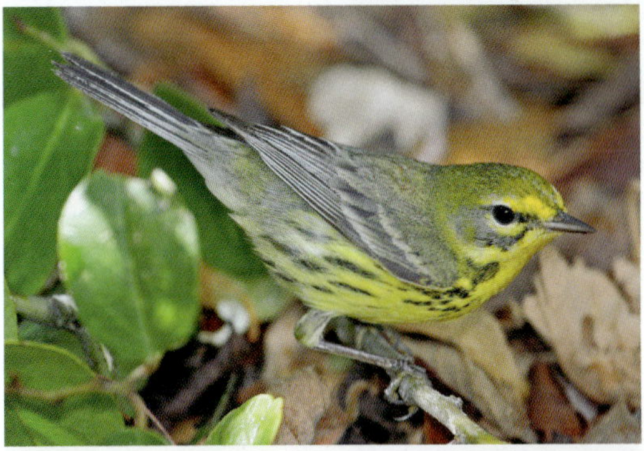
Adult female. Cayman Islands, April.

Range Breeds in North America. Winters south to the Bahamas, West Indies and Central America.
Status in Jamaica Common winter visitor (July–May).

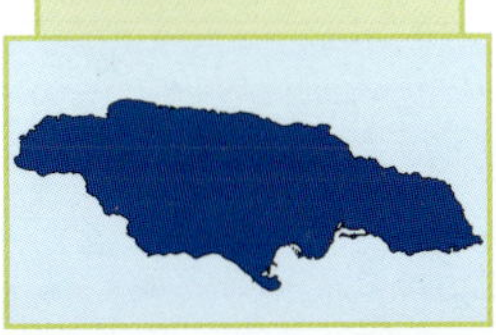

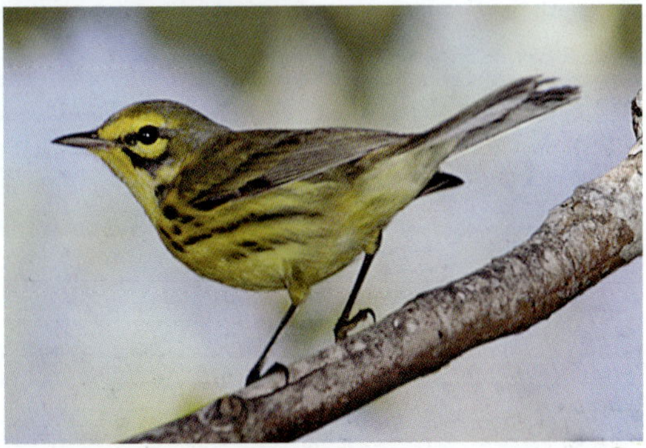
Adult male. Note white undersides to tail feathers. Cayman Islands, April.

Palm Warbler
Dendroica palmarum

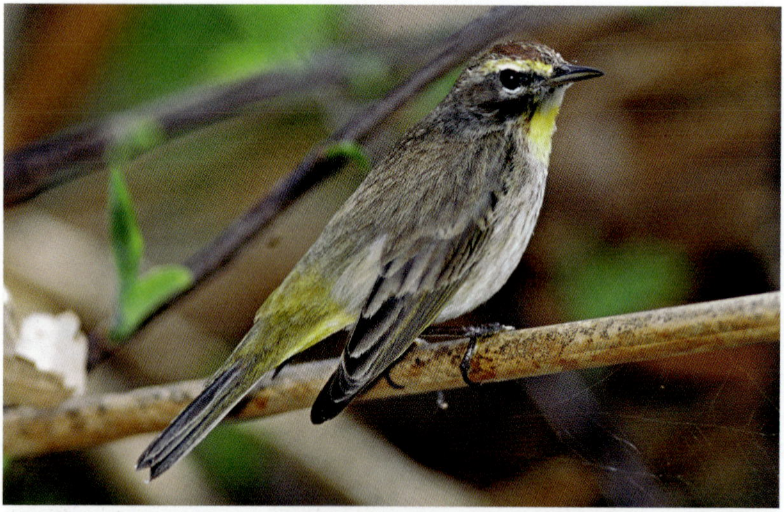

Adult. Note brown crown and yellowish supercillium, and absence of wing-bars. Florida, February.

Taxonomy Polytypic (2). Races in Jamaica are *D. p. palmarum* and *D. p. hypochrysea.*
Description L 12–14cm (5–6in). A dull, brownish, faintly-streaked warbler. In winter has faint light yellowish supercilium and darker eye-stripe. Crown brownish. Upperparts grey-brown, rump greenish. Underparts greyish with darker streaks. Undertail-coverts lemon-yellow. In spring begins to develop reddish-brown crown, rufous-streaked yellowish underparts and greenish-yellow rump.
Similar species No similar species flick their tails up.
Voice Various chips.
Habitat and behaviour Coastal scrub, woodland and cays, grassland, gardens chiefly in the lowlands and foothills, usually seen hopping about on or near the ground and in low shrubs. Constantly flicks its tail up and down. Feeds on insects.

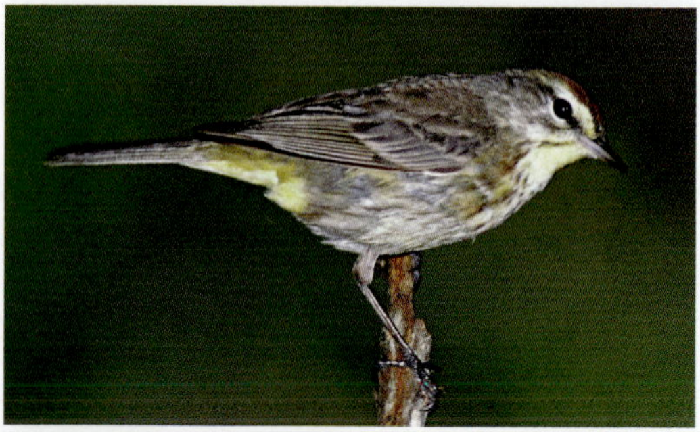

Adult non-breeding. Florida, February

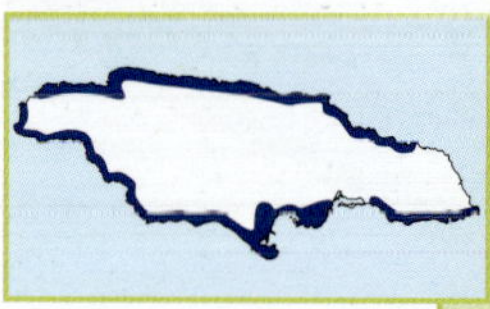

Range *D. p. palmarum* breeds in northern North America. Winters south to the Caribbean. *D. p. hypochrysea* breeds northeast United States, winters south to Florida.
Status in Jamaica Common winter visitor (September–May).

Bay-breasted Warbler
Dendroica castanea

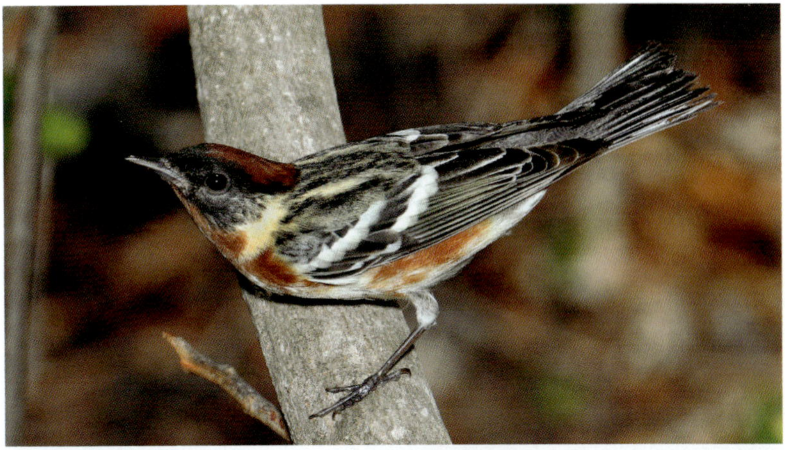

Adult male. Note chestnut crown, throat and sides, and black mask. Cayman Islands, April.

Taxonomy Monotypic.
Description L 13–15cm (5–6in). Large, with two bright white wing-bars. Adult male in breeding plumage has deep chestnut-brown crown, throat and flanks. Broad black mask. Buff patch on neck. Adult female and non-breeding male duller with chestnut wash on flanks. Non-breeding female and first-winter male greenish above and lemon-yellow to white below, vent washed buffy. There are several intermediate plumages.
Similar species First-winter Blackpoll has greyish neck-sides and white vent. Vagrant first-winter Pine Warbler drabber, lacks greenish tones.
Voice Chips.
Habitat Wet limestone forest.

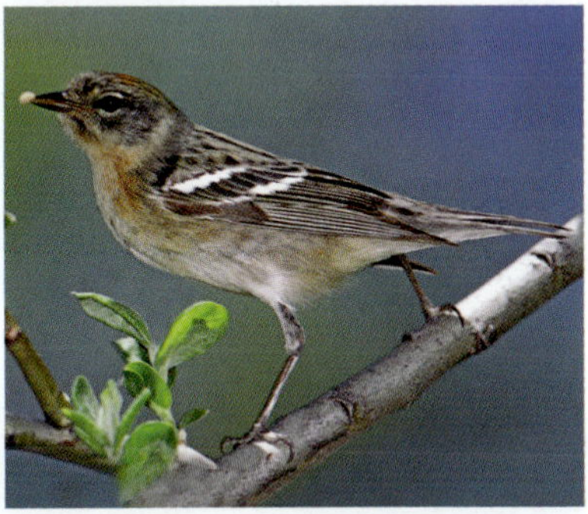

Adult female. May.

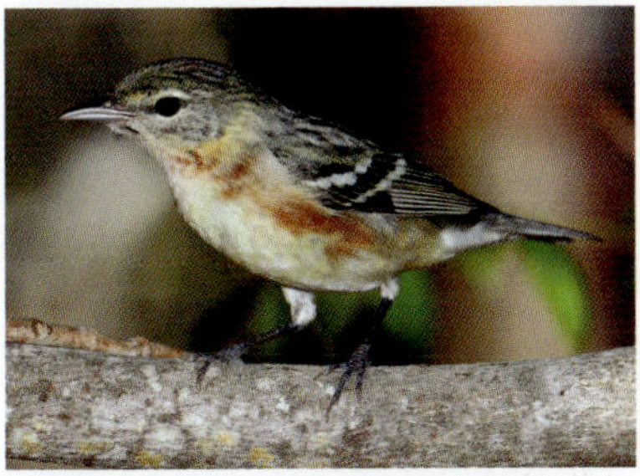

Adult male non-breeding. Cayman Islands, April.

Range Breeds in North America. Winters in South America via Cuba, Jamaica, Central America.
Status in Jamaica Regular passage migrant (October-November, April-May).

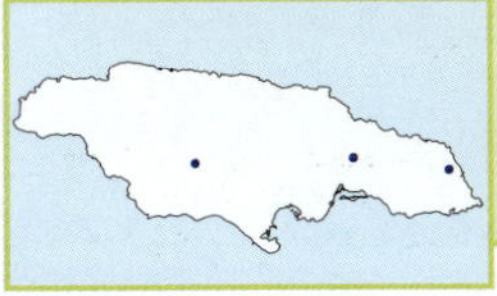

Arrow-headed Warbler

Dendroica pharetra

Local names Ants Bird, Ants Picker.

Taxonomy Monotypic. Endemic to Jamaica.

Description L 13cm (5in). Adult male head and upperparts heavily streaked black-and-white, more finely on head. Throat to abdomen spotted and streaked with dark grey arrowheads pointing to bill, smaller on face and throat. Lower underparts are white streaked buff on undertail-coverts. Iris dark brown, eye-ring white, dark eye-stripe. Bill dark grey, paler below. Wings brownish olive with two narrow white wing-bars. Outer tail feathers have small white tips. Legs light grey. Adult female similar to male but arrowheads grey on white. Juvenile on nest has no arrowheads; head and back olive with yellow flecks, two indistinct wing-bars; iris brown, eye-ring yellowish, black eye-stripe. Underparts yellow with grey smudges.

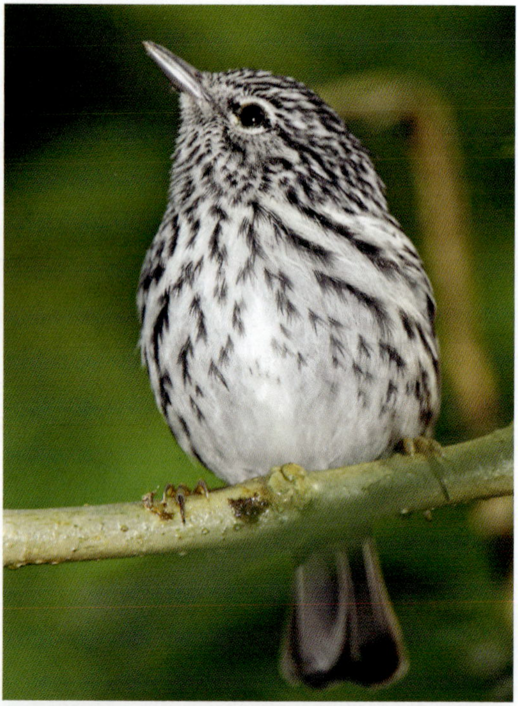

Adult male. Port Royal Mountains, Jamaica, November.

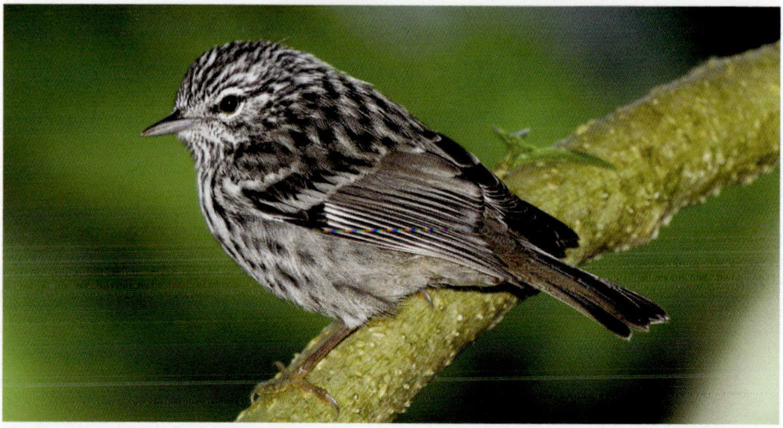

Immature. Note grey flanks and indistinct arrowheads, olive rump and wings.Port Royal Mountains, Jamaica, November.

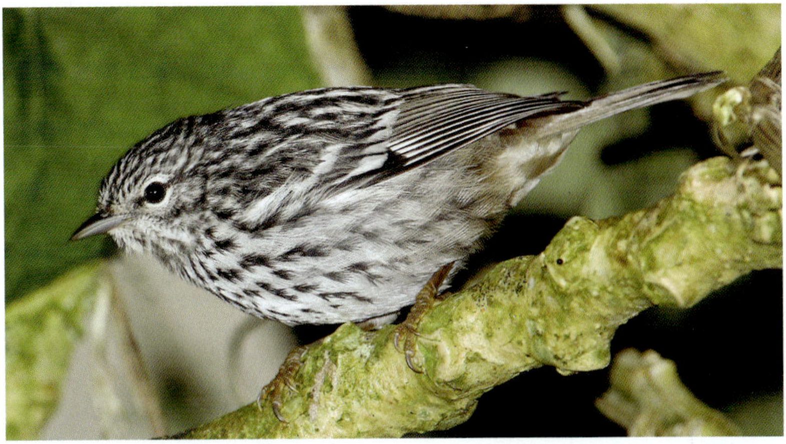

Immature. Port Royal Mountains, Jamaica, November.

White tail-spots smaller than adult. Immatures grey where adults are white.

Similar species Adult Black-and-white Warbler has three broad white stripes on head and (unlike Arrow-headed Warbler) usually feeds on the trunks of trees. Immature Arrow-headed resembles Jamaican Vireo, which has plain grey head with no eye line, pinkish bill and no white in tail.

Voice A high metallic *tick, tick, tick,* and a high-pitched squeaky song.

Habitat and behaviour Humid forest at all elevations, in dense vegetation. Feeds on insects, probing actively under leaves, on branches and in dense hanging vines (but not on tree trunks), singly or in family parties. Nest (May–June) is a cup of grasses and ferns, well concealed in moss.

Range Jamaican endemic.
Status in Jamaica Locally fairly common resident.

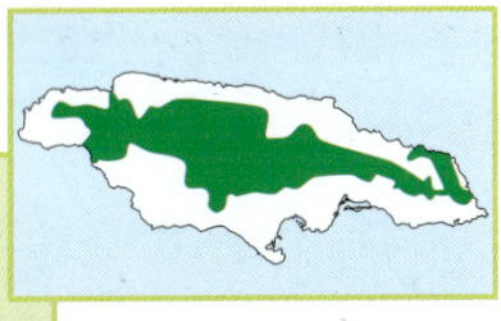

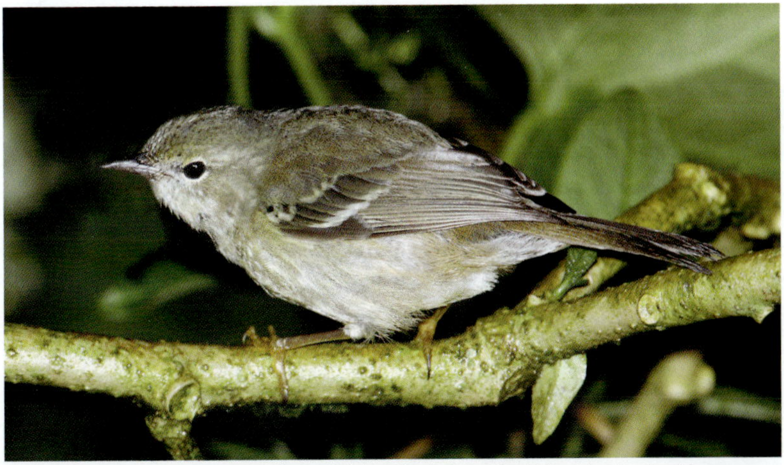

Juvenile. Note lack of streaks or arrowheads. Two indistinct wing-bars. Port Royal Mountains, Jamaica, February.

Blackpoll Warbler
Dendroica striata

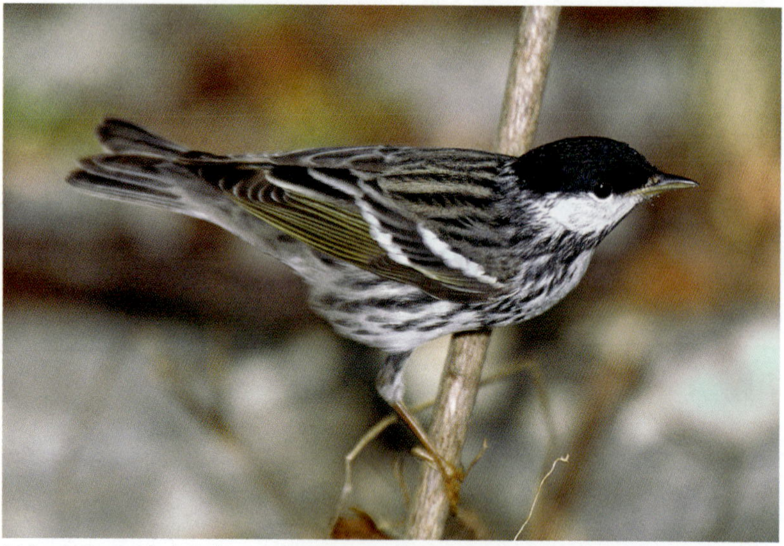

Adult male. Cayman Islands, April.

Taxonomy Monotypic.
Description L 13cm (5.25in). Usually seen in non-breeding plumage but may be in breeding plumage by late spring. Non-breeding male is olive-green above, streaked darker. Breast and sides are lemon-yellow, flanks streaked. Belly and undertail-coverts are white. Wings are dark with two white wing-bars. First-winter has indistinct streaking, greyish neck sides and white vent. Legs are pale.
Similar species None.
Habitat and behaviour Often seen in dry coastal scrub. Usually feeds on insects, spiders and small fruits, gleaning them from twigs close to the ground in thick undergrowth. Usually arrives in autumn in groups that quickly move on southwards.

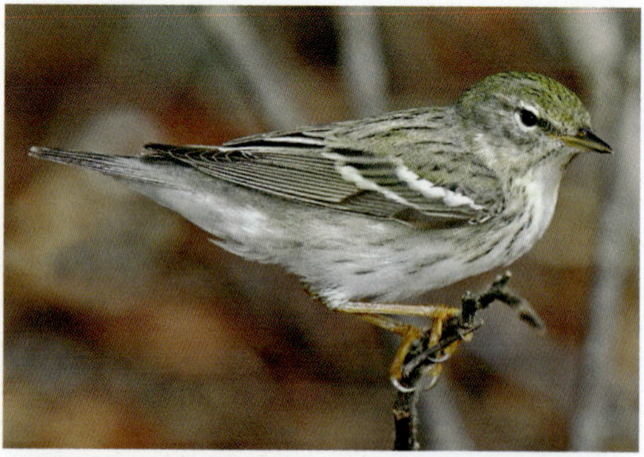

Adult female. Cayman Islands, April.

Range Breeds in northern North America. Winters south to the Lesser Antilles and Central America.
Status in Jamaica Rare but irruptive passage migrant.

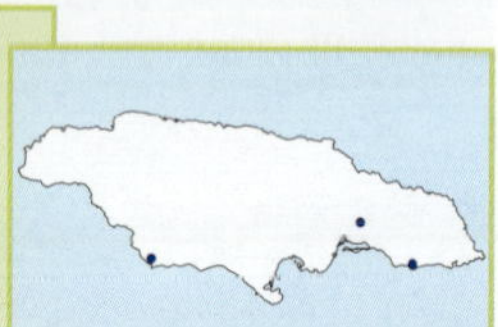

Black-and-white Warbler

Mniotilta varia

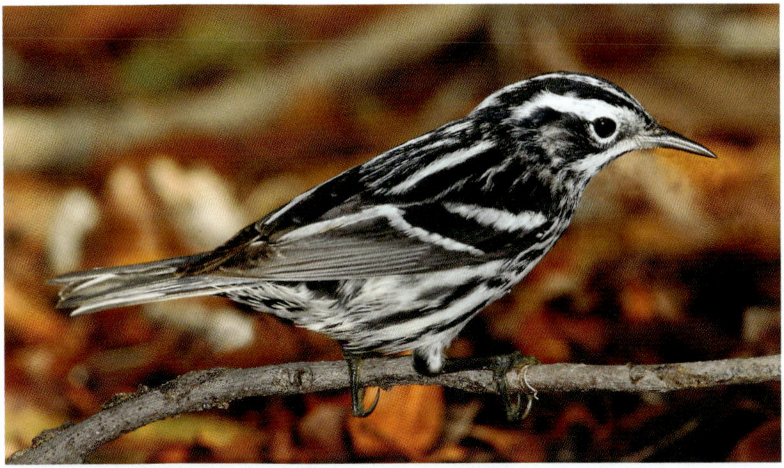

Adult male. Cayman Islands, April.

Local names Ants Bird, Ants Picker.
Taxonomy Monotypic.
Description L 12cm (4.5in). A boldly striped, black-and-white warbler with black-and-white head-stripes, a long, thin bill and two white wing-bars. Undertail-coverts white with black spots. Breeding male has black cheeks and throat. Non-breeding male lacks black throat. Female and first-winter have buffy-grey cheeks, white throat, grey-streaked belly and buffy flanks and undertail-coverts.
Similar species Arrow-headed Warbler lacks the three white head-stripes and is not seen on trunks and branches of trees.
Voice Generally chips. Males may start to sing occasionally in March–April before departure.
Habitat and behaviour All types of forests, woods, scrub and secondary growth. Usually seen creeping up and down trunks and along branches of trees using its long bill to extract insects (including ants, hence its local name) and spiders from bark.

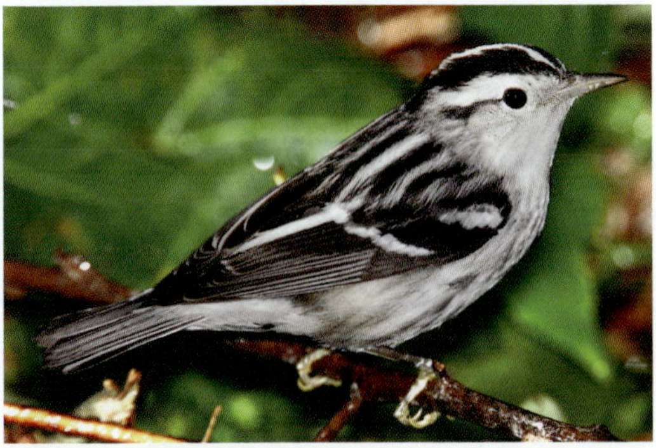

Adult female. Cayman Islands, April.

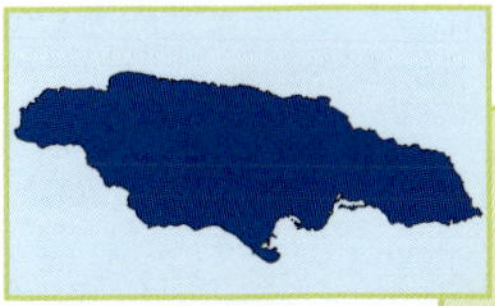

Range Breeds in North America. Winters south to the West Indies, Bahamas, Central America and northern South America.
Status in Jamaica Common and widespread winter visitor (July–May)

American Redstart
Setophaga ruticilla

Local name Butterfly Bird, Christmas Bird.
Taxonomy Monotypic.
Description L 12cm (4.5in). Adult male black on back, head, upper breast and tail, with very bright orange patches on wings, breast-sides and tail-sides. Belly white. Adult female has brownish head, olive upperparts and lemon-yellow to orange-yellow patches on wings, underwing-coverts, breast-sides and tail. First-winter male resembles female. Over its first winter it develops black feathers on back and sides, and the breast patch becomes orange-yellow, contrasting with the lemon-yellow underwing-coverts. First-winter female has little yellow on the wing and tail.

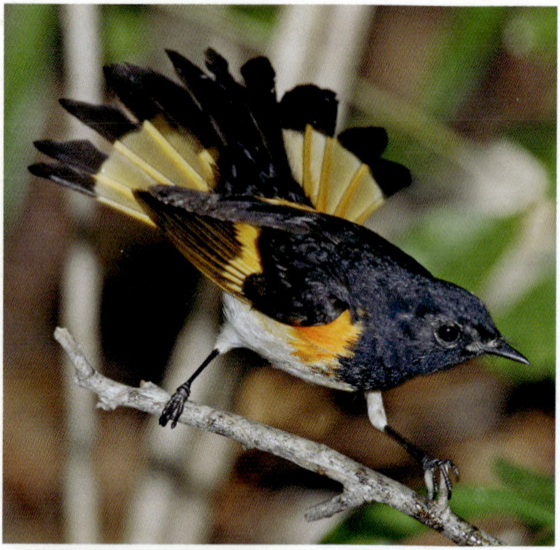

Adult male. Cayman Islands, April.

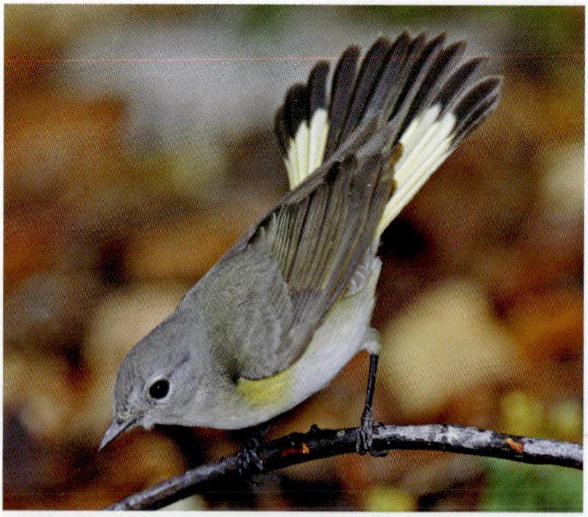

Adult female. Cayman Islands, April.

Similar species None.
Voice A loud *chip*.
Habitat and behaviour Found in all habitat types – swamps, gardens, mangroves, woodland and forests. Less abundant at high elevations. Males tend to occupy optimal habitats in forests while juveniles and females are found in poorer quality habitats in scrub. Very active, flits about in the canopy with its tail fanned, and sallies out into the open to feed on flying insects. One of the first migrants to arrive. Has been recorded as early as July.

Range Breeds in North America. Winters south to the West Indies, Central America and northern South America.
Status in Jamaica Very common winter visitor (August–May). One of the most common winter visitors to Jamaica.

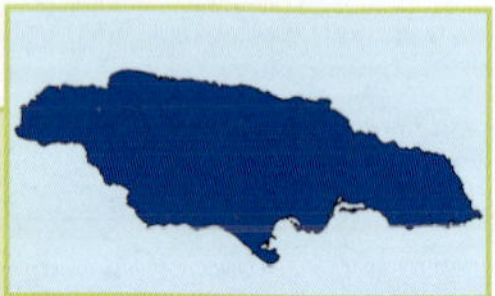

Prothonotary Warbler

Protonotaria citrea

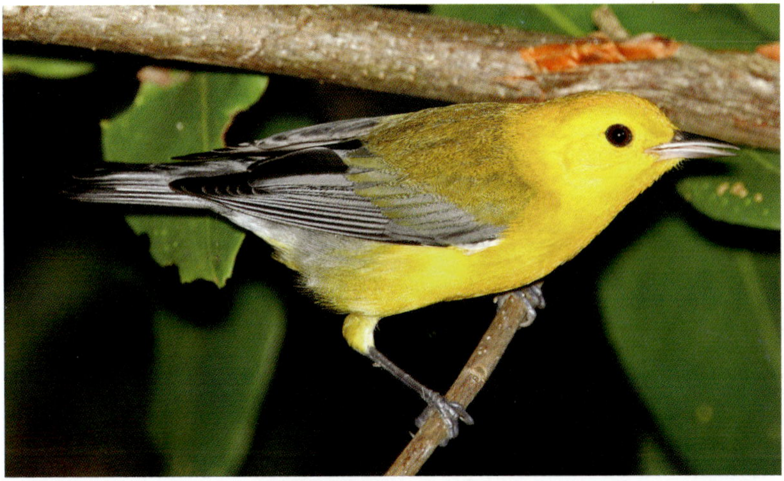

Adult male. Note yellow head and bluish-grey wings. Costa Rica, February.

Taxonomy Monotypic.
Description 12cm (5in). Adult male is a chunky warbler with bright yellow head and underparts and green upperparts contrasting strongly with the bluish-grey wings. Vent and undertail-coverts white. Dark eye and long black bill contrast with yellow head. Adult female and first-winter are duller yellow with olive wash on head showing little contrast with greyish wings.
Similar species Yellow Warbler is greenish-yellow above.
Voice Usually chips.
Habitat and behaviour Usually in scrubby woodlands near mangroves. Feeds mainly on insects.

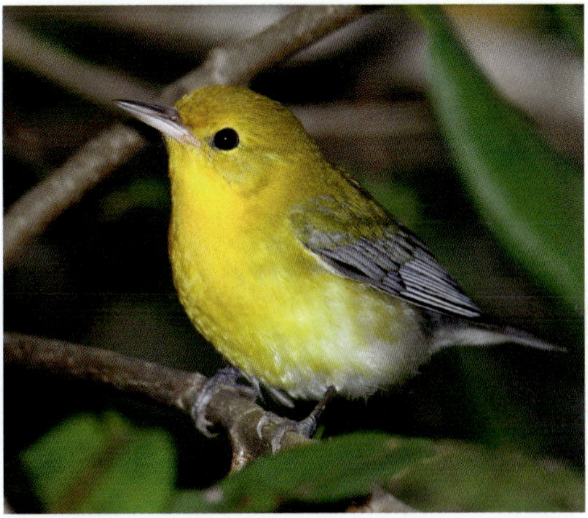

Adult female. Note olive wash on head. Costa Rica, February.

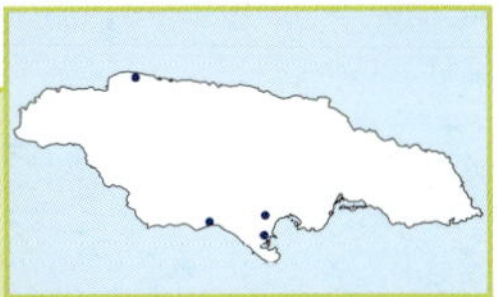

Range Breeds in North America. Winters in Central America and northern South America via the West Indies.
Status in Jamaica Regular passage migrant (October and March) and occasional winter visitor, usually in coastal areas.

Worm-eating Warbler
Helmitheros vermivorum

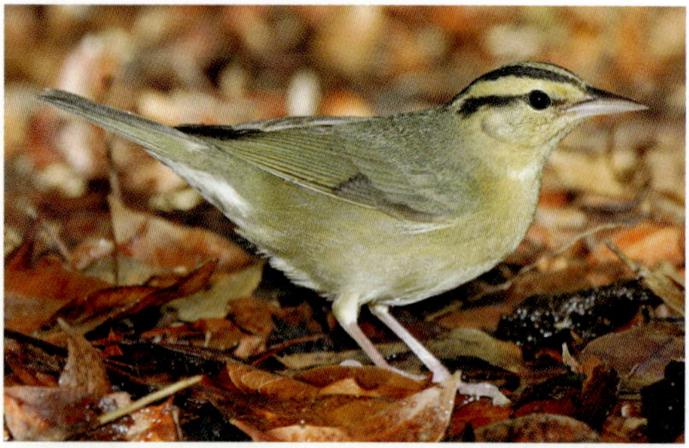

Adult. Cayman Islands, April.

Taxonomy Monotypic.
Description L 11–13cm (5–6in). A light coloured warbler with buffy-yellowish head with bold blackish stripes through eye and on either side of crown. Bill pale. Olive-brown back and wings. Unstreaked underparts are yellowish-olive.
Similar species Swainson's only has eye-stripe, has rufous cap and usually forages on ground.
Voice Usually chips in flight.
Habitat and behaviour Forests at all elevations. Feeds on insects. Forages in shrubs and dense plants (including clusters of dry leaves) from ground level to upper canopy. Secretive and hard to detect.

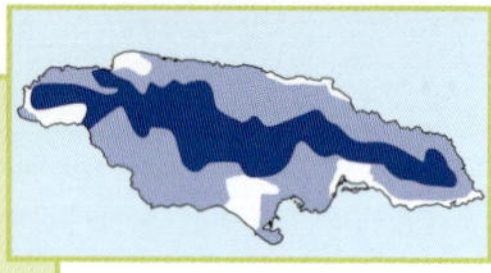

Range Breeds North America, wintering in Central America, the Bahamas and Greater Antilles.
Status in Jamaica Fairly common passage migrant and winter visitor (September–May).

Swainson's Warbler
Limnothlypis swainsonii

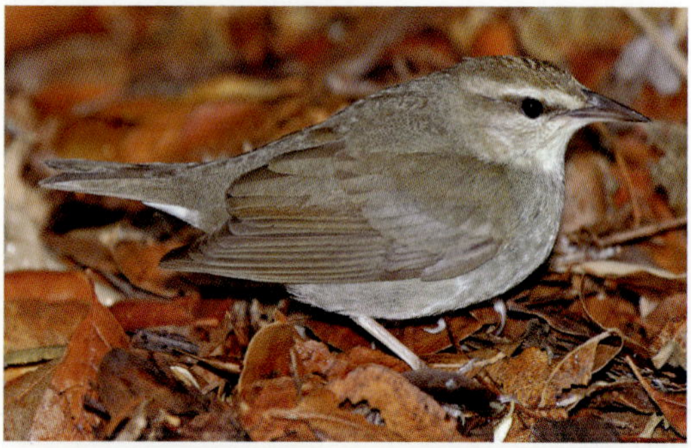

Adult. Cayman Islands, April.

Taxonomy Monotypic.
Description L 13–14cm (5–6in). Plain warbler, brown above and whitish below, with greyish flanks. Has dark eye-stripe and pale supercilium. Crown rich brown. Bill long, thick and pointed.
Similar species Worm-eating Warbler has striped head and yellower underparts.
Voice Loud chips.
Habitat and behaviour Mangrove margins, lowland woodlands but mainly occurs in montane forest. A secretive and hard-to-see species. Feeds on insects and spiders, usually foraging on the ground in the leaf litter.

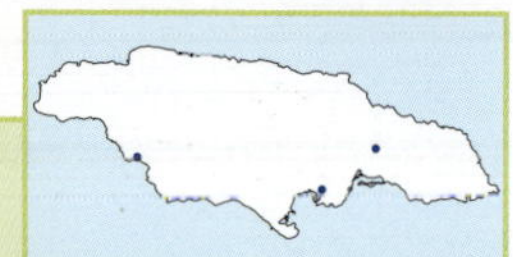

Range Breeds in North America. Winters in the Bahamas, Greater Antilles and Central America.
Status in Jamaica Uncommon winter visitor (September–April).

Ovenbird
Seiurus aurocapilla

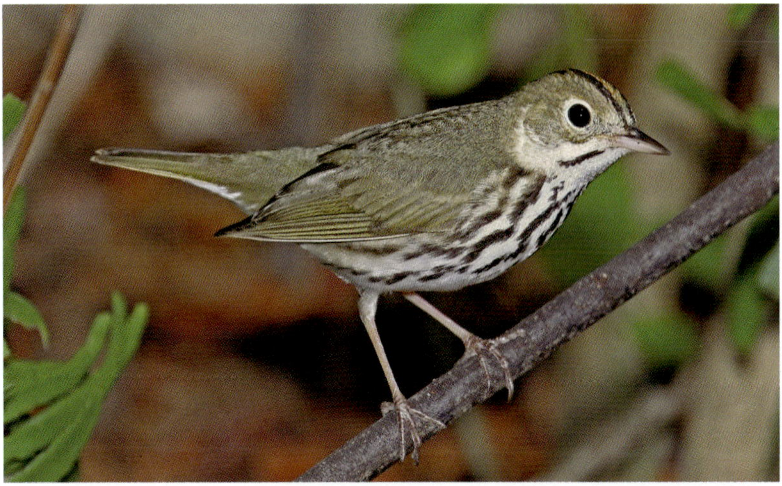

Adult. Note heavily streaked breast. Cayman Islands, April.

Local name Betsy Kick-up.
Taxonomy Polytypic (3). Race on Jamaica is *S. a. furvivor.*
Description L 14–16cm (6–7in). Olive above. White below, heavily streaked black on breast and flanks. Centre of crown rich orange-brown, bordered black. Bold white eye-ring. Legs and lower mandible pink.
Similar species Crown-stripe and distinctive walk diagnostic.
Voice Chips loudly.
Habitat and behaviour Mainly lowland and mid-level woodland. Walks (rather than hops), jerking its tail upwards. Feeds mainly on the ground on ants, beetles and seeds in leaf litter or dense understorey.

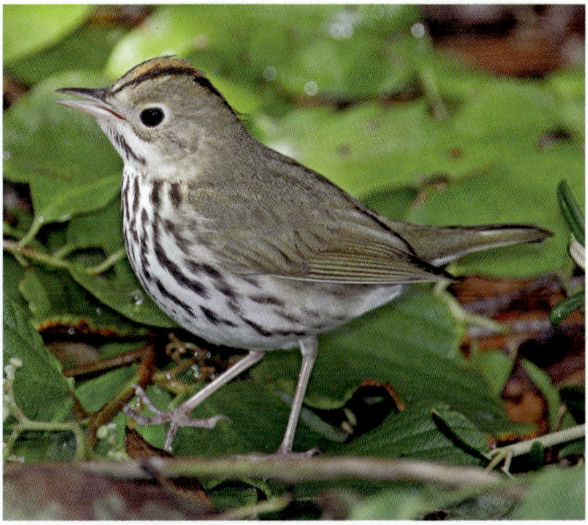

Adult. Note the crown stripes. Cayman Islands, April.

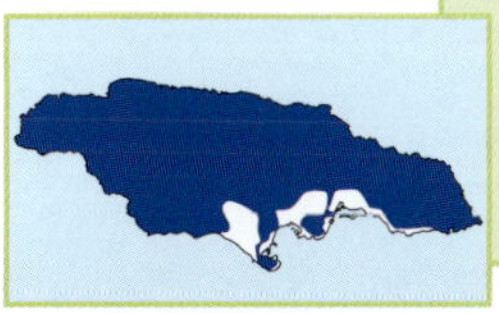

Range *S. a. furvivor* breeds in Newfoundland. Winters in Bahamas, Cuba and the West Indies south to Panama. Other races in Canada and the United States.
Status in Jamaica Common passage migrant and winter visitor (September–May).

Northern Waterthrush

Seiurus noveboracensis

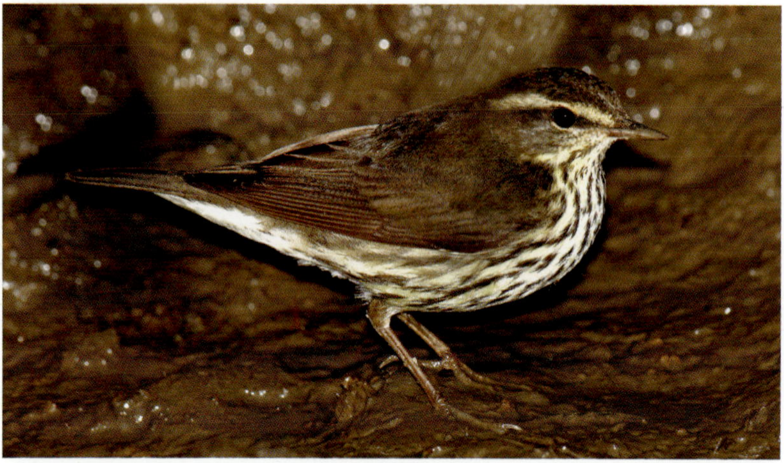

Adult. Note the streaked breast and pinkish-brown legs. Costa Rica, March.

Taxonomy Monotypic.
Description L 13–15cm (5–6in). Dark above and pale buffy-yellow or white below, breast and flanks streaked brown. Throat finely spotted. Dark eye-stripe. Prominent whitish or yellowish supercilium of uniform width throughout. Upper mandible dark brown, lower mandible paler at sides. Legs pinkish brown. First-winter underparts and supercilium more buffish.
Similar species Louisiana Waterthrush (which see).
Voice Very loud, incessant chipping.
Habitat and behaviour Mostly in mangrove forests and south coast wetlands and associated habitats. Pumps tail while walking on the ground. Bobs and teeters. Forages actively in understorey or on the ground, walking and bobbing tail. Feeds on insects, snails and other small aquatic invertebrates picked from the surface of the mud.

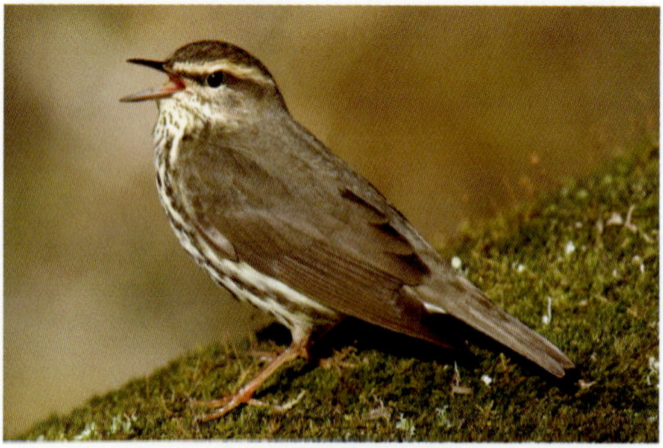

Adult. Note the plain, untapered supercillium. May.

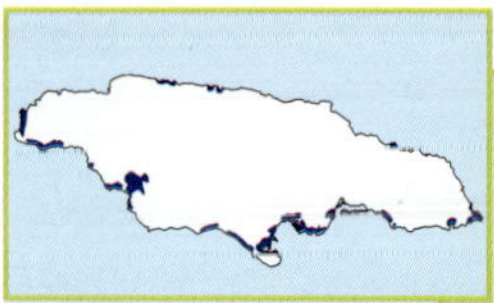

Range North America. Winters south to northern South America via Central America, the Bahamas and West Indies.
Status in Jamaica Locally abundant winter visitor and passage migrant (August–April).

Lousiana Waterthrush
Seiurus motacilla

Taxonomy Monotypic.
Description L 13–15cm (5–6in). Closely resembles more abundant Northern Waterthrush but supercilium buff before eye and white to the rear; broadens behind the eye. Also throat unspotted; belly white but flanks and vent buffish, underpart streaks more diffuse. Legs bright pink.
Similar species Northern Waterthrush.
Voice Loud, incessant chipping.
Habitat and behaviour Seen singly or in small numbers on stream banks in wooded areas. On migration may sometimes be seen in gardens. Walks about on the ground looking for insect prey in the leaf litter.

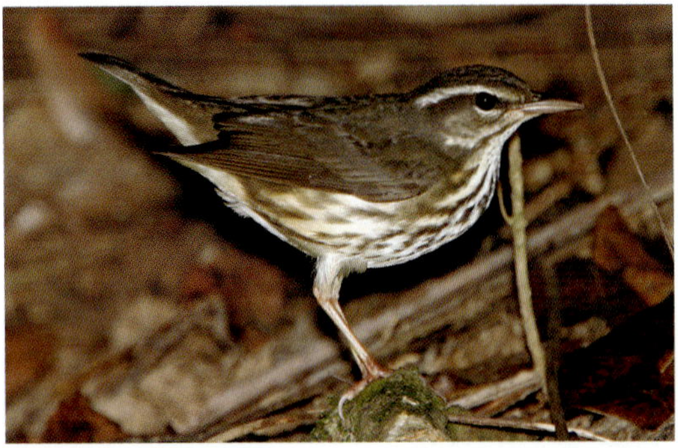
Adult. Note white chin, bright pink legs and buff wash on flanks. Cuba, Nov.

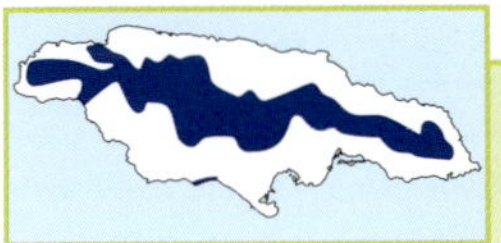

Range North America, wintering in Bermuda, Bahamas, Central America, the West Indies and northern South America.
Status in Jamaica Regular winter visitor (August–April). Less abundant than Northern Waterthrush.

Hooded Warbler
Wilsonia citrina

Taxonomy Monotypic.
Description L 12–14cm (6–7in). Olive-brown with bright yellow face; underparts yellow. Outer tail feathers white edged with black. Adult male has black hood around yellow forehead and face. Adult female black mainly on crown. First-winter female lacks any black on hood.
Similar species Yellow Warbler lacks white in tail; more prominent dark eye.

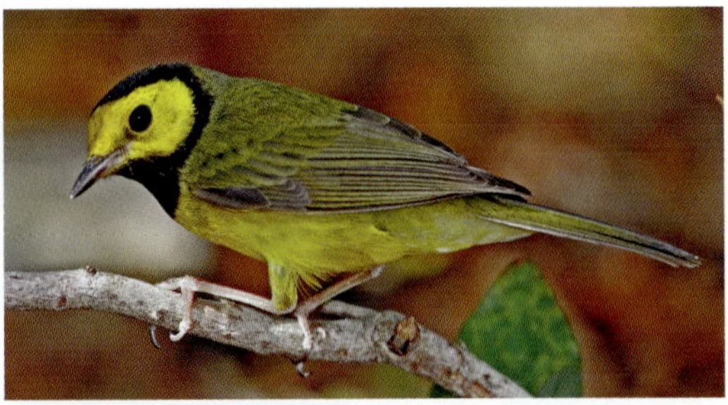
Adult male. Cayman Islands, April.

Voice Loud chips.
Habitat and behaviour Dense woods and mangroves. Feeds on insects close to the ground.

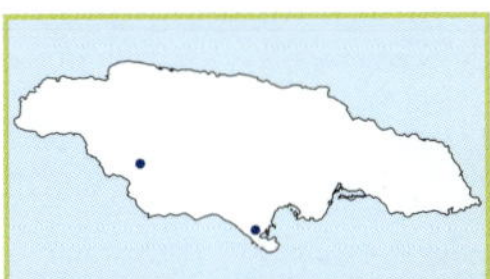

Range North America. Winters Caribbean and Central America.
Status in Jamaica Rare winter visitor or passage migrant (October–April).

Common Yellowthroat
Geothlypis trichas

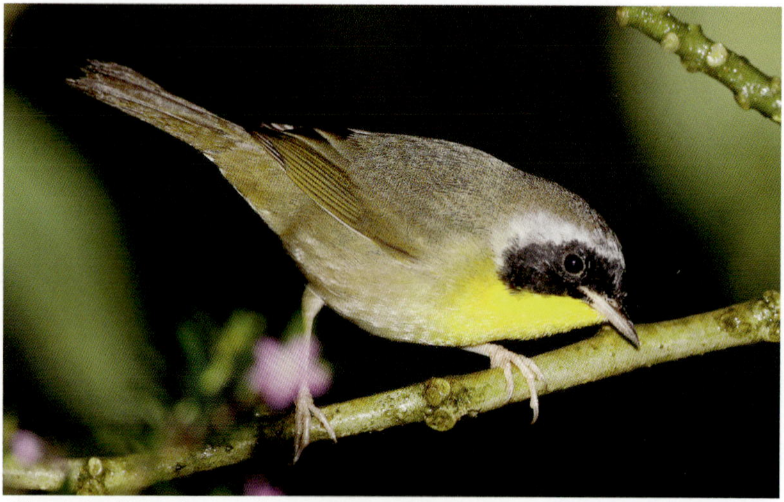

Adult male. Note black mask. Port Royal Mountains, Jamaica, February.

Taxonomy Polytypic (13). Race on Jamaica is *G. t. trichas*.
Description L 11–13cm (4–5in). Adult male unmistakeable; black mask with grey upper margin. Throat, breast and undertail-coverts bright yellow. Lower belly white. Back and wings plain olive-greenish. Adult female duller and lacks mask, cheeks grey. Has an indistinct eye ring. First-winter female very pale and brownish. Lacks eye ring. First-winter male as first-winter female but has indistinct dark mask.
Similar species None.
Voice Loud chipping.
Habitat and behaviour Most abundant along wetland edges but also found in bushes and hedges, from lowland swamps to the mountains. Feeds on insects and spiders. Tends to forage close to the ground in deep undergrowth.

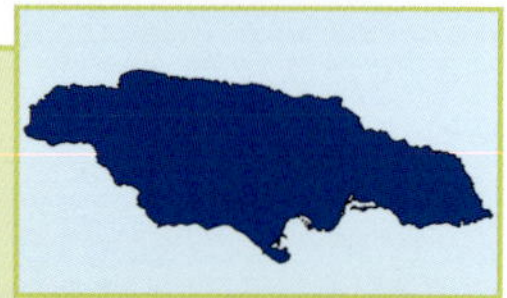

Range *G. t. trichas* breeds in southeast United States. Winters in Bahamas, West Indies and northern South America.
Status in Jamaica Common winter visitor and passage migrant (September–May)

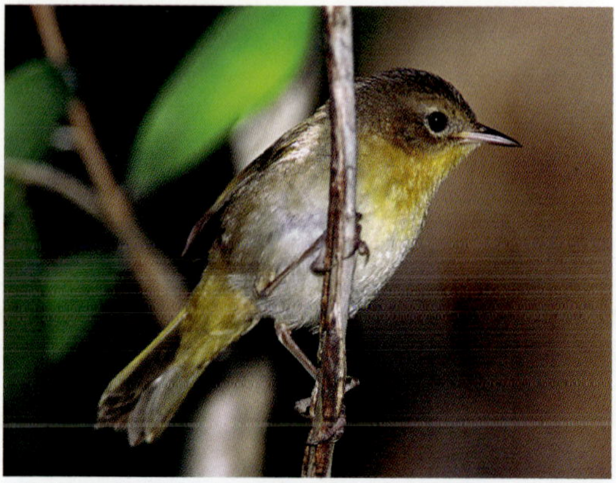

Adult female. Note dull yellow breast. Cayman Islands, April.

Scarlet Tanager
Piranga olivacea

Taxonomy Monotypic.
Description L 16cm (6in). Breeding male in spring is red overall with black wings and tail. Non-breeding male greenish with black wings: some are still in breeding plumage when they first arrive in autumn. Female and juvenile olive green with dark wings.
Similar species Male breeding Summer Tanager has red wings. Female and juvenile Summer Tanagers have paler wings.
Habitat Usually seen in wooded gardens.

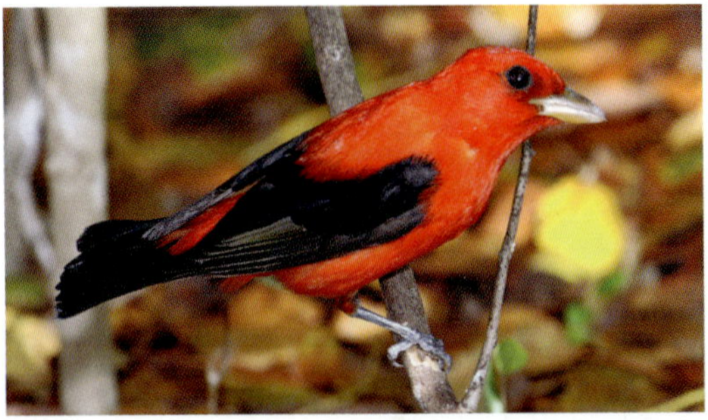
Adult male in breeding plumage. Cayman Islands, April.

Range Breeds in North America, wintering in Central America south to Panama.
Status in Jamaica Rare passage migrant (October–May).

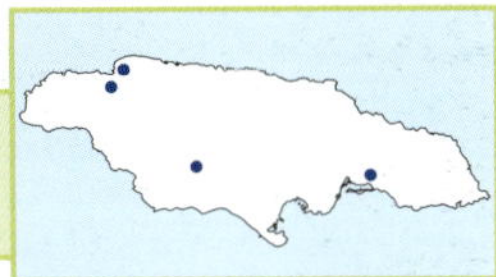

Summer Tanager
Piranga rubra

Taxonomy Polytypic (2).
Description L 17cm (6.5in). Bill is long and thick and always pale in colour. Breeding male wholly bright red. Female and juvenile olive-green. Immature males show some reddish feathers early in their second year.
Similar species Scarlet Tanager has darker wings in all plumages.

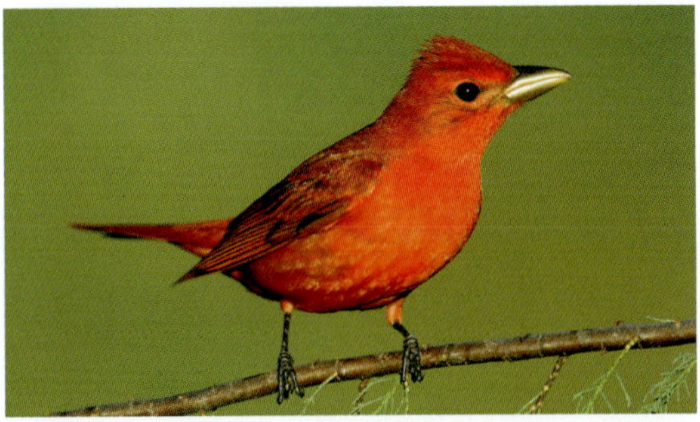
Adult male in breeding plumage. Unmistakable red body and black wings. April.

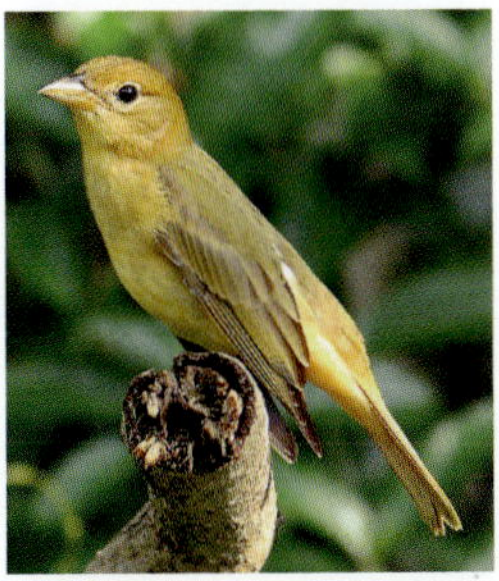
Adult female. Florida, September.

Habitat and behaviour Secretive. Often seen in wooded gardens.

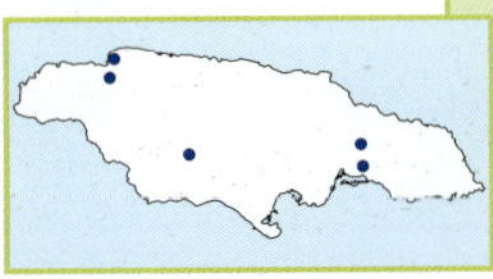

Range Breeds in North America, wintering in Central America south to Bolivia.
Status in Jamaica Rare passage migrant (Oct–April).

Jamaican Spindalis
Spindalis nigricephala

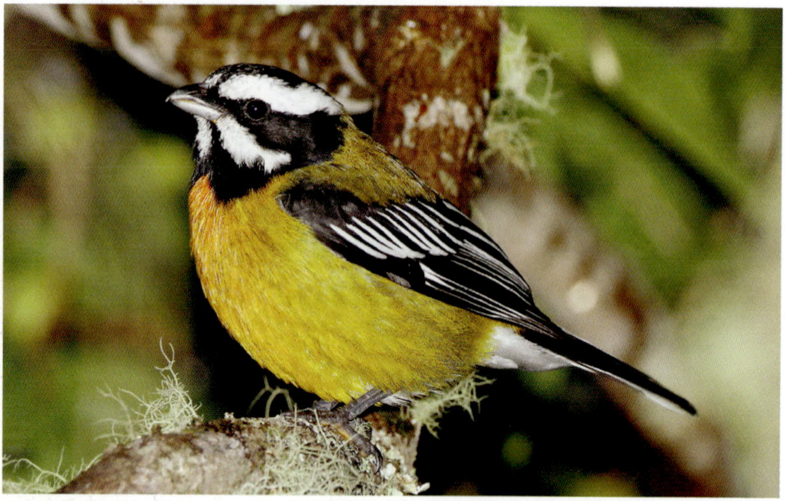

Adult male. Port Royal Mountains, Jamaica, February.

Local names Mark Head, Goldfinch, Champa Beeza. **Alternative name** Jamaican Stripe-Headed Tanager. **Taxonomy** Monotypic. Endemic to Jamaica. **Description** L 18cm (7in). Distinctive chunky yellow, black and white bird with a short bill. Adult male has black head with broad white superciliary and malar stripes and a black-bordered white chin. Bill is thick, dark grey; lower mandible is silver with a dark tip. Breast is orange, fading to a mixture of orange and yellow on belly. Back is greenish-yellow becoming orange on rump. Flight feathers and upperwing-coverts are black, edged white. Tail black, outer feathers tipped white. Undertail-coverts are white. Legs grey. Female and juvenile have greyish heads, with indistinct head markings mottled grey and white. Breast and belly are greenish-yellow

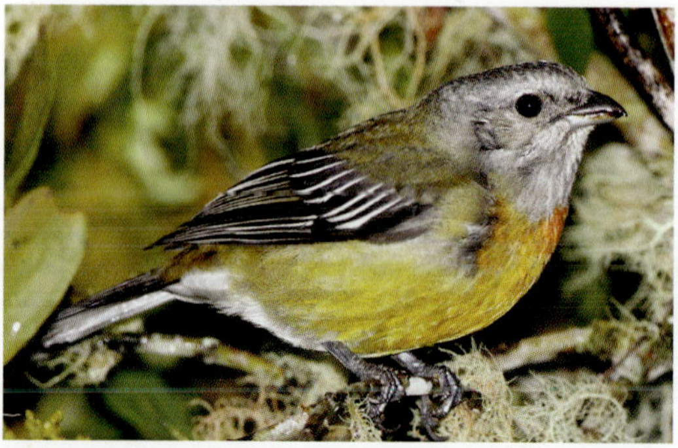

Adult female. Port Royal Mountains, Jamaica, February.

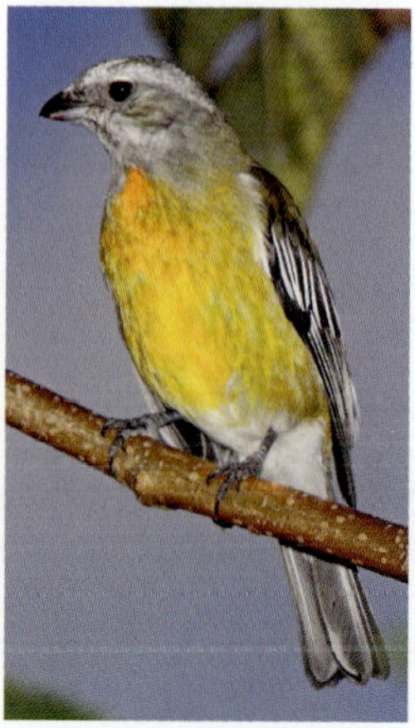

Imm. male molting to adult. Nov.

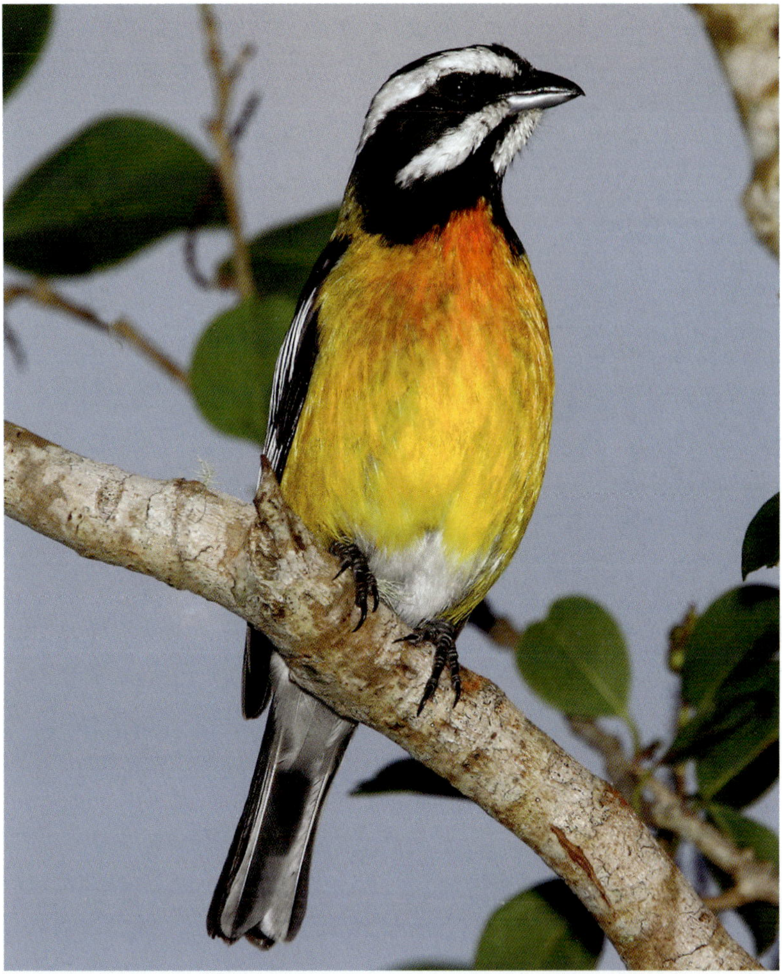

Adult male. Port Royal Mountains, Jamaica, May.

with a small orange patch at centre of breast. Back is greenish-grey becoming yellowish on rump. Tail is dark grey, wings are dark grey, edged white in adult females and edged yellow in juveniles.
Similar species Bananaquit is much smaller and has a long, decurved bill.
Voice In flight often gives a very soft, weak *seep*. Males often sing from exposed perches, giving a series of high, fast *chi-chi-chi-chi-chi* notes.
Habitat and behaviour Widespread in moist and wet forests from sea level to the high mountains. Usually seen in pairs or family groups. Feeds, often in loose flocks, on berries, flowers and leaves. Nest (April–July) is a loosely built cup.

Range Jamaican endemic.
Status in Jamaica Common and widespread.

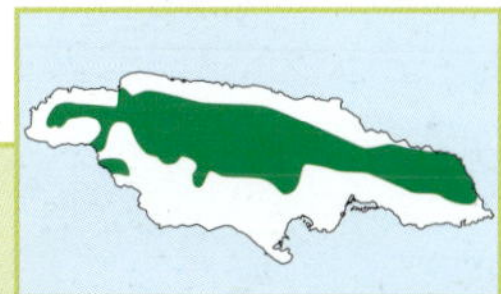

Bananaquit
Coereba flaveola

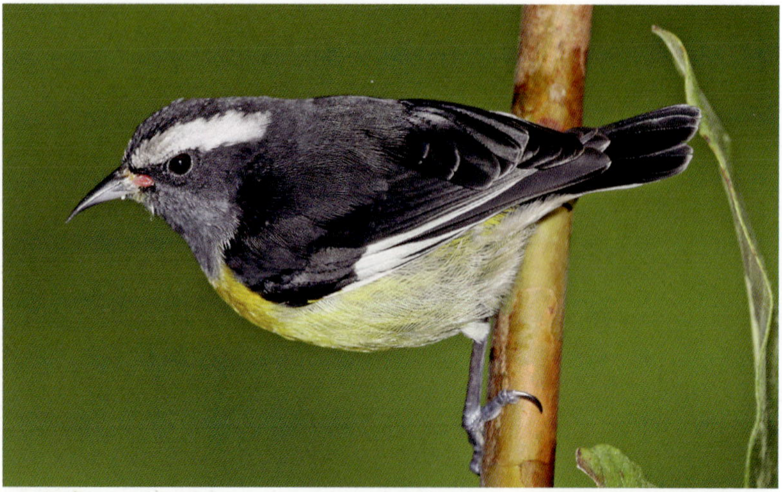

Adult. Rocklands, Jamaica, November.

Local names Teasy, Sugar Bird, Beeny Bird.
Taxonomy Polytypic (41). *C. f. flaveola* is endemic to Jamaica.
Description L 10cm (4in). A small black, yellow and white bird with decurved bill, bright yellow rump and short black tail. Bright white supercilium is broader in males than in females; dusky towards rear in latter. Bill shiny black with a crimson spot at its base. Iris dark brown. Throat is dark grey. Breast bright yellow, contrasting with the white lower belly and greyish-tinged flanks. Wings dark grey; bases of primaries white, forming a white wing-patch. Bend of wing yellow. Three outer tail feathers are tipped white. Legs grey. Juvenile greyish above and greyish-yellow below, with yellowish throat and indistinct supercilium.
Similar species Female Black-throated Blue Warbler has some resemblance to juvenile Bananquit but has a straight bill without any red spot at the base.
Voice A rapid *zizizizizizizizizi* on one tonal level, and many other high, buzzy calls.
Habitat and behaviour Ubiquitous wherever there are flowering trees, shrubs or herbs, in gardens, woodlands and forests from sea level to the highest mountains. Feeds

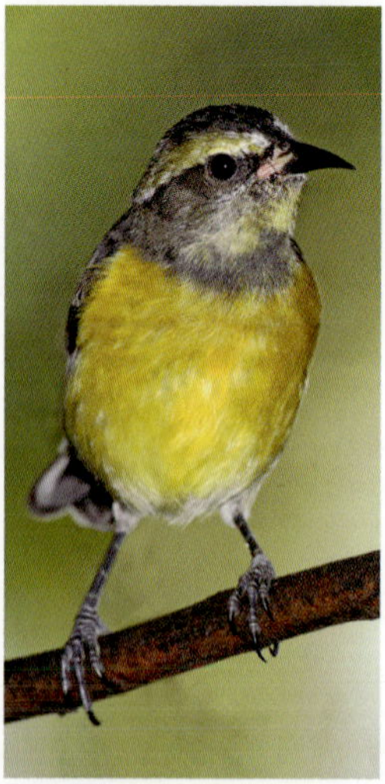

Juvenile. Rocklands, November.

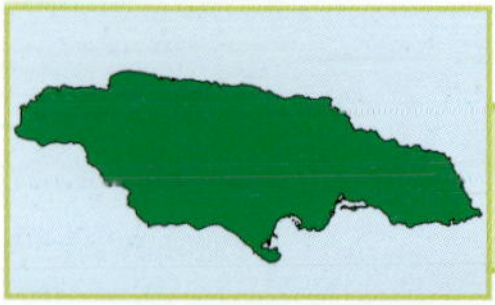

Range *C. f. flaveola* endemic to Jamaica. Other races on other Caribbean islands (not Cuba), and in Central and South America.
Status in Jamaica Abundant and widespread resident.

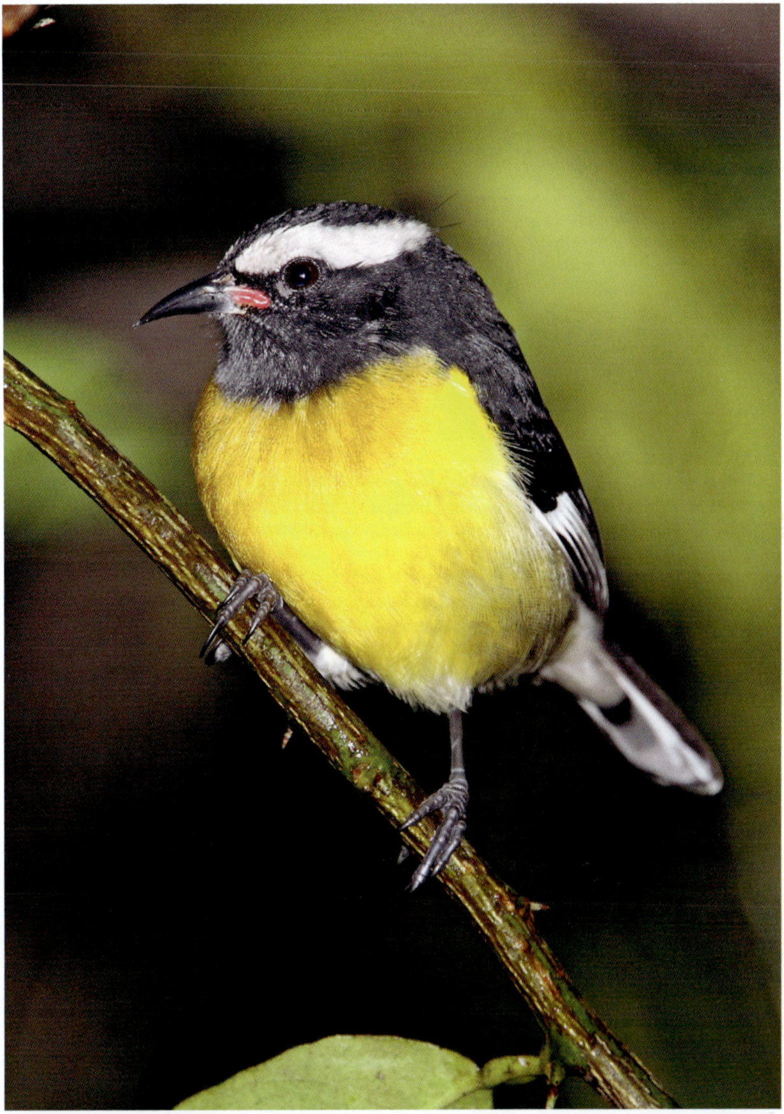

Adult. Rocklands, Jamaica, May.

on nectar (often puncturing the base of flowers), insects and small berries. Comes to hummingbird feeders or basins of sugar. Nest (chiefly March–June but often several times a year, triggered by rainfall) is dome-shaped, loosely woven of long grasses, with a side entrance. It is usually constructed in dense shrubbery e.g. Bougainvillea, often close to a wasps' nest (which has been shown to provide protection from predators). Lays 3–4 heavily spotted eggs. Males also build separate nests for sleeping in.

Yellow-faced Grassquit
Tiaris olivaceus

Taxonomy Polytypic (5). Race in Jamaica is *T. o. olivaceus.*
Description L 10cm (4in). A small olive-brown finch. Adult male has bright yellow supercilia that make a V-shape when viewed from the front. Bright yellow throat-patch in black bib. Yellow crescent below eye. Legs grey. Adult female has cream supercilium, chin and crescent below eye; throat-patch pale yellow. Juvenile pale brown with facial markings indistinct or absent.
Similar species Black-faced Grassquit female and juvenile resemble juvenile Yellow-faced Grassquit but have pinkish legs. Munias and bishops in non-breeding plumage lack the distinctive facial marks.
Voice A series of high-pitched trills and ticks.
Habitat and behaviour Gardens, grasslands (especially with Guinea-grass), edges of forests, woods and cleared areas, chiefly in the lowlands. Feeds by hopping on to a grass seed-head and weighing it down to the ground where it can be stripped of seeds. May roam in flocks around the island depending on food availability. Nest (year-round but chiefly January–May) domed with a side entrance. It is made of woven grasses and attached to a shrub or small tree, such as a lime tree.

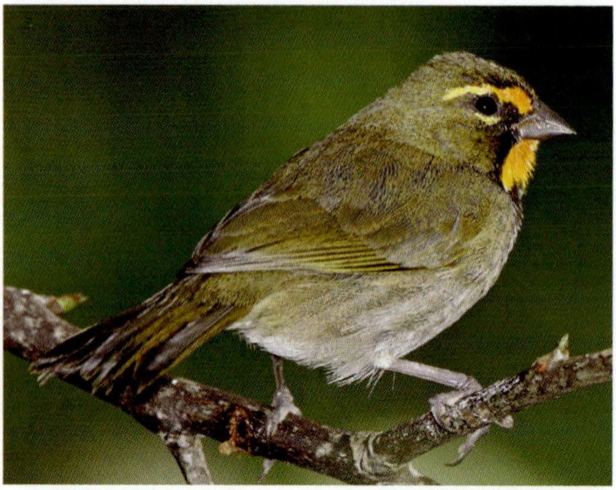
Adult male. Rocklands, Jamaica, November.

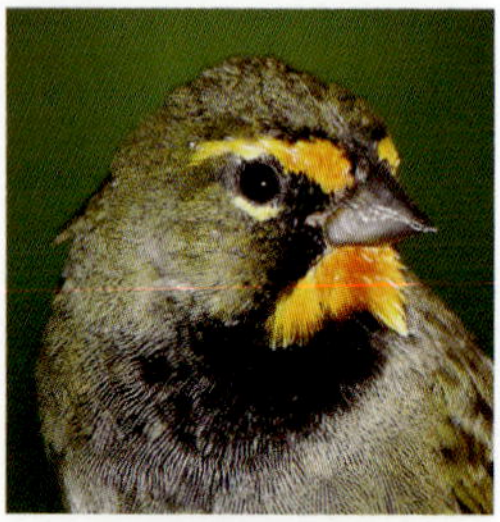
Adult male. Rocklands, Jamaica, May.

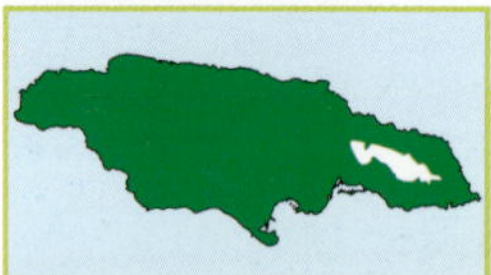

Range *T. o. olivacea* resident on Jamaica, Cuba, Hispaniola and islands, and the Cayman Islands. Other races in Puerto Rico, Cozumel Island and Central America to northern South America.
Status in Jamaica Common and widespread.

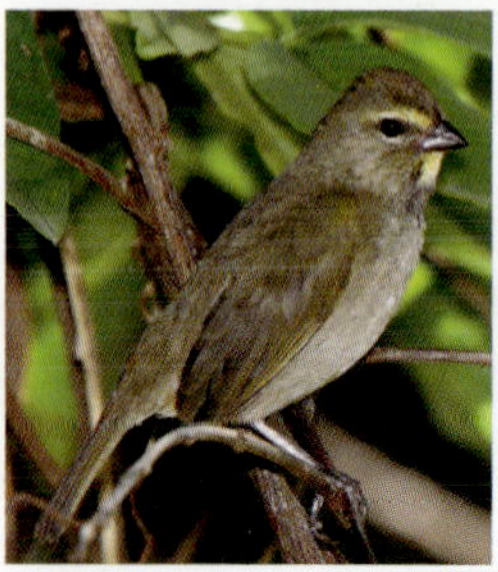
Adult female. Rocklands, May.

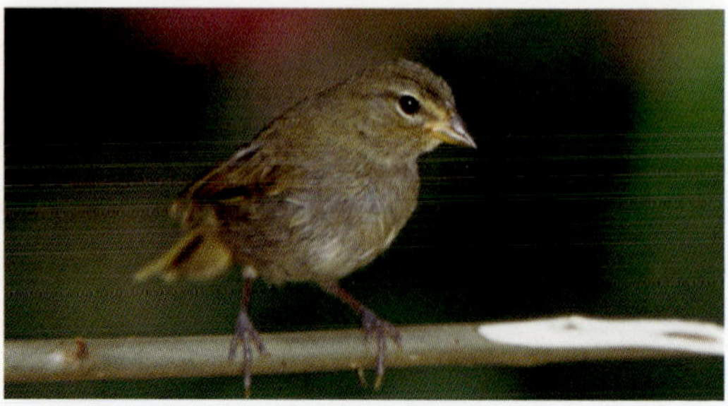
Juvenile. Rocklands, Jamaica, November.

Black-faced Grassquit
Tiaris bicolor

Local names Black Sparrow.
Taxonomy Polytypic (8). Race in Jamaica is *T. b. marchii*.
Description L 10cm (4in). A small finch with dull brown back, greyish lower underparts and buffy undertail-coverts. Legs pinkish grey. Adult male has black face and breast. Upper mandible darker than the lower, gape red. Female and juvenile have grey heads and breasts. After about three months young males start to develop the black breast.
Similar species Juvenile Yellow-faced Grassquit has dark legs. Munias and bishops in non-breeding plumage have larger bills and browner, more contrasting plumage. Bishops are heavily-streaked above.
Voice *whichi? whichi-chi-chi.*

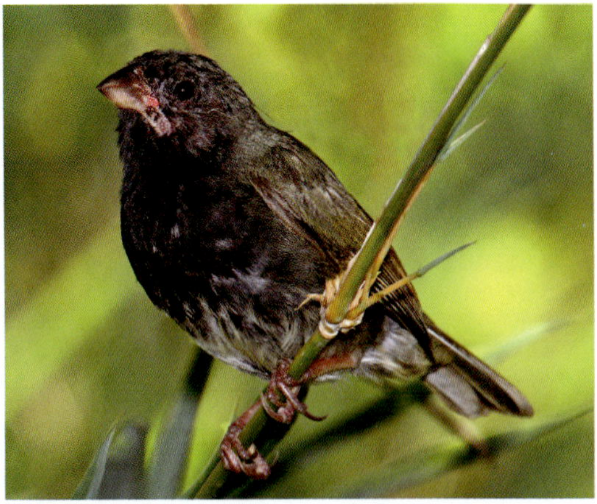

Adult male breeding. Note black on throat and belly. Rocklands, May.

Habitat and behaviour Usually seen in small flocks or family parties in gardens and along roadsides. More common around human habitations than Yellow-faced Grassquit. Feeds mainly on seeds of grasses and weeds. Hops on the ground or feeds in grasses and shrubs. Comes to bird feeders to eat cracked rice. Nest (year-round but chiefly April-August) domed, made of grass; built in vines, trees or bushes, or in pots on verandahs. Lays three eggs, which are cream, streaked rufous especially at the round end.

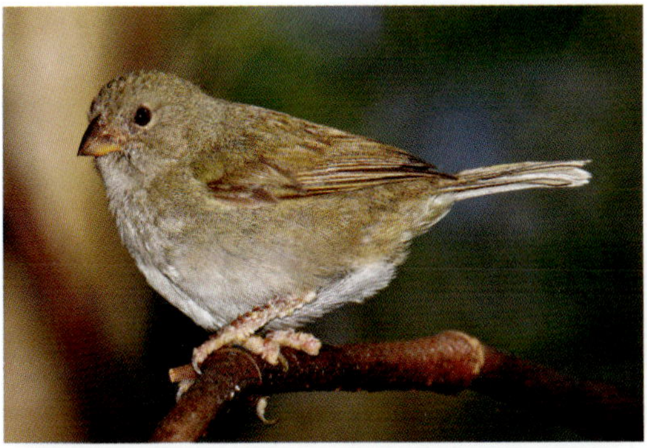

Adult female. Rocklands, November.

Range *T. b. marchii* on Jamaica, Hispaniola and adjacent islands. Other races in Bahamas and throughout the Caribbean (except mainland Cuba) and North and South America.
Status in Jamaica Common and widespread resident.

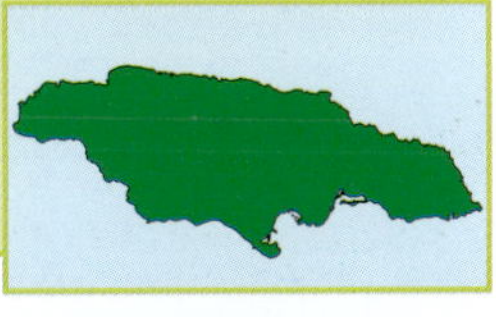

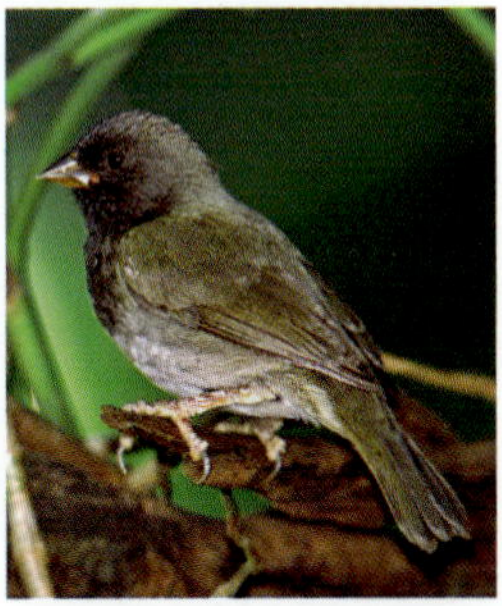

First-year male. Rocklands, Nov.

Yellow-shouldered Grassquit
Loxipasser anoxanthus

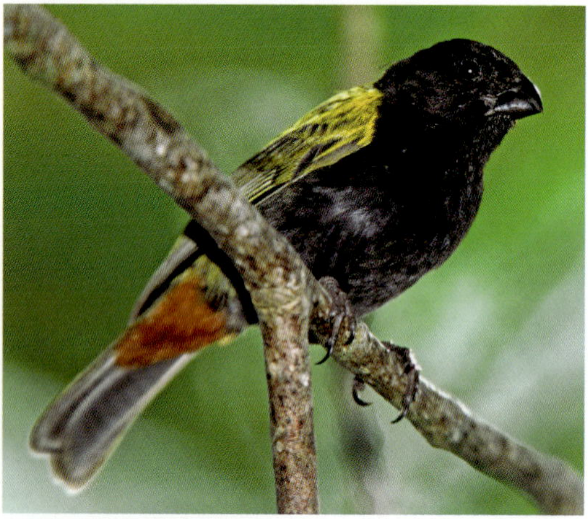

Adult male. Montego Bay, Jamaica, May.

Local names Yellow-Backed Finch, Yellow-Shouldered Finch, Yellow-Back.
Taxonomy Monotypic genus. Endemic to Jamaica.
Description L 10cm (4in). A small, brownish-olive finch. Male has black head and breast, dark grey belly and flanks, and rusty undertail-coverts. Upper back and wing-coverts are bright yellow, becoming greenish-yellow on rump. Wings and tail dark grey, edged yellow. Legs dark brownish-pink to black. Female has olive head, finely spotted with dark grey. Back and wing-coverts are paler yellow than male, carpal patch bright yellow. Underparts grey, lightly washed olive, with rusty undertail-coverts. Juvenile resembles the female but has less yellow in wing, undertail-coverts pale rufous.
Similar species Black-faced and Yellow-faced Grassquits are smaller, lack yellow on wing.
Voice A descending *chi-chi-chi-chi-chi*, sounding a bit like beads shaken in a jar. Has at least four local dialects.
Habitat and behaviour Common in hills and mountains, including roadsides and garden shrubbery, ranging down to sea level. Apparently has an altitudinal movement to the lowlands in winter. Often seen in family groups but does not flock. Behaves more like a bullfinch than a grassquit. Feeds on small fruits and seeds, such as Spanish Needle, Prickly

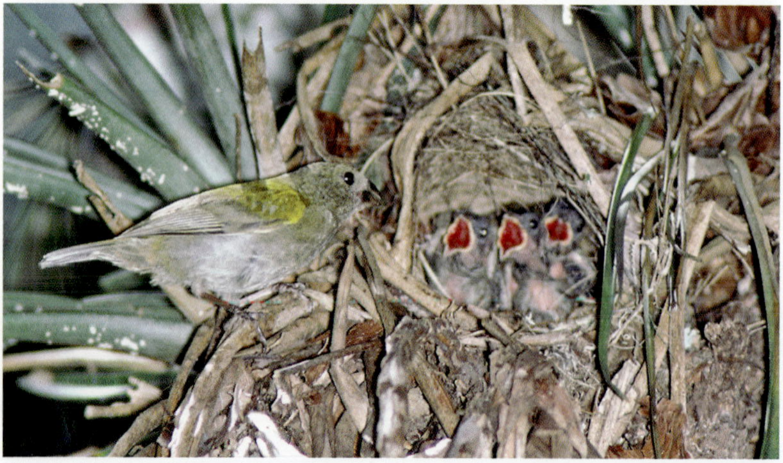

Although this bird is feeding young, it is second-year bird breeding in sub-adult plumage. Port Royal Mountains, May.

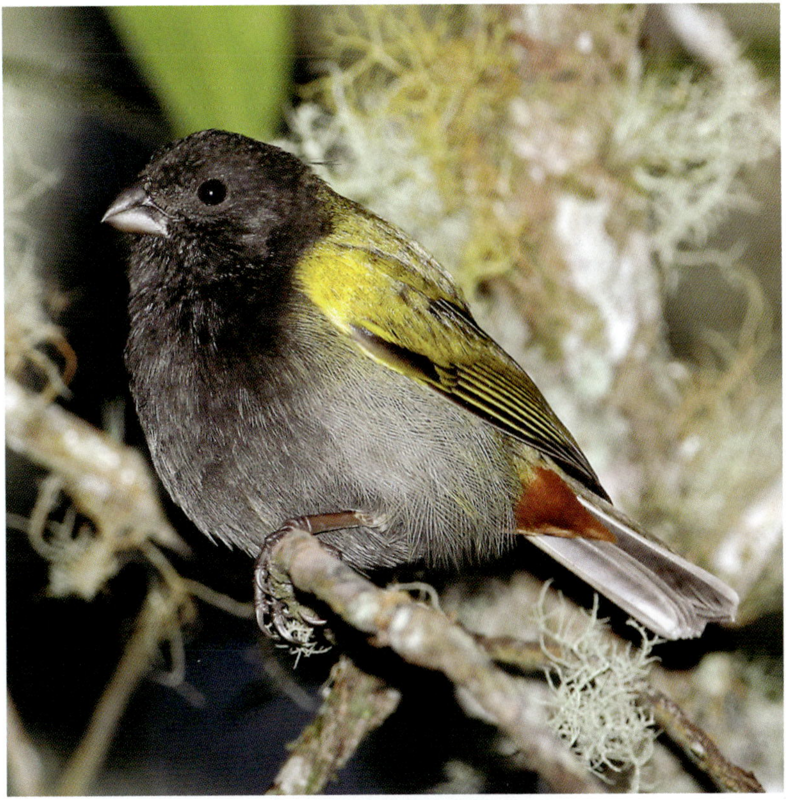

Adult male. This is a rather pale individual. Port Royal Mountains, Jamaica, November.

Yellow, Maiden Plum and Fiddlewood. Nest (March-July) is an open bowl made of grasses and twigs, under the canopy in cedar, citrus bushes and under bromeliads such as *Tillandsia*.

Range Jamaica.
Status Locally common resident.

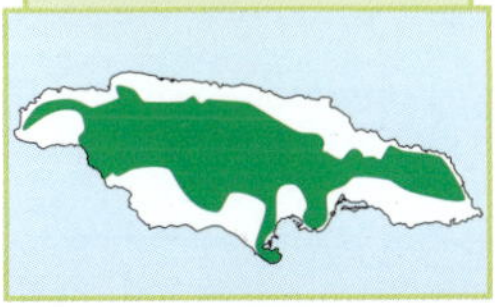

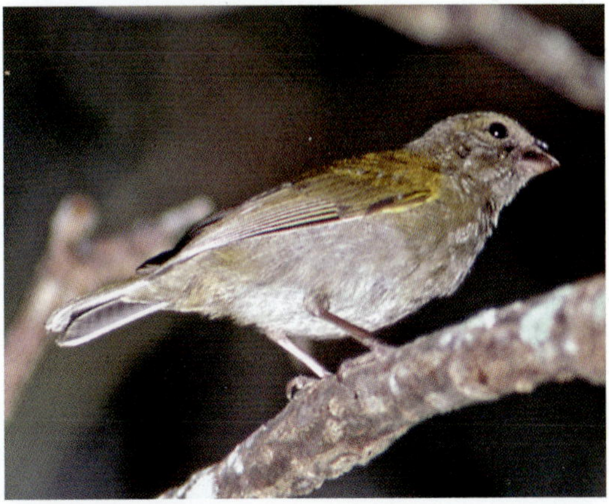

Immature male. Port Royal Mountains, Jamaica, May.

Orangequit
Euneornis campestris

Local names Blue Baize, Long-Mouth Bluequit, Blue Badas.
Taxonomy Monotypic genus. Endemic to Jamaica
Description L 14cm (5.5in). Male is a small, slender, blue-black bird. Appears bright blue in sunlight, with a bright rust-coloured, rectangular throat-patch and a slightly decurved black bill. Dark eye-stripe, iris dark russet. Legs black. Female and juvenile brown, head bluish-grey washed olive-brown which is more pronounced in the female. Wings and tail olive-brown, edged fawn. Underparts pale grey with indistinct whitish streaking. Immature male takes two years to acquire blue adult plumage, going through a patchy blue and brown transitional stage in its second year (July–September).
Similar species First-year Greater Antillean Bullfinch resembles female or juvenile Orangequit but much larger, more erect in posture and has a finch-like bill. Blue Mountain Vireo resembles female but has a browner back and yellower underparts.
Voice A soft, high-pitched *tseet*, *tsit* or *swee*, sounding a bit like a squeaky wheel. Bursts of intense calling also associated with groups in dense vegetation during the breeding season.
Habitat and behaviour Open woodland, roadside bushes in wet mid-level and mountain areas. Feeds on fruit, seeds and flowers, from low to mid-canopy in trees and shrubs.

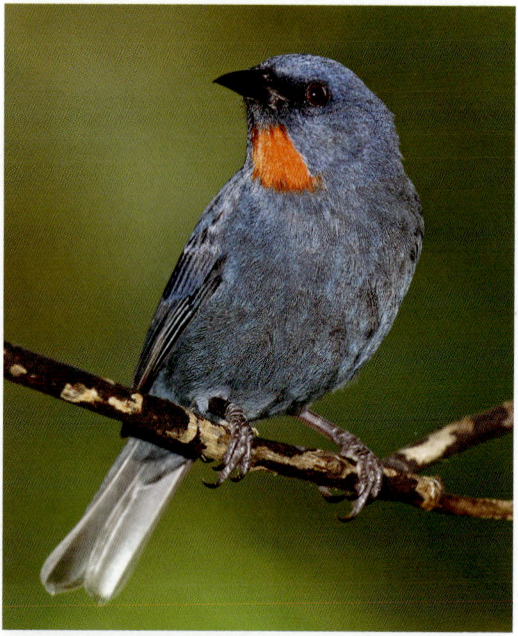

Adult male. Rocklands, Jamaica, November.

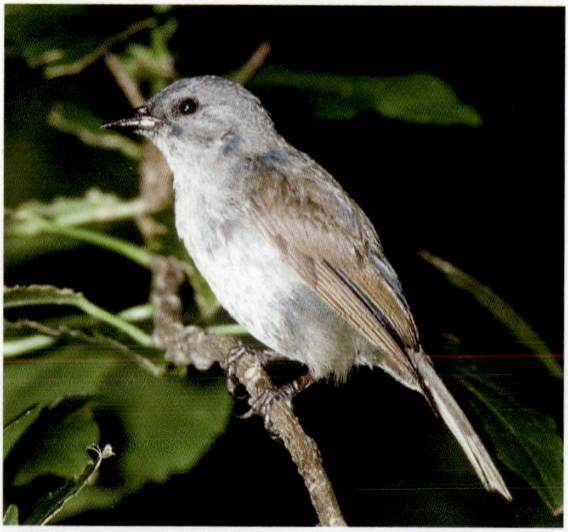

Second-year male moulting to adult plumage. Rocklands, Jamaica, May.

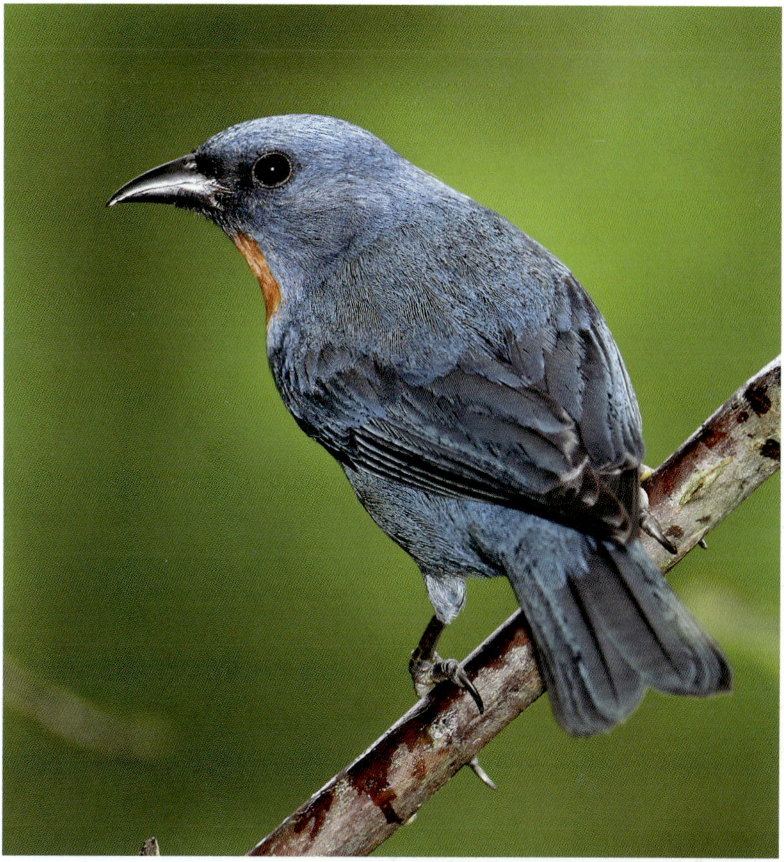

Adult male. Rocklands, Jamaica, November.

Nest (April–June) is a very well concealed, deep, roughly-built cup; sometimes in bunches of seeds or in a large leaf.

Range Jamaica.
Status in Jamaica Locally common resident.

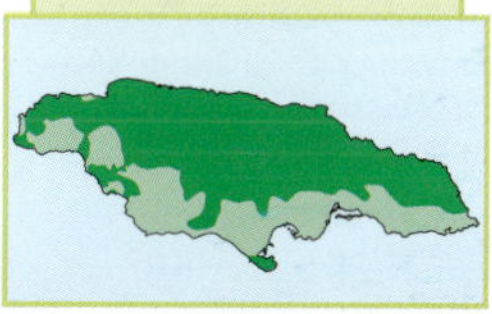

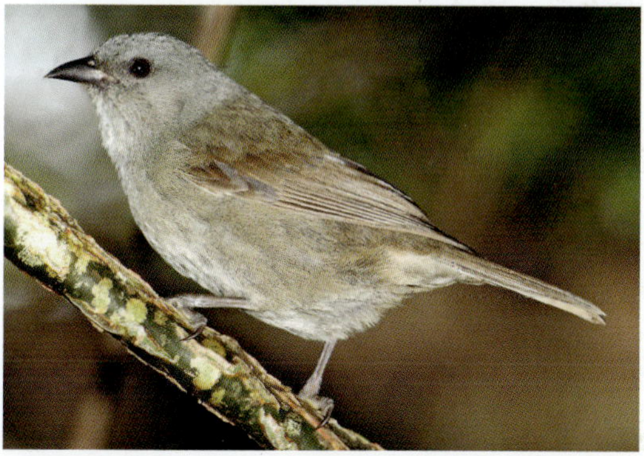

Immature. Rocklands, Jamaica, November.

Greater Antillean Bullfinch
Loxigilla violacea

Local name Black Sparrow, Jack Sparrow, Cotton Tree Sparrow.
Taxonomy Polytypic (5). *L. v. ruficollis* is endemic to Jamaica.
Description L 16cm (6in). A stout finch with a very thick, short black bill. Adult male glossy black above and dull black below. Supercilium, throat and undertail-coverts orange-rufous. Legs dusky brown, iris reddish-brown. Adult female duller black than male. Juvenile is olive-brown above, paler below; rusty markings on head, undertail-coverts pale, lower mandible pale in centre. May appear mottled during transition to adult plumage (July-Sept.).

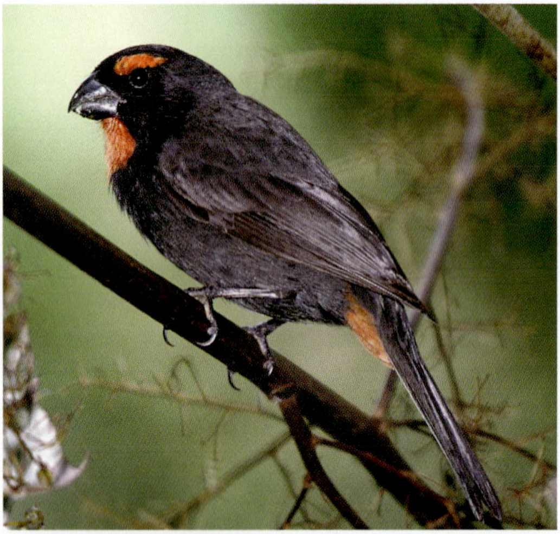
Adult male. Port Royal Mountains, Jamaica, May.

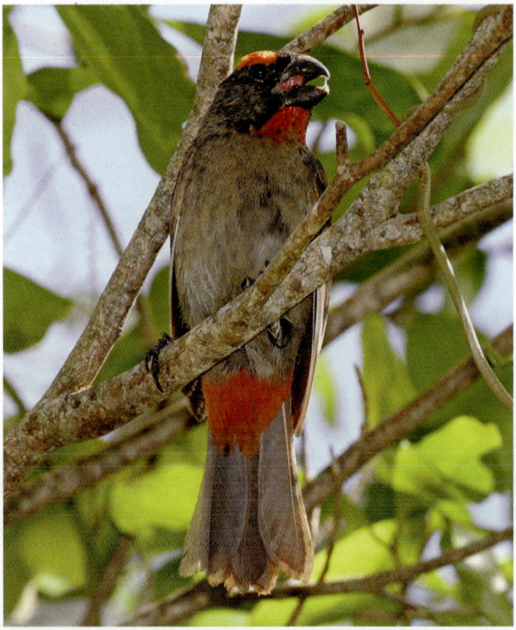
Second-year male moulting to adult plumage. Rocklands, Jamaica, May.

Similar species None. Much larger than other similar species (e.g. cowbirds, munias and bishops). First-year can resemble juvenile or female Orangequit but size, posture and bill all diagnostic.
Voice A wheezy insect-like call but also capable of loud squawks.
Habitat and behaviour Bushy areas and forest undergrowth at all elevations, but chiefly in the mountains. Feeds on the seeds of a variety of wild and cultivated plants, as well as on the buds and petals of flowers; also fruit, such as green plantain, coffee and peppers. Uses its large bill skillfully to remove the hard husks of seeds such as four-o'clocks. Nest (March–June) in holes in dead tree limbs, lined with *Tillandsia* down or paper-thin tree bark.

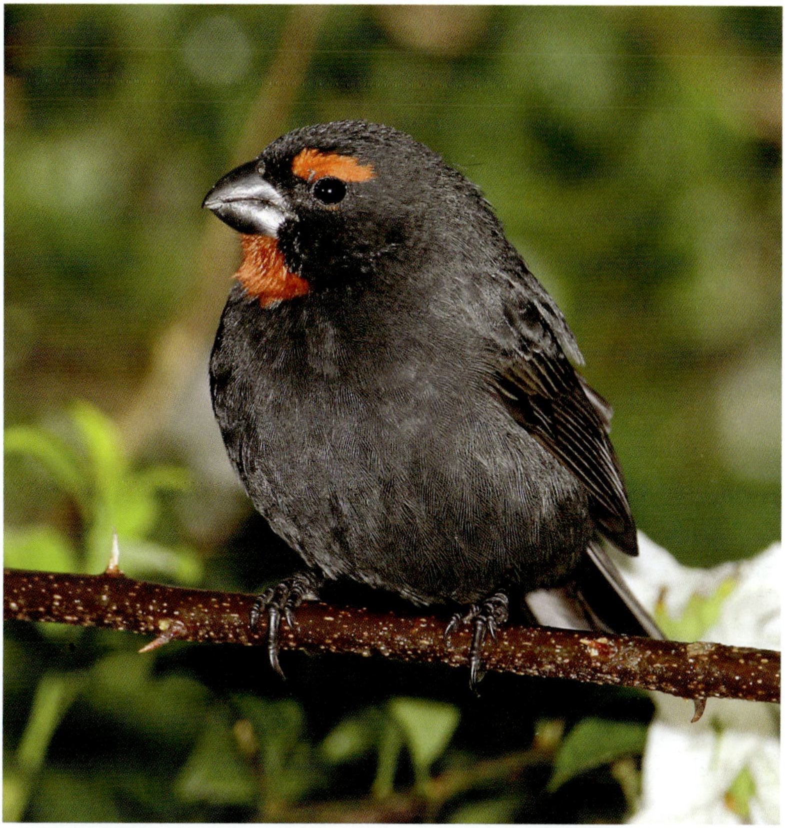

Adult male. Marshall's Pen, Jamaica, November.

Range *L. v. ruficollis* endemic to Jamaica. Other races in the Bahamas, Hispaniola and adjacent small islands and Saona Islands off Puerto Rico.
Status in Jamaica Common and widespread resident.

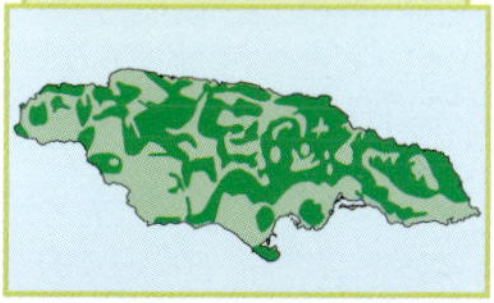

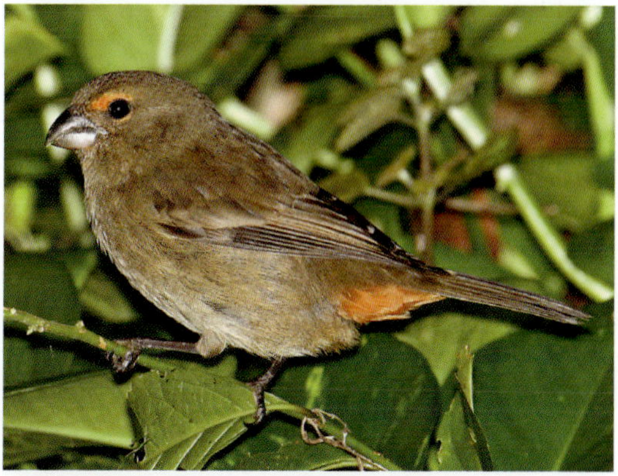

Juvenile. Cockpit Country, Jamaica, November.

Saffron Finch
Sicalis flaveola

Local name Canary.
Taxonomy Polytypic (4).
Description L 14cm (5.3in). Male is a very distinctive bright yellow bird with an orange crown. Rest of the upperparts buffy-yellow, streaked brownish on the back. Wings and tail brown, edged yellow. Face and underparts orange-yellow, yellower towards tail. Bill is conical; upper mandible brown, lower mandible cream. Legs buff. Female is paler yellow. Juvenile green-grey above streaked brown, white below with a pale yellow breast-band.
Similar species None.
Voice Call note a loud chip, and squeaky song (beginning in March) *chup-chip-chup-zeeee, chup-zeeee chup-zeeee chup-zeeee.*
Habitat and behaviour Open grassy areas, gardens, towns and chicken farms from sea level to the mountains (except the highest). May occur in flocks. Roosts communally in small trees. Feeds on a variety of grain and seeds, also nibbles hibiscus leaves. Nest (year-round but chiefly March–July) under cover in holes in trees, at the base of palm fronds, in Night-blooming Cereus or under the eaves of houses. May compete for nest sites with House Sparrow.

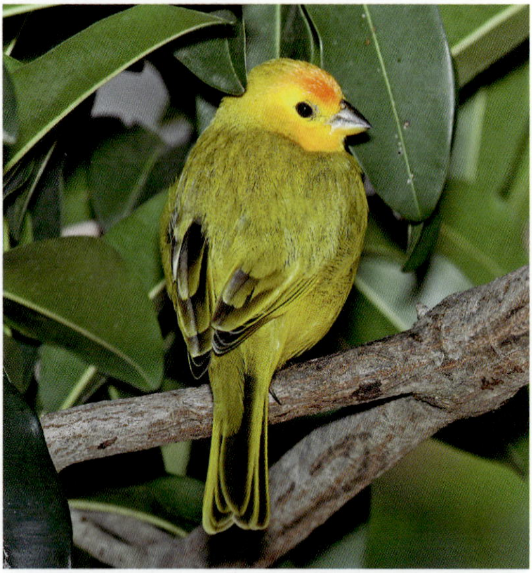

Adult male. Rocklands, Jamaica, February.

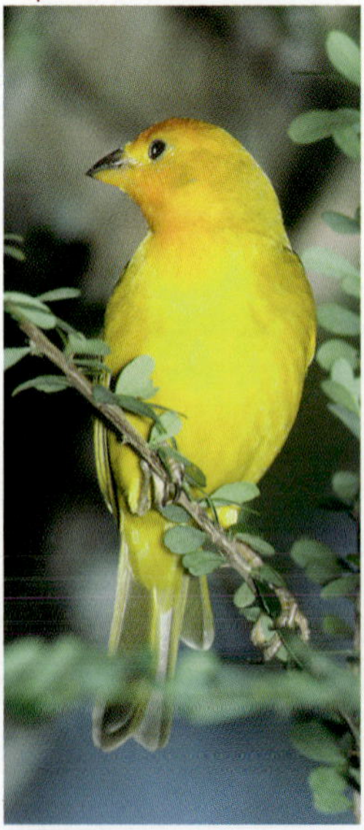

Adult male. Rocklands, Jamaica, May.

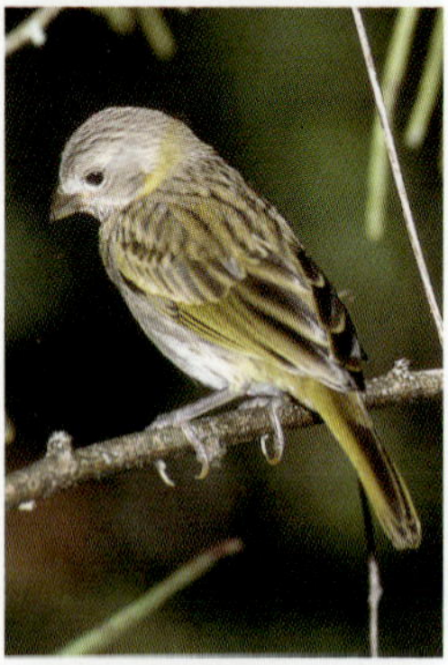

Juvenile. Rocklands, May.

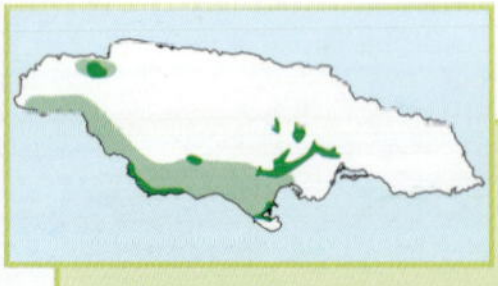

Range South America. Introduced to Jamaica in the 1820s, when a few pairs were released at the Rectory in Black River, St. Elizabeth.
Status in Jamaica Fairly common and widespread. May be declining.

Grasshopper Sparrow

Ammodramus savannarum

Local names Savanna Bird, Grass Dodger, Grass Pink, Tichicro.

Taxonomy Polytypic (11). *A. s. savannarum* is endemic to Jamaica.

Description L 13cm (5in). Adult head dark brown with a median white crown-stripe. Face is brownish with an ill-defined supercilium and rufous supra-loral spot. Bill large, greyish and finch-like, upper mandible darker. Back brownish-grey, washed with buff. Wings have dark spots. Breast buffy, unstreaked. Lower underparts buff, washed with grey. Legs pink. Juvenile resembles adult but white below and breast finely streaked.

Similar species Non-breeding Yellow-crowned and Orange Bishops similar but have yellowish supercilia and lack median crown-stripe.

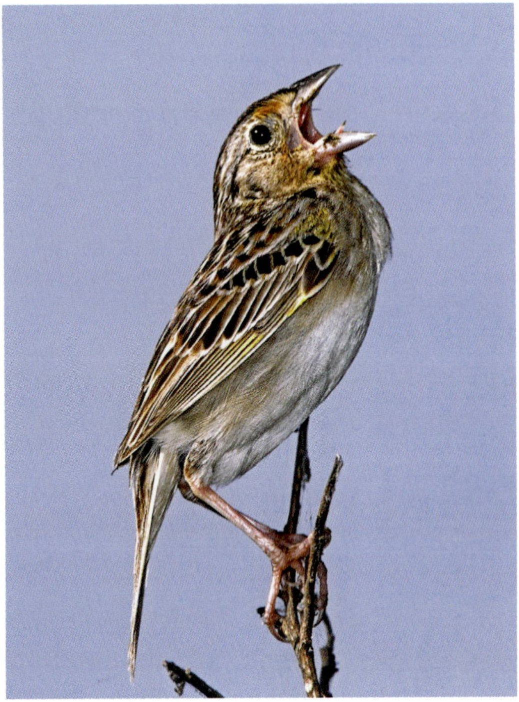

Adult male. Black River, Jamaica, May.

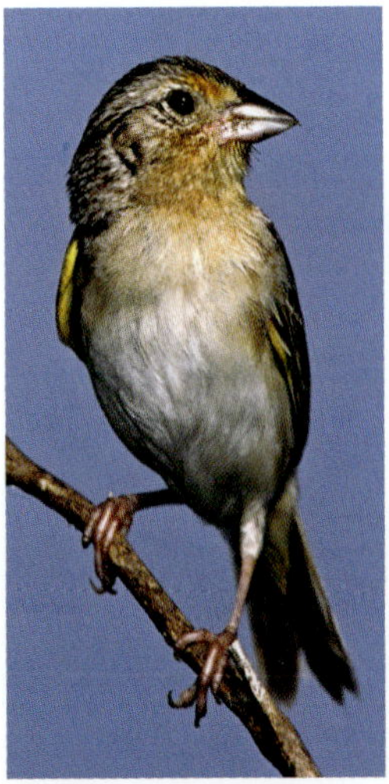

Adult female Black River, May.

Voice A high-pitched, insect-like *zeeeee-tick-zeeeeeee*.

Habitat and behaviour Grassy fields, pastures and savannas, particularly with Pangola, African Star and Guinea-grass. Usually seen perched close to the ground on fences or clumps of grass or flying short distances between clumps. Feeds on grass seeds and insects. Nest (April–June) is domed, built close to the ground under the creeping stems of grass. Lays three spotted eggs.

Range *A. s. savannarum* endemic to Jamaica. Other races inhabit Hispaniola and Puerto Rico, North, Central and northern South America.

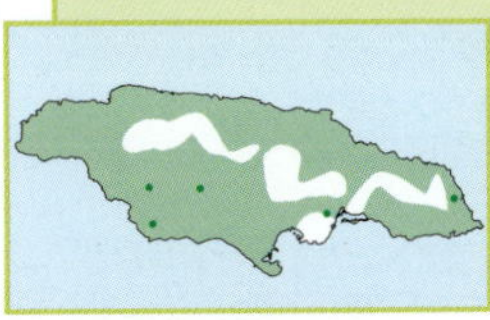

Status in Jamaica Locally common resident. Numbers may be increased by passage migrants in winter.

Lincoln's Sparrow
Melospiza lincolnii

Taxonomy Polytypic (3).
Description L 14–15cm (5–6in). A brown sparrow with white underparts, lightly streaked with black above and below. Flight feathers rufous-edged. Crown brown with grey central stripe. Broad grey supercilia contrast with dark ear-coverts and buffy malar stripes. Light eye-ring. Yellowish lower mandible contrasts with dark upper mandible.
Similar species None.
Voice Silent in Jamaica.
Habitat and behaviour Shy. Skulks in dense thickets, therefore probably overlooked. Difficult to call into view.

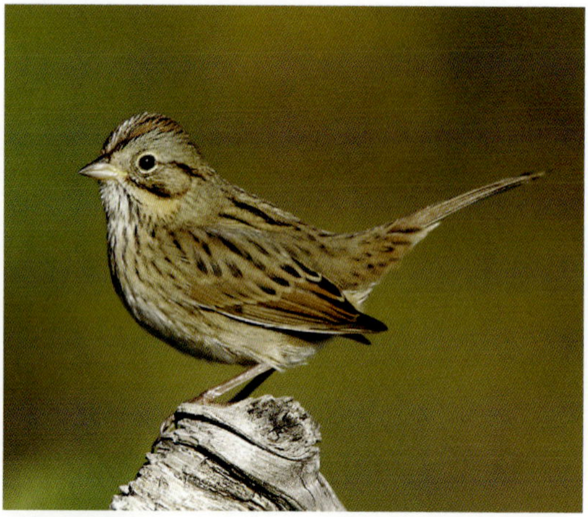

Adult. October.

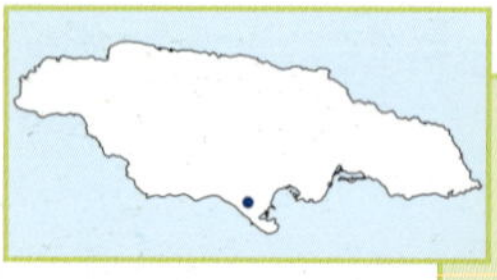

Range Breeds North America. Winters in Bahamas and Greater Antilles, southern United States and Central America.
Status in Jamaica Probably a regular but uncommon winter visitor (December–April).

Rose-breasted Grosbeak
Pheucticus ludovicianus

Taxonomy Monotypic.
Description L 19cm (7in). Large and long-winged, with a thick pale bill. Breeding male has a black head and back with a bright rose-coloured breast. Rump is white, wings and tail are black with conspicuous white patches on the wings. Underparts are white. Non-breeding male has brown tips to the upperpart feathers. Immature male brown above, white streaked with brown below. May have a pink spot in the centre of the breast. Female is brown, with whitish

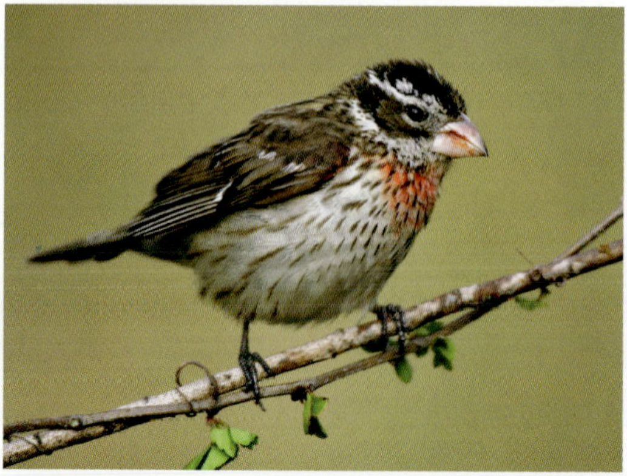

Juvenile male. May.

supercilia and dark-streaked underparts.

Similar species The male in particular is unmistakable. Female is sparrow-like but thick bill help separation.

Voice On Jamaica mostly calls; a sharp *chink*. Also a melodious song.

Habitat and behaviour
Mid-level forest and margins of cultivations; occurs mainly in highland forests in winter. Feeds mainly on fruits and seeds.

Range North America, wintering south to Peru via the Bahamas, Mexico, Greater Antilles and islands in the western Caribbean.

Status in Jamaica
Regular but rare winter visitor and brief passage migrant in small numbers (October and March). A few remain in winter, chiefly in the mountains.

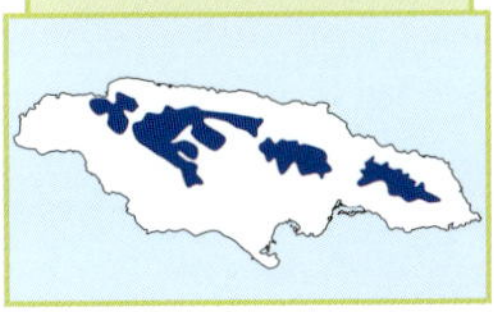

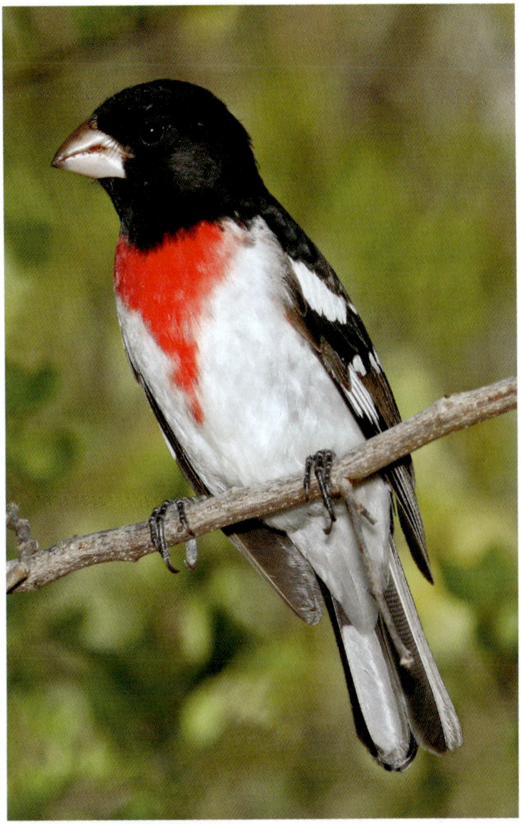

Adult male in breeding plumage. Cayman Islands, April.

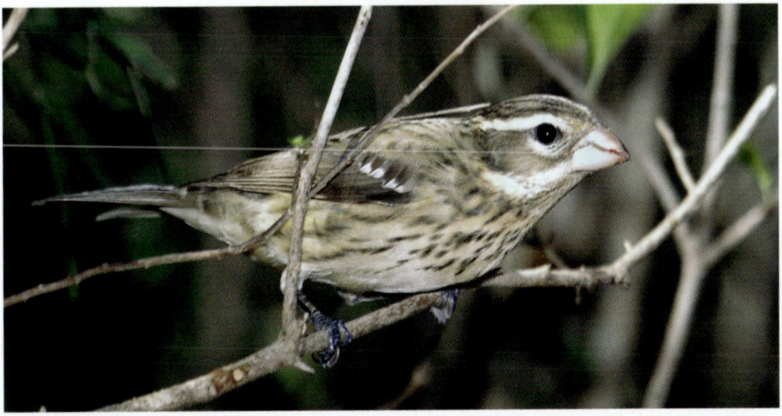

Adult female. Cayman Islands, April.

Indigo Bunting
Passerina cyanea

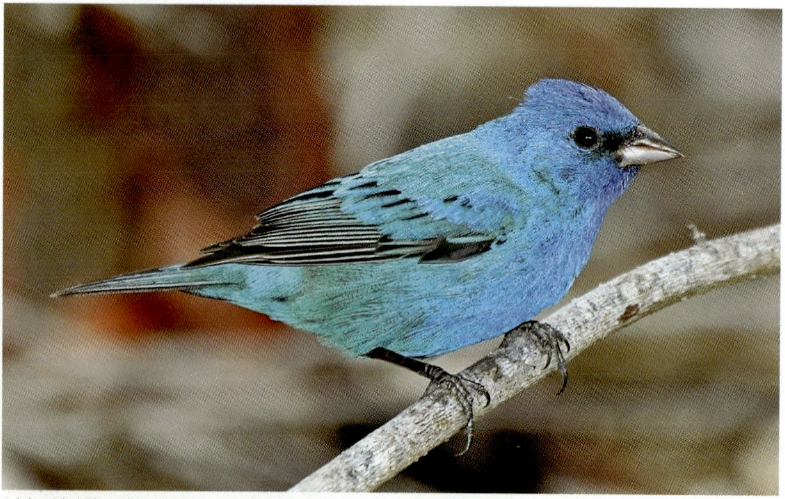

Adult male in breeding plumage. Cayman Islands, April.

Taxonomy Monotypic.
Description L 12cm (5in). Breeding male is bright indigo blue. Non-breeding male is brown with patches of blue, acquiring full breeding plumage in April or early May before leaving the winter territory. Female and juvenile in winter plumage are cinnamon-brown, paler below with indistinct flank streaks. First-winter male brown with blue blotches.
Similar species Blue Grosbeak (Appendix 2) is similar but larger, and very rare in Jamaica.
Voice Only sings when massing to migrate.
Habitat and behaviour Gardens, roadsides. May be seen in small flocks. Feeds on fruits and seeds, often in mixed flocks with Yellow-faced and Black-faced Grassquits. Readily comes to bird-feeding stations for seed and water.

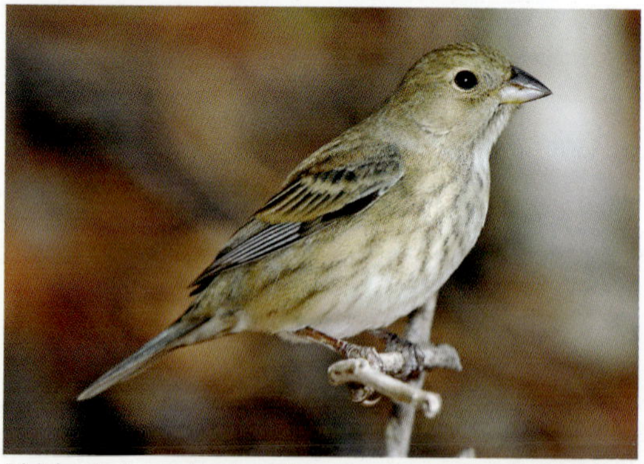

Adult female. Cayman Islands, April.

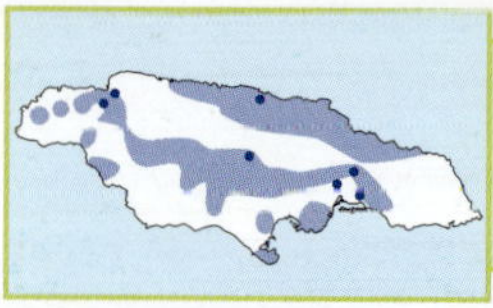

Range Breeds in North America, winters south to Peru via Central America, the Bahamas and Greater Antilles.
Status in Jamaica Regular winter visitor and spring passage migrant (January–May).

Bobolink
Dolichonyx oryzivorus

Local names October Pink, Butter Bird, Rice Bird.
Taxonomy Monotypic.
Description L 16cm (6in). Finch-like, with sharply pointed tail feathers. Non-breeding adult and juvenile brownish buff with dark streaks on mantle and flanks. Wings dark, edged paler. Dark stripes on sides of crown and behind eye. Males in breeding plumage in spring are black, with white rump and scapulars and yellow nape-patch.
Similar species In autumn they are in non-breeding plumage and could be mistaken for cowbirds or sparrows, but pointed tail feathers distinctive.
Voice In spring groups occasionally sing in a loud chorus, reminiscent of parrotlets but with many bell-like notes; in autumn they merely *pink!*
Habitat and behaviour Flocks are usually seen on golf courses, in marshy reeds, vegetable patches gone to seed, ricefields and grasslands; occasionally in trees in spring.

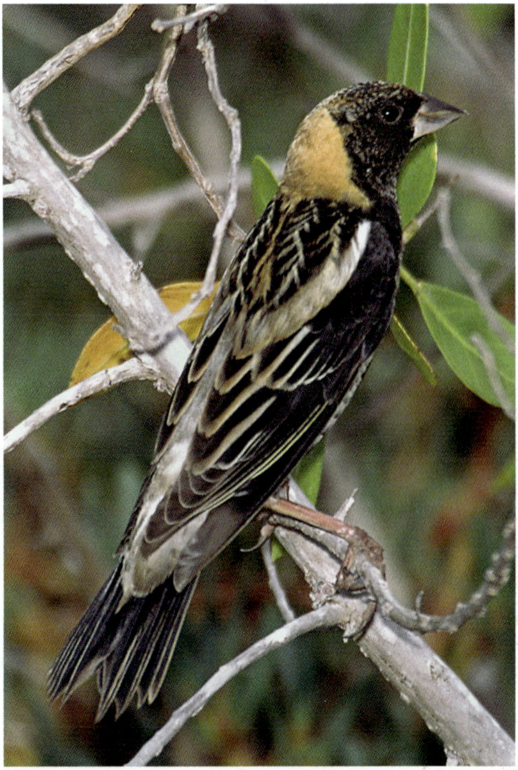
Adult male in breeding plumage. Cayman Islands, April.

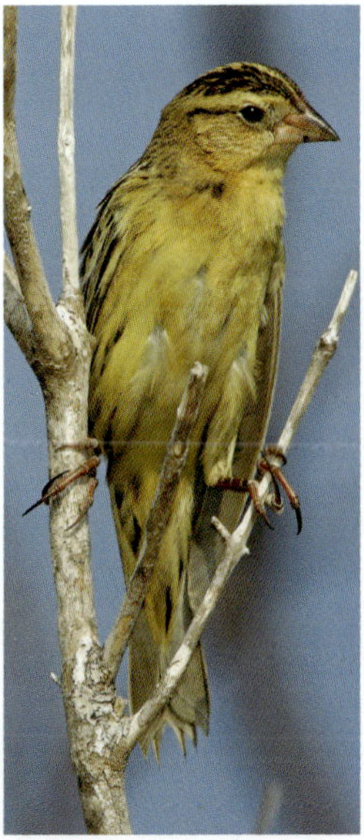
Adult female. Cayman Islands, Apr.

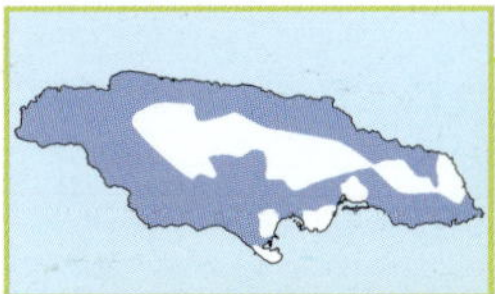

Range Breeds in North America. Winters in South America via the West Indies.
Status in Jamaica Regular passage migrant September–October and March–May, usually staying at a site for about two weeks. Probably declining as less rice is being grown locally.

Jamaican Blackbird
Nesopsar nigerrimus

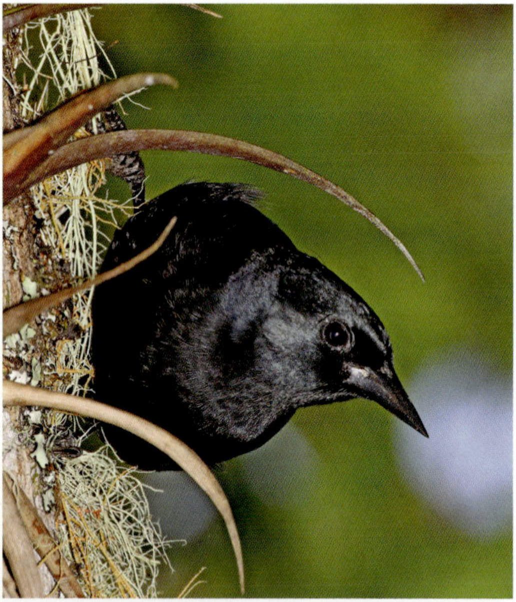

Adult. Port Royal Mountains, Jamaica, November.

Local names Wildpine Sergeant, Black Banana Bird.
Taxonomy Monotypic genus. Endemic to Jamaica.
Description L 18cm (7in). Shiny black icterid with a sharply pointed black bill and a short, slightly forked, rounded tail. Iris dark brown. Legs black. Juvenile slightly browner but inseparable in the field.
Similar species Jamaican Crow is much larger with a thicker bill. Greater Antillean Grackle has a yellow iris. Male Shiny Cowbird has a thicker bill and purplish iridescence on head and back. Male Jamaican Becard has a thick heavy bill.
Voice A wheezy, tuneless *zwheezoo-whezoo whe,* or single note.
Habitat and behaviour Sheltered gullies and valley bottoms in mature rain forest in the mountains at approximately

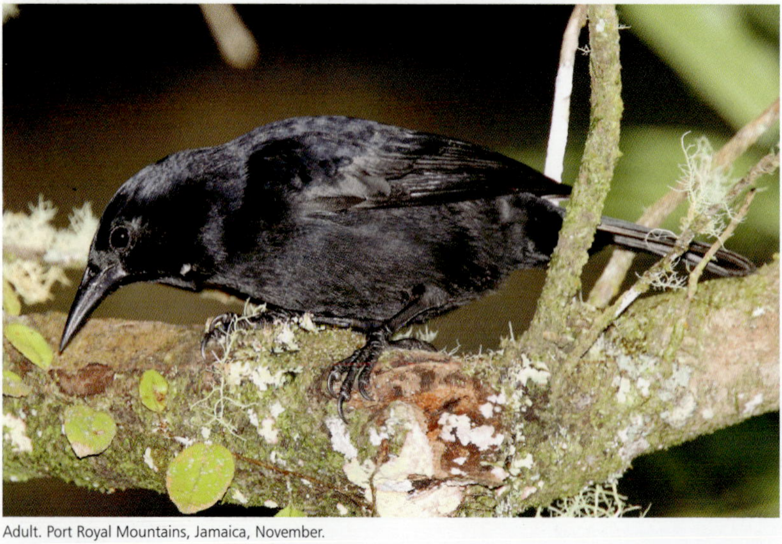

Adult. Port Royal Mountains, Jamaica, November.

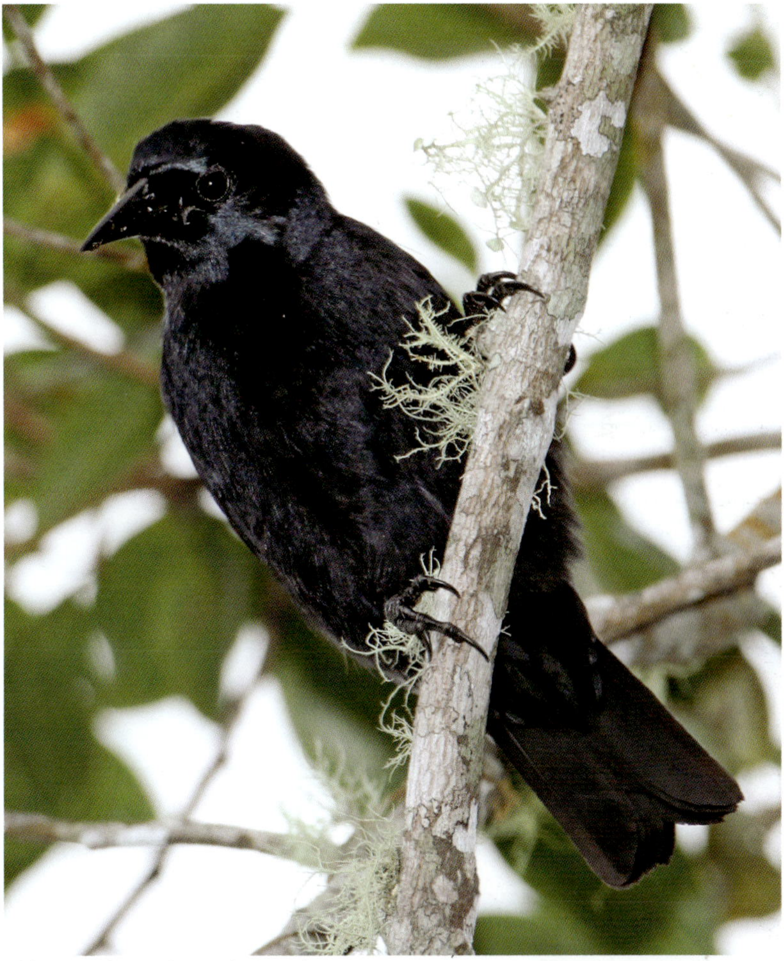

Adult. Port Royal Mountains, Jamaica, November.

500–2200m. Arboreal. Never found in flocks but family parties are sometimes seen after the nesting period. Feeds on small invertebrates, foraging silently in bromeliads and moss or at the base of tree ferns, tossing out dead leaves and sticks. The sound of disturbed leaves and a rain of debris often first attracts attention to its whereabouts Nest (May-July) is a cup, well-hidden in foliage. May descend to lower altitudes in the cooler months.

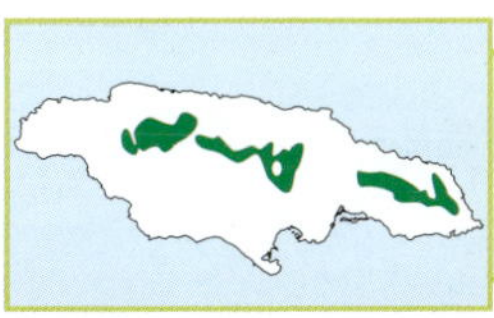

Range Endemic to Jamaica. Global conservation status: Endangered.
Status in Jamaica Uncommon and very local resident. May be declining as a result of habitat destruction.

Greater Antillean Grackle
Quiscalus niger

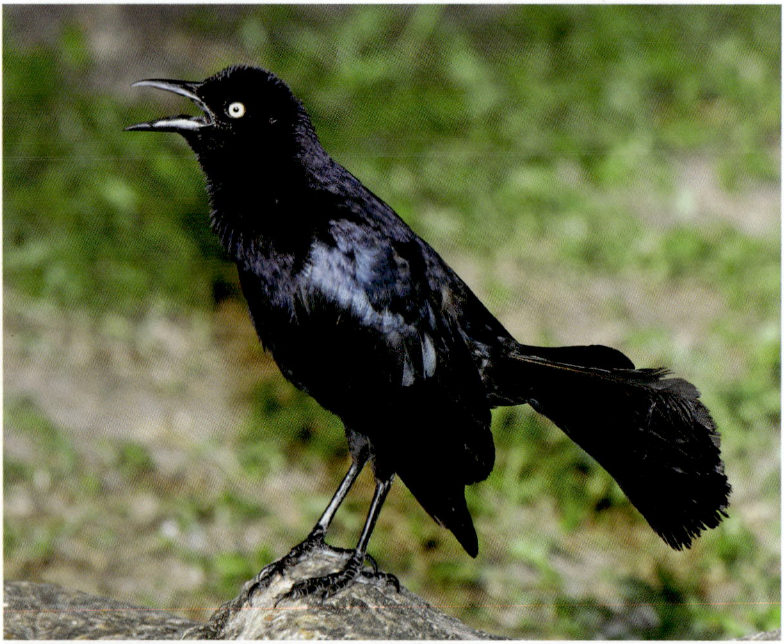

Adult in display. Hope Gardens, Jamaica, May.

Local name Cling-Cling.
Taxonomy Polytypic (7). *Q. n. crassirostris* is endemic to Jamaica.
Description L 26cm (10in). Black with sharply pointed conical black bill, long keel-shaped tail and bright yellow iris. Adult male is glossy black with pale yellow eyes. Adult female is slightly smaller and duller with a less deeply keeled tail. Juvenile lacks tail keel; iris is light brown.
Similar species Jamaican Blackbird never occurs in disturbed habitats, has dark iris and lacks keel-shaped tail.
Voice Has a variety of bell-like notes. Call note sounds like local name – *cling-cling-cling*. The food call of the young is a raucous *cheef*, *chef* or *chee*.
Habitat and behaviour Cultivated land and around human habitations in the lowlands and at mid-levels. Not usually seen in undisturbed habitats. Roosts in large noisy flocks in trees near the sea or on buildings, especially power-plants. Flocks feed inland during the day and fly to the coast in the evening. Bold and fearless, often walking around hotel restaurants and feeding on scraps at occupied tables. Nests (April–June) in colonies in tall trees, the cup-shaped nests wedged between the trunk and the branches. Young are fledged by July–August.

Range *Q. n. crassirostris* endemic to Jamaica. Other races are found in Cuba, the Cayman Islands, Hispaniola, Puerto Rico and adjacent islands.
Status on Jamaica Common resident.

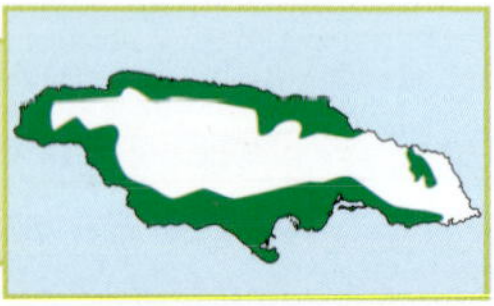

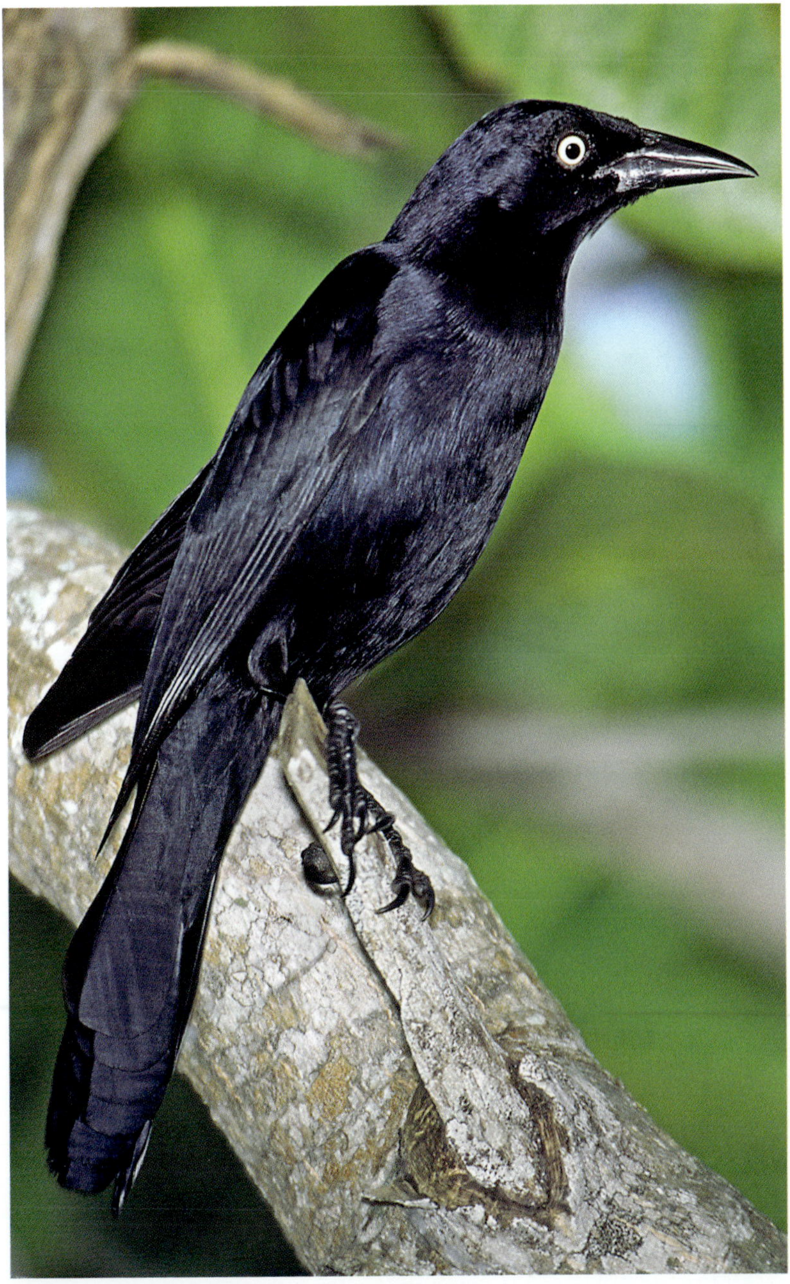

Adult. Montego Bay, Jamaica, May.

Great-tailed Grackle
Quiscalus mexicanus

Taxonomy Polytypic (8).
Description L male 43–46cm (17–18in); female 32–34cm (13in). Male is black with a greenish gloss. Iris is yellow. Tail is keeled and wedge-shaped. Female is brown with a buffy supercilium, iris yellowish to brown; tail is shorter. Juvenile is brown, streaked darker below, iris brown.
Similar species Greater Antillean Grackle is smaller.

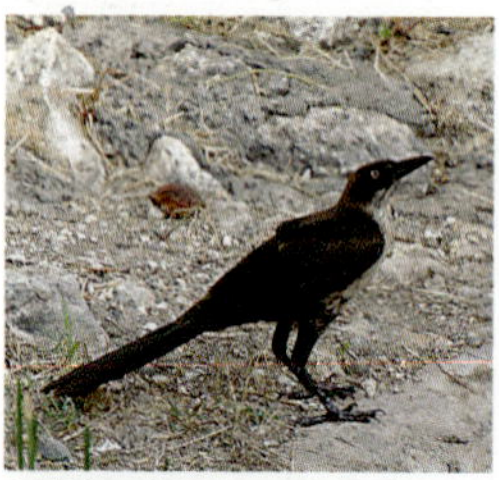

Adult female. Portmore Sewage Ponds, Jamaica, June.

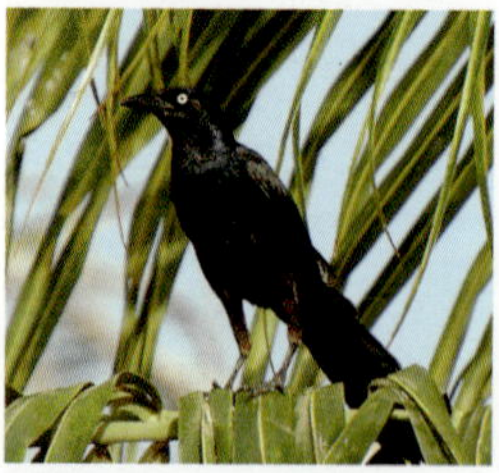

Adult male. Portmore Sewage Ponds, Jamaica, June.

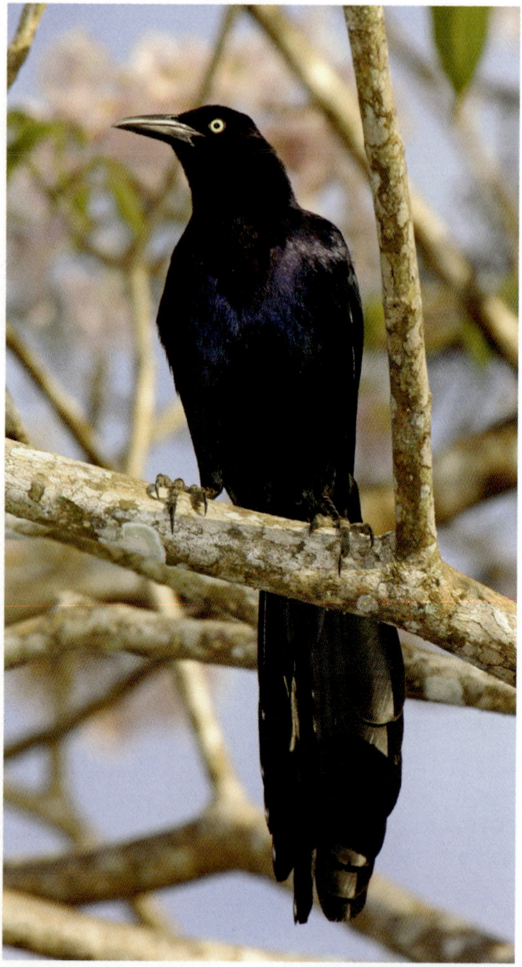

Adult male. Costa Rica, February.

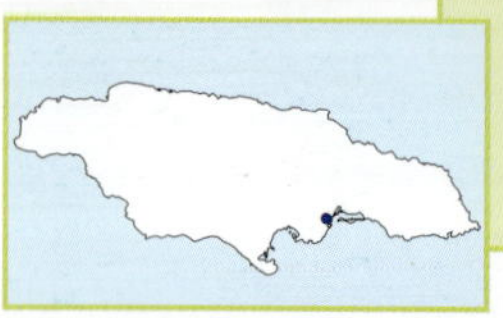

Range North and Central America. Introduced to Jamaica.
Status in Jamaica Probably accidentally introduced with cargo; a few individuals were regularly seen at Norman Manley Airport in 2005. May be established and spreading. Bred at Portmore Sewage Ponds in 2007.

Shiny Cowbird
Molothrus bonariensis

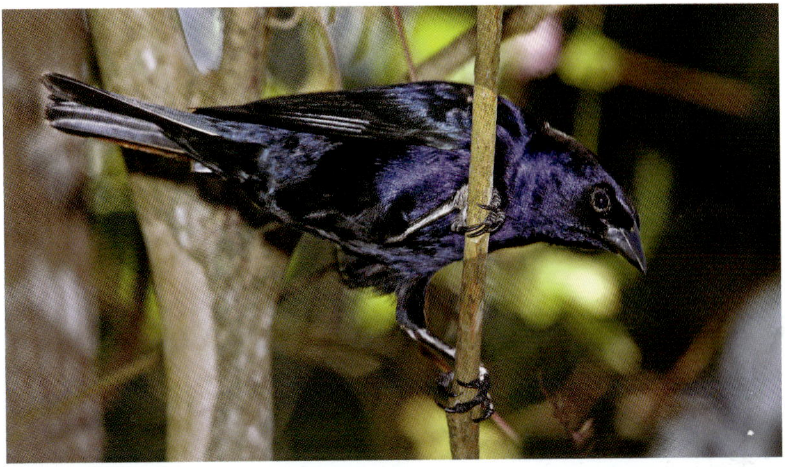

Adult male. Rocklands, Jamaica, May.

Taxonomy Polytypic (7).
Description L 19cm (8in). Adult male is glossy black with iridescent purple on head and back. Bill black, broad at the base and sharply pointed. Adult female and juvenile buffy-grey with slightly paler supercilia, throat slightly paler than breast.
Similar species Jamaican Blackbird lacks iridescence and never forms flocks.
Voice Includes single *tcuck* and various rattles and whistles.
Habitat and behaviour Cultivated land and around human habitations, chicken- and fish-farms. Not usually seen in undisturbed habitats. In winter large flocks are seen feeding on the ground on spilt grain around farms or roosting in thousands on trees. In the breeding season, individuals spread out to breed in dry, open (usually disturbed) forests and woodland across the island. This species is a brood parasite (mainly April–June), laying its eggs in nests of other species, such as Jamaican Oriole. The host species incubates the egg and raises the chick, which outgrows the host chicks; they may starve or be ejected from the nest.

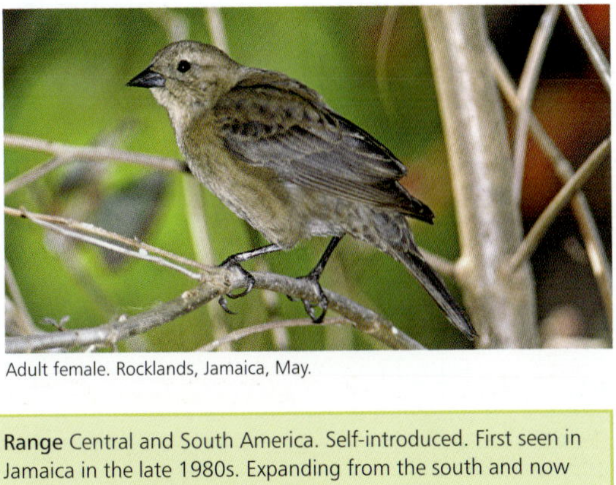

Adult female. Rocklands, Jamaica, May.

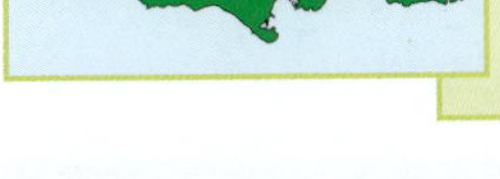

Range Central and South America. Self-introduced. First seen in Jamaica in the late 1980s. Expanding from the south and now resident on most West Indian islands.
Status in Jamaica Common resident. Probably increasing.

Jamaican Oriole
Icterus leucopteryx

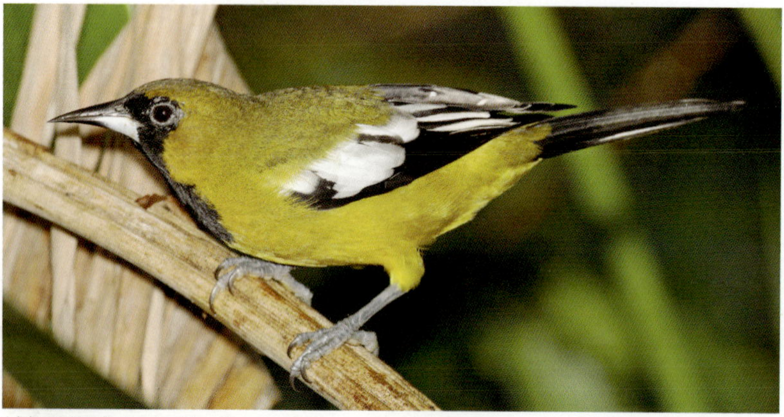

Adult. Rocklands, Jamaica, November.

Local names Auntie Katie, Banana Katie.
Taxonomy Polytypic (3). *I. l. leucopteryx* is endemic to Jamaica.
Description L 20cm (8in). A long-tailed greenish bird with a black mask and bib and white wing patches. Adult is bright greenish-yellow on head, nape and mantle, brighter yellow below. Sharply pointed black bill. Wings and tail black, edged white. There is a large white patch in the wings. Juvenile is duller, with cinnamon tips to the wing-coverts forming two wing-bars and cinnamon spotting on head and back; tail is greenish-yellow.
Similar species None.
Voice *you-cheat, you-cheat*, sometimes *you-cheat-you*. Song (October–June) is a melodious *Auntie-katie* or *if-you-want-it free, don't-ask-me!*
Habitat and behaviour Seen year-round in gardens and forests at all elevations. Sometimes uses communal roosts. Feeds by prising-out insects from bromeliads, pods (e.g. *Poinciana*) and from under tree bark. Also eats fruits and flowers. Nest (March–June) is long and pendulous, loosely woven from dried strands of *Tillandsia*, grass or plastic threads.

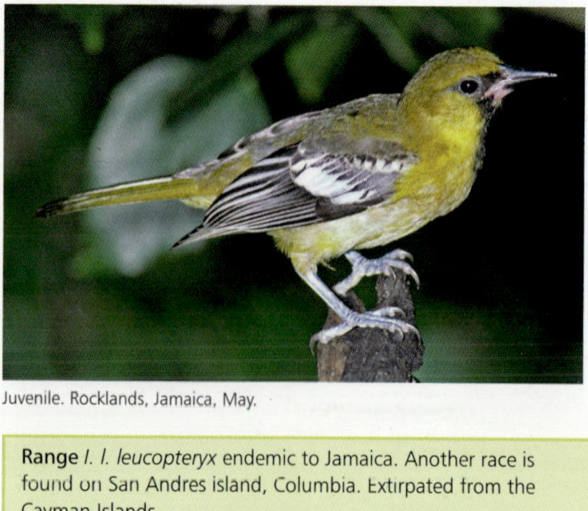

Juvenile. Rocklands, Jamaica, May.

Range *I. l. leucopteryx* endemic to Jamaica. Another race is found on San Andres island, Columbia. Extirpated from the Cayman Islands.
Status in Jamaica Common and widespread resident. May be at risk from brood parasitism by the Shiny Cowbird.

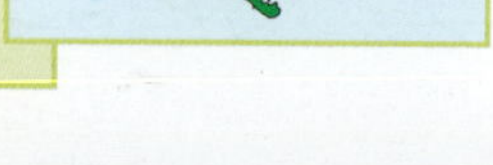

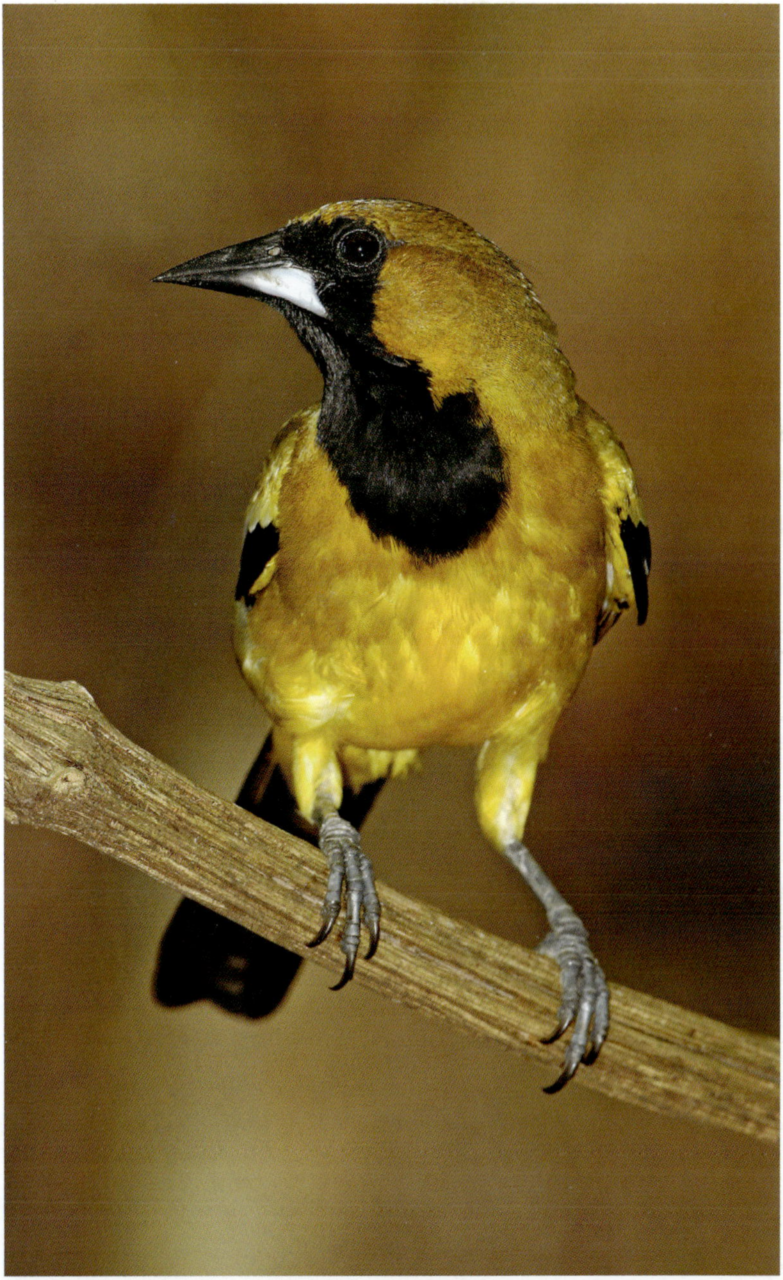

Adult. Rocklands, Jamaica, May.

Baltimore Oriole

Icterus galbula

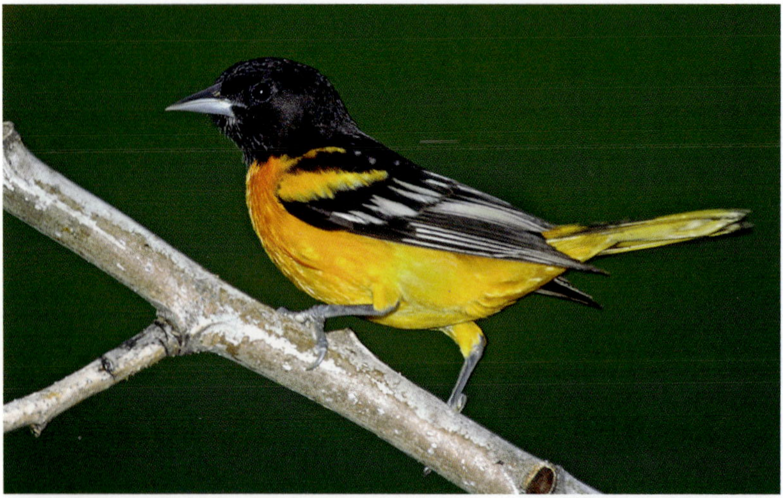

Adult male. May.

Taxonomy Monotypic.
Description L 18cm (7in). A dark-winged bird with an orange tail. Male has a black head, bib and mantle. Rump and underparts are bright orange. Wings dark with one white wing-bar. Tail black with orange corners. Female has brownish upperparts with no bib, two white wing-bars and orange-yellow rump, tail and underparts.
Similar species Jamaican Oriole lacks the black head, and is yellow not orange.
Voice Sings a sweet song, quite different from that of the Jamaican Oriole.
Habitat and behaviour While in Jamaica seen in hedges and large or small trees e.g. ackee, privet, *Erythrina* and silk cotton trees. Feeds on fruit.

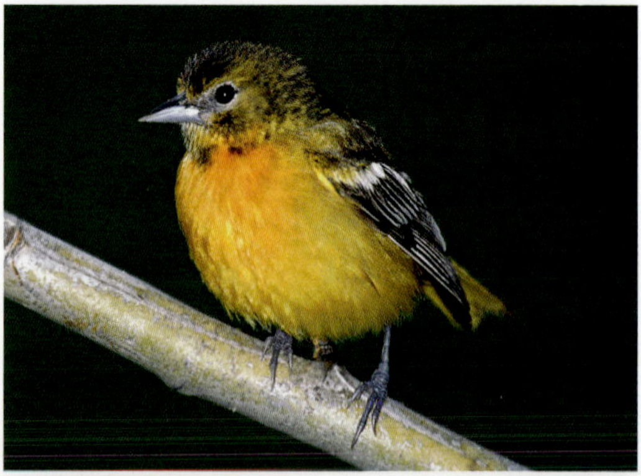

Adult female. May.

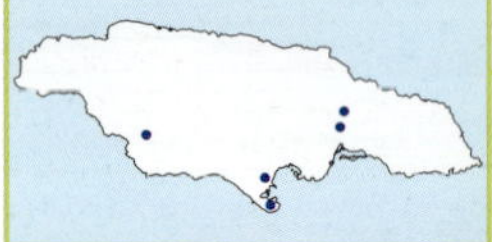

Range Breeds in North America. Winters in South America via Central America and Greater Antilles.
Status in Jamaica Uncommon but regular passage migrant (October–November and January–April). A few overwinter.

APPENDIX 1
Probably extinct species

A number of species are known to have become extinct in the 400 years since Europeans first visited Jamaica; these include two species of macaw and perhaps a hummingbird, and there may have been others, including parrots and a ground dove. Two probably extinct species are rather better known, and were last recorded in the late nineteenth century, with the last record of an endemic subspecies a few decades later. A map for each of these birds is included in the accounts below; this shows the approximate former breeding range.

Jamaican Petrel
Pterodroma caribbaea

Local name Blue Mountain Duck.
Taxonomy Monotypic. Endemic to Jamaica.
Description L 40cm (16in). Dark, mainly uniform brown petrel with whitish rump and uppertail-coverts. Tail wedge-shaped. Legs and feet black.
Similar species Black-capped Petrel of Cuba and Hispaniola is blacker with white underparts and neck.
Voice Not described but probably produced eerie sounds heard at night in mountain valleys. Most vocal when breeding.
Habitat and behaviour Probably pelagic during non-breeding period. Flight patterns were probably similar to Black-capped Petrel: fast wingbeats followed by high arching glides. Fed far offshore, probably off the Carolinas. Nests (probably starting October, fledging by May) in burrows in shale, pockets of soil in limestone and holes under trees, probably above 1,000m in the Blue and John Crow Mountains, entering and leaving nests only at night.

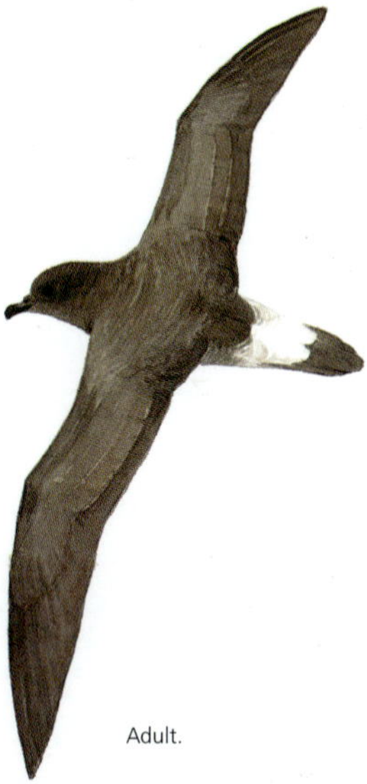

Adult.

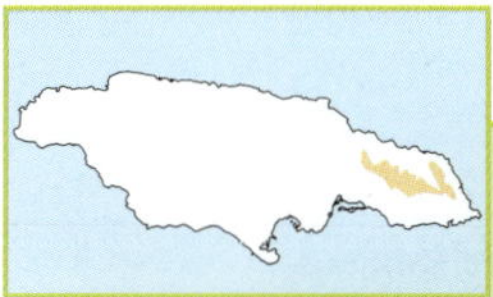

Range Jamaican endemic species; may also have occurred on Guadeloupe and Dominica. Global conservation status: Critically Endangered (Possibly Extinct).
Status in Jamaica Probably extinct. Last collected in 1879. Possibly extirpated c.1936 by hunting for food, combined with predation by mongooses (introduced 1872) and feral pigs. A small population could still survive in remote and inaccessible areas but recent searches have been unsuccessful. All dark petrels seen in or near Jamaica or surrounding waters should be carefully observed and photographed.

Uniform Crake
Amaurolimnas concolor

Taxonomy Polytypic (3). *A. c. concolor* was endemic to Jamaica.
Alternative name Jamaican Wood Rail.
Description L 26cm (10in). Plain reddish brown body and legs. Bill short and yellowish-green.
Similar species Clapper Rail is greyer and smaller. Its bill is longer, dark above, and reddish brown at base.
Voice Not described from Jamaica. In Central America reportedly gives clear whistles and a nasal *kek*.
Habitat and behaviour Swamp forests and wooded banks of rivers and streams. Terrestrial and sluggish.

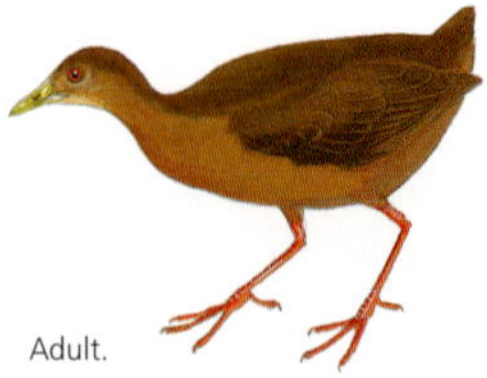

Adult.

Range *A. c. concolor* endemic to Jamaica. Other races extant, from Mexico south to South America.
Status in Jamaica Not reported since 1911. Probably extirpated but there have been no extensive searches.

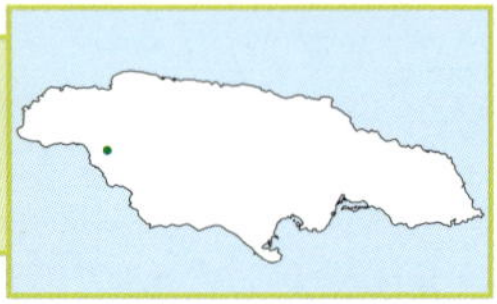

Jamaican Pauraque
Siphonorhis americana

Taxonomy Monotypic. Endemic to Jamaica.
Description L 24cm (9in). A speckled nightjar with brown upperparts streaked with darker brown and grey. Rufous collar on hind-neck. White collar on lower throat. Belly lighter brown with dark arrow-shaped marks. Tail long and brown extending well beyond wings when perched. Male has tail feathers with broad white tips (like a cuckoo's tail). Female tail feathers are edged buff.
Similar species Antillean Nighthawk is larger, with white band in wing and characteristic call; tail does not extend beyond wings when perched.
Voice Not described.
Habitat and behaviour Probably dry forests. Historical specimens were collected from Great Salt Pond (near Hellshire Hills), Bluefields (Westmoreland), Freeman's Hall (near Albert Town) and near Linstead.

Range Jamaican endemic. Global conservation status: Critically Endangered (Possibly Extinct).
Status in Jamaica Very rare (<50 individuals) or extinct. Not collected since 1860. A few sightings of unidentified night birds have encouraged the hope that this species may persist in low numbers in remote, inaccessible areas.

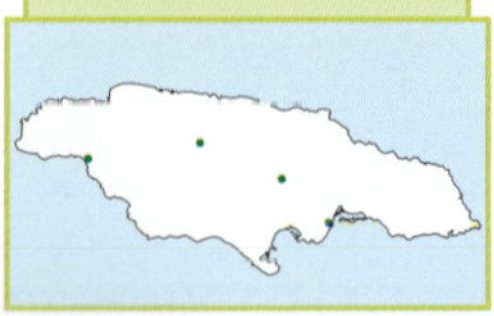

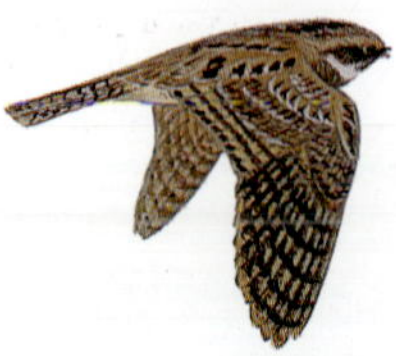

Adult.

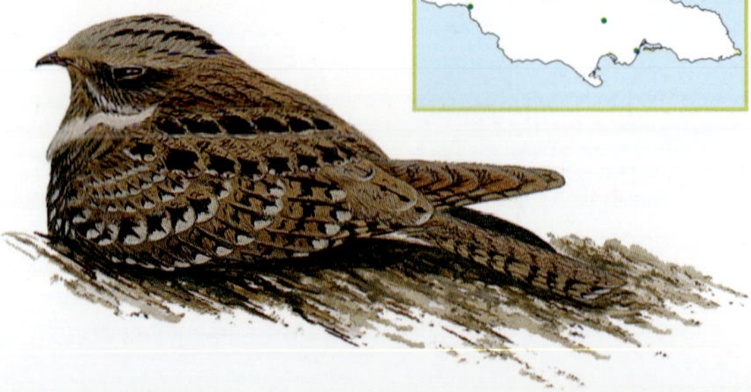

APPENDIX 2
Vagrants

In this section information on similar species, voice, habitat and behaviour is only included where it assists field identification. Subspecies is stated where known. Recent introductions and escapes are also discussed.

Canada Goose *Branta canadensis*
Taxonomy: Polytypic (7). **Description:** L 64–113cm (25–45in). Large brownish-grey goose with dark neck and white cheeks and throat. **Habitat:** Wetland margins. **Range:** Widespread in the northern hemisphere. **Status:** Vagrant (October–April). Mentioned in the literature but not recently reported.

Gadwall *Anas strepera*
Taxonomy: Polytypic (2). **Description:** L 53cm (21in). Medium-sized dark diving duck with a white speculum. Male has a black stern. Female has orange along the side of the bill. **Habitat:** Freshwater. **Range:** *A. s. strepera* breeds in central North America. Winters south to Guatemala. A rare non-breeding visitor to the Bahamas and Cuba. **Status:** Vagrant. Mentioned in the literature but not recently reported.

Mallard *Anas platyrhynchos*
Taxonomy: Polytypic (5). **Description**: L 41cm (16in). Breeding male has a green head with a white ring around base of neck; bill yellow; breast rufous-brown. Female mottled brown with an orange and black bill. Eclipse male is like female but has blackish-green bill. **Similar species**: Male Northern Shoveler also has a green head, but has chestnut sides and a very long spatulate bill. **Habitat and behaviour**: Fresh or saline ponds. Feeds from surface on pond vegetation, seeds and snails. **Range**: Worldwide. *A. p. platyrhynchos* has been introduced to many West Indian islands. **Status**: Vagrant. A common domesticated duck.

Northern Pintail *Anas acuta*
Taxonomy: Monotypic. **Description:** L male 69–74cm (23–30in), female 54–56cm (21–22in). An elegant duck with a long, slender neck, long wings and a long pointed tail. Speculum green in males, brown in females, with a white trailing edge. Breeding male has a chocolate-brown head contrasting with white neck and grey body. Female and non-breeding male are grey-brown mottled and scalloped dark brown and grey. **Habitat and behaviour:** Fresh, saline or brackish ponds. **Range:** Breeds in North America and Eurasia. Extends further south in winter. **Status:** Vagrant or very rare winter visitor.

White-cheeked Pintail *Anas bahamensis*
Taxonomy: Polytypic (3). *A. b. bahamensis.* **Description:** L 38–48cm (15–19in). A slender brown duck with a long neck and long, whitish, pointed tail. Dark brown cap and nape contrast with white cheek. Bill thin and grey with red base. Speculum green, fringed brown. **Habitat and behaviour:** Fresh, saline or brackish ponds. **Range:** Breeds on Caribbean islands (including Cuba, Hispaniola and Puerto Rico) as well as in South America. Dispersive after breeding. **Status:** Vagrant. Not recently reported.

Canvasback *Aythya valisineria*
Taxonomy: Monotypic. **Description:** L 53cm (21in). Medium-sized diving duck with distinctive sloping forehead, long black bill and angled back of head. Breeding male has chestnut head, black breast and tail, white sides and grey back. Female, immature and male in eclipse brownish with darker head. **Similar species:** See Redhead. **Range:** Breeds in North America. Rare non-breeding resident in Cuba. Winters south to Mexico. **Status:** Vagrant. Not recently reported.

Redhead *Aythya americana*
Taxonomy: Monotypic. **Description:** L 51cm (20in). Medium-sized diving duck. Male and female similar to Canvasback but have rounded heads and grey bills tipped with black. **Similar species:** Canvasback has sloping forehead and longer black bill. **Habitat:** Wetlands. **Range:** Breeds in North America. Winters south to Guatemala. **Status:** Vagrant. Not recently reported.

Greater Scaup *Aythya marila*
Taxonomy: Polytypic (2). *A. m. mariloides*. **Description:** L 46cm (18in). Medium-sized, diving duck. Very similar to more common Lesser Scaup, but head is rounded and white wing-bar extends to the primaries. **Similar species:** See Lesser Scaup. **Habitat:** Salt ponds. **Range:** Breeds North America and northern Eurasia. Winters south to southern United States and southern Eurasia. **Status:** Vagrant. Only reported once in last 20 years.

Bufflehead *Bucephala albeola*
Taxonomy: Monotypic. **Description:** L 36cm (14in). Small diving duck. Male mainly black above and white below, with large white patch on rear of black head. Female browner, with a white flash behind the eye. **Habitat:** A sea duck, seen in saline bays and lagoons. **Similar species:** See Hooded Merganser. **Range:** Breeds in northern North America. Winters south to Mexico and Greater Antilles. **Status:** Vagrant. Only reported once since 2000, at Montego Bay sewage ponds in February 2008.

Hooded Merganser *Lophodytes cucullatus*
Taxonomy: Monotypic. **Description:** L 46cm (18in). Medium-sized diving duck with a short, thin bill, crested head. Breeding male has large white patch on black crest. Eclipse male and female browner, with brown crest. **Habitat:** Inland – mostly ponds and lagoons. **Range:** Very rare non-breeding visitor to the Bahamas, Greater Antilles (except Jamaica) and some Lesser Antilles. Breeds in central North America. Winters in coastal southern states. **Status:** Vagrant. Only reported once since 2000, at lower Rio Grande, Portland, in February 2007.

Ring-necked Pheasant *Phasianus colchicus*
Taxonomy: Polytypic (31). **Description:** L 76–92cm (30–36in). Large gamebird with a long, pointed tail. Male has short crest, red skin on face, green neck and white neck-ring. Female drab. **Habitat:** Usually seen on or close to the ground. **Range:** Native to Asia. Widespread in captivity and as an escaped exotic. **Status:** Escape; possibly established in the wild on Long Mountain near Kingston.

Sooty Shearwater *Puffinus griseus*
Taxonomy: Monotypic. **Description:** L 40–46cm (16–18in). All-dark shearwater with contrasting pale underwings. Tail rounded. Wings are narrow and long. **Habitat and behaviour:** Flies close to surface of the sea, flapping and gliding. **Range:** Southern South America, Australasia, North Atlantic. Reported offshore through much of the Caribbean. **Status:** Vagrant.

Leach's Storm Petrel *Oceanodroma leucorhoa*
Taxonomy: Polytypic (5). **Description:** L 20cm (8in). Small and blackish with a paler wing-bar, rump white with dark centre, tail long and forked. In flight always close to water surface; zig-zags. Flaps alternate with glides. **Similar species:** Wilson's Storm-Petrel *Oceanites oceanicus* and Band-rumped Storm-Petrel *Oceanodroma castro* have been reported from waters adjacent to Jamaica and also seem quite likely to occur at times. **Range:** Worldwide. **Status:** Vagrant. Reported at least once in last 20 years. Probably overlooked.

American White Pelican *Pelecanus erythrorhynchos*
Taxonomy: Monotypic. **Description:** L 158cm (62in). Unmistakeable, very large white bird with enormous yellow to pink bill. Flight feathers black. **Habitat and behaviour:** Fishes from the surface in small groups. Unlike Brown Pelican does not dive. **Range:** Breeds in North America. Winters south to Costa Rica. **Status:** Vagrant. Not recently reported.

Neotropical Cormorant *Phalacrocorax brasilianus*
Taxonomy: Polytypic (2). *P. b. mexicanus.* **Description:** L 66cm (26in). Slender cormorant with longish tail and relatively short bill. Adult black with white-edged, yellowish throat pouch. Immature is greyer. **Similar species:** Double-crested Cormorant is larger, has an orange throat pouch and is shorter-tailed. **Range:** Central America to South America. Common year-round resident on Cuba. **Status:** Vagrant.

Little Egret *Egretta garzetta*
Taxonomy: Polytypic (4). **Description:** L 55–65cm (22–25in). White egret with black bill and lores, legs black, feet bright yellow. **Similar species:** Snowy Egret has yellow to orange lores. Juvenile Little Blue Heron has a bicoloured bill and pale lores. **Range:** Widespread in Old World. Also breeds on Barbados. **Status:** Vagrant. Only reported once.

Western Reef Heron *Egretta gularis*
Taxonomy: Polytypic (2). **Description:** L 55–65cm (22–25in). Blackish with white throat or entirely white. Heavy bill brownish with yellow base. Legs greenish-black, feet greenish-yellow. **Similar species:** Little Blue Heron has pale greenish legs and a bicoloured bill. **Range:** Africa and Asia, occasionally strays to the West Indies. **Status:** Vagrant. Only reported once.

American Bittern *Botaurus lentiginosus*
Taxonomy: Monotypic. **Description:** L 61cm (24in). Bulky, very long neck often hunched up. Brown above, paler below with bold chestnut stripes extending from throat to flanks. **Similar species:** Juvenile night-herons are shorter-necked, greyer and less boldly streaked below. **Habitat and behaviour:** Thick vegetation in freshwater swamps. Perches with bill pointing up. Hard to detect unless flushed. **Range:** Breeds in North America and Mexico. Winters south to Central America and rarely to the West Indies. **Status:** Vagrant (October–March). Not recently reported.

Wood Stork *Mycteria americana*
Taxonomy: Monotypic. **Description:** L 102cm (40in). Very large white bird with black tail and black flight feathers. Dark, bare head and heavy dark bill. Legs dark. **Habitat:** Coastal mangroves, mud-flats, ponds and inland wetlands and pond margins. **Range:** Southeast United States to northern Argentina. Rare breeding species and year-round resident in Cuba and Hispaniola. **Status**: Vagrant in Jamaica. Mentioned in the literature but not recently reported.

Scarlet Ibis *Eudocimus ruber*
Taxonomy: Monotypic. **Description:** L 58cm (23in). Adult is an unmistakable red ibis with black wing tips. Juvenile is brown, underparts and rump white, sometimes tinged pink. **Similar species**: Immature White Ibis resembles juvenile but lacks any red on back or rump. **Range:** Resident in Central and South America and Trinidad. **Status:** Vagrant. Previously bred in Jamaica but has been extirpated.

Roseate Spoonbill *Platalea ajaja*
Taxonomy: Monotypic. **Description:** L 66–81cm (26–32in). Very large, pink spoonbill with a white head and neck and a distinctive large spoon-shaped bill. **Habitat and behaviour:** Waves bill back and forth to feed in shallow water, filtering out fish and other small animals and detritus from the mud. **Range:** Southern United States, Central and South America and the West Indies. Locally common resident in Hispaniola and Cuba. **Status**: Vagrant. Not recently reported.

Mississippi Kite *Ictinia mississippiensis*
Taxonomy: Monotypic. **Description:** L 36cm (14in). Falcon-like with long wings and a notched dark tail. Adult is grey with a pale head and secondaries. Juvenile (August–March) has a series of pale tips to the upperpart feathers, a brown-streaked breast and a barred tail. **Range:** Breeds in southern United States. Winters in South America. **Status:** Vagrant. Mentioned in the literature but not recently reported.

Northern Harrier *Circus hudsonius*

Taxonomy: Monotypic. **Description:** L 46cm (18in). Long-winged and long-tailed with a white rump. Adult male is pale grey above and mainly white below. Adult female is dark brown above and paler below, with barred flight feathers and tail. Juvenile resembles the female but has rufous-orange underparts and wing-linings. **Behaviour:** Often glides close to the ground, with wings held above the horizontal. **Range:** Breeds in North America, wintering south to Venezuela. Winters in the Bahamas and in Cuba, also rarely in other Greater and Lesser Antilles. **Status:** Vagrant or very rare passage migrant.

Sharp-shinned Hawk *Accipiter striatus*

Taxonomy: Polytypic (8). Endemic races occur on Cuba, Hispaniola and Puerto Rico. **Description:** L 25–35cm (10–14in). Small, with short, rounded wings. Long, narrow, white-tipped tail, barred dark. Underparts rufous, finely barred white. Upperparts are steely blue. Juvenile browner, streaked and barred brown below. **Similar species:** See Broad-winged Hawk. **Range:** North and Central America and the Greater Antilles. Migrates south to South America. **Status:** Vagrant.

Broad-winged Hawk *Buteo platypterus*

Taxonomy: Polytypic (6). Endemic races occur on Cuba, Puerto Rico and Lesser Antilles. **Description:** L 35–41cm (14–16in). Mid-sized, with broad, rounded wings and short, rounded tail. Typically dark brown above and white below with rufous-barred body. Adult has one white band across black tail. Juvenile has evenly-barred tail. **Similar species:** Red-tailed Hawk is larger and has a red-brown tail and white breast with dark streaks on belly. **Range:** North and Central America wintering south to Brazil. Also Greater and Lesser Antilles. **Status:** Vagrant.

American Oystercatcher *Haematopus palliatus*

Taxonomy: Polytypic (2). *H. p. palliatus.* **Description:** L 43–54cm (17–21in). Adult has black head and neck. Back and wings brown, striking white wing-bar. Underparts white. Bill long, thick and bright orange. Legs pinkish. Immature duller. **Habitat:** Rocky outcrops by the sea and stony beaches. **Range:** Occasionally breeds in the Bahamas, Cuba and Lesser Antilles. Also seen year-round in Puerto Rico. **Status:** Vagrant. Not recently reported.

American Avocet *Recurvirostra americana*

Taxonomy: Monotypic. **Description:** L 43–46cm (15in). Large, long-legged, black-and-white shorebird with a long, thin, upturned, black bill and grey legs. **Similar species:** Black-necked Stilt is more slender, with longer red legs and straight bill. **Habitat:** Shallow wetlands. **Feeds** on aquatic crustaceans and insects, sweeping bill from side to side underwater. **Range:** Breeds in North America. Winters south to Cuba and Central America. **Status:** Vagrant. Only reported once in the last 25 years.

American Golden Plover *Pluvialis dominica*

Taxonomy: Monotypic. **Description:** L 26cm (10in). Stocky with a short bill. When birds arrive in August they are in breeding plumage; underparts black, back blackish mottled with golden flecks. Non-breeding plumage mainly grey, with a dark cap contrasting with white supercilium. **Similar species:** Grey Plover has distinct black wing-pit mark; also paler above and lacks golden speckling. **Range:** Breeds in North America. Winters in South America. **Status:** Very rare vagrant or passage migrant. More frequently seen August–November.

Piping Plover *Charadrius melodus*

Taxonomy: Monotypic. **Description:** L 18cm (7in). Pale grey plover with short bill and orange-yellow legs. In breeding plumage bill orange with black tip. Narrow black breast band. Non-breeding adult has a black bill and breast-band (if present) is pale. Tail black-tipped, uppertail-coverts white. **Similar species:** Semipalmated Plover is a much darker brown above. Snowy Plover has dark legs. **Habitat and behaviour:** Sandy margins of saline water bodies or the sea (e.g. cays). Finds food by sight. Runs a few steps then feeds and runs on. **Range:** Breeds in North America. Winters south to southeastern United

States, Bahamas and West Indies. Global conservation status: Threatened. **Status**: Vagrant. Mentioned in the literature but not recently reported (despite searches).

Snowy Plover *Charadrius alexandrinus*

Taxonomy: Polytypic (5). *C. a. nivosus.* **Description:** L 14–15cm (6 in). Very small, pale plover with a thin black bill and dark legs. In breeding plumage has a dark patch behind eye, a black forehead and a black side-patch on neck. These marks turn to grey when non-breeding. **Similar species:** Piping Plover is larger and has yellow-orange legs; in breeding plumage has orange bill with black tip. **Habitat and behaviour**: Salt-flats and beaches. Probes for insects. **Range:** Breeds in southern United States and the West Indies (Bahamas, Cuba, Hispaniola, Puerto Rico and Lesser Antilles). Winters south to Panama. May be threatened by habitat loss and introduced predators. Other races worldwide. **Status**: Vagrant. Mentioned in the literature but not recently reported.

Marbled Godwit *Limosa fedoa*

Taxonomy: Polytypic (2). **Description:** L 40–51cm (16–20in). **Description:** Large, plain, brownish or buffy shorebird with a long, slightly upturned, dark-tipped, pinkish bill. Underwing cinnamon. **Range:** Breeds North America. Winters California to South America. **Status:** Vagrant. One record since 2000; Old Harbour Bay, November 2006.

Upland Sandpiper *Bartramia longicauda*

Taxonomy: Monotypic. **Description:** L 28–32**cm** (11–13in). Long neck with small, squarish head, large dark eye. Bill and feet yellow. When standing the wings appear much shorter than the relatively long tail. **Range:** Breeds in North America. **Status:** Vagrant but may be under-reported. At least one record from the last 20 years.

Wilson's Phalarope *Phalaropus tricolor*

Taxonomy: Monotypic. **Description:** L 22–24cm (9in). Slender-looking sandpiper with long, needle-like bill. In breeding plumage female has a broad dark stripe through eye and down neck, male less strongly marked; legs are black. In non-breeding plumage grey above, white below. Legs are yellow. In flight shows contrast between dark upperparts and white rump. **Similar species:** Red-necked Phalarope has dark smudge through eye, dark legs. **Habitat and behaviour:** Ponds and lagoons. Feeds rapidly while swimming, plucking insects from surface, often spinning in water to stir up food. **Range:** Breeds in northern North America. Winters south to Argentina. Recorded from most islands in the West Indies (mostly August–April). **Status**: Vagrant. Mentioned in the literature but not recently reported.

Red-necked Phalarope *Phalaropus lobatus*

Taxonomy: Monotypic. **Description**: L 18–20cm (7–8in). Delicate-looking small sandpiper with a long, thin, black bill. Legs dark. Black eye-patch and crown. Dark above with red neck when breeding. In non-breeding plumage grey above and white below. **Similar species:** Wilson's Phalarope has grey crown, legs yellow in winter. **Habitat and behaviour:** Ponds and lagoons. Feeds rapidly while swimming, plucking insects from surface, often spinning in water to stir up food. **Range**: Circumpolar breeder. Winters in the Atlantic off South America. Reported from the Greater Antilles (mostly November–March). **Status:** Vagrant. Only one record in last 20 years.

Black-legged Kittiwake *Rissa tridactyla*

Taxonomy: Polytypic (2). **Description:** L 38–41cm (15–16in). Medium-sized gull with white head, body and tail. Wings grey above, white below, with neat black tips. Legs black. Bill yellow. Non-breeding adult has dark spot behind eye and grey nape. Juvenile shows dark M-shaped pattern on upperwings, black nape-bar and black tail-tip. **Range:** Breeds around northern Atlantic and Pacific. Winters south. **Status:** Reported in literature off east coast (December–March) but not recently recorded.

Black Noddy *Anous minutus*

Taxonomy: Polytypic (7). *A. m. americanus.* **Description:** L 33cm (13in). Very dark, slender noddy. Bill thin and as long as head. Tail paler than back. **Similar species:** Brown Noddy more chunky, with longer tail and shorter bill; browner, but this is hard to detect except in good light. **Range:** Breeds on tropical islands off Central America and Venezuela. **Status**: Vagrant. Only reported once, from the Pedro Cays in the 1990s.

Pomarine Skua *Stercorarius pomarinus*

Alternative name Pomarine Jaeger. **Taxonomy:** Monotypic. **Description:** L 65–78cm (26–31 in). Large chunky dark brown and white seabird usually seen in flight. Twisted, spoon-shaped central tail feathers extend beyond tail. Primaries blackish with white bases. Light phase adult has white belly with mottled band across breast. Dark phase adult all dark below. Immature heavily barred, lacks tail-spoons. **Similar species:** Much bulkier than other jaegers. **Habitat:** Usually far from land. **Range:** Breeds Arctic to northern Eurasia, ranges to northern S America and the Mediterranean. Recorded throughout the W Indies. In Hispaniola probably a passage migrant in November–December and April–May, with some remaining throughout the winter. **Status**: Vagrant. Only seen from boats at sea, therefore probably under-reported.

Arctic Skua *Stercorarius parasiticus*

Alternative name Parasitic Jaeger. **Taxonomy:** Monotypic. **Description**: L 46–67cm (18–26). Large brownish-grey and white seabird usually seen in flight, when shows projecting long, thin, central tail feathers. Primaries dark grey with white bases. Light-phase adult has mottled white belly with narrow darker band across breast. Dark-phase adult has uniformly dark underparts. Immature dark with pale, usually rufous feather-tips. **Habitat and behaviour:** Generally far offshore. May harass other seabirds for food. **Similar species:** Other skuas. **Range**: Breeds in Arctic, winters at sea in southern oceans. Recorded throughout the West Indies. In Hispaniola thought to be mostly a passage migrant, southbound in November–December, northbound April–May. **Status:** Vagrant or very rare passage migrant. Only seen from boats at sea therefore probably under-reported.

Long-tailed Skua *Stercorarius longicaudus*

Alternative name Long-tailed Jaeger. **Taxonomy:** Polytypic (2). *S. l. pallescens.* **Description**: L 50–58cm (20–23in). Slender dark seabird usually seen in flight when very long, thin, central tail feathers extend well beyond tail. Primaries very pointed, only outer feathers have white bases. Cap black. Underparts whitish with grey belly. **Immature** occurs in light, intermediate and dark phases, heavily barred. **Similar species:** Other skuas are bulkier. **Range**: Breeds in Arctic and Siberia, winters at sea off South America. Recorded throughout the West Indies. **Status:** Very rare vagrant. Only seen from boats at sea therefore probably under-reported.

Black-billed Cuckoo *Coccyzus erythropthalmus*

Taxonomy: Monotypic. **Description**: L 28–31cm (10–12in). Slender cuckoo, grey-brown above and white below, with a short, decurved dark bill; long brownish tail with small white and black tips. Eye-ring reddish to yellowish. **Similar species:** Yellow-billed Cuckoo has more distinct tips to tail feathers, lower mandible yellow, rufous on wings. Mangrove Cuckoo has black mask, yellow lower mandible and buffy underparts. **Range**: Breeds in North America, winters south to Bolivia. Recorded as a passage migrant mainly from the Greater Antilles and Bahamas. **Status:** Vagrant or rare passage migrant.

Common Nighthawk *Chordeiles minor*

Taxonomy: Polytypic (8). **Description:** L 22–24cm (in). Very similar to Antillean Nighthawk, from which it is reliably distinguished only by call – but Common Nighthawk is mostly silent during September and October. **Similar species:** Antillean Nighthawk. **Voice:** Usually detected by call. A single, nasal note; *neet* or *peet*. **Habitat and behaviour:** Feeds on flying insects over open fields and similar grassy areas at dawn and dusk. **Range:** Breeds in North America, winters in South America. **Status:** Very rarely recorded vagrant or passage migrant, reported several times since 2000 (Marshall's Pen, Mandeville and Windsor Trelawny in May 2002). Probably under-reported due to similarity to Antillean Nighthawk.

Whip-poor-will *Caprimulgus vociferus*

Taxonomy: Polytypic (6). Probably *C. v. vociferus.* **Description**: L 23–26cm (9–10in). Medium-sized, brown nightjar with pale grey, black-edged 'braces' across inner-wing. Tail long and dark, with large white corner-patches in males. Usually only detected by its call. Flies close to the ground, like a moth. **Similar species:** Identified from other nightjars by call. **Voice:** *Whip-poor-will.* **Range**: *C. v. vociferus* breeds in eastern North America, winters south to Cuba and Panama. **Status:** Vagrant.

Chimney Swift *Chaetura pelagica*

Taxonomy: Monotypic. **Description:** L 12–14cm (5–6in). Small, dark swift with short, rounded tail and paler throat. **Similar species:** Black Swift is larger and has a longer, slightly forked tail. **Habitat:** Over open fields. **Range:** Breeds in North America, winters in the Amazon. Reported throughout the West Indies and Bahamas. **Status**: Vagrant or very rare passage migrant. Only reported twice since 2000; Windsor, Trelawny 2002, Pedro Cays 2007.

Willow Flycatcher *Empidonax traillii*

Taxonomy: Polytypic (4). Probably *E. t. traillii.* **Description:** L 15cm (6in). Dark greyish brown above, upper breast greyish. Chin and underparts dull white. Two pale wing-bars. **Similar species:** Both Western and Eastern Wood-Pewees have grey throats. **Range:** Breeds in North America, winters to northwest South America. **Status:** Vagrant or rare passage migrant (mid-September–November and April).

Western Wood-Pewee *Contopus sordidulus*

Taxonomy: Polytypic (4). **Description**: L 14–16cm (5–6in). **Description:** Dark brown above, with two pale wing bars. Base of lower mandible orange or all dark. **Similar species:** Eastern Wood Pewee is paler, base of lower mandible pale yellow. **Behaviour:** Unlike Jamaican Pewee, does not shiver tail on landing. **Range:** Breeds western North America. Migrates south to Venezuela. **Status:** Vagrant. Only reported once since 2000, at Portland Ridge, 2000.

Eastern Wood-Pewee *Contopus virens*

Taxonomy: Monotypic. **Description:** L 16cm (6in). Dark brown above, with two pale wing bars. Base of lower mandible pale yellow. No eye-ring. **Similar species:** Difficult to distinguish (except by voice) from Western Wood Pewee, which is darker. Base of lower mandible orange or all-dark. **Habitat:** Woods and their margins. **Range:** Breeds eastern North America. Winters central South America. Rare passage migrant through the western Caribbean. **Status**: Vagrant. Only reported once since 2000; a bird netted at Cornwall Barracks, Portland, in November 2003.

Eastern Kingbird *Tyrannus tyrannus*

Taxonomy: Monotypic. **Description:** L 22cm (9in). Upperparts very dark grey. Black tail tipped white. Concealed crown-patch scarlet. Underparts white, with dusky lower throat and breast. Immature has pale feather edging on wings. **Similar species:** Grey Kingbird is paler. Loggerhead Kingbird has orange-yellow crown patch and yellow wash on belly and undertail coverts. **Range:** Breeds in eastern North America. Winters in the Amazon. **Status**: Vagrant. Only reported once since 2000, at Pedro Cays in 2006.

Fork-tailed Flycatcher *Tyrannus savana*

Taxonomy Polytypic (4). **Description** L 33–41cm (13–16in). A large kingbird, mainly dark above and white below with very long black tail streamers. **Similar species** Loggerhead Kingbird is smaller and lacks tail streamers. **Habitat and behaviour** Found in lowland scrub and mangroves. **Range** Breeds in Central America and winters in South America. **Status** Very rare winter visitor or vagrant.

Purple Martin *Progne subis*

Taxonomy: Polytypic (3). **Description:** L 20–22cm (8–8.5in). Large, broad tail very slightly forked. Male all dark, glossed purple. Female and juvenile pale below, head and throat greyish. **Similar species:**

Caribbean Martin has white belly. **Range:** Breeds in North America. Winters in Brazil. **Status:** Vagrant or rare passage migrant in spring and autumn.

Cliff Swallow *Petrochelidon pyrrhonota*

Taxonomy: Polytypic (4). **Description:** L 12–15cm (5–6in). Chunky swallow with blackish-brown wings, pale buff forehead and rump, dark crown and rufous chin and throat which may include dark patch. Underparts dirty white. **Similar species:** Possibly overlooked because of close resemblance to Cave Swallow, which has a reddish-brown forehead and rump. **Behaviour:** Feeds on insects; may occur in flocks of Cave Swallows. **Range:** Breeds in North America and winters south to Argentina. Widely reported from the West Indies. **Status:** Vagrant. Mentioned in the literature but not recently reported.

Buff-bellied Pipit *Anthus rubescens*

Alternative name: American Pipit. **Taxonomy:** Polytypic (4). *A. r. rubescens.* **Description:** L 17cm (7in). Slender body and long tail with pale outer feathers. Plain grey-brown above, buffy below with streaky breast. Supercilium buff. Legs dark. **Habitat and behaviour:** Terrestrial. Bobs tail when walking. **Range:** Breeds in North America. Winters to Central America and Bahamas. Also Eurasia. **Status:** Vagrant. Mentioned in the literature but not recently reported.

Ruby-crowned Kinglet *Regulus calendula*

Taxonomy: Polytypic (3). Probably *R. c. calendula.* **Description:** L 9–11cm (3–4in).Tiny, grey, with bold white eye-ring; dark wings with two white wing bars. Male has concealed red crown patch. **Habitat and behaviour**: Scrub. Flicks wings. Feeds on insects. **Range:** Breeds in North America, winters south to Bahamas, Cuba and Guatemala. **Status:** Vagrant. Mentioned in the literature but not recently reported.

Pearly-eyed Thrasher *Margarops fuscatus*

Taxonomy: Polytypic (4). **Description:** L 28–29cm (11in). Thrush-like with a long, white-tipped tail. Brown above. White below, heavily streaked brown. Iris white, bill yellow, legs dark. **Range:** Bahamas, Hispaniola, Puerto Rico, Virgin Islands and Lesser Antilles. **Status**: Vagrant. Mentioned in the literature but not recently reported.

Wood Thrush *Hylocichla mustelina*

Taxonomy: Monotypic. **Description:** L 19–21cm (8–9in). Large spotted thrush, with bright rufous crown and nape, contrasting with slightly browner back, tail and wings. Bright white eye-ring. Underparts white with bold black spots on cheeks, throat breast and flanks. **Similar species:** Ovenbird is smaller and has striped crown. **Habitat and behaviour:** Understorey of mature forests. Feeds on the ground on insects and fruit. Very secretive and probably overlooked. Detected through ringing. **Range:** Breeds in North America from Canada to Florida. Winters south via Central America to Columbia and rarely the West Indies. **Status**: Vagrant or rare passage migrant or (September, October, April). Has been netted at Windsor, Trelawny.

American Robin *Turdus migratorius*

Taxonomy: Polytypic (7). **Description:** L 23–28cm (9–11in). Large, dark thrush, dark grey above and orange-red below. Chin and throat are white with dark streaks. Bill yellow, legs dark. Male has blackish head. **Habitat and behaviour:** Feeds on ground in open areas. **Range:** Breeds in North America. Winters south to Mexico, Bahamas and Cuba. **Status:** Vagrant. Mentioned in the literature but not recently reported.

Blue-headed Vireo *Vireo solitarius*

Taxonomy: Polytypic. **Description:** L 14cm (5in). Blue-grey head contrasts with green back. Throat white. Bold white spectacles. Two bold white wing bars. Flanks yellowish. **Range:** Breeds in eastern North America. Winters to Central America. **Status**: Vagrant. Mentioned in the literature but not recently reported.

Warbling Vireo *Vireo gilvus*
Taxonomy: Polytypic (10). **Description:** L 12–13cm (5in). Very plain, pale vireo; unmarked except for whitish supercilium and indistinct dark eye-stripe. Iris dark. Greyish green above, white below with yellow wash on neck and flanks, slightly buffy neck. Juvenile has more yellow on flanks. **Similar species:** Red-eyed and Black-whiskered Vireos are larger, darker and have red irises. **Range:** Breeds in North America. Winters south to Central America. In West Indies also reported from Cuba and Hispaniola. **Status:** Vagrant. Mentioned in the literature (in October) but not recently reported.

Philadelphia Vireo *Vireo philadephicus*
Taxonomy: Monotypic. **Description:** L 13cm (5in). Upperparts unmarked greyish, tinged olive. Distinct whitish supercilium. Yellowish throat, Belly and undertail coverts white in adult, yellowish in juvenile. Iris brown. **Similar species:** Black-whiskered Vireo has red iris and dark-edged supercilium. **Habitat and behaviour:** Very secretive. Apparently feeds on insects, spiders and fruits mainly in upper canopy. **Range:** Breeds in eastern North America. Winters south to northern Columbia. Rare passage migrant through Bahamas and Cuba. **Status:** Vagrant or rare passage migrant (October–November, February). Mentioned in the literature but not recently reported.

White-winged Crossbill *Loxia leucoptera*
Taxonomy: Polytypic (3). *L. l. megaplaga.* **Description:** L 15cm (5.75in). Heavy brown bill has crossed tips. Wings black with two broad, white, wing bars. Male is reddish. Female brown with dark streaks, yellow rump. Juvenile is browner than female. **Habitat and behaviour:** Pine forests. Feeds on cones. **Range:** Subspecies is endemic to Hispaniola. **Status:** Vagrant. Only reported once, from Bellevue in the Port Royal Mountains, St. Andrew.

Blue-winged Warbler *Vermivora pinus*
Taxonomy: Monotypic. **Description:** L 11–12cm (5in).Medium-sized. Head, throat and underparts yellow. Blue-grey wings with two white wing-bars. Undertail-coverts and outer tail feathers white. Dark eye-stripe. **Similar species:** Prothonotary Warbler is brighter yellow and lacks wing-bars. **Habitat:** Woodlands. **Range:** Breeds eastern United States, winters from Mexico to Panama. **Status:** Vagrant or rare winter visitor.

Golden-winged Warbler *Vermivora chrysoptera*
Taxonomy: Monotypic. **Description:** L 13cm (5in). Medium-sized blue-grey warbler with yellow wing patch. Undertail-coverts and outer tail feathers white. Male has black throat and broad black eye-patch, with white above and below. Forecrown bright yellow. Female duller, with grey eye-patch and greenish-yellow crown. **Habitat and behaviour:** Gardens and woodlands. In tree canopy, often feeds by probing for insects in dead leaves. **Range:** Breeds in the United States. Winters Central America, Bahamas and Greater Antilles (mainly Cuba, Puerto Rico and Cayman Islands). **Status:** Vagrant or rare passage migrant.

Tennessee Warbler *Vermivora peregrina*
Taxonomy: Monotypic. **Description:** L 11cm (4in). Non-breeding adult pale greenish above with pale yellow breast and white undertail-coverts. Thin bill, pale supercilium. Juvenile is yellower below. **Similar species:** Orange-crowned Warbler has broken eye ring, thin streaks on breast and yellow undertail-coverts. **Habitat:** Lowland gardens and mountain forest. **Range:** Breeds in North America. Winters south through Central America and the West Indies to northern South America. **Status:** Vagrant or rare winter visitor (October–March).

Orange-crowned Warbler *Vermivora celata*
Taxonomy: Polytypic (4). **Description:** L 12–14cm (5–6in). Plain olive green above, greyish-yellow below with faint grey streaks on breast. Yellow undertail-coverts. Short supercilium, broken yellow eye-ring. **Similar species:** Does not resemble any common species. **Range:** Breeds in North America. Winters in the northern Bahamas and Central America. **Status:** Vagrant. Not recently reported.

Nashville Warbler *Vermivora ruficapilla*

Taxonomy: Polytypic (2). **Description:** L 12–13cm (5–6in). Head and mantle plain grey. Wings greenish. Throat, breast and undertail coverts yellow. Belly white. White eye-ring. **Similar species:** Northern Parula has white wing-bars. **Range:** Breeds in North America, winters south to Central America. In West Indies recorded from Bahamas and Greater Antilles. **Status:** Vagrant.

Pine Warbler *Dendroica pinus*

Taxonomy: Polytypic (4). **Description:** L 13–14cm (6in). Pale neck-patch contrasts with darker cheeks. Male olive-green above, throat and breast yellow, faint dark streaks on breast-sides. Undertail-coverts white. Wings greyish with two white wing bars. Female and juvenile are duller. **Similar species:** Bright adults are sometimes mistaken for Yellow-throated Vireo but the latter has clearly defined yellow spectacles. **Range:** Breeds in eastern North America. Winters south to Texas. There are resident subspecies on Bahamas and Hispaniola. **Status:** Vagrant or rare passage migrant.

Cerulean Warbler *Dendroica cerulea*

Taxonomy: Monotypic. **Description:** L 10–13cm (4–5in). Blue (male) or blue-green (female) above, with two white wing-bars. White below, flanks streaked darker. Juvenile female dullest; olive-green above, yellow below. **Similar species:** Black-throated Blue Warbler darker with white patch at base of primaries and no wing-bars. Male has a black throat. **Range:** Breeds North America. Winters in the mountains of Central and northern South America. Global conservation status: Vulnerable. **Status:** Vagrant. Only recorded twice in Jamaica (Hardwar Gap 1964, Rocklands 1985).

Kentucky Warbler *Oporornis formosus*

Taxonomy: Monotypic. **Description:** L 12–14cm (5–7in). Unmarked dark olive green above, yellow below. Yellow eyebrow and incomplete yellow eye-ring resemble spectacles. Legs pinkish. Male crown, eye-patch and breast-sides black. Female and juvenile paler. **Similar species:** Yellow-throated Vireo is chunkier and has white wing-bars. **Range:** Breeds in northern United States, wintering south to Venezuela. In West Indies recorded from Bahamas, Greater Antilles, US Virgin Islands, Antigua and Guadeloupe. **Status:** Vagrant.

Mourning Warbler *Oporornis philadelphia*

Taxonomy: Monotypic. **Description:** L 13–15cm (5–6in). Male has blue-grey hood. Scaly black bib. Wings and back greyish-olive. No wing bars. Underparts bright yellow. Legs and feet pinkish. May have broken white eye-ring. Female and juvenile duller, lacks black bib. **Range:** Breeds in North America. Winters Central and northern South America. **Status:** Vagrant or rare passage migrant through the Greater Antilles.

Wilson's Warbler *Wilsonia pusilla*

Taxonomy: Polytypic (3). **Description:** L 12cm (4.75in). Greenish above and bright yellow below, with striking dark eye and yellow supercilium. Male has neat black cap. **Range:** Breeds in North America. Winters south to Panama. **Similar species:** Yellow Warbler is larger and lacks yellow supercilium. Female and immature Hooded Warblers have white tail-patches. **Range:** Breeds North America, winters Central America south to Panama. Rare migrant in Bahamas and Cuba. **Status:** Vagrant. Only reported once since 2000 (from Hardwar Gap).

Canada Warbler *Wilsonia canadensis*

Taxonomy: Monotypic. **Description:** L 13cm (5in). Plain blue-grey above and yellow below. Undertail-coverts white. Male has a breast-band of bold black streaks. Female and juvenile duller with less distinct breast-band. **Range:** Breeds North America. Winters South America. **Status:** Very rare vagrant. Only reported once since 2000, on Pedro Cays in May 2007.

Yellow-breasted Chat *Icteria virens*

Taxonomy: Polytypic (2). **Description:** L 18–19cm (7in). Large ,very chunky warbler with heavy black bill and long, rounded tail. Lores black. White spectacles and supercilium. Greenish-olive above, throat and breast yellow, lower belly and undertail coverts white. **Range:** Breeds in North America. Winters south to Panama including Cuba and the Bahamas. Also recorded In West Indies from Cayman Islands and Hispaniola. **Status:** Vagrant or very rare winter visitor. Probably overlooked and mainly recorded through mist-netting (near Windsor several times since 2000).

Lark Sparrow *Chondestes grammacus*
Taxonomy: Polytypic (2). **Description:** L 15cm (6in). Adult greyish with chestnut, black-and-white stripes on head, chestnut cheeks. Small black spot in centre of buffy breast. Tail is long and rounded, dark with white margin. Juvenile has dull head-markings. **Habitat and behaviour:** Feeds on seeds on the ground in open, scrubby dry areas. **Range:** Breeds in North America. Winters south to Mexico. Vagrant in Bahamas, Cuba and Jamaica. **Status:** Vagrant. Mentioned in the literature but not recently reported.

White-crowned Sparrow *Zonotrichia leucophrys*
Taxonomy: Polytypic (5). *Z. l. leucophrys.* **Description:** L 18cm (7in). Adult has black-and-white crown-stripes. Neck, rump, tail and underparts plain grey. Wings rufous-edged, streaked, brown, buff and grey. Bill and legs are yellowish. Juvenile has brown and buff crown-stripes. **Range:** Breeds in Canada. Winters south to southeastern United States, Cuba and Jamaica (October–April). **Status:** Vagrant or very rare winter resident. Mentioned in the literature but not recently reported.

Blue Grosbeak *Passerina caerula*
Taxonomy: Polytypic (7). **Description:** L 16–18cm (6–7in). Stout-billed, with long, rounded tail. Breeding male dark blue. Wings blackish wings with rufous wing-bars. Dark upper mandible contrasts with grey lower. Female and juvenile are brownish-rufous with buffy median coverts. **Behaviour:** Frequently spreads and flicks tail up. **Range:** Breeds in southern United States and Central America. Winters south to Panama. In West Indies reported from Bahamas, Greater Antilles and Virgin Islands. **Status:** Vagrant. Only reported once recently; Stony Hill, St. Andrew in October 2006.

Painted Bunting *Passerina ciris*
Taxonomy: Polytypic (2). *P. c. ciris.* **Description:** L 13cm (5in). Bill heavy. Male unmistakeable with green back, blue head, red underparts. Female and juvenile greenish above and yellowish below with conspicuous dark eyes. **Habitat:** Usually in low, dense scrub. **Range:** *P. c. ciris* breeds in southeast United States, winters Bahamas, Cuba, Jamaica and Yucatán. **Status**: Vagrant. Mentioned in the literature but not recently reported.

Dickcissel *Spiza americana*
Taxonomy: Monotypic. **Description:** L 15cm (6in). Grey with a long, yellow supercilium. Wings and mantle boldly streaked buff and black. Rufous shoulder-patch. Male has grey bib boldly edged black, yellow upper breast. Female less strongly marked. Juvenile greyer. **Habitat and behaviour:** Observations elsewhere suggest that Dickcissel is most likely to be seen in flocks over rice or sugar cane but there is no information from Jamaica. **Range:** Breeds in North America. Winters Mexico to northern South America. Irruptive. **Status**: Vagrant. Mentioned in the literature but not recently reported.

Orchard Oriole *Icterus spurius*
Taxonomy: Monotypic. **Description:** L 15–18cm (6–7in). A small oriole. Breeding male has a black head, wings and tail. Two wing-bars; upper one is broad and deep reddish-brown, lower bar is thin and white. Underparts are deep reddish-brown. Female has an olive-brown head and tail, with grey mantle. Two thin white wing-bars. Underparts are dull greenish-yellow. First-summer male resembles female with a black bib. **Similar species:** Juvenile Jamaican Oriole is yellower with brighter wing-bars. **Range:** Breeds North America. Winters south to northern South America. In West Indies reported as a vagrant from Cuba, Bahamas and Hispaniola. **Status**: Vagrant. Mentioned in the literature but not recently reported.

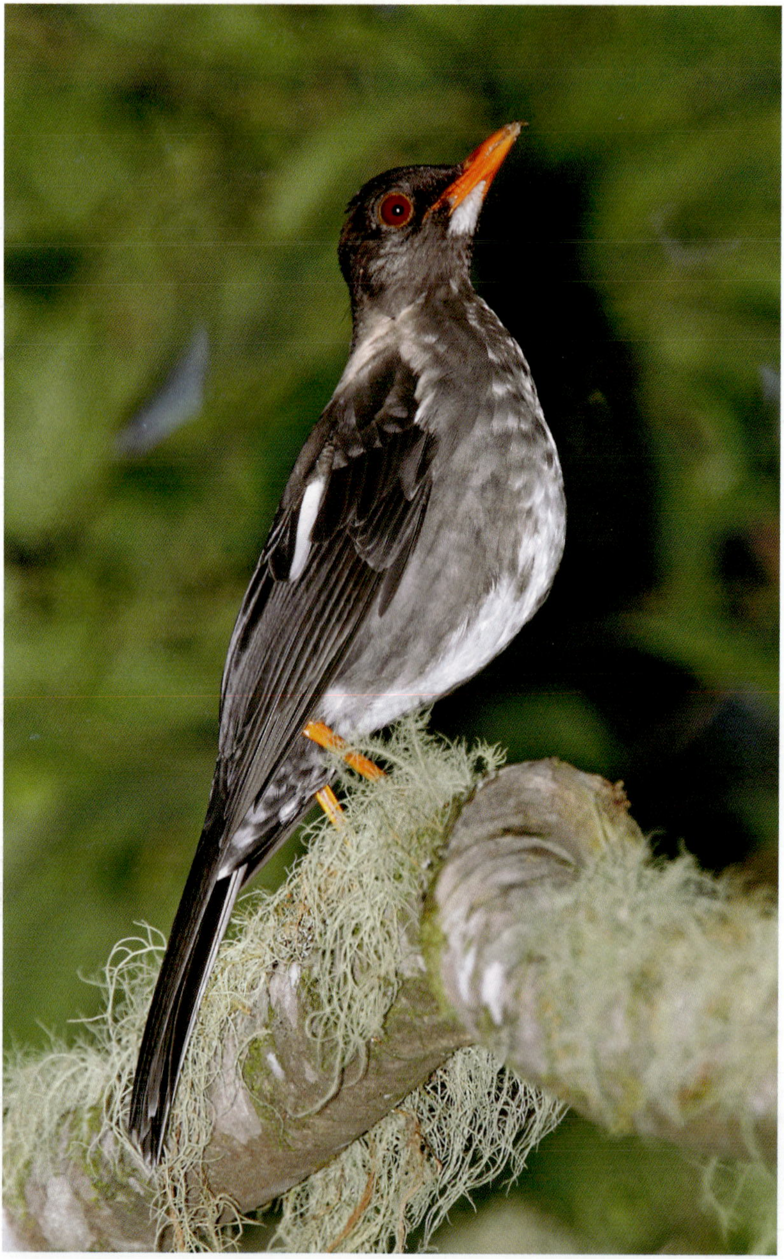

Adult. Port Royal Mountains, Jamaica, November.

APPENDIX 3: SPECIES ENDEMIC TO JAMAICA

There are a total of 30 species that are endemic to Jamaica.

COMMON NAME	SCIENTIFIC NAME	NOTES
Jamaican Petrel *	*Pterodroma caribbaea*	
Ring-tailed Pigeon	*Patagioenas caribaea*	
Crested Quail-Dove	*Geotrygon versicolor*	
Yellow-billed Parrot	*Amazona collaria*	
Black-billed Parrot	*Amazona agilis*	
Jamaican Lizard Cuckoo	*Coccyzus vetula*	
Chestnut-bellied Cuckoo	*Coccyzus pluvialis*	
Jamaican Owl	*Pseudoscops grammicus*	
Jamaican Pauraque *	*Siphonorhis americana*	
Jamaican Mango	*Anthracothorax mango*	
Red-billed Streamertail	*Trochilus polytmus*	
Black-billed Streamertail	*Trochilus scitulus*	
Jamaican Tody	*Todus todus*	
Jamaican Woodpecker	*Melanerpes radiolatus*	
Jamaican Elaenia	*Myiopagis cotta*	
Jamaican Pewee	*Contopus pallidus*	
Sad Flycatcher	*Myiarchus barbirostris*	
Rufous-tailed Flycatcher	*Myiarchus validus*	
Jamaican Becard	Pachyramphus niger	
White-eyed Thrush	*Turdus jamaicensis*	
White-chinned Thrush	*Turdus aurantius*	
Jamaican Crow	*Corvus jamaicensis*	
Jamaican Vireo	*Vireo modestus*	
Blue Mountain Vireo	*Vireo osburni*	
Jamaican Euphonia	*Euphonia jamaica*	
Arrow-headed Warbler	*Dendroica pharetra*	
Jamaican Spindalis	*Spindalis nigricephala*	
Yellow-shouldered Grassquit	*Loxipasser anoxanthus*	
Orangequit	*Euneornis campestris*	
Jamaican Blackbird	*Nesopsar nigerrimus*	

* Extremely rare or extinct.

APPENDIX 4: SUBSPECIES ENDEMIC TO JAMAICA

There are a total of 19 subspecies that are endemic to Jamaica.

COMMON NAME	SCIENTIFIC NAME	NOTES
Uniform Crake *	*Amaurolimnas concolor concolor*	
Northern Jacana	*Jacana spinosa violacea*	
Plain Pigeon	*Patagioenas inornata exigua*	
Common Ground Dove	*Columbina passerina jamaicensis*	
Caribbean Dove	*Leptotila jamaicensis jamaicensis*	
Olive-throated Parakeet	*Aratinga nana nana*	
Northern Potoo	*Nyctibius jamaicensis jamaicensis*	
Vervain Hummingbird	Mellisuga minima minima	
Greater Antillean Elaenia	*Elaenia fallax fallax*	
Loggerhead Kingbird	*Tyrannus caudifasciatus jamaicensis*	
Stolid Flycatcher	*Myiarchus stolidus stolidus*	
Golden Swallow *	*Tachycineta euchrysea euchrysea*	
Cave Swallow	*Petrochelidon fulva poeciloma*	
Bahama Mockingbird	*Mimus gundlachii hillii*	
Rufous-throated Solitaire	*Myadestes genibarbis solitarius*	
Bananaquit	*Coereba flaveola flaveola*	
Grasshopper Sparrow	*Ammodramus savannarum savannarum*	
Greater Antillean Grackle	*Quiscalus niger crassirostris*	
Jamaican Oriole	*Icterus leucopteryx leucopteryx*	

* Extremely rare or extinct.

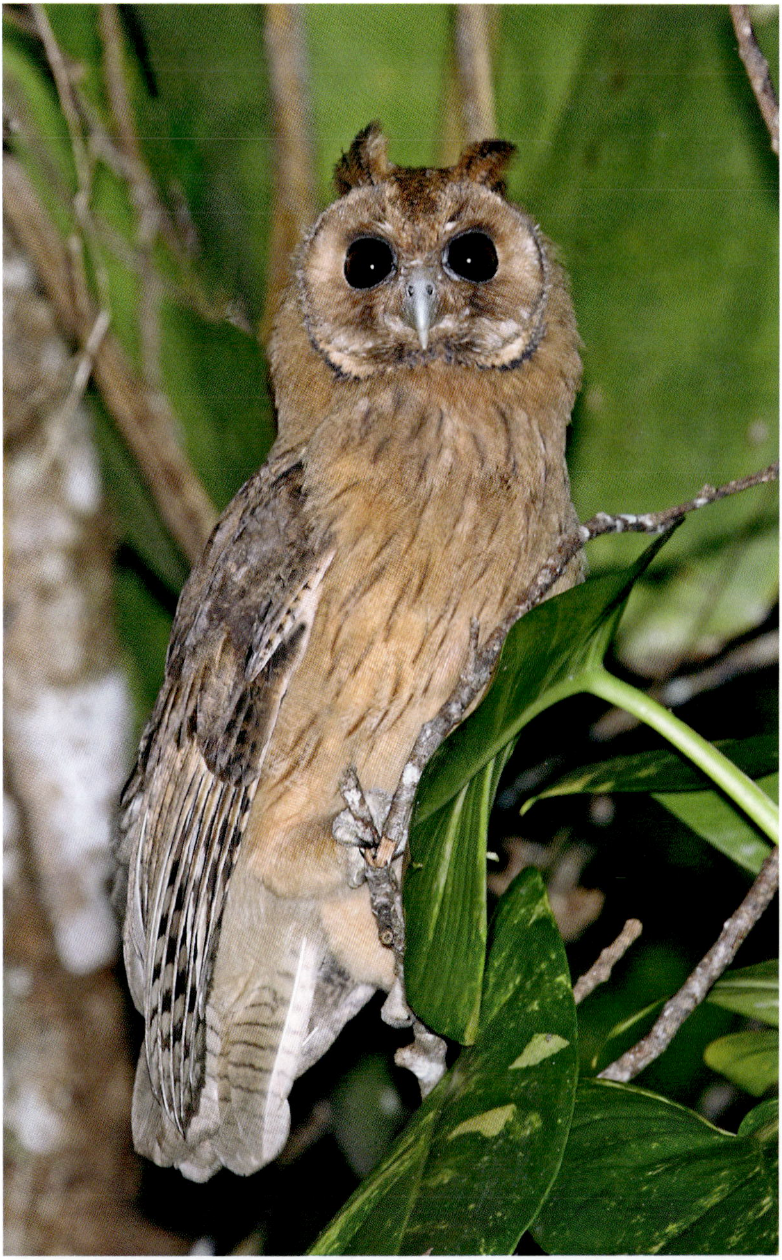

Adult Jamaican Owl. Port Antonio, Jamaica, November.

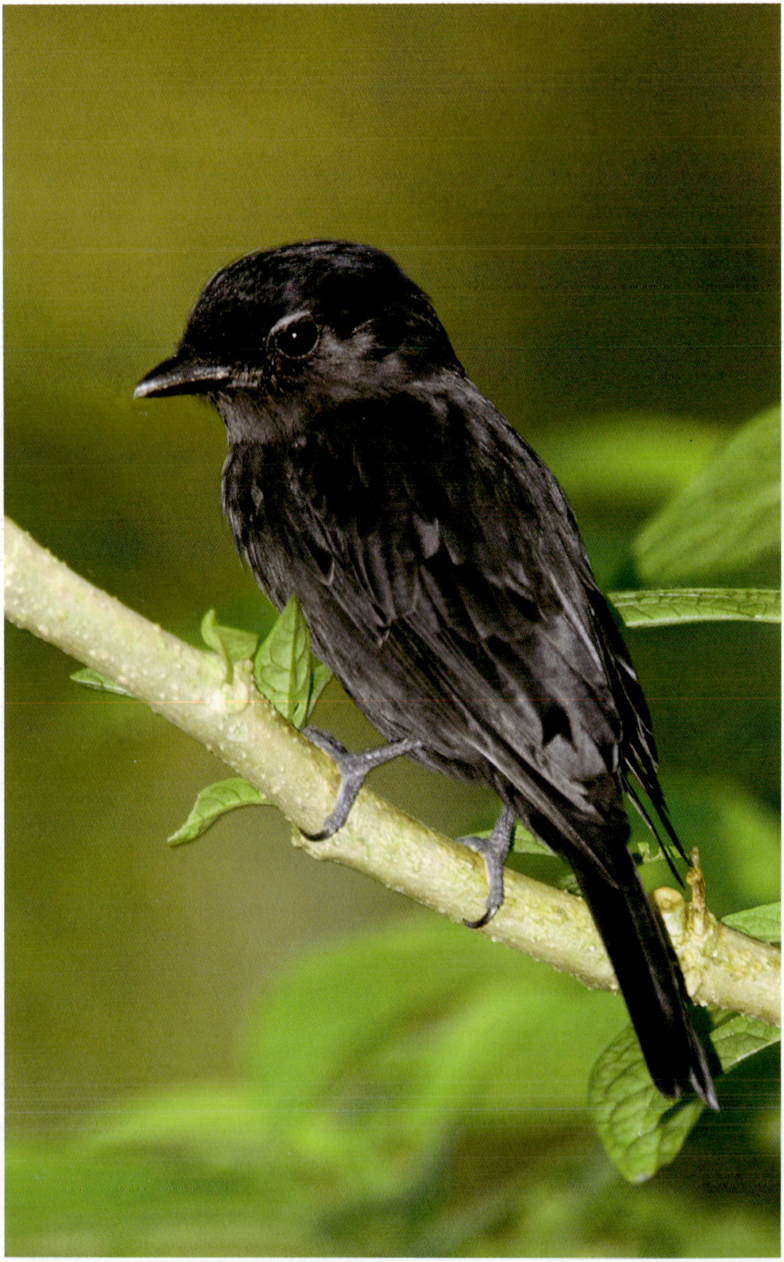

Adult male Jamaican Becard. Port Royal Mountains, Jamaica, November.

APPENDIX 5: CARIBBEAN ENDEMIC SPECIES AND SUBSPECIES RECORDED IN JAMAICA

There are a total of 18 species and subspecies endemic to the Caribbean that occur in Jamaica.

COMMON NAME	SCIENTIFIC NAME	NOTES
West Indian Whistling Duck	*Dendrocygna arborea*	
Pied-billed Grebe	*Podilymbus podiceps antillarum*	
White-tailed Tropicbird	*Phaethon lepturus catesbyi*	
Brown Pelican	*Pelecanus occidentalis occidentalis*	
Reddish Egret	*Egretta ruficollis ruficollis*	
Osprey	*Pandion haliaetus ridgwayi*	
Red-tailed Hawk	*Buteo jamaicensis jamaicensis*	
American Kestrel	*Falco sparverius sparveroides/dominicensis*	
Clapper Rail	*Rallus longirostris caribaeus*	
Yellow-breasted Crake	*Porzana flaviventer gossii*	
Spotted Rail	*Pardirallus maculatus maculatus*	
Common Moorhen	*Gallinula chloropus cerceris*	
Scaly-naped Pigeon	*Patagioenas squamosa*	
Zenaida Dove	*Zenaida aurita zenaida*	
Barn Owl	*Tyto alba furcata*	
White-collared Swift	*Streptoprocne zonaris pallidifrons*	
Antillean Palm Swift	*Tachornis phoenicobia*	
Yellow Warbler	*Dendroica petechia eoa*	

APPENDIX 6: SCIENTIFIC NAMES OF PLANTS MENTIONED IN THE TEXT

FAMILY	COMMON NAME	SCIENTIFIC NAME
Aquifoliaceae	Winter Berry	*Ilex macfadyenii*
Agavaceae	Maypole	*Agave sobolifera*
Anacardiaceae	Burnwood	*Metopium brownii*
	Cashew	*Anacardium occidentale*
	Maiden Plum	*Comocladia* spp.
Annonaceae	Soursop	*Annona muricata*
Bignoniaceae	Spathodea/African Tulip	*Spathodea campanulata*
Bombacaceae	Silk Cotton	*Ceiba pentandra*
Bromeliaceae	Bromeliads	*Tillandsia, Hohenbergia, Guzmania* spp.
	Spanish Moss	*Tillandsia usneoides*
Cactaceae	Night-blooming Cereus	*Cereus hexagonus*
	Prickly Pear, Seaside Tuna	*Opuntia dillenii*
	Dildo	*Cephalocereus swartzii*
	Prickly Pear Tree	*Opuntia spinosissima*
	Jamaican Mistletoe	*Rhipsalis baccifera*
	Torchwood Dildo	*Harrisia gracilis*
Caesalpiniaceae	Poinciana	*Delonix regia*
	Logwood	*Haematoxylum campechianum*
	Tamarind	*Tamarindus indica*
	Wild Bauhinia, Bull-Hoof	*Bauhinia divaricata*
Cannabaceae	Ganja	*Cannabis sativa*
Casuarinaceae	Casuarina	*Casuarina equisetifolia*
Clusiaceae	Strangler Fig	*Clusia* spp.
Combretaceae	Black Mangrove	*Avicennia germinans*
	Broadleaf	*Terminalia latifolia*
	Buttonwood Mangrove	*Conocarpus erectus*
	White Mangrove	*Laguncularia racemosa*
Compositae	Spanish Needle	*Bidens pilosa* var. *radiata*
Cucurbitaceae	Cho-cho	*Sechium edule*
Graminae	Bamboo	*Bambusa vulgaris*
	Indian Corn, Corn	*Zea mays*

FAMILY	COMMON NAME	SCIENTIFIC NAME
	Sorghum, Guinea Corn	*Sorghum saccharatum*
Lauraceae	Sweetwoods	*Ocotea* spp.
Malvaceae	Blue Mahoe	*Hibiscus elata*
	Hibiscus	*Hibiscus rosa-sinensis*
Meliaceae	West Indian Cedar	*Cedrela odorata*
Mimosaceae	Acacia	*Acacia* spp.
	Cashaw	*Prosopis juliflora*
	Privet	*Pithecellobium unguis-cati*
Moraceae	Fig	*Ficus* spp.
	Santa Maria	*Calophyllum calaba*
	Trumpet Tree	*Cecropia peltata*
Musaceae	Plantain	*Musa* spp.
Myricaceae	Waxwood	*Myrica cerifera*
Myrtaceae	Guava	*Psidium guajava*
	Pimento	*Pimenta dioica*
Nyctaginaceae	Bougainvillea	*Bougainvillea* spp.

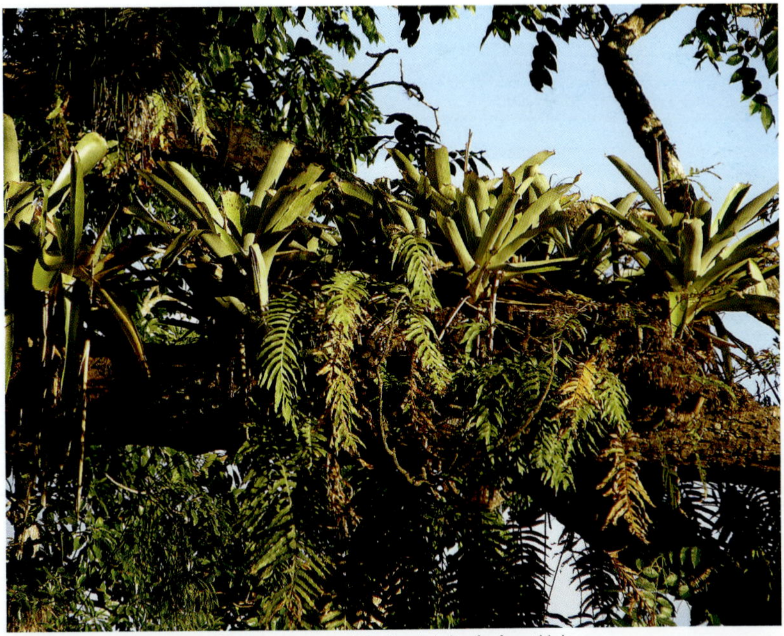

Bromeliads at Marshall's Pen – an important source of food and shelter for forest birds.

FAMILY	COMMON NAME	SCIENTIFIC NAME
Orchidaceae		*Broughtonia sanguinea*
		Oncidium tetrapetalum
Palmae	Thatch Palms	*Calyptronoma, Sabal, Thrinax, Coccothrinax* spp.
	Coconut	*Cocos nucifera*
	Royal Palm	*Roystonea* spp.
Papilionaceae	Erythrina	*Erythrina* spp.
Pittosporaceae	Mock Orange	*Pittosporum undulatum*
Rhizophoraceae	Red Mangrove	*Rhizophora mangle*
Rosaceae	Peach	*Prunus persica*
Rubiaceae	Coffee	*Coffea* spp.
	Pentas	*Pentas* spp.
Rutaceae	Mountain Pride	*Spathelia sorbifolia, S. glabrescens*
	Lime	*Citrus aurantifolia*
	Orange	*Citrus sinesis*
	Prickly Yellow	*Xanthoxylum martinicensis*
Sapindaceae	Ackee	*Blighia sapida*
	Wild Ackee	*Cupania glabra*
Sapotaceae	Bullet	*Sideroxylon* spp.
	Naseberry	*Manilkara zapota*
Simaroubaceae	Bitterwood	*Picrasma excels*
	Red Birch	*Bursera simaruba*
Solanaceae	Bird-Pepper	*Capsicum frutescens*
	Peppers	*Capsicum* spp.
Sterculiaceae	French Peanut	*Sterculia apetala*
Verbenaceae	Chinese Hat	*Holmskioldia sanguinea*
	Fiddlewood	*Citharexylum caudatum*
	Vervain, Vervine	*Stachytarpheta jamaicensis*
Vitaceae	Pudding Withe	*Cissus sicyoides*
Zingiberaceae	Ginger lilies	*Hedychium* spp.
Cyatheaceae	Tree Ferns	*Cyathea spp*
Podocarpaceae	Yacca	*Podocarpus urbanii*

APPENDIX 7: JAMAICAN BIRD SOUNDS

The songs of many Jamaican birds (including most of the endemics) are available on *Bird Songs in Jamaica*, a set of CDs by George Reynard and Robert Sutton. This is available from the Library of Natural Sounds, Cornell Laboratory of Ornithology (http://birds.cornell.org). This appendix lists the species that appear on the CD. Listening to the sounds and calls of the Jamaican forest beforehand will be of great benefit to birders, especially first-time visitors to the island.

COMMON NAME	SCIENTIFIC NAME
West Indian Whistling-Duck	*Dendrocygna arborea*
Blue-winged Teal	*Anas discors*
Masked Duck	*Nomonyx dominicus*
Least Grebe	*Tachybaptus dominicus dominicus*
Pied-billed Grebe	*Podilymbus podiceps antillarum*
Least Bittern	*Ixobrychus exilis exilis*
Great Egret	*Ardea alba egretta*
Cattle Egret	*Bubulcus ibis ibis*
Green Heron	*Butorides virescens virescens*
Black-crowned Night-heron	*Nycticorax nycticorax bancrofti*
Yellow-crowned Night-heron	*Nyctanassa violacea hoactli*
Red-tailed Hawk	*Buteo jamaicensis jamaicensis*
American Kestrel	*Falco sparverius*
Black Rail	*Laterallus jamaicensis jamaicensis*
Clapper Rail	*Rallus longirostris caribaeus*
Yellow-breasted Crake	*Porzana flaviventer gossii*
Common Moorhen	*Gallinula chloropus cerceris*
American Coot	*Fulica americana americana*
Caribbean Coot	*Fulica caribaea*
Limpkin	*Aramus guarauna pictus*
Grey Plover	*Pluvialis squatarola*
Wilson's Plover	*Charadrius wilsonia wilsonia*
Killdeer	*Charadrius vociferous ternominatus*
Black-necked Stilt	*Himantopus mexicanus mexicanus*

COMMON NAME	SCIENTIFIC NAME
Northern Jacana	*Jacana spinosa violacea*
Greater Yellowlegs	*Tringa melanoleuca*
Willet	*Tringa semipalmata semipalmatus*
Least Tern	*Sternula antillarum antillarum*
Royal Tern	*Thalasseus maximus maximus*
Sandwich Tern	*Thalasseus sandvicensis acuflavidus*
White-crowned Pigeon	*Patagioenas leucocephala*
Plain Pigeon	*Patagioenas inornata exigua*
Ring-tailed Pigeon	*Patagioenas caribaea*
White-winged Dove	*Zenaida asiatica asiatica*
Zenaida Dove	*Zenaida aurita aurita*
Mourning Dove	*Zenaida macroura macroura*
Common Ground Dove	*Columbina passerine jamaicensis*
Caribbean Dove	*Leptotila jamaicensis jamaicensis*
Crested Quail-dove	*Geotrygon versicolor*
Ruddy Quail-dove	*Geotrygon montana montana*
Olive-throated Parakeet	*Aratinga nana nana*
Green-rumped Parrotlet	*Forpus passerinus*
Yellow-billed Parrot	*Amazona collaria*
Black-billed Parrot	*Amazona agilis*
Mangrove Cuckoo	*Coccyzus minor*
Chestnut-bellied Cuckoo	*Coccyzus pluvialis*
Jamaican Lizard Cuckoo	*Coccyzus vetula*
Smooth-billed Ani	*Crotophaga ani*

COMMON NAME	SCIENTIFIC NAME
Barn Owl	*Tyto alba furcata*
Jamaican Owl	*Pseudoscops grammicus*
Antillean Nighthawk	*Chordeiles gundlachii gundlachii*
Northern Potoo	*Nyctibius jamaicensis jamaicensis*
Black Swift	*Cypseloides niger niger*
Antillean Palm Swift	*Tachornis phoenicobia phoenicobia*
Jamaican Mango	*Anthracothorax mango*
Red-billed Streamertail	*Trochilus polytmus*
Black-billed Streamertail	*Trochilus scitulus*
Vervain Hummingbird	*Mellisuga minima minima*
Jamaican Tody	*Todus todus*
Belted Kingfisher	*Ceryle alcyon*
Jamaican Woodpecker	*Melanerpes radiolatus*
Yellow-bellied Sapsucker	*Sphyrapicus varius*
Jamaican Elaenia	*Myiopagis cotta*
Greater Antillean Elaenia	*Elaenia fallax fallax*
Eastern Wood-Pewee	*Contopus virens*
Jamaican Pewee	*Contopus pallidus*
Sad Flycatcher	*Myiarchus barbirostris*
Rufous-tailed Flycatcher	*Myiarchus validus*
Stolid Flycatcher	*Myiarchus stolidus stolidus*
Grey Kingbird	*Tyrannus dominicensis dominicensis*
Loggerhead Kingbird	*Tyrannus caudifasciatus jamaicensis*
Jamaican Becard	*Pachyramphus niger*
Jamaican Vireo	*Vireo modestus*
Blue Mountain Vireo	*Vireo osburni*
Black-whiskered Vireo	*Vireo altiloquus altiloquus*
Jamaican Crow	*Corvus jamaicensis*
Caribbean Martin	*Progne dominicensis*
Golden Swallow	*Tachycineta euchrysea euchrysea*
Cave Swallow	*Hirundo (Petrochelidon) fulva poeciloma*
Rufous-throated Solitaire	*Myadestes genibarbis solitarius*

COMMON NAME	SCIENTIFIC NAME
White-eyed Thrush	*Turdus jamaicensis*
White-chinned Thrush	*Turdus aurantius*
Grey Catbird	*Dumetella carolinensis*
Northern Mockingbird	*Mimus polyglottos orpheus*
Bahama Mockingbird	*Mimus gundlachii hillii*
European Starling	*Sturnus vulgaris*
Northern Parula	*Parula americana*
Yellow Warbler	*Dendroica petechia eoa*
Magnolia Warbler	*Dendroica magnolia*
Black-throated Blue Warbler	*Dendroica caerulescens caerulescens*
Prairie Warbler	*Dendroica discolor discolor*
Palm Warbler	*Dendroica palmarum palmarum*
Arrow-headed Warbler	*Dendroica pharetra*
Black-and-white Warbler	*Mniotilta varia*
American Redstart	*Setophaga ruticilla*
Worm-eating Warbler	*Helmitheros vermivorum*
Ovenbird	*Seiurus aurocapilla furvivor*
Northern Waterthrush	*Seiurus noveboracensis*
Common Yellowthroat	*Geothlypis trichas trichas (brachydactylus)*
Bananaquit	*Coereba flaveola flaveola*
Jamaican Spindalis	*Spindalis nigricephala*
Yellow-faced Grassquit	*Tiaris olivacea olivacea*
Black-faced Grassquit	*Tiaris bicolor marchii*
Yellow-shouldered Grassquit	*Loxipasser anoxanthus*
Greater Antillean Bullfinch	*Loxigilla violacea ruficollis*
Orangequit	*Euneornis campestris*
Saffron Finch	*Sicalis flaveola*
Grasshopper Sparrow	*Ammodramus savannarum savannarum*
Jamaican Blackbird	*Nesopsar nigerrimus*
Greater Antillean Grackle	*Quiscalus niger crassirostris*
Jamaican Oriole	*Icterus leucopteryx leucopteryx*
Jamaican Euphonia	*Euphonia jamaica*

SELECTED BIBLIOGRAPHY

Adams, C. D. 1972. *Flowering Plants of Jamaica*. University of the West Indies, Mona, Jamaica.

American Ornithologists's Union 1998. *Check-list of North American Birds*. 7th Edition. Allen Press, Lawrence, Kansas.

BirdLife International. 2008. *Species fact sheets.* http://www.birdlife.org

Bond, J. 1936. *Birds of the West Indies.* 1st edition. Collins, London.

Bradley, P. E. 2000. *The Birds of the Cayman Islands – an annotated checklist*. BOU Checklist No. 19. British Ornithologists' Union, The Natural History Museum, Tring.

Clements, J. 2007. *The Clements Checklist of the Birds of the World.* Christopher Helm Publishers, London.

Downer, A. & Sutton, R. *Birds of Jamaica – a Photographic Field Guide.* Cambridge University Press, Cambridge.

Gosse, P. H. 1847. *The Birds of Jamaica*. Van Voorst, London.

Gosse, P. H. 1849. *Illustrations of the Birds of Jamaica*. Van Voorst, London.

Gosse, P. H. 1851. *A Naturalist's Sojourn in Jamaica*. Longman, London.

Hallett, B. 2006. *Birds of the Bahamas and the Turks and Caicos Islands.* Macmillan Caribbean, Oxford.

Hawkes, A. 1974. *Wild Flowers of Jamaica*. Collins/Sangster, London and Kingston.

Keith. A., Wiley, J., Latta, S. & Ottenwalder, J. 2003. *The Birds of Hispaniola – Haiti and the Dominican Republic – an annotated checklist.* BOU Checklist No 21. British Ornithologists' Union, The Natural History Museum, Tring.

Lack, D. 1976. *Island Biology, Illustrated by the Land Birds of Jamaica*. Blackwell Scientific Publications, Oxford.

Latta, S., Rimmer, C., Keith, A., Wiley, J., Raffaele, H., McFarland, K. & Fernandez, E. 2006. *Birds of the Dominican Republic and Haiti.* Christopher Helm Publishers, London.

Raffaele, H., Wiley, J., Garrido, O., Keith, A. & Raffaele, J. 1998. *Birds of the West Indies*. Christopher Helm Publishers, London.

Robbins, C., Bruun, B. & Zim, H. 1966. *Birds of North America.* Golden Press, New York.

Senior, Olive. 1983. *A–Z of Jamaican Heritage.* Heinemann Educational Books (Caribbean).

Sibley, D. 2000. *The North American Bird Guide*. Christopher Helm Publishers, London.

Sloane, H. 1707. *A Voyage to the Islands Madera, Barbados, Nieves, S. Christophers and Jamaica*. Privately printed, London.

INDEX

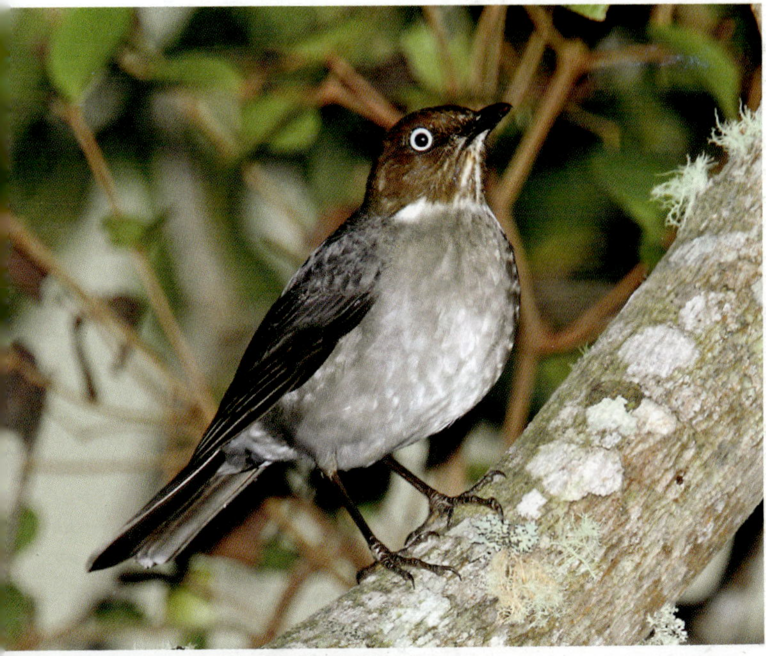

lult White-eyed Thrush. Port Royal Mountains, Jamaica, November.